D0079065

2014

Geology of the Moon

REVISED EDITION 1972 · PRINCETON UNIVERSITY PRESS, PRINCETON, NEW JERSEY

GEOLOGY OF THE MOON

A Stratigraphic View

Thomas A. Mutch

L.C. Card 70-38387

ISBN 0-691-08110-7

Revised Edition, 1972

Printed in the United States of America
By Princeton University Press, Princeton, New Jersey

This book has been composed in Times Roman

End paper and frontispiece photograph by Ansel Adams
Autumn Moon, The High Sierra from Glacier Point

Preface

WORK on this manuscript began in the fall of 1967 when I was a visitor at the U.S. Geological Survey Center of Astrogeology in Flagstaff, Arizona. At that time I had just become familiar with the large amount of geologic analysis and mapping of the Moon that had been accomplished, chiefly by U.S. Geological Survey personnel. My initial reasons for attempting a synthesis of these studies in an accessible format were two-fold. First, I wanted to direct the attention of scientists—geologists, in particular—to the large body of work that properly should serve as a background for future studies of the Moon. Most of this work is not published in professional journals and certainly not in journals regularly read by geologists. For that reason there are, even today, wide-spread misconceptions about the novelty of lunar geologic analysis.

My second initial goal was more personal. As a stratigrapher I am particularly interested in the history of the Earth and, by extension, in the history of science as well. When first introduced to lunar geology I was struck by similarities between the development of terrestrial stratigraphy in the nineteenth century and geologic analysis of the Moon, chiefly in the last decade. It is only natural that scientists most closely involved with a new field, lunar geology in this case, will stress the novelty of their work and their conclusions. Adopting a different point of view, I thought it might be instructive to explore possible similarities between our evolving understanding of the Earth's history and unravelling of the Moon's history. In that connection I have made frequent use of analogies between terrestrial and lunar features throughout the text.

It was clear from the very start that the value of this book would rest in large part on the illustrations. Surprisingly, very few representative collections of pictures taken on the Ranger, Surveyor, and Orbiter missions are available. Pictures appearing in the popular press are limited to a small collection of "standard" views (including, for example, the famous Orbiter II oblique view of the crater Copernicus). Pictures used in connection with scientific articles have generally been poorly reproduced and incompletely described. Recognizing this situation I have tried to provide the reader with a comprehensive review of the several pre-Apollo missions and to acquaint him with the diversity of features shown in the photographs. Even so, I make no claim for completeness. I suspect that anyone who has worked extensively with the entire collections will find that many of his favorite photographs are missing.

It is something of a tradition for authors of science books to include a few prefatory remarks about their intended audiences, as if in this way the authors could reassure themselves that their books will in fact be read. In my own case I admit to having no gift for prophecy. The book was first envisioned as an auxiliary text to be used either in advanced undergraduate or elementary graduate level courses. To that end I have presented brief reviews of the fundamentals as well as more detailed discussions of "space age" studies. Perhaps, however, the book will find its chief use among professional geologists wishing an introduction to lunar geology or among nongeological planetary scientists who wish to acquaint themselves with the geologist's point of view. With these latter two audiences in mind I have tried to take note, however briefly, of major publications through the spring of 1969—essentially up to the flight of Apollo XI. My intent in liberally sprinkling the text with references is to encourage additional readings in a variety of primary sources.

Since my own scientific discipline is stratigraphy the emphasis of this book is stratigraphic and historical. However, at many points I have felt the need to discuss topics in related fields—for example, petrology, geomorphology, and geophysics. Inevitably this has led me away from the field in which I might claim some competency. I recognize that these side trips make me vulnerable to criticisms of misstatement, oversimplification, omission, and imbalance. Nonetheless this approach seemed preferable to entrenching myself behind an easily defended but tightly constraining wall of specialization.

Although my approach has been historical, I should emphasize that this is *not* intended to be a comprehensive review of all former studies of the Moon. For example, the pioneering work of Harold C. Urey has not received the attention it deserves. Decisions to include or exclude references to published work were based on the applicability of that work to our developing geologic and stratigraphic understanding of the Moon. I apologize in advance to those readers who will be dismayed to find that some unrelated but important publications have not been discussed.

I have made no effort to revise the first eleven chapters of the book in the light of Apollo XI results. Even in the few months since these chapters were written the receipt of new information precisely dates many passages as "pre-Apollo–post-Surveyor." The recent analysis of rocks brought back by the Apollo XI astronauts contradicts some previously held views and makes other earlier speculations look like axiomatic truths. Rather than updating a major part of the text—an exhausting exercise at best—I thought that the reader might be better entertained by doing his own mental editing, determining what changes must be made and what conclusions are still valid.

In balance, and perhaps in contrast with prevailing views, I am not sure that many changes are necessary. Administrators of the space program are in the habit of billing each mission as *the* critical step in the advancement of the total program. In a similar vein scientists closely connected with the program have sometimes been too cooperative in publicizing pre-mission controversies which, they imply, will be settled once and for all when the first data of a particular mission are received. This simplistic point of view preceded the Ranger missions with the first detailed pictures of the Moon's surface,

Preface

WORK on this manuscript began in the fall of 1967 when I was a visitor at the U.S. Geological Survey Center of Astrogeology in Flagstaff, Arizona. At that time I had just become familiar with the large amount of geologic analysis and mapping of the Moon that had been accomplished, chiefly by U.S. Geological Survey personnel. My initial reasons for attempting a synthesis of these studies in an accessible format were two-fold. First, I wanted to direct the attention of scientists—geologists, in particular—to the large body of work that properly should serve as a background for future studies of the Moon. Most of this work is not published in professional journals and certainly not in journals regularly read by geologists. For that reason there are, even today, wide-spread misconceptions about the novelty of lunar geologic analysis.

My second initial goal was more personal. As a stratigrapher I am particularly interested in the history of the Earth and, by extension, in the history of science as well. When first introduced to lunar geology I was struck by similarities between the development of terrestrial stratigraphy in the nineteenth century and geologic analysis of the Moon, chiefly in the last decade. It is only natural that scientists most closely involved with a new field, lunar geology in this case, will stress the novelty of their work and their conclusions. Adopting a different point of view, I thought it might be instructive to explore possible similarities between our evolving understanding of the Earth's history and unravelling of the Moon's history. In that connection I have made frequent use of analogies between terrestrial and lunar features throughout the text.

It was clear from the very start that the value of this book would rest in large part on the illustrations. Surprisingly, very few representative collections of pictures taken on the Ranger, Surveyor, and Orbiter missions are available. Pictures appearing in the popular press are limited to a small collection of "standard" views (including, for example, the famous Orbiter II oblique view of the crater Copernicus). Pictures used in connection with scientific articles have generally been poorly reproduced and incompletely described. Recognizing this situation I have tried to provide the reader with a comprehensive review of the several pre-Apollo missions and to acquaint him with the diversity of features shown in the photographs. Even so, I make no claim for completeness. I suspect that anyone who has worked extensively with the entire collections will find that many of his favorite photographs are missing.

It is something of a tradition for authors of science books to include a few prefatory remarks about their intended audiences, as if in this way the authors could reassure themselves that their books will in fact be read. In my own case I admit to having no gift for prophecy. The book was first envisioned as an auxiliary text to be used either in advanced undergraduate or elementary graduate level courses. To that end I have presented brief reviews of the fundamentals as well as more detailed discussions of "space age" studies. Perhaps, however, the book will find its chief use among professional geologists wishing an introduction to lunar geology or among nongeological planetary scientists who wish to acquaint themselves with the geologist's point of view. With these latter two audiences in mind I have tried to take note, however briefly, of major publications through the spring of 1969—essentially up to the flight of Apollo XI. My intent in liberally sprinkling the text with references is to encourage additional readings in a variety of primary sources.

Since my own scientific discipline is stratigraphy the emphasis of this book is stratigraphic and historical. However, at many points I have felt the need to discuss topics in related fields—for example, petrology, geomorphology, and geophysics. Inevitably this has led me away from the field in which I might claim some competency. I recognize that these side trips make me vulnerable to criticisms of misstatement, oversimplification, omission, and imbalance. Nonetheless this approach seemed preferable to entrenching myself behind an easily defended but tightly constraining wall of specialization.

Although my approach has been historical, I should emphasize that this is *not* intended to be a comprehensive review of all former studies of the Moon. For example, the pioneering work of Harold C. Urey has not received the attention it deserves. Decisions to include or exclude references to published work were based on the applicability of that work to our developing geologic and stratigraphic understanding of the Moon. I apologize in advance to those readers who will be dismayed to find that some unrelated but important publications have not been discussed.

I have made no effort to revise the first eleven chapters of the book in the light of Apollo XI results. Even in the few months since these chapters were written the receipt of new information precisely dates many passages as "pre-Apollo–post-Surveyor." The recent analysis of rocks brought back by the Apollo XI astronauts contradicts some previously held views and makes other earlier speculations look like axiomatic truths. Rather than updating a major part of the text—an exhausting exercise at best—I thought that the reader might be better entertained by doing his own mental editing, determining what changes must be made and what conclusions are still valid.

In balance, and perhaps in contrast with prevailing views, I am not sure that many changes are necessary. Administrators of the space program are in the habit of billing each mission as *the* critical step in the advancement of the total program. In a similar vein scientists closely connected with the program have sometimes been too cooperative in publicizing pre-mission controversies which, they imply, will be settled once and for all when the first data of a particular mission are received. This simplistic point of view preceded the Ranger missions with the first detailed pictures of the Moon's surface, the Surveyor missions with the first chemical analysis of lunar material, and most recently the Apollo XI mission with the first return of lunar material for analysis on Earth. So much emphasis is being placed on each dramatic "first" that there is a real danger scientists will themselves begin to believe that study of the Moon is somehow exempt from a requirement for orderly collection of data and gradual sharpening of insights. Knowing the tortuous complexity of the Earth's geologic record it is—at the very least—unsettling to hear some scientists publicly state that analysis of the first few rocks hastily collected from an area on the Moon some tens of meters in size will settle questions of the Moon's origin and history.

Faced with the precipitously rapid investigation of the Moon, I would maintain that there is a special need to provide a permanent and accessible record of the many geologic studies which preceded the Apollo flights. Of course, some of these are now invalidated or will be contradicted by subsequent collection of data, but many others will provide, either in principle or in fact, important first steps on which additional work must be based.

October, 1969 THOMAS A. MUTCH

NOTE TO THE REVISED EDITION

The principal change in the revised edition is a complete rewriting of Chapter XII, which deals with the results of the Apollo missions. More than forty new figures have been included, and an attempt has been made to sift through the numerous publications that analyze Apollo results. Revisions elsewhere in the book are minor, although readers should note that Appendix A has been updated.

My own feeling is that most of the data and interpretations within the first eleven chapters remain current and, in large part, confirmed by the Apollo explorations. The brief essay of Chapter XI, "Lunar Stratigraphy Reconsidered," obviously could, itself, be reconsidered. I leave it unchanged, because I think it is of some historical value to preserve this pre-Apollo analysis of the value of geologic interpretations *before* the first lunar landing.

June, 1972 THOMAS A. MUTCH

the Surveyor missions with the first chemical analysis of lunar material, and most recently the Apollo XI mission with the first return of lunar material for analysis on Earth. So much emphasis is being placed on each dramatic "first" that there is a real danger scientists will themselves begin to believe that study of the Moon is somehow exempt from a requirement for orderly collection of data and gradual sharpening of insights. Knowing the tortuous complexity of the Earth's geologic record it is—at the very least—unsettling to hear some scientists publicly state that analysis of the first few rocks hastily collected from an area on the Moon some tens of meters in size will settle questions of the Moon's origin and history.

Faced with the precipitously rapid investigation of the Moon, I would maintain that there is a special need to provide a permanent and accessible record of the many geologic studies which preceded the Apollo flights. Of course, some of these are now invalidated or will be contradicted by subsequent collection of data, but many others will provide, either in principle or in fact, important first steps on which additional work must be based.

October, 1969 THOMAS A. MUTCH

NOTE TO THE REVISED EDITION

The principal change in the revised edition is a complete rewriting of Chapter XII, which deals with the results of the Apollo missions. More than forty new figures have been included, and an attempt has been made to sift through the numerous publications that analyze Apollo results. Revisions elsewhere in the book are minor, although readers should note that Appendix A has been updated.

My own feeling is that most of the data and interpretations within the first eleven chapters remain current and, in large part, confirmed by the Apollo explorations. The brief essay of Chapter XI, "Lunar Stratigraphy Reconsidered," obviously could, itself, be reconsidered. I leave it unchanged, because I think it is of some historical value to preserve this pre-Apollo analysis of the value of geologic interpretations *before* the first lunar landing.

June, 1972 THOMAS A. MUTCH

Acknowledgments

I WISH to thank the personnel of the U.S. Geological Survey Branch of Astrogeologic Studies for their generous cooperation in preparation of this book. Eugene M. Shoemaker was instrumental in stimulating my interest in lunar geology and in making it possible for me to pursue these interests. My introduction to the geology of the Moon came during the summer and fall of 1967 when I was a visitor at the U.S. Geological Survey Center of Astrogeology in Flagstaff, Arizona. During that time and on several occasions since then the Geological Survey scientists have generously shared with me a great deal of information concerning lunar geology. The proof of my indebtedness is found in the large number of references to their reports. I am particularly indebted to Harold Masursky, John F. McCauley, and Don E. Wilhelms for reading several versions of the manuscript and offering many helpful suggestions.

George W. Colton and Ronald S. Saunders also made numerous suggestions for improvement of the manuscript. In preparing the revised edition, James W. Head, III provided a great deal of help with regard to the geology of Apollo sites. John A. Creaser played a major role in drafting and arranging most of the figures. Additional help was provided by Virginia Washburn. Margaret T. Cummings and Cynthia C. Downing cheerfully typed countless versions of the final manuscript. Editorial help was provided by Sara Lou Johnson, and Dorothy Hollmann and Frank Mahood of Princeton University Press. Numerous individuals and publishers have permitted me to use figures previously published or privately circulated. Credits are noted in the figure captions.

Finally, it will be clear to all readers that I am greatly indebted to the National Aeronautics and Space Administration. Without the program of lunar exploration sponsored by NASA, we would have none of the magnificent photographs which make up a major part of this book.

Although I have profited greatly from the assistance of others I freely acknowledge my own responsibility for the selective inclusion—and exclusion—of data as well as interpretation of these data. The final manuscript is a personal evaluation of lunar geology and its underlying principles.

Contents

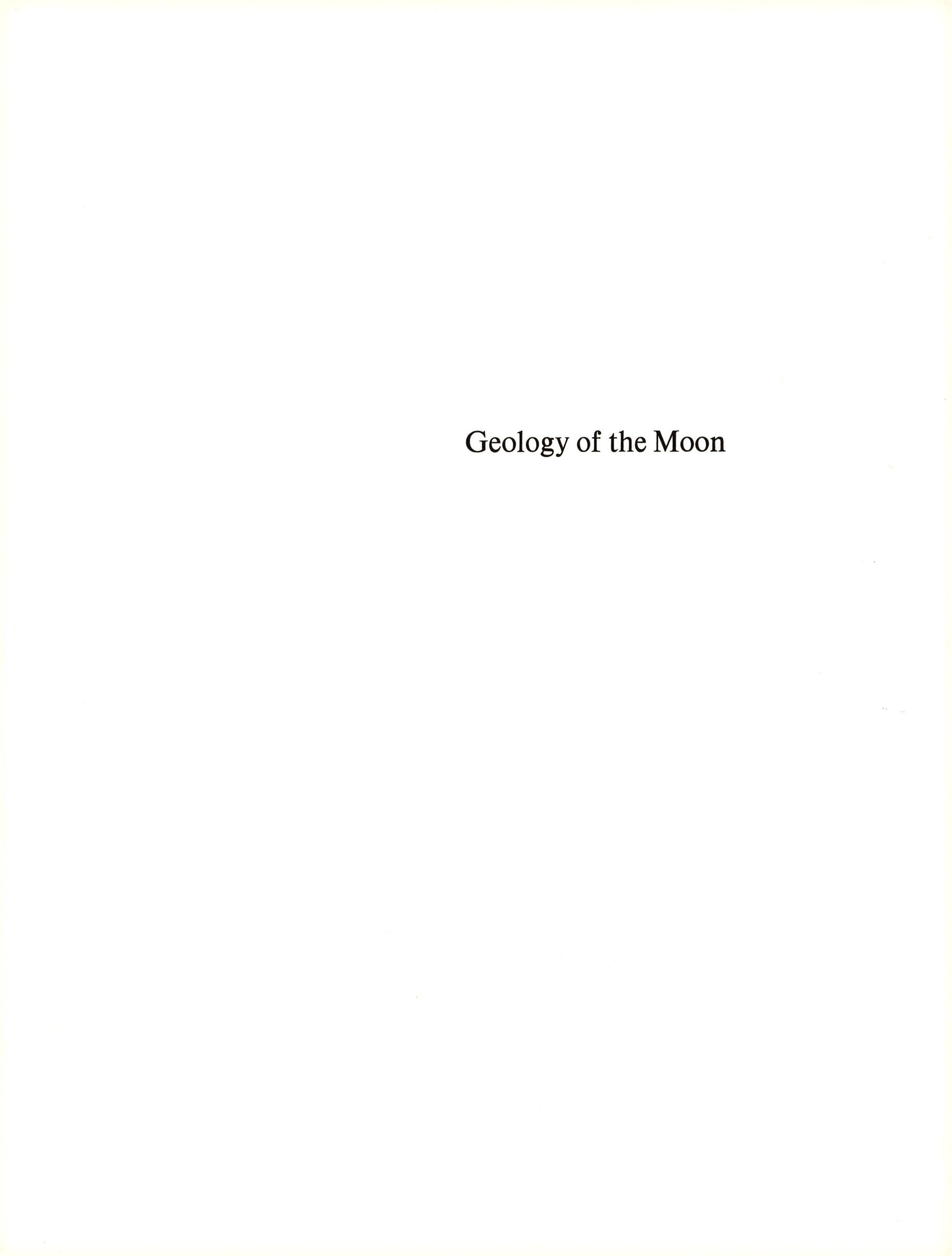

Geology of the Moon

History's Lessons

CHAPTER I

AT the very outset it is appropriate to point out the irony inherent in what we often refer to as the development of scientific thought. Few people would contend that men have become progressively more intelligent over the past centuries. But at the same time one often encounters the attitude that scientists of today will never repeat those errors of observation and interpretation made in previous generations. Unfortunately the scientists who present this point of view often are the ones so involved in "original" research that they have no opportunity to review the history of thought in their own fields and to become familiar with the origins of ideas recorded in this history.

The Moon is an excellent proving ground for the thesis that scientists profit from the experiences of their predecessors. A great deal of time and effort, especially during the past two hundred years, has been spent in study of the Earth's geologic history. Now, quite suddenly accomplishments of the "space age" have provided opportunity to construct a reasonably detailed geologic history of the Moon. The proponents of this venture argue that geologic principles worked out from examples on the Earth can be applied equally well on the Moon. It is as if we have been allowed to practice on one planet, freely drawing and erasing patterns as our knowledge and technical skills increase. Then, for a final version, we are presented a second planet[1] virtually unblemished by the pencil marks of previous geologists.

However, as we start this final effort have we really taken the trouble to review the principles of terrestrial geology and to reflect upon the limits of the subject? This is an intriguing question which will repeatedly come to the surface throughout the book.

GEOLOGIC STUDIES OF THE EARTH

In order to set the stage properly, a brief historical review of geologic work on the Earth and on the Moon is necessary. One disclaimer must be inserted. Here and elsewhere in the book geology is defined as that science which deals with the history of a planet revealed in its rocks, especially layered rocks. Some scientists will protest that this definition unfairly favors the historical and stratigraphic aspects. They might point out that significant work in many geologic subfields is carried on outside this historical

[1] The Moon here is included among the planets. Though this is not strictly accurate, it results in simpler sentences.

framework. For example, experimental synthesis of minerals or theoretical study of fault generation are problems not directly involving an appreciation of geologic time. Nonetheless, the geologic importance of such problems is fully realized only when they are considered in an historical context.

In any brief review of the history of science one is inclined simply to tabulate names, dates, and discoveries. However, critical inspection of the historical record demonstrates how hazardous it is to fabricate a black and white story of ignorance and knowledge, all based on the accomplishments of a few selected scientists. Instead, the development of thought is demonstrably circuitous, fortuitous, and repetitive. The work of the historian is further complicated by the fact that the accomplishments of many ancient scientists are known only through second-hand report. We have access to the published results of more recent scientists, but many of these publications raise more questions than they answer. To what extent was a particular writer influenced by preceding articles or by conversations with other scientists? To what extent did he himself influence the development of others? Did he really understand the importance of a prophetic conclusion inconspicuously sandwiched between more prominent irrelevant observations? Of course we cannot hope to answer all those questions, but at least we can be properly sceptical about any quick tour through great moments in science—including, perhaps, the one that follows.

The eighteenth century provides a convenient dividing point in the history of geology. Before then observations were numerous but generally unsystematic. There were a few incisive discussions of the superposition of strata, such as that by Steno (1638–1687), but no general scheme of evolution for the Earth and its inhabitants was widely considered. The inhibiting influence of the Church was probably the chief reason for this neglect. Although churchmen openly encouraged scientific effort in many other areas, geology, with its strong historical content, was disturbingly at odds with that history which had been constructed from Biblical accounts of the creation.[2]

The concept of a geologic succession of strata was developed in considerable detail by both Arduino (1713–1795) and Lehmann (1719–1767). However, their contributions were overshadowed by the teachings of a German professor, Werner (1749–1817), who proposed that a single worldwide flood had systematically deposited most of the rocks visible on the Earth's surface. "Proposed" is perhaps too mild a word, for Werner commonly made no distinctions between hypothesis and fact. His writings abound with such phrases as "it is obvious," "we are certain," "we know," and "we are convinced." Werner was an eloquent and superbly persuasive lecturer. He had a truly messianic influence on his students, who apparently accepted without question much which he chose to present as fact.

[2] Although heated controversies about the reality of biologic evolution generally belong to the past, it is possible that a bias against gradual geologic change continues to affect the work of many contemporary scientists. Surely it is undeniable that several recent theories of a catastrophic history for the Earth have a great attraction for many laymen (e.g., Velikovsky, 1950).

Opposed to Werner's views were those of a Scottish geologist, Hutton (1726–1797). Hutton was a keen observational scientist, content to let his theories follow naturally from his observations. An examination of sediments and rocks coupled with a study of the recent effects of running water led him to the conclusion that processes of sedimentary erosion and deposition operating in his day had operated throughout the past at approximately the same rate. This concept implied a vastness of geologic time that stood in contrast to the brief period of accumulation of rocks during Werner's proposed catastrophic flood. As Hutton himself maintained (1788, p. 304): "The result, therefore, of our present enquiry is that we find no vestige of a beginning, no prospect of an end."

Hutton's realization that presently observable processes have gradually changed the surface of the Earth has earned him a place as one of the first proponents of "uniformitarianism." In addition, his demonstration that some rocks have solidified from a molten state, intruding previously formed rocks, makes him one of the early advocates of "plutonism."

In some historical reviews the early nineteenth century is pictured as a time of heated controversy between Werner and his neptunist-catastrophic followers on one hand and Hutton and his plutonist-uniformitarian supporters on the other. In fact, Hutton's career slightly preceded Werner's, and there is no indication that the two ever met face to face. In addition, Hutton was absorbed more in pursuing his own investigations than in publicizing them. His few publications received generally favorable attention, but were also vehemently attacked by some clergymen and scientists. Unfortunately, the progress of geology in Hutton's native land was effectively retarded by one of Werner's students, Jameson (1774–1854), who taught at the University of Edinburgh. He seems to have inherited much of Werner's charisma and dogmatism.

Jameson notwithstanding, the ultimate resolution of the conflict between Hutton and Werner was destined to end in Hutton's favor. With Werner's death his personal attraction dissipated, and his students quickly realized that actual relations among rocks contradicted his theories. Hutton's views, on the other hand, were demonstrably correct. They were first published in 1795 in a weighty and opaquely worded two-volume treatise, *Theory of the Earth, with Proofs and Illustrations.* His views became widely known, however, only when his friend Playfair published a briefer and clearer exposition in 1802 entitled *Illustrations of the Huttonian Theory of the Earth.* The strength of its content and the clarity of its style made an immediate and lasting impact.

The early years of the nineteenth century saw a rapid development of modern historical geology and stratigraphy. A number of French geologists, including Lamarck and Cuvier, contributed to an understanding of the historical succession of strata and their contained fossils. But the work which most directly influenced the character of modern stratigraphy was done in England. Utilizing the principles of superposition, faunal succession, and correlation of rocks with similar aspect, William Smith produced a sur-

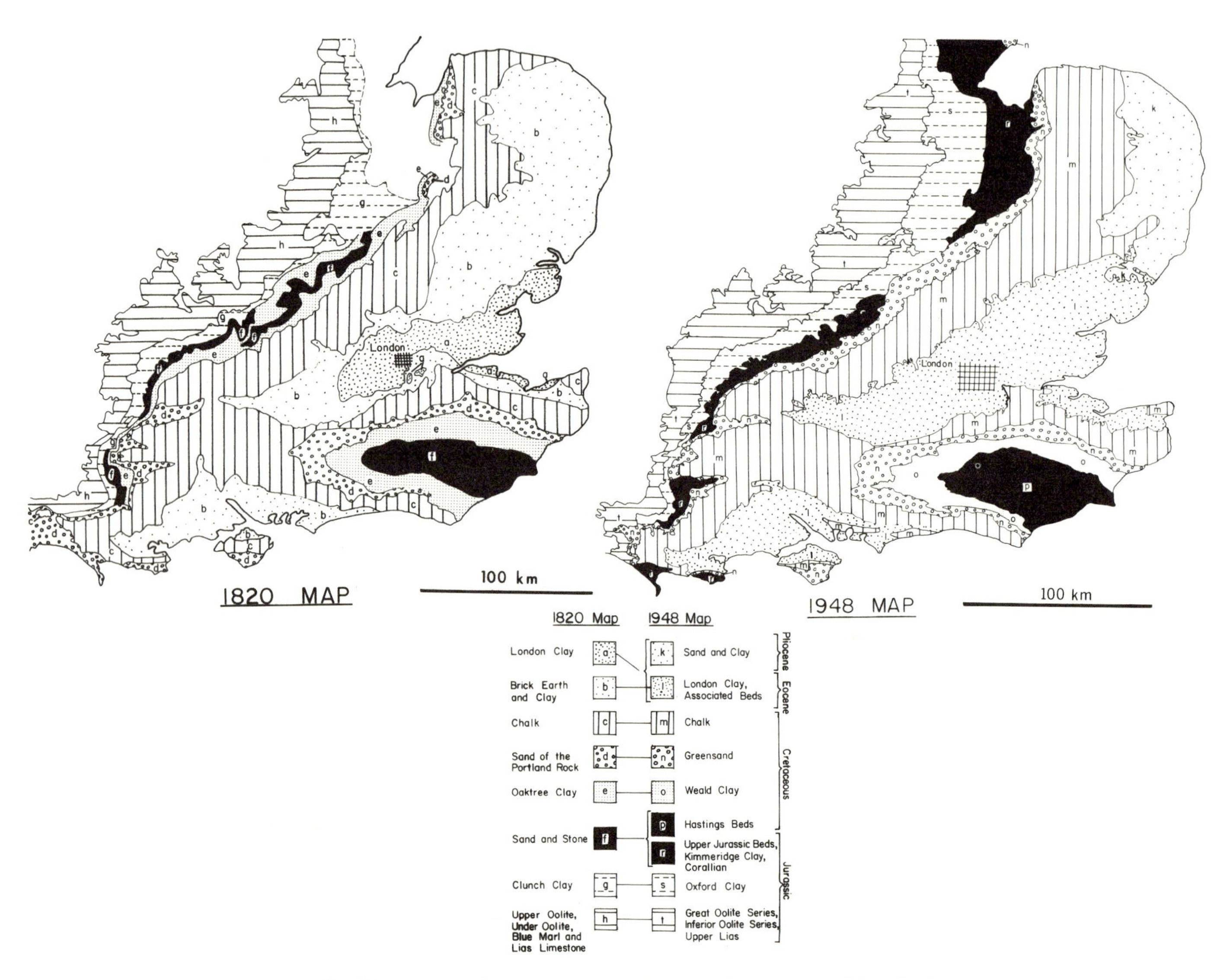

FIGURE I-1. Part of a geologic map of England published by William Smith in 1820 compared with a map published by the Geological Survey of Great Britain in 1948. Both maps have been slightly modified. Similarities occur not only on the maps but also in the description and arrangement of geologic units in the legends.

prisingly detailed map of England in 1820 (Fig. I-1). Refinements in subdivision of rocks in England and Europe came quickly following the work of other geologists, especially Murchison, Sedgewick, and Lyell. Lyell played a particularly important role in advocating subdivision of Tertiary rocks according to their changing faunas. More than any other person he is responsible for establishing the rudiments of modern biostratigraphy.

By the latter half of the nineteenth century the geologic time scale had been cast in a form closely resembling that presently accepted (Fig. I-2). It is interesting to note that recent radioactive measurements support the original positioning of system boundaries. The durations of most of the periods are shown to be approximately the same.

The sudden—and fairly late—appreciation of Earth history following centuries of ignorance and apathy represents a type of scientific explosion more commonly associated with the present day. Equally impressive is the fact that the revolution was engineered by a small group of geologists utilizing field observations made almost exclusively in Europe and the British Isles. Seen in retrospect, the intuitive brilliance of this early stratigraphic work is unmistakable. However, a contemporary of Lyell might have argued very convincingly that Lyell's classification of rocks was essentially "European," having no relevance for other continents and other geologic ages. At that time refutation of the criticism would have been difficult, if not impossible. The certifying tests have been necessarily long and extensive.

FIGURE 1-2. The geologic time scale.

*Radiometrically determined ages (millions of years)**	*Duration of periods (millions of years)*	*Era*	*Period*	*Derivation of name*	*Time and place of pioneering work*
	2	CENOZOIC	QUATERNARY	An addition to earlier threefold divisions of layered rocks	Sicily; middle 19th century
2	63	CENOZOIC	TERTIARY	Holdover from earlier threefold divisions of layered rocks	London Basin, Paris Basin, Po Valley; middle 19th century
65	70	MESOZOIC	CRETACEOUS	Latin word for chalk	Europe, England and Russia; early and middle 19th century
135	55	MESOZOIC	JURASSIC	Jura Mts., Europe	
190	35	MESOZOIC	TRIASSIC	Threefold division of rocks of this age in Germany	
225	45	PALEOZOIC	PERMIAN	Province of Perm in Russia	Chiefly in England; early 19th century
270	50	PALEOZOIC	CARBONIFEROUS: PENNSYLVANIAN	Coal-bearing: One of the United States	
320	20	PALEOZOIC	CARBONIFEROUS: MISSISSIPPIAN	Coal-bearing: Region adjacent to upper Mississippi River	
340	60	PALEOZOIC	DEVONIAN	Devonshire, England	
400	30	PALEOZOIC	SILURIAN	The Silures, an ancient British tribe	
430	70	PALEOZOIC	ORDOVICIAN	The Ordovices, an ancient British tribe	
500	100	PALEOZOIC	CAMBRIAN	British name for Wales	
600; > 3400†		PRECAMBRIAN		Before Cambrian	Canada; late 19th century

* After Faul (1966), pp. 59-61.

† The oldest dated crustal rocks are 3,400 m.y. (Cloud, 1968). The age of the Earth is approximately 4,500 m.y. (Patterson, Tilton, and Inghram, 1955). It is probable that a significant period of time elapsed between formation of the Earth and formation of a permanently solid crust on the Earth. A review of the most ancient mineral dates and a discussion of early Earth evolution is contained in Cloud (1968).

Geologists of the nineteenth and twentieth centuries have had remarkable success in mapping and interpreting rock records according to the initial stratigraphic concepts. As more facts have been assembled, an increasingly detailed picture of the flooding of continents, the rise and fall of mountains, and the evolution of life has emerged. Some completely unsuspected chapters in Earth history have been uncovered—for example, the drifting apart of continents that once were connected. Giving proper credit to these advances one might still maintain that the dominant accomplishment of modern geologists has been nothing more than the diligent application of stratigraphic principles correctly stated more than 150 years ago.

ASTRONOMICAL STUDIES OF THE MOON

Ancient studies of the Moon—indeed, almost all studies before the twentieth century—can be conveniently divided into two categories, astronomical on one hand and geographic or geologic on the other. Astronomical studies have dealt mostly with the Moon's size, mass, distance from Earth, and motion. Well before the time of Christ, Greek scientists had correctly evaluated some of the chief astronomical features. Thales (ca. 624–547 B.C.) used as the basis for abstract deductive science certain geometric concepts that the Egyptians had used for practical purposes. Although none of his writings have survived, he reportedly understood something of the recurrent motions of planetary bodies and was able to predict a solar eclipse. Following him, both Anaxagoras (ca. 500–428 B.C.) and Aristotle (384–322 B.C.) recognized that the progression of the Moon's phases results from its illumination by sunlight.

Aristarchus (ca. 310–230 B.C.) devised an ingenious method to measure relative distances to the Sun and Moon. In brief, he reasoned that when the Moon is precisely half full, the Earth, Moon, and Sun define a right triangle with the right angle at the Moon. Measurement of the acute angle whose vertex is at the Earth permits calculation of relative lengths for the sides of the right triangle (Fig. I-3). Although several critical assumptions regarding the Moon's orbit were in error, the results of the calculation demonstrated that the Sun was much more distant from Earth than was the Moon.

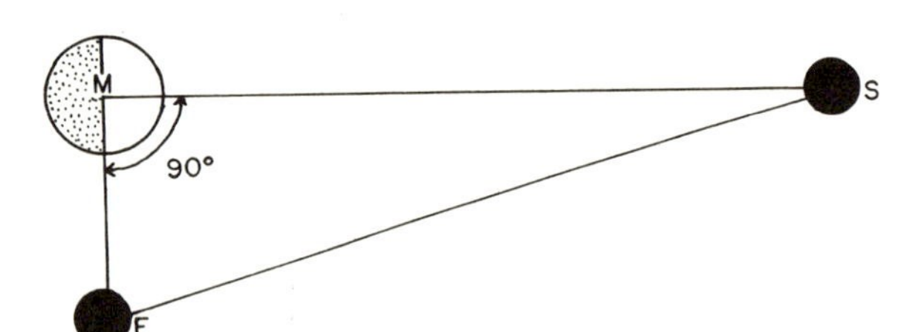

FIGURE I-3. Aristarchus' method of measuring the relative Earth–Moon and Earth–Sun distances. When the Moon is half full the angle SME is considered to be 90°. Measurement of the angle MES permits calculation of relative lengths of sides in triangle MES.

Aristarchus was able to determine relative sizes of the three bodies by measuring angular cross sections of the Sun and Moon and comparing these to the size of the Earth's shadow cast on the Moon during a lunar eclipse. Later measurements of the Earth's diameter by Eratosthenes (ca. 275–194 B.C.) permitted conversion of Aristarchus' ratios to absolute dimensions (Fig. I-4).

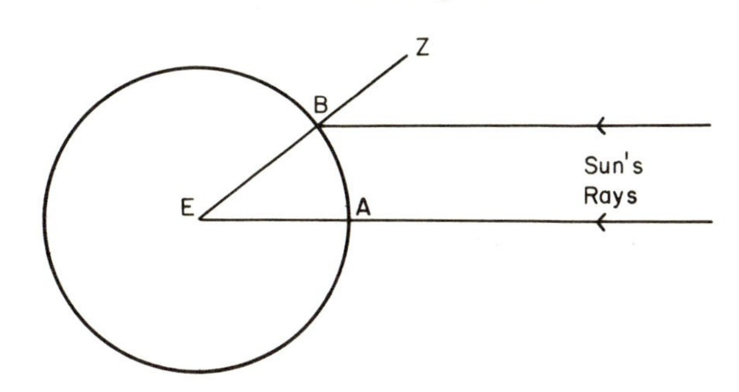

FIGURE I-4. Eratosthenes' method of measuring the size of the Earth. When the Sun is directly overhead at A the angle between the Sun's rays and the local zenith, Z, is measured at B. Since the Sun's rays which intersect the Earth are essentially parallel, the measured angle is equal to the angle BEA. Measurement of the distance from A to B, coupled with knowledge of the equivalent arc, permits calculation of the Earth's circumference and radius.

The Moon's distance from the Earth was more accurately calculated by Ptolemy (ca. A.D. 150). He made use of the fact that when the Moon is directly overhead for an observer at one point on the Earth, it will be angularly displaced from the local zenith at a second observational position. Measurement of this displacement angle and knowledge of the Earth's size and the position of the two observational points permits a geometric solution of the Earth-Moon distance relative to the Earth's diameter (Fig. I-5).

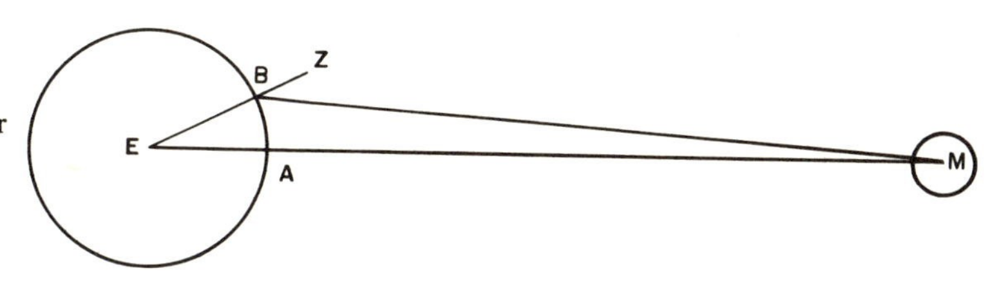

FIGURE I-5. Ptolemy's method of measuring the distance to the Moon. Two observers at A and B simultaneously view the Moon when it is directly overhead of A. The observer at B measures angle ZBM. EB, radius of the Earth, is known from Eratosthenes' calculations. Measurement of distance between A and B allows calculation of angle BEA. Knowledge of two angles, BEA and EBM, and the included side, EB, permits determination of other sides in the triangle BEM.

In reviewing studies of the Moon it is appropriate to include a brief history of ideas relating to general planetary motion. The Moon did not figure prominently in the development of these ideas, but the final result was renewed interest in all bodies of the solar system, including the Moon.

Understanding of planetary motion evolved slowly and uncertainly.[3] It is all too easy to oversimplify the history of thought by contrasting Ptolemy's "incorrect" cosmology with Copernicus' "correct" interpretations. Ptolemy devised a complex mathematical model which explained the great majority of astronomical observations. According to his scheme, the planets and the Sun travel about the Earth along paths defined by a complex of circles piled on circles. Some fourteen hundred years later Copernicus (1473–1543) described a Sun-centered geometry for all the planets, including Earth. This idea was by no means new. Both Heraclides (340–310 B.C.) and Aristarchus conceived analogous systems, and the central position of the Sun was probably considered reasonable by more than a few of Copernicus' contemporaries. Furthermore, his break with

[3] A fascinating account of this evolution is contained in Koestler (1959). Much of the present discussion is adapted from this source.

the Ptolemaic system is not as dramatic as many writers suggest. Like Ptolemy, he was convinced that the secret of planetary motion involved some combination of that perfect geometrical form, the circle. Hemmed in by this constraint his final geometrical model became even more complex than Ptolemy's. Finally, it seems likely that Copernicus considered his new cosmology only an abstract hypothesis to explain the observations, not a description of physical reality.

The ideas of Copernicus were tested and refined by Kepler (1571–1630). Unlike Copernicus, who was an indifferent astronomical observer, Kepler took account of many precise measurements of successive planetary positions made by his contemporary Tycho. After many false starts he finally realized that the planets travel along elliptical rather than circular paths. In addition he determined that both the position of the orbits and the motion of the planets in their orbits could be described by simple mathematical constructions. Blessed with hindsight, we realize the tremendous importance of these conclusions, related as they are to the dynamics of the solar system. But for Kepler, working with little knowledge of gravitational forces, the discoveries held no singular attraction.

Historical distortion of the achievements of Galileo (1564–1642) is even more common than oversimplification of the lives of Copernicus and Kepler. He is best remembered by many as an enlightened Renaissance scholar who bravely championed the ideas of modern astronomy, even when faced with trial for heresy. The actual situation is considerably more complicated. It is sufficient to note that Galileo was an arrogantly outspoken man, that he did not even deign to mention the ideas of Kepler although these were well known to him, that he presented some patently incorrect evidence in support of Copernicus, and that many contemporary churchmen treated him with both tolerance and understanding.

The last observation raises again a question regarding the oft-cited axiom that the Church attempted to stifle all scientific thought throughout the Middle Ages and the Renaissance. Certainly this simple view is not adequate to explain the slow, halting development of astronomy. Many people within the Church supported and even encouraged new models for the solar system and the universe. Astronomers were free to pursue any line of inquiry as long as they did not go out of their way to assert dogmatically the dominance of their views over the teachings of the Church fathers. The inability of so many scientists to break away from the Aristotelian view of nature advocated by the Church has causes deeper and more complex than simple censorship.

GEOGRAPHIC STUDIES OF THE MOON

Geographic studies of the Moon begin with the first telescopic observations at the start of the seventeenth century. Before then most speculations were dictated as much by emotion as by reason. On the credit side, Democritus (ca. 460–370 B.C.) considered that the markings were caused by great mountains, and we may assume that a number of Greek astronomers shared this view.

More romantic visions are recorded by Plutarch (A.D. 46–after 120) in a dialogue entitled "Concerning the Face Which Appears in the Orb of the Moon." One of the participants in this dialogue proposed that the bright and dark spots on the Moon are a reflected image of the ocean and land areas present on Earth. His friend presents the alternate hypothesis that the dark areas result from blackening of air above a region of fire. This idea is rejected chiefly for two reasons: first, that such an unsightly mixture of soot and smoke ill befits an object otherwise known for its beauty; secondly, that one should not expect to find air on the Moon since it is located in the region of a "superior substance."

In Dante's *Divine Comedy* Beatrice argues that the light and dark regions on the Moon are the light of God variably filtered through the Moon as "mingled virtue shineth through the body." The poetic eloquence is momentarily impressive, but provides little substance for the more pedestrian scientist.

Galileo was the first to publish an account of the Moon's surface as seen through a telescope. He reported his findings in a brief treatise entitled *The Starry Messenger*, published in 1610. Galileo's clear and simple descriptions of his observations stand in vivid contrast to the typically florid writing style of most of his contemporaries. His prose is direct, unencumbered with any specialized "scientific" vocabulary. Indeed, the clear factual style of the article serves as an excellent standard for twentieth-century science writers.

Among the features verbally noted by Galileo were rough topography shown by the irregular line of the terminator, the dominance of numerous circular depressions, and the shadowing effects of crater walls and mountains. On a broader scale he described the dark spots and light areas as being analogous to terrestrial oceans and land respectively, although it is not certain that he meant this analog literally. His drawings of the Moon are rudimentary but nonetheless better than most critics are willing to acknowledge. Maria Humorum, Imbrium, and Serenitatis are unmistakable. The crater Ptolemaeus is clearly shown, but other large craters—notably Copernicus—are difficult to identify (Fig. I-6).

Scores of observers quickly followed in Galileo's footsteps, and by the end of the seventeenth century more than twenty-five maps of the Moon's surface had been produced. Among the more important were those of Langrenus (1645), Rheita (1645), Hevelius (1647), Riccioli (1651), and Cassini (1680).

Langrenus' map was one of the first reconstructions of the entire Earthside hemisphere and clearly shows more than 250 of the most remarkable of all lunar features, the craters. Anybody who has suffered through a grade school geography course as many of them were taught some few years ago scarcely needs to be reminded of the fact that explorers, cartographers, and politicians revel in the naming of every promontory and depression. The Moon with its vast number of distinctive craters offered wonderful opportunities along these lines, a point which Langrenus was quick to appreciate. Since he was a scientist in the court of King Philip II of Spain, he no doubt realized the value in naming craters for assorted kings and noblemen. Few of these names survive, although one elegant crater still bears the name he assigned it, his own.

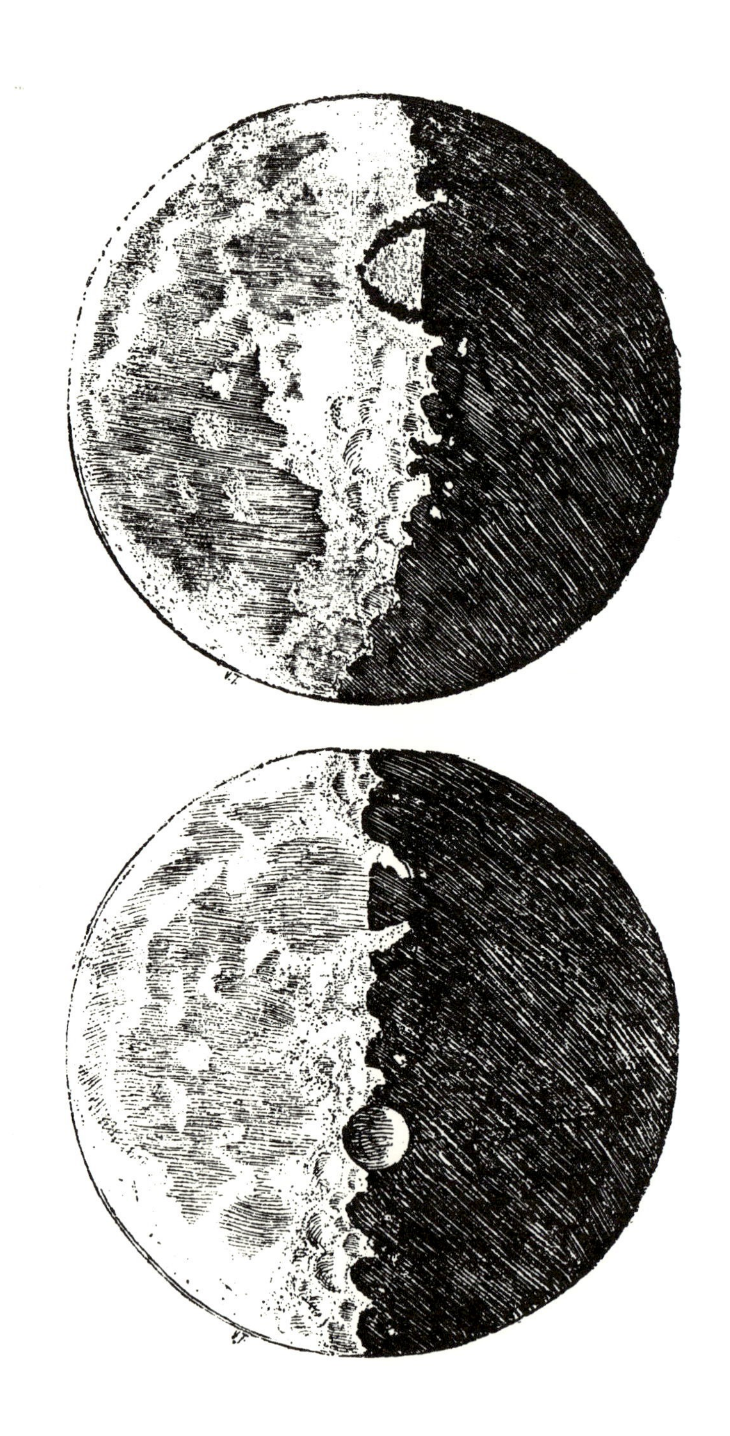

FIGURE I-6. Drawings of the Moon by Galileo (1610).

FIGURE I-7. A map of the Moon by Rheita (1645).

In Rheita's map the prominent ray system which radiates from the crater Tycho is portrayed, if somewhat diagrammatically (Fig. I-7). Numerous other craters are identifiable, including Copernicus, Kepler, Timocharis, Aristarchus, Langrenus, and Euclides. Rheita's map provides one particularly interesting contrast with Langrenus'. In the latter case the Moon is shown with north at the top just as one would view it with the unaided eye. In Rheita's map the image is rotated 180° so that north appears at the bottom. The reason is that several of the early telescopes, patterned after Galileo's, consisted of a convex objective lens and a concave eyepiece. This lens arrangement produced an erect image. It soon became apparent that much more efficient telescopes could be produced with two convex lenses. In these the image is reversed. Return to an erect position is possible with an additional lens, although loss of light results. This seemingly trivial difference in image orientation has been a thorn in the side of twentieth-century students of the Moon. Astronomers have historically insisted on putting north at the bottom, while cartographers and geologists have campaigned for a restoration to its rightful position at the top. Anyone who thinks it makes no difference should try examining a map of a familiar Earthly region upside down.

Hevelius produced the first Moon map which is reasonably accurate by modern standards. In addition, he was the first to map the peripheral regions intermittently revealed by lunar librations (Fig. I-8). Hevelius was successful in recognizing subtle variations in brightness and texture not previously depicted. Examples include the dark annulus around Tycho and a bright region around Copernicus, distinct from the crater's ray system. Throughout the maria he recognized slight variations in brightness, although he did not record them very accurately.

Riccioli is responsible for introducing the system of nomenclature which is still followed today. Rejecting the heraldic names of Langrenus he used instead the names of scholars and scientists who had studied the Moon. This has proved a reasonably satisfactory arrangement but is not without some inconsistencies. The crater Copernicus is rightfully impressive as is Kepler on a smaller scale. Galileo, however, is awarded a crater on the limb of the Moon, only nine miles in diameter and without any notable features. The crater Newton is unobtrusively tucked away near the South Pole.

Cassini's map, the last of the pioneering seventeenth-century efforts, exceeds in accuracy all those which preceded it (Fig. I-9). Central peaks are shown within many craters. The rays surrounding Copernicus are accurately delineated. Especially noteworthy is a ray loop north of Copernicus, a feature which was recently noted by Shoemaker (1962) in a ballistic analysis of crater evacuation processes. The prominent lineation which cuts across the central part of the lunar disc from lower left to upper right is an exaggerated portrayal of a single prominent ray from the crater Tycho.

Observers of the eighteenth and nineteenth centuries progressively improved the early maps, but they continued to be handicapped by the need to render an artistic facsimile of what was seen visually. Even though micrometric measuring techniques were introduced and the geography was referred to a standard set of coordinates, mappers were

FIGURE I-8. A map of the Moon by Hevelius (1647).

FIGURE I-9. A map of the Moon by Cassini (1680).

still limited to an approximation of major features. The trend toward increasing detail culminated with a map produced by Schmidt in 1878 (Fig. I-10). Printed as twenty-five separate sections, the composite map has a diameter of over 6 feet. More than 33,000 individual craters are shown.

The discovery of the photographic process ushered in a new era of lunar study. Ironically the importance of this tool was very early recognized but was not fully exploited in subsequent years. The first photograph of the Moon was obtained in 1840, only one year after the development of a workable photographic technique. Several photographic atlases were published toward the end of the nineteenth century, to be joined by more in the present century. Even so, the first atlas sufficiently comprehensive and versatile to be widely used was that edited by Kuiper in 1960.

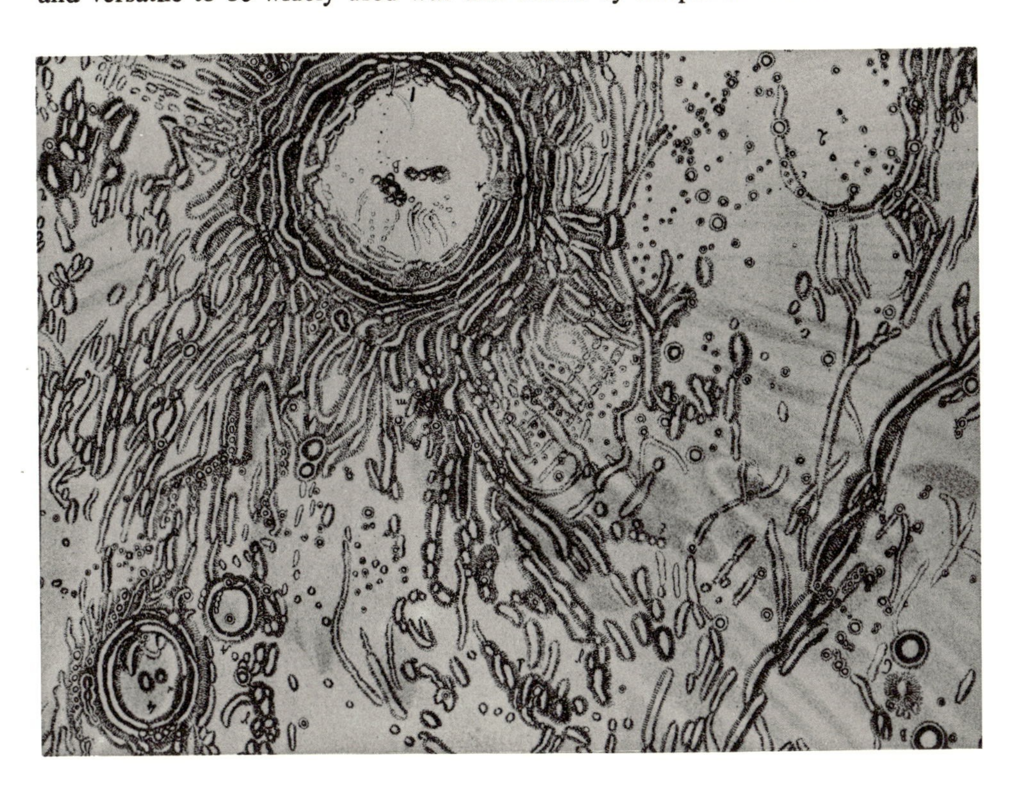

FIGURE I-10. Part of a map of the Moon by Schmidt (1878) showing the region around the crater Copernicus.

Notwithstanding the accessibility of photographs, three major efforts to construct detailed, line-drawn maps principally from visual observations have been carried out in the twentieth century by Goodacre (1910), Wilkins (1933), and Fauth (1964). These projects can be justified on the grounds that visual observations under ideal conditions have a resolving power about four times greater than the best photographs (Kuiper, 1960). But drawings, no matter how conscientiously constructed, still are subjective interpretations. In addition, few line drawings can reproduce the subtleties of tone and texture present in a photograph. It is certain that emphasis on visual observations and line drawings, coupled with limited circulation of photographs, inhibited study of the Moon by many geologists during the first half of this century. Without ready access to the original scene they were deprived opportunity for independent judgments.

The Aeronautical Chart and Information Center (ACIC) of the U.S. Air Force is making integrated use of visual observations, thousands of telescopic photographs, and—more recently—Orbiter photographs to construct topographic maps of the entire Moon at a scale of 1:1,000,000. These are air-brush portrayals with subtle textures much more clearly shown than in previous line drawings by other workers. Topographic charts for most of the near side are completed, and eventually the entire Moon may be mapped at this scale, a total of 144 quadrangles. These maps are constructed on a Mercator projection in the equatorial region, on a Lambert projection in the mid-latitudes, and on a polar stereographic projection in the Polar regions. They do not replace good Orbiter photographs as a portrayal of the lunar scene, but they provide an excellent base for geologic information (Figs. I-11 and I-12). In this way they serve the same purpose as do quadrangle maps on the Earth.

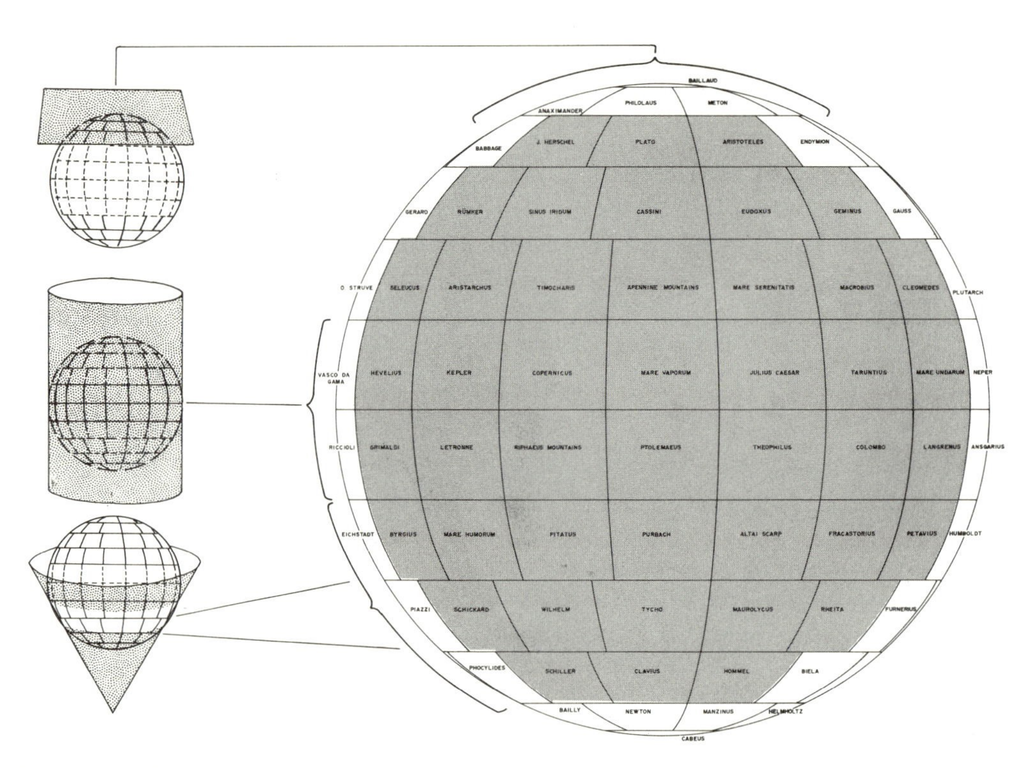

FIGURE I-11. Mapping projections employed by the Aeronautical Chart and Information Center, U.S. Air Force, in preparation of 1:1,000,000 topographic maps. Mercator projections are used for the equatorial region, Lambert conformable for the mid-latitudes, and polar stereographic for the polar regions. Geologic maps, published by the U.S. Geological Survey, are available for those quadrangles with a gray background. For more information, see Appendix A.

The sudden surge in photographic information following the success of three impacting spacecraft (Rangers), five soft-landing spacecraft (Surveyors), five orbiting satellites (Orbiters), several manned orbiting and landed spacecraft (Apollos), and various Russian spacecraft (Zond and Lunas) is truly unprecedented. Almost the entire Moon has been photographed with a resolution of 100 meters or better, and overlapping photographs provide stereoscopic coverage for many areas. The far side is revealed for the first time. Regions near the limbs, which appear badly distorted through the telescope,

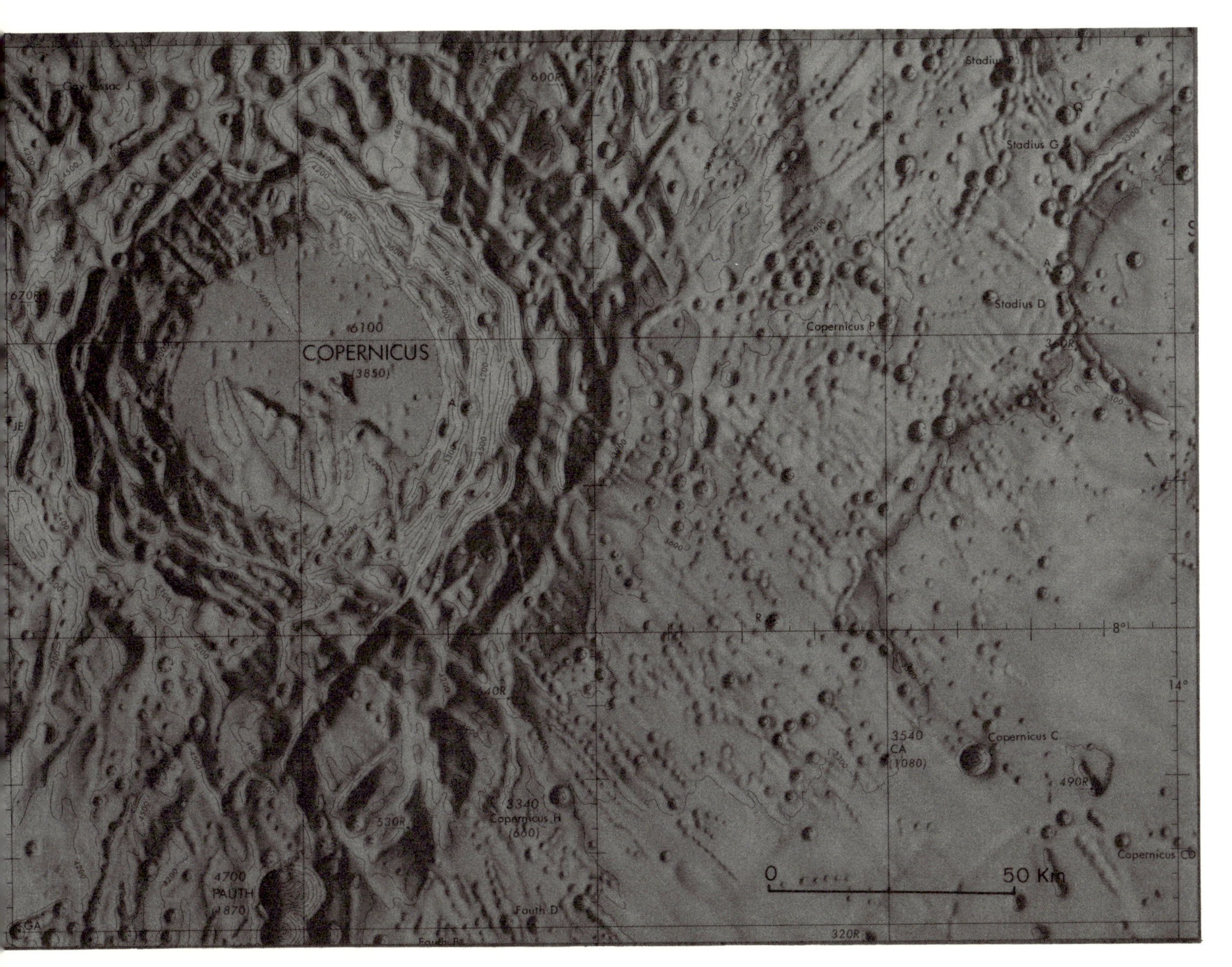

FIGURE I-12. Part of a topographic chart of the Copernicus quadrangle published by the Aeronautical Chart and Information Center, U.S. Air Force (1964). The map is published at a scale of 1:1,000,000. The figure above is slightly reduced. Contour lines are approximate. The contour interval is 300 feet.

have been photographed from points directly overhead. Finally, many small and important details are displayed in regions where telescopic observations reveal none. As one might expect the receipt of these photographs has dramatically stimulated interest in the geography and geology of the Moon.

GEOLOGIC STUDIES OF THE MOON

Geologic interpretations of the Moon prior to the space age can be divided into four groups. First, the origin of the Moon as an Earth-satellite has been a matter of serious discussion since the latter part of the nineteenth century when G. H. Darwin (1879) suggested that the Moon escaped from the Earth owing to the tidal action of the Sun. His theory was generally rejected as a result of more detailed studies of fission mechanisms, but recently has been put forward again by Wise (1963) who has proposed that fission may be related to increased rate of rotation of a primitive Earth following segregation of the Earth's core. Another theory, proposed by Kuiper (1951) and Ruskol (1962) among others, is that the Moon formed by accretion relatively close to the Earth while the Earth was still growing. Still another group of theories develops the idea that the Moon was captured by the Earth during a close approach (e.g., Gerstenkorn, 1955; Urey, 1962; and Alfvén, 1963). Incorporating elements of several theories, MacDonald (1964) has speculated that many relatively small Moons were captured by the Earth early in its history and that these coalesced into a single Moon as recently as 1.5 billion years ago.

These several reconstructions rest heavily on explanations of the orbital and bulk properties of the Moon, on chemical and mineralogical information contained in meteorites, and—more generally—on accompanying interpretations of solar system evolution. All of the theories have elements of interest for the historically oriented geologist. For example, study of present tidal effects and extrapolation of similar effects in the past leads to the conclusion that the Moon was very close to the Earth some 1.5 to 2 billion years ago (MacDonald, 1964). If the two bodies were so close, it is reasonable to expect that the event might be recorded in terrestrial sedimentary rocks which contain primary structures compatible with unusually strong tides.

A second type of lunar study concentrated attention on the physical and inferred chemical properties of surficial materials. An extensive literature on photometry of scattered Moonlight, and thermal emission, electromagnetic properties, and luminescence of the lunar surface is summarized by Kopal (1966). Inasmuch as these studies consider the character of lunar materials they might be termed geologic. However, the majority of publications are of limited use to a geologist because they do not adequately differentiate between probably different materials existing side by side on the surface of the Moon.

A third body of literature is restricted to interpretation of lunar topographic features, notably craters. Two contrasting origins for craters have been repeatedly advocated. Numerous people have suggested that the craters are volcanic depressions, either calderas or maars. The idea was first proposed by Dana in 1846. A modern endorsement notable both for length and single-mindedness is provided by Spurr (1944, 1945, 1948, 1949).

The opposing view is that the craters are results of meteoritic, asteroidal, or cometary impacts. Gilbert is commonly remembered as the author of the impact interpretation. In fact, several people anticipated him, but his analysis of the problem (1893) is particularly thorough and eloquent. Considering his commitment to the impact hypothesis, it is ironic that he failed to recognize the origin of a structure which is among the best preserved and most thoroughly studied of all impact craters—Meteor Crater, Arizona. In a paper entitled "The Origin of Hypotheses, illustrated by the Discussion of a Topographic Problem" (1896) he describes how he first thought the crater to be the scar of a "falling star." He rejected the idea in favor of a volcanic steam explosion, chiefly because experiments with the deflection of a magnetic needle failed to reveal any large iron body beneath the crater.

Certainly one of the most influential publications on the Moon in the past decade is *The Measure of the Moon* by R. B. Baldwin (1963). A number of subjects are discussed, but the central thesis of the book is that lunar craters bear close resemblance to meteoritic craters.

The case for impact origin of craters has been stated repeatedly in the modern literature. If one judges the opposing views only on the bulk of publications, the impact argument is clearly weightier than the volcanic one. Arguments between adherents of the two positions have been—and continue to be—vehement and colorful. The literature abounds with hyperbole, some in absolute earnest and some tongue-in-cheek. A more detailed discussion is saved for later chapters. Suffice it to note here that present controversy has something of the flavor of arguments between Neptunists and Plutonists concerning the nature of the crust of the Earth. A more apt comparison, well known to modern geologists, is the controversy about the origin of granite which reached an emotional peak some fifteen or twenty years ago. One group followed the traditional view, first articulated by Hutton, that all granites form from a molten state. A second group maintained that virtually all granites are the product of solid-state transformation of previously formed rocks, a process known as granitization. The ultimate resolution of this verbal battle was predictable if not very satisfactory to the protagonists. Most geologists presently believe that some granites are formed in the first way, some in the second. The crater argument is now approaching a similar rational, if equivocal, conclusion.

Many interpretations of lunar features have been limited to random observations and inadequately supported speculations. These ideas can be assembled in a fourth and final group, entertaining but not very edifying. For example, the presence of massive glaciers and accompanying vegetation patterns controlled by glacial changes has been a recurring vision. Contrary to what one might expect, this idea was not advocated solely by "outsiders" unfamiliar with the Moon's appearance. Fauth, one of the last great visual observers of the twentieth century, dogmatically asserted the presence of ice 200 miles thick (1908). Throughout a career of meticulous observation and mapping, his certainty never wavered.

In the wake of solid—sometimes a little too solid—landings of various Surveyor and Ranger spacecraft it would be difficult to maintain that the maria are actually oceans. However, several workers have maintained that the dark maria surfaces are floors of relic seas. A proposed analog between craters and atoll-like coral reefs (Beard, 1925) provides the ultimate in one form of circular reasoning.

Evidence for "human" activity has not gone unnoticed. Gruithuisen, an otherwise distinguished observer, recorded a walled and fortressed city standing in the maria (1824). Ocampo (1949) reasoned that we are looking at an atomic bomb target range.

This brief review of geologic studies of the Moon would not be complete without mention of the numerous influential contributions of H. C. Urey. His interests have ranged widely over a number of geochemical problems relating not only to the Moon, but also to the other planets. His book *The Planets: Their Origin and Development* (1952) laid the groundwork for the modern chemical-physical approach to this problem. Since then he has published many articles, dealing especially with the composition and early history of the Moon.

LUNAR STRATIGRAPHY

During the many years that geologists were unraveling Earth history by deciphering the sequence of events recorded in stratified rocks, very few scientists gave serious thought to a similar study of the Moon. Obviously this neglect stemmed partially from its inaccessibility. But, as we shall see shortly, a reasonable sequence of geologic events can be deduced from telescopic photographs. Though not widely circulated, photographs with the necessary degree of resolution have been in existence for most of the twentieth century.

The first modern attempt to recognize a geologic history preserved in surficial rocks came in 1962 with the publication of an article entitled "Stratigraphic Basis for a Lunar Time Scale" by Shoemaker and Hackman. Within the brief span of twelve pages, the authors proposed that the Moon could be mapped geologically, that a sequence of events was clearly visible, and that a stratigraphic succession comparable to that on Earth could be defined. These are tremendously exciting possibilities.

For the sake of dramatic emphasis one is tempted to state that the ideas of Shoemaker and Hackman were completely new. However, they had been anticipated by at least two other famous geologists. Gilbert (1893) described the ejection of material from the Imbrium basin following a giant impact and suggested that a geologic history of the Moon might be constructed around this event. A more specific statement was supplied by Barrell (1927), who discussed lunar superposition relations in a manner conceptually identical to that of Shoemaker and Hackman.

Spurr (1948) established a series of lunar time divisions and attempted to decipher a sequence of structural, igneous, and sedimentary events within each division. A Russian scientist, Khabakov (1960), proposed dividing lunar history into seven time periods. Significantly, though, he defined periods of formation of topographic relief, not periods of deposition. In this sense his argument was not developed along stratigraphic lines.

After giving proper credit to these several efforts to interpret the Moon geologically, the fact remains that Shoemaker and Hackman were the first to demonstrate specifically that the Moon's crust could be interpreted stratigraphically. Mapping a region in the vicinity of Copernicus, the authors divided the stratigraphic sequence into several systems, each deposited during a particular period of lunar history. Certain of their ideas are expressed implicitly and casually, but the similarity between the stratigraphic framework for the Moon and that for the Earth was clearly intentional. As we shall see later, when a stratigrapher talks about a "system" or a "period" he is considering a category with specific limits, material and temporal. Shoemaker and Hackman had these same limits in mind (p. 298). In their own words:

> Reconnaissance studies indicate that the stratigraphic systems recognized in the Copernicus region can be correlated and mapped over most of the visible hemisphere of the Moon. . . . A lunar time scale corresponding to the stratigraphic systems described and extending from the beginning of lunar history to the present is therefore proposed . . . .

Following this pioneering effort there has been a systematic attempt to map the entire Earthside hemisphere utilizing the initial stratigraphic assumptions and many of the original stratigraphic divisions. Geologic maps at a scale of 1:1,000,000 have been published or are available as open-file reports for all 44 Earthside quadrangles (Fig. I-11, Appendix A). These maps, prepared by the U.S. Geological Survey, constitute a continuing test of the original stratigraphic framework. In later chapters there will be opportunity to consider the interim results of this test. Can the postulated systems actually be recognized throughout the Moon? Do the periods as originally conceived and defined have decipherable importance as divisions of time? On an even more fundamental level one might ask by what processes these stratigraphic sections are produced. Are we justified in treating them as sedimentary sequences analogous to those on Earth?

The work of Shoemaker, Hackman, and those few other geologists involved in the lunar mapping program provides a rationale for interpretation of the Moon's history comparable to that provided for the Earth by Smith, Lyell, and their contemporaries. Suddenly we find ourselves on another planet retreading the paths of nineteenth-century stratigraphy and historical geology on Earth. A logical assumption is that these century-old paths should now be familiar, that all the turns should be marked and the dead ends closed to further traffic. But logic, especially that based on past experience, is a fragile quantity—which brings us full circle back to the first paragraphs of this chapter.

tographs taken with cameras by astronauts orbiting the Moon will probably be more suitable for stereoscopic viewing.

basalt in composition (Turkevich and others, 1968*a*). Analyses obtained at the Surveyor VII site, just north of Tycho, reveal significant differences, suggesting rocks more anorthositic than basaltic (Turkevich and others, 1968*b*).

The Moon's Shape and Motion

CHAPTER

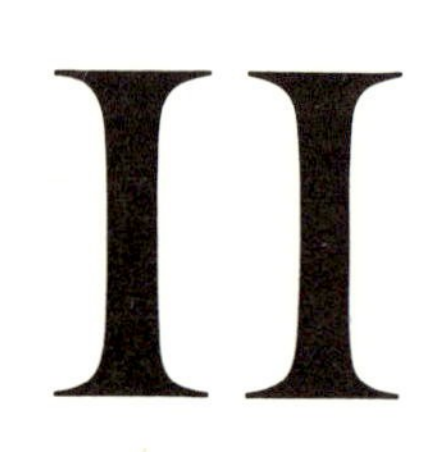

COMPREHENSIVE and detailed discussions of the Moon's shape and motion are found in a number of books (e.g. Arthur, 1963; Kopal, 1966). Expectably, these are traditional subjects of interest for astronomers, who have viewed the Moon from a distance. However, our purpose here is only to consider several phenomena which affect visual observations and which have geologic significance. Geometric and mathematical treatments are included more for explanatory clarity than for any fundamental rigor. This approach may appear oversimplified to those with astronomical training, but it is adequate for the task at hand. On Earth, for example, stratigraphic studies have been independent of all but the most rudimentary knowledge of our own planet's large-scale shape and motion. On the other hand, one of the chief tasks of a field geologist is spatial analysis of geologic features: vertical successions of strata, orientation of fault planes, or trend directions of fold axes. In order to make equivalent analyses of lunar features it is important to have some understanding of how information from photographs can be utilized to locate features relative to a standard coordinate system.

MOTION OF THE MOON

The form of the Moon's course about the Earth is not a circle but an ellipse. The mean distance between the centers of the two bodies is 384,402 km. However, the distance at perigee (closest approach) is only 356,410 km, and the distance at apogee (farthest separation) is 406,697 km. These values have recently been determined by reflection of radar waves and are known to an accuracy of less than 2 km (Bruton and and others, 1959). As a consequence of the elliptical orbit the apparent size of the lunar disc viewed from Earth varies systematically. The perigee diameter is approximately 12 percent greater than the apogee diameter.

Knowing the Earth–Moon distance and the angular diameter of the Moon, it is a simple matter to determine the Moon's linear diameter as very nearly 3,476 km (Fig. II-1). A noteworthy comparison is that the Moon's diameter is approximately one-quarter that of the Earth's. Since the surface area of a sphere varies with the square of the diameter, the Moon's area is only one-sixteenth that of the Earth's. Put a little differently, the surface of the entire Moon covers an area only slightly larger than Africa.

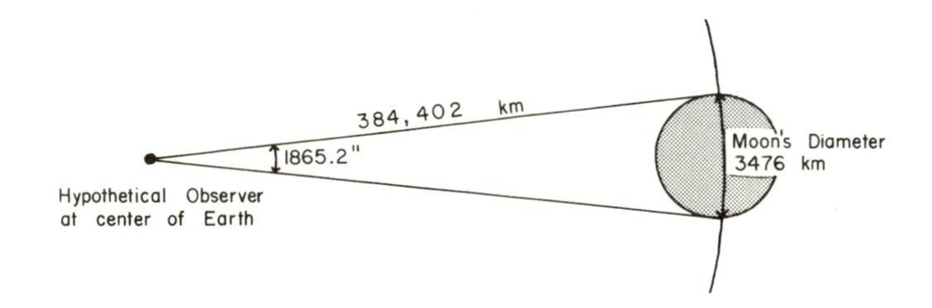

FIGURE II-1. The Moon's diameter is a fraction of the great circle with radius 384,402 km, corresponding to the fractional value that 1,865.2″ (approximately ½°) is of 360°. Because of the small angular diameter of the Moon, the measured arc is nearly identical with the actual straight diameter.

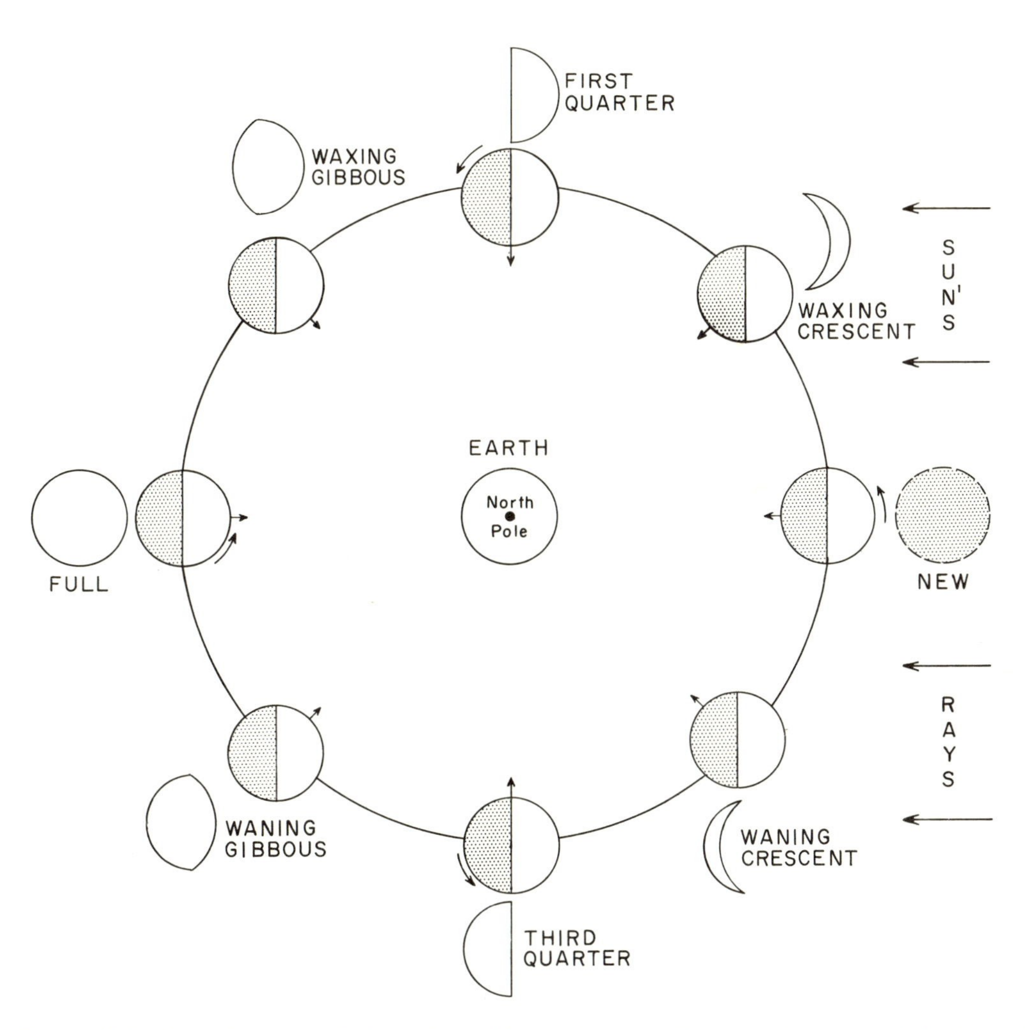

FIGURE II-2. Approximate revolution of the Moon about the Earth during one period (approximately 27 days). Note that the Moon rotates as it revolves so that the same hemisphere always faces the Earth.

The approximate geometry of the Moon's rotation about its axis and revolution in its orbit about the Earth is shown in Figure II-2. In general, the result of these motions is that one-half of the Moon is never visible. A more precise statement is possible after considering two empirical laws stated by Kepler in 1609:

(i) Each planet moves about the Sun in an elliptical orbit. (For the present problem we can substitute "Moon" for "planet" and "Earth" for "Sun.")
(ii) The straight line joining a planet (Moon) and the Sun (Earth) sweeps over equal areas in equal times.

and three empirical laws stated by Cassini toward the end of the seventeenth century:

(i) The Moon rotates eastward about a fixed axis with constant angular velocity and in a period equal to that for one complete revolution about the Earth.
(ii) The planes of the Moon's equator and the Earth's orbit (ecliptic) meet at a fixed angle, namely 1°32′.
(iii) The normal to the ecliptic, the normal to the plane of the Moon's orbit, and the axis of rotation of the Moon all lie in the same plane.

Some of these relations are shown in Figure II-3.

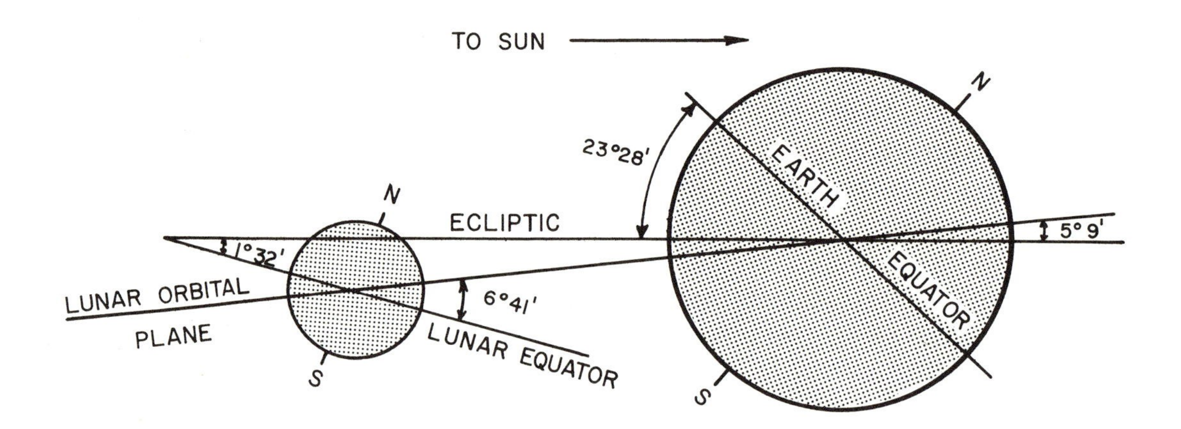

FIGURE II-3. Schematic drawing showing relative orientation of Earth, Moon, and ecliptic.

Brief study of these several laws leads to the conclusion that the relative motion of the Earth and Moon will allow the observer over a period of time to see considerably more than one-half of the lunar surface. These phenomena are known as optical librations.

The first is termed a libration in longitude. Although the Moon rotates at a constant rate, its angular velocity of revolution in its elliptical orbit varies. The average rates are the same; that is, the Moon rotates once during a complete revolution. However, since the two get temporarily out of step in the course of a single period because of the Moon's elliptical orbit, we are allowed to see first more of one side, then more of the other. The maximum angular displacement is nearly 8°. The geometry is diagrammatically shown in Figure II-4.

FIGURE II-4. Schematic drawing showing libration in longitude due to elliptical orbit of Moon. At positions 1 (perigee) and 3 (apogee) the subterrestrial point "S" lies approximately on the prime meridian. At position 2 the subterrestrial point is displaced to the right of the prime meridian (positive libration in longitude), and at position 4 the subterrestrial point is displaced to the left of the prime meridian (negative libration in longitude).

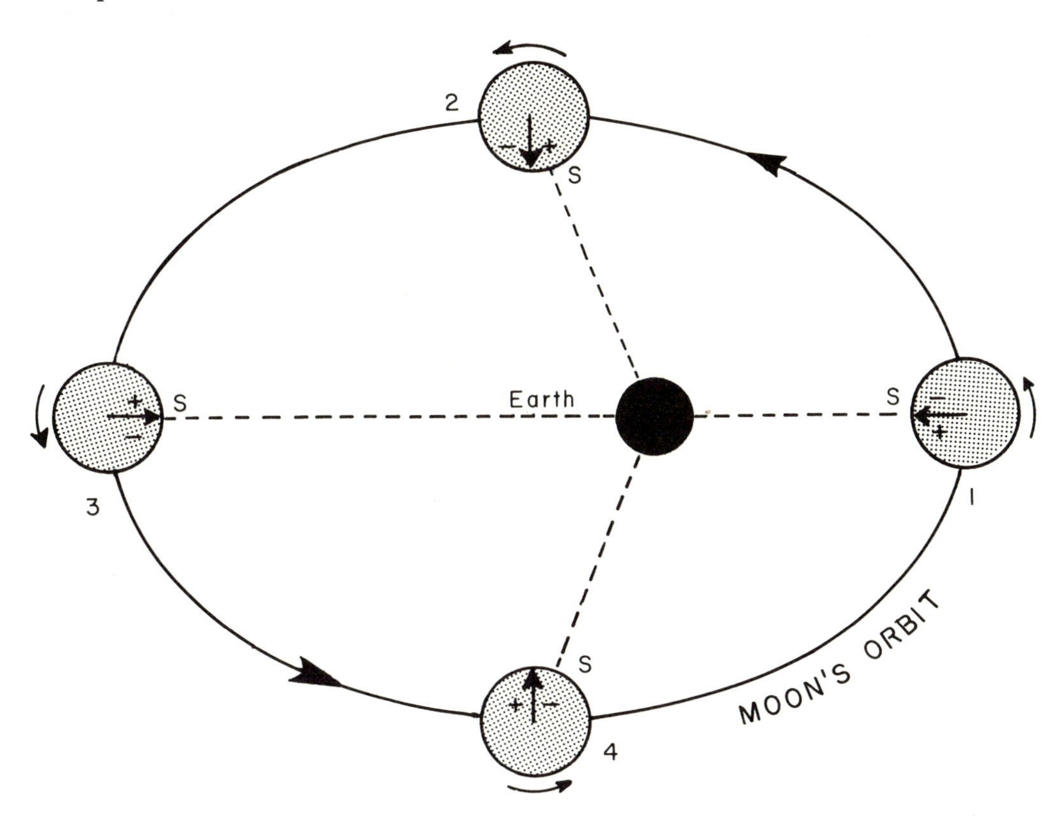

The second libration is one in latitude. This results from the fact that the Moon's axis of rotation is tilted 6°41′ from a perpendicular to its orbital plane. Because this orientation remains constant during the Moon's revolution about the Earth we are permitted to see alternately more of each of the polar regions. This relationship is shown in Figure II-5.

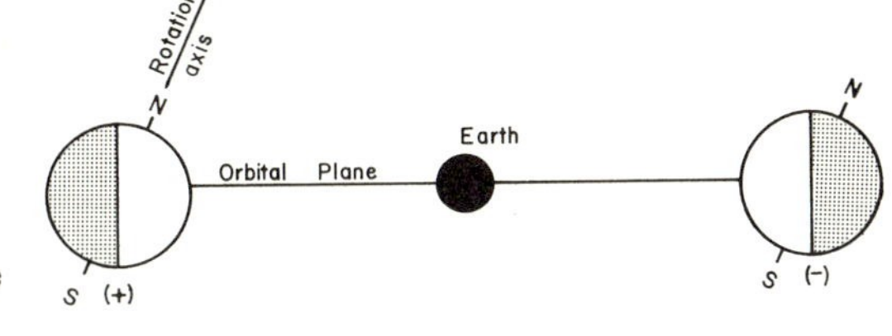

FIGURE II-5. Schematic drawing showing libration in latitude. Because the Moon's rotational axis is tilted to its orbital plane, an observer on Earth can see alternately more of the north polar regions (positive libration in latitude) and south polar regions (negative libration in latitude).

Finally, there is a diurnal libration, actually the result of changing terrestrial perspective rather than a consequence of the Moon's motion. As we observe the Moon from its rising to setting we are transported across a base line equivalent in length to the

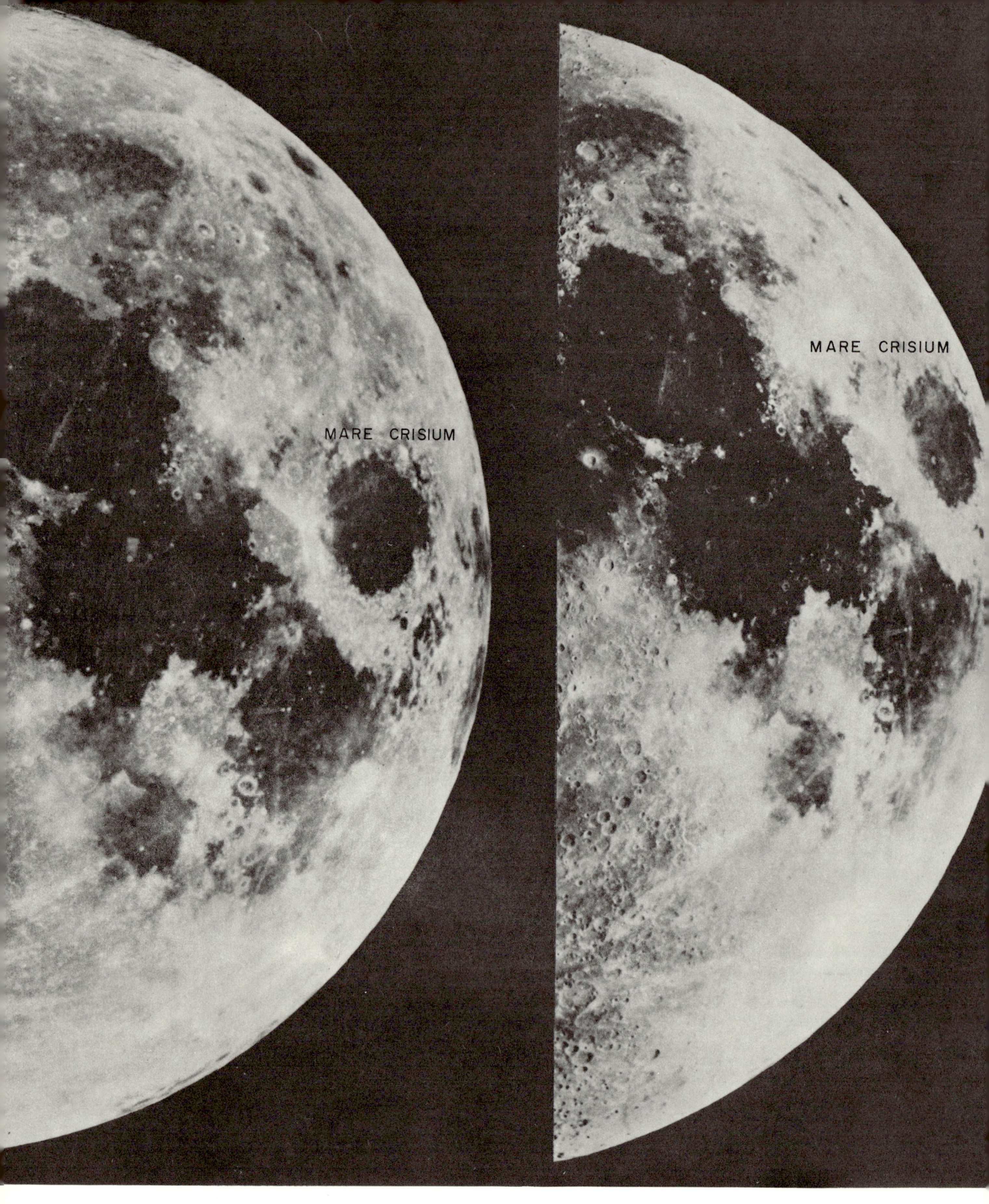

FIGURE II-6. The effect of librations in latitude and longitude on the eastern limb of the Moon. A positive libration in longitude produces a less distorted view of Mare Crisium (from Kopal, 1965).

Earth's diameter. This motion allows us to peer successively around the western and eastern edges. This libration has a value of about 1°.

As a consequence of these combined librations we can see from Earth, at one time or another, almost 60 percent of the entire lunar surface. The magnitude of the effect is suggested visually in Figure II-6. Appreciation of optical librations is clearly important for the geologist. A particular feature near the limb might be completely absent, or at best strongly distorted, on one telescopic photograph. Another photograph, taken at a time of more favorable libration, will show this same feature quite clearly.

COORDINATES OF LUNAR FEATURES

In locating features on the Moon, the most basic problem is to decide which way is up. In telescopic images the lunar disc is rotated 180° relative to the image seen with the naked eye. Accordingly the bottom appears at the top, and the right side on the left. Since this reversed image is the one most familiar to astronomers they have traditionally published their plates with the same orientation, indicating north at the bottom of the plate. Some of these reversed plates have their left side labelled west to coincide with west in our sky; other plates have the east and west labels reversed.

The convention used in most recent cartographic and geologic publications, and adopted by the International Astronomical Union in 1961, is to place the cardinal points in positions consistent with the image of the Moon as seen with the unaided eye. North is at the top, south at the bottom, east on the right, and west on the left. A man standing on the Moon would see the Sun rising in the east just as he would on Earth. This convention, the so-called astrônautical convention, is the one used throughout this book.

The axes of the lunar coordinate system are defined by the principal axes of inertia of the Moon. This is a coordinate system comparable to that for the Earth, where the defining planes are marked by the equator and a prime meridian arbitrarily chosen to pass through Greenwich Observatory. As on the Earth, so on the Moon, distances from the reference axes are measured in latitude and longitude. One convention is to measure latitude (β) positively northward and negatively southward; longitude (λ) positively eastward and negatively westward. A more familiar notation for geologists is simply 10°N latitude or 25°E longitude.

As one might expect, the point of zero longitude and latitude is not marked by any particular feature. Historically, the positions of many structures have been measured relative to Mösting A, a small bright crater close to this intersection. Precise determinations of latitude and longitude for Mösting A allow final transformation to absolute coordinates for other features.

The actual measurement of coordinates for particular lunar features is not exceedingly complicated in principle, involving only definition of azimuth and distance from a reference point. The measurement can be accomplished either visually or on a photograph.

SHAPE OF THE MOON

Determination of the Moon's detailed shape is much more difficult than the determination of coordinates. In the latter problem the Moon is treated as a perfect sphere; no account is taken of the fact that there are small but significant deviations from sphericity. In order to solve the shape problem it is necessary to determine precisely the radial distance of surface points from the Moon's center.

There are three principal ways of determining the exact position of lunar surface points. The first two, here called the limb and terminator techniques, are somewhat specialized. The third approach, stereoscopic in principle, is more generalized.

The limb technique involves measurement of the lunar globe seen in projection. Obviously, the portion of the Moon which can be so examined from the Earth is controlled by the libration magnitudes and is extremely small. The terminator technique relies on a striking phenomenon first noted by Galileo. On a smooth spherical globe the terminator should define a smoothly curving great circle. But the actual terminator is extremely irregular, reflecting the rough surface illuminated by the low Sun. In a sense, then, each one of the terminator lines defines a topographic cross section, and analysis of successively displayed terminators should yield a continuous three-dimensional map. The chief difficulty lies in accurate determination of the precise terminator line. What is actually seen is a broad, diffuse zone, the character of which can be radically changed simply by altering the photographic exposure time.

The stereoscopic approach involves analysis of several photographs taken at times of different libration. If the Moon were perfectly spherical the coordinates of a point calculated from two photographs would be the same. In fact the coordinates are different, due both to the large-scale nonsphericity of the Moon and to local irregularities in relief. Analysis of differences in measurement between the photographs allows one to calculate absolute elevations. The mathematics is not simple, but neither is it uncertain. The difficulties lie in two other areas. First, measurements and calculations are so time-consuming that the number of available points is limited. Most of the nineteenth- and early twentieth-century contour maps were based on less than 100 points for the entire visible hemisphere. Recent analyses by Baldwin (1963) and Mills (1968) include 696 and 917 points respectively, still sparse control for an area the approximate size of the United States. A second problem is more qualitative. It concerns the problem of making critical point determinations on lunar craters and peaks that are, in fact, massive structures, which often look completely different from different perspectives. In addition, location of identical points on photograph pairs is hindered by the poor resolution inherent in any telescopic view.

One of the more interesting questions relating to global form is whether the heavily cratered terrae, or "highlands," actually stand higher than the dark maria. Clearly this is true directly along the contact between the two; it is the relative heights some distance away from the contact that are difficult to determine. Baldwin (1963) concludes that the highlands do, indeed, stand higher than the maria (Fig. II-7a). This conclusion is qualitatively confirmed by limb profiles which reveal Maria Humboldtianum, Marginus, and Smythii as depressed and flattened regions. A more complex relationship is shown by Mills (1968) (Fig. II-7b). The mean level of maria points is distinctly below that of the terrae points. However, a harmonic analysis of these points by Mills indicates the contours do not closely parallel the margins of the maria. A prominent bulge occurs near the center of the Earthside hemisphere.

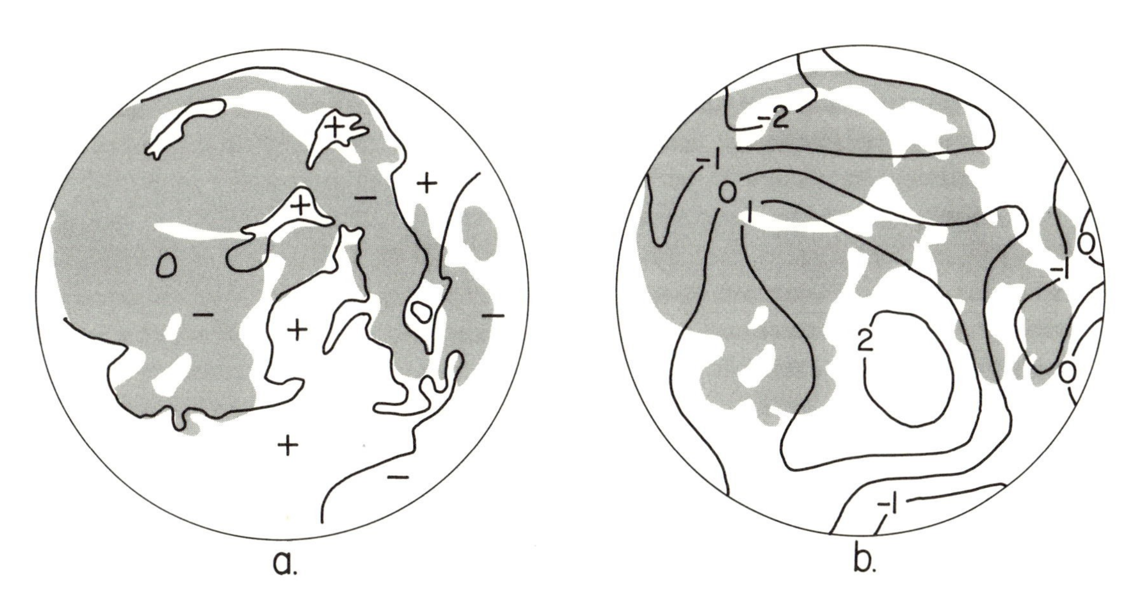

FIGURE II-7. Global form of the Moon according to (a) Baldwin (1963) and (b) Mills (1968). The single contour in (a) separates lower (−) areas from higher (+) areas. The contour interval in (b) is 1 km. Stippled areas are maria.

The calculation of relative elevations for neighboring points is simpler than the delineation of the global form, principally because the smaller angles permit simpler calculations and greater tolerances in measurements. Of particular interest is the local height measurement based on shadow length. Over the years a great deal of effort has been expended in this type of analysis. It has been particularly valuable in determining crater depths and profiles.

Figure II-8 shows a special case of shadow measurement for a lunar mountain. The length of the shadow AM and the projected distance to the terminator BT can be measured. ABM and OBT are similar triangles so that

$$\frac{BM}{AM} = \frac{BT}{OB}$$

Since OB is the known radius of the Moon, the unknown height of the mountain BM

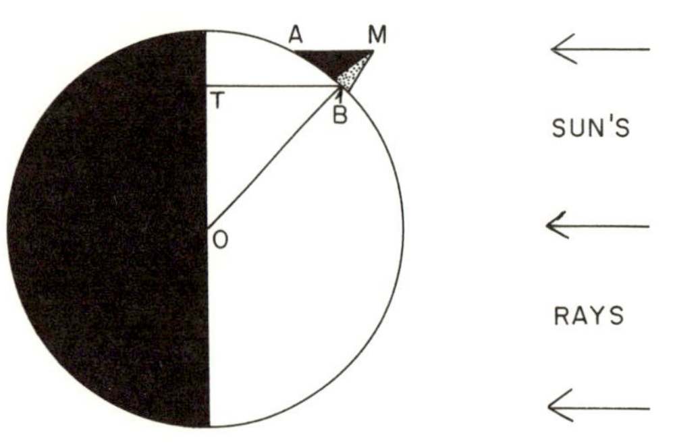

FIGURE II-8. Measuring the height of a lunar mountain. The section is drawn parallel to the incident sunlight and through the center of the Moon.

can be calculated. For actual computational purposes this relationship is expressed differently. In the general case, of course, it is not feasible to measure directly the distance to the terminator. An important point to note is that calculation of height by shadow length for this general case requires knowledge not only of the lunar phase but also of the selenographic latitude and longitude for the promontory whose height is being determined.

TOPOGRAPHIC PROFILES

Of much more value than point determinations of height are continuous profiles which show areal distributions of slopes. There are at least three ways in which this problem can be attacked. First, the shadow method can be extended. It is apparent that a single application of this technique yields the elevation difference between only two points, the peak top and the spot occupied by the tip of the shadow. As the Sun rises or sets, the shadow length will continuously change so that the tip will occupy successive points along a line. Analysis of each of these points will permit construction of a profile. This is not a very economical technique. It requires numerous photographs or visual observations and suffers from uncertainties in shadow position already mentioned.

Stereoscopic construction of local profiles is a more promising approach. Orbiter photographs incorporate overlapping coverage with different angular perspectives. These photographs can be handled in much the same way as a mosaic of aerial photographs of the Earth. Good qualitative results are obtained by examination of photograph pairs with a portable stereoscope. Quantitative profiles are obtainable with standard plotters developed for construction of topographic maps from terrestrial aerial photographs.

Difficulties lie in two areas. First, Orbiter photographs are mosaics built up from many individual framelets (see Fig. III-9). Imperfections in joining produce a spurious discontinuity along the join line. Secondly, the base-height ratio in many photographic pairs is undesirably small; because of flight constraints the angular displacement between many stereographic pairs is at the lower limit for acceptable quantitative work. These difficulties can be minimized by special photographic processing and by specialized plotting equipment. However, it is unlikely that topographic maps based on Orbiter photographs will ever be produced routinely by photogrammetric techniques. The photographs taken with hand-held cameras by astronauts orbiting the Moon will probably be more suitable for stereoscopic viewing.

A third profile technique exploits an unusual empirical fact: namely, that the brightness of any point on the lunar surface is dependent only on the phase angle, brightness longitude, and albedo of the surface material (Minnaert, 1961). Albedo is defined as the reflecting power of the surface at full Moon. Phase angle is the angle measured between the incident ray and the emergent ray (Fig. II-9). Brightness longitude is the angle measured in the phase plane between the emergent ray and a line perpendicular to the intersection of the ground with the phase plane. Albedo values can be determined from the full Moon image. Variations in brightness vary systematically with phase angle as shown in Figure II-10.

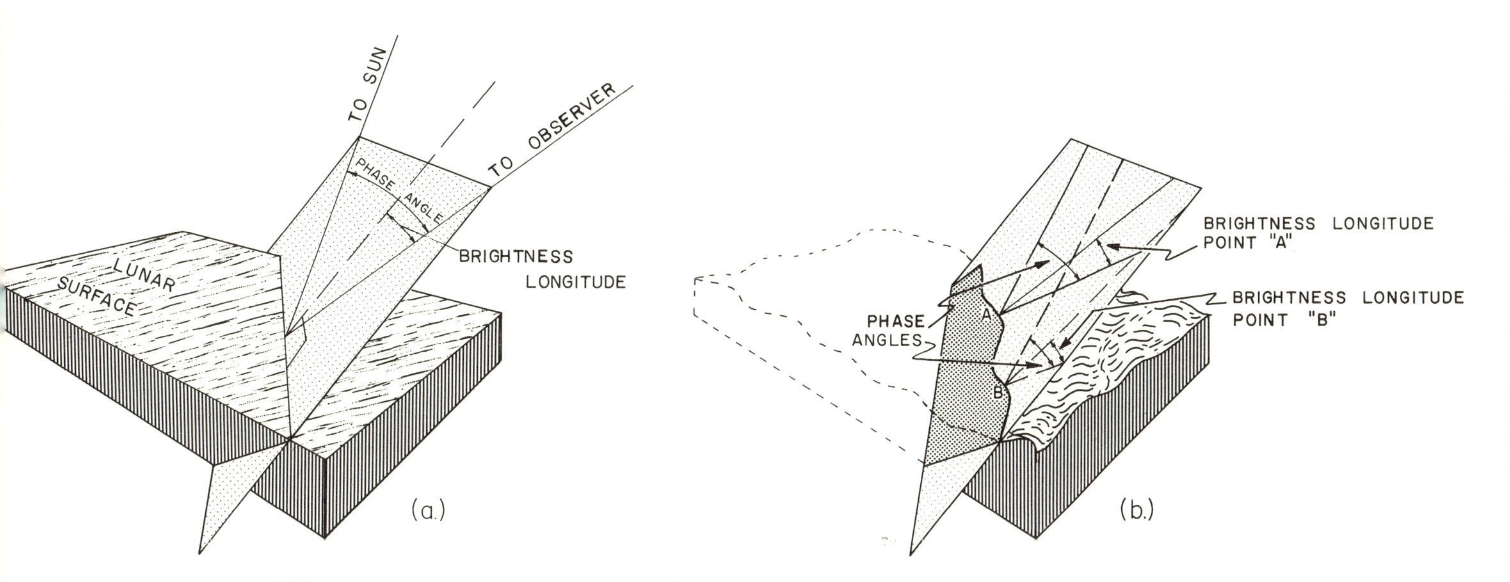

FIGURE II-9. Relation between brightness longitude, phase angle, and incident and emergent rays. (a) depicts situation for level surface. (b) depicts situation for undulating surface. Phase angles at points "A" and "B" are identical, but brightness longitudes have different values (after Watson, 1968).

A complete brightness determination follows the geometry shown in Figure II-9. If the surface block is rotated about its intersection with the phase plane, the brightness remains constant, since neither the phase angle nor the brightness longitude change in value. But, if the block is tilted along an axis perpendicular to the intersection in the plane of the block, then the brightness longitude changes, with a resultant change in visual brightness. Consideration of this relationship will reveal that brightness variations along the intersection of the surface block and the phase plane indicate changes in

FIGURE II-10. Variation of global light intensity of the Moon with phase. Note that, moving away from full Moon, the brightness falls off more sharply than does the percentage of the sunlit side visible from Earth.

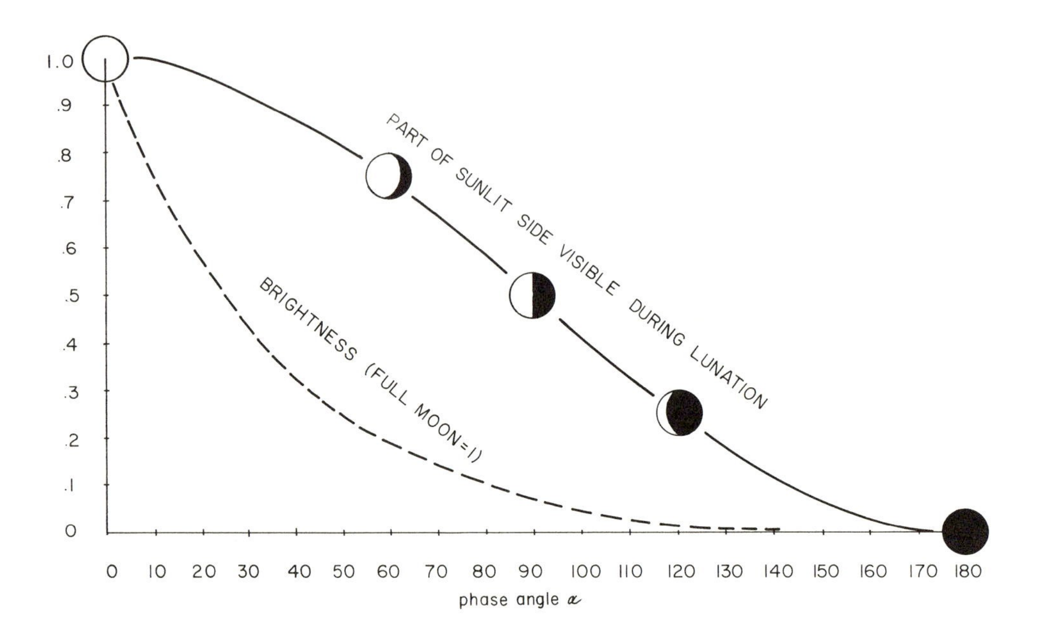

slope component measured in the phase plane. Varying brightness, then, will reveal an irregular topographic profile.

This technique for obtaining slope information is termed photoclinometry (Watson, 1968). Of all the schemes for measuring slopes it is the most intriguing—and probably the most frustrating in its application. One obvious limitation is that profiles can be constructed only in the phase plane. A second constraint is that the critical variable, brightness, is usually measured as emulsion density on a photographic negative. Disastrous "artificial" changes in brightness can be caused by minor irregularities in film manufacture, by variations in exposure procedures, and by imprecise film development. Unfortunately, Orbiter photographs share all these deficiencies. The problem can be partially obviated by working directly from the electronic signal returned to Earth rather than from the regenerated photograph. Unfortunately this requires an unusually sophisticated information retrieval system.

Actual results of the photoclinometric process are so far limited. It is likely that large areas will be mapped by automated programs in the near future. However, the great cost of data reduction and uncertainties due to imprecision in brightness value determinations will pose continuing problems.

Conflicting impressions of the aggregate roughness of the lunar surface have been published. A large program of shadow measurements supervised by Kopal indicates that "there appear to be no steep slopes anywhere on the Moon of any appreciable size" (Kopal, 1966, p. 251). In the equatorial region he records 90 percent of the slopes less than 2°, and in the highland 90 percent less than 10°. This contrasts with Pohn's determination of minimum slope values ranging between 13° and 48° for the inner walls of nineteen prominent craters (1963). Surveyor III landed in a crater with slopes ranging from 10° to 15° (Surveyor III Mission Report, 1967*a*). The crater appears no more precipitous than many which surround it. Several photogrammetric profiles measured in the same region, each extending about 10 km, indicate 30 percent of the slopes greater than 10°. Surveyor V landed in a ten-meter-diameter crater with interior slopes as steep as 20° (Surveyor V Mission Report, 1967*b*). Kopal reasons that the steep slopes measured by Pohn and others occupy only a minute fraction of the lunar surface and so do not contradict his conclusion of general flatness (1966, p. 253). As more and more examples of steep slopes are documented, Kopal's interpretation of the data becomes less tenable.

Remote Sensing Techniques

CHAPTER

MUCH of what we know about the Moon comes through the use of remote sensors. For the most part this phrase is a vaguely impressive synonym for the words "telescope" and "camera," but the category of remote sensors also includes those instruments which detect ultraviolet, infrared, and radar waves.

In any scientific effort comprehensive knowledge of the data available for analysis should precede interpretation and speculation. A brief review of data available for geologic study of the Moon is especially appropriate since much of it is quite different from conventional geologic data. In this chapter we will emphasize distinctive characteristics of images obtained by the several types of remote sensor. Although a number of geologic examples will be introduced, comprehensive review of data interpretation is reserved for successive chapters.

One could be excused for thinking that remote sensing techniques provide information which quickly becomes obsolete when more direct analysis is possible and which, therefore, is of little long-term value. Certainly there is a tendency for the results—scientific and otherwise—of each space mission to be instantly billed as "spectacular" and "unique," only to be forgotten in the wake of the next mission's "spectacular" results. Nonetheless, for some years to come scientists will depend on remote sensing techniques in planetary exploration, notably in the geologic exploration of Mars, Mercury, and Venus. Venus provides an especially good example of the value of remote sensing. Its surface is inhospitably hot and permanently shrouded with dense clouds. Most of what we learn about the topography of this surface will come through the use of radar imagery.

TELESCOPIC PHOTOGRAPHS

Telescopic photographs of the Moon have been obtained chiefly at those observatories listed in Figure III-1. The majority of photographs are not published but are privately circulated on request. Unquestionably this poses a serious problem for the casual student. There is no easy way of determining, in advance, the number, quality, and repository of photographs showing a particular feature.

The problem is intensified by the fact that a diagnostic pattern may be revealed on only one or, at the most, several photographs among the many that are available. A

FIGURE III-1. Telescopes frequently used for observation of the Moon. The diffraction limit is a measure of theoretical resolution, varying inversely with telescope aperture, and calculated for 0.5 microns wavelength. Owing to atmospheric turbulence, resolution of better than 0.2 km on the Moon is almost never realized.

Observatory	*Location*	*Type*	*Focus*	*Aperture (cm)*	*Diffraction Limits (km on Moon)*
Palomar (Hale telescope)	California	Refl.	Prime. Cass. Coudé	508	.048
Lick	California	Refl.	Prime. Coudé	305	.080
Mt. Wilson	California	Refl.	Newt. Cass. Coudé	254	.096
Kitt Peak (National observatory)	California	Refl.	Cass. Coudé	213	.115
McDonald	Texas	Refl.	Prime. Cass. Coudé	208	.117
Perkins	Arizona	Refl.	Newt. Cass.	175	.140
U.S. Naval	Arizona	Refl.		155	.157
Catalina	Arizona	Refl.		155	.157
Mt. Wilson	California	Refl.	Newt. Cass.	152	.160
Pic du Midi	France	Refl.	Prime. Cass.	109	.222
Lowell	Arizona	Refl.	Cass.	107	.225
Yerkes	Wisconsin	Refr.		102	.240
Lick	California	Refr.		91	.265
U.S. Geol. Survey	Arizona	Refl.	Cass.	76	.320
Pic du Midi	France	Refr.	Newt.	61	.400
Lowell	Arizona	Refr.		61	.400
Lowell	Arizona	Refr.	Cass.	51	.460

critic without access to the relevant photographs has no way of assessing another person's interpretations. Comparison of the photographs in Kuiper's 1960 atlas with one of the 1:1,000,000 geologic maps published by the U.S. Geological Survey provides an excellent demonstration of this dilemma. A great deal of geologic detail shown on the map will not be visible on the photographs. The explanation is—hopefully—not that the geologist had an unusually vivid imagination, but rather that he had access to a much larger suite of photographs and also made visual observations. (It is true that some sceptics might not completely discount the imaginative content.)

Variations between individual photographs of the same area are largely a function of terminator position. When features are close to the terminator (lunar sunrise and

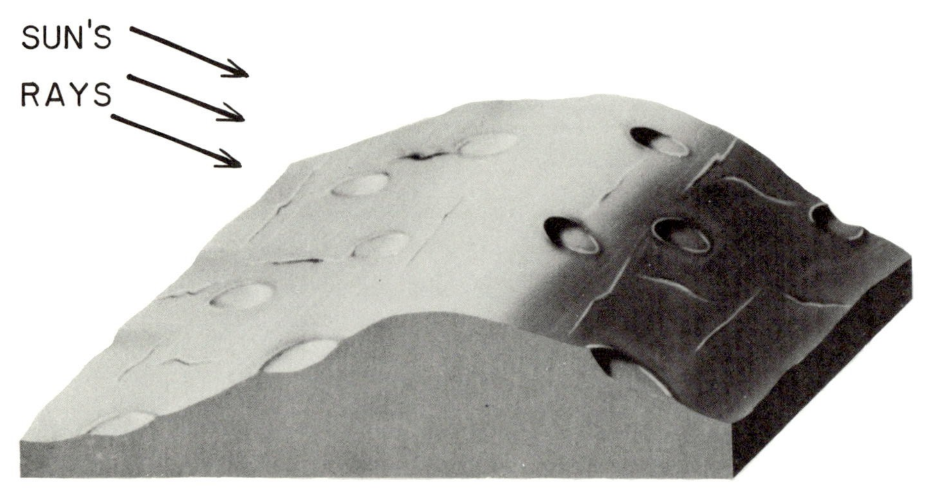

FIGURE III-2. Schematic diagram showing variation in shadowing definition on opposite-facing slopes.

sunset) even the subtlest protuberances and depressions are strongly shadowed (Figs. III-2 and 3). Figures III-4 illustrates the additional information gained by opposite illumination as the same scene is shadowed first by a rising Sun in the east and then by a setting Sun in the west. When the Sun is directly overhead (lunar midday) many gentle topographic features are invisible.

The image produced when the phase angle is close to zero is unique. In such a full Moon photograph, dark maria, brilliant rays, and luminous highlands are distinguished by their different albedos (Fig. III-5). At the zero position (the precise zero phase can never be obtained from Earth stations, since the Moon would be eclipsed by the Earth) the lunar surface shows a brightness largely independent of slope and geographic position. Instead, brightness is determined by the small-scale structure, or possibly the chemical composition, of the surficial material. The problem is by no means as simple as these brief sentences might suggest and will be considered in more detail later. Suffice it to point out here that different portions of the Moon do have different brightnesses at full Moon and that the brightness values are termed albedos.

Resolution of detail in photographs is extremely variable, being a consequence of telescope optics, local seeing conditions, and photographic plate reproduction. On the best photographs it is possible to resolve some objects 200 meters across, but identification of features smaller than 5 km is difficult on average photographs. The question of resolution is further complicated by the fact that features with certain geometries are most easily identified. For example, narrow clefts can be distinguished even though their widths are less than the diameter of the smallest recognizable crater (Wilkins and Moore, 1955, p. 349).

Visual images invariably provide more detail than a photograph taken at the same instant through the same telescope. For this reason the proliferation of telescopic photographs and even the receipt of Orbiter photographs do not preclude the need for visual observations. The time involved in obtaining a single good look through a telescope is great. Consequently the chief value of visual work is for study of particular features rather than for regional surveys.

FIGURE III-3. Sunset over Petavius. The Moon as shown in (b) is less than a half lunar day older than in (a), but this small difference in Sun elevation results in distinctive shadowing differences. Note particularly the mare region in the upper left corner of both figures (from Kopal, 1965).

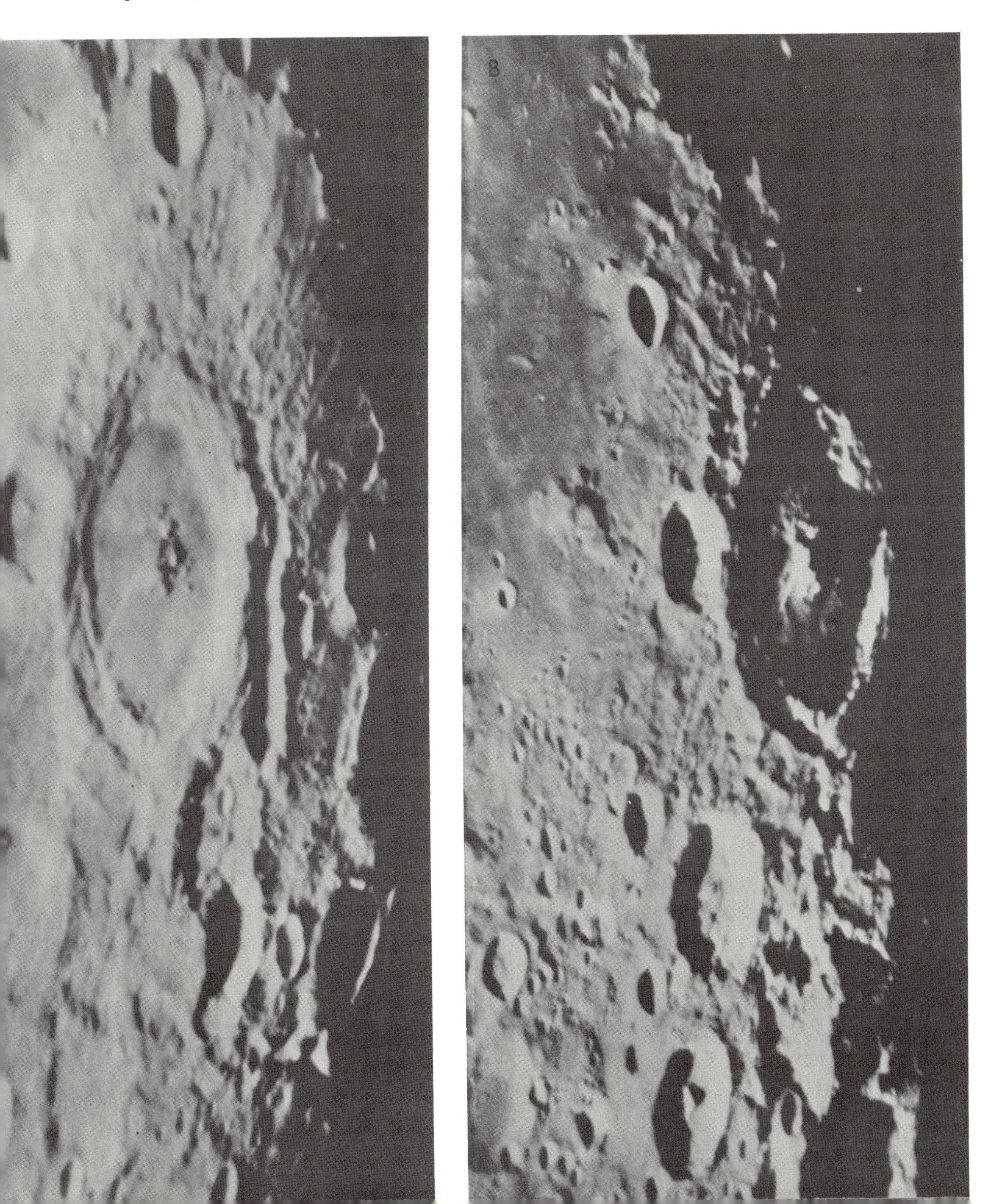

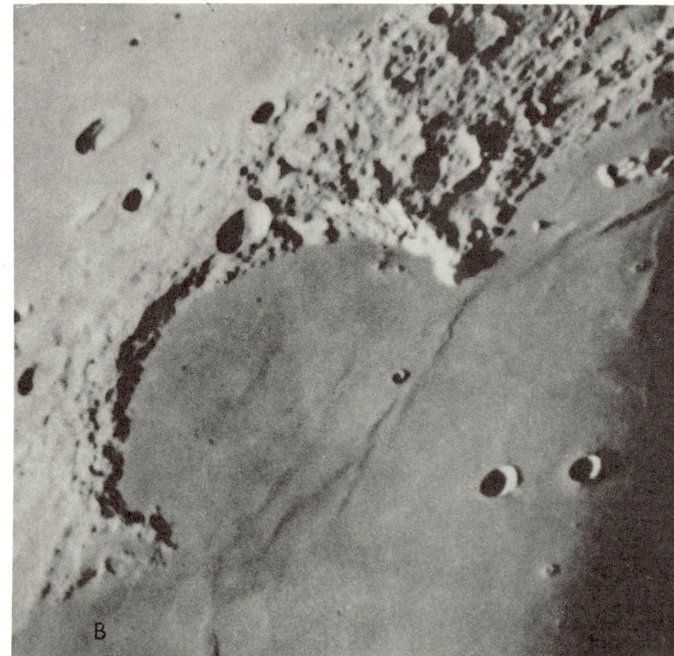

FIGURE III-4. Sunrise (a) and sunset (b) over Sinus Iridum. Note particularly the different perspectives of the mare ridges and the structures in the crater wall which forms the boundary of Sinus Iridum (from Kopal, 1965).

ORBITER PHOTOGRAPHS

By all odds, the greatest contribution of the space program to the lunar geologist interested in regional mapping has been the suite of photographs taken by satellites orbiting the Moon. This has been an outstandingly successful effort with five photographic missions completed in little more than a year, from August 1966 to August 1967. The first three Lunar Orbiter spacecraft were placed in near-equatorial orbits in order to photograph possible Apollo landing sites (Figs. III-6 and 7a). Orbiters IV and V were placed in near-polar orbits in order to gain more complete coverage outside the equatorial belt. Orbiter IV photographed the entire nearside (Figs. III-6 and 7b); Orbiter V returned pictures from selected sites of scientific interest on the nearside and completed coverage of the farside (Figs. III-6, 7c).

A total of 1,950 photographs has been received, almost every one showing details previously unknown. It is surely no exaggeration to say that many of these photographs have not yet been thoroughly studied. Doubtless, a large number of obvious features have gone unnoticed. For example, it is entirely possible that certain small craters photographed in the later missions were formed some time after early flights and would not appear on the first photographs. In order to demonstrate this possibility one would

FIGURE III-5. The full Moon.

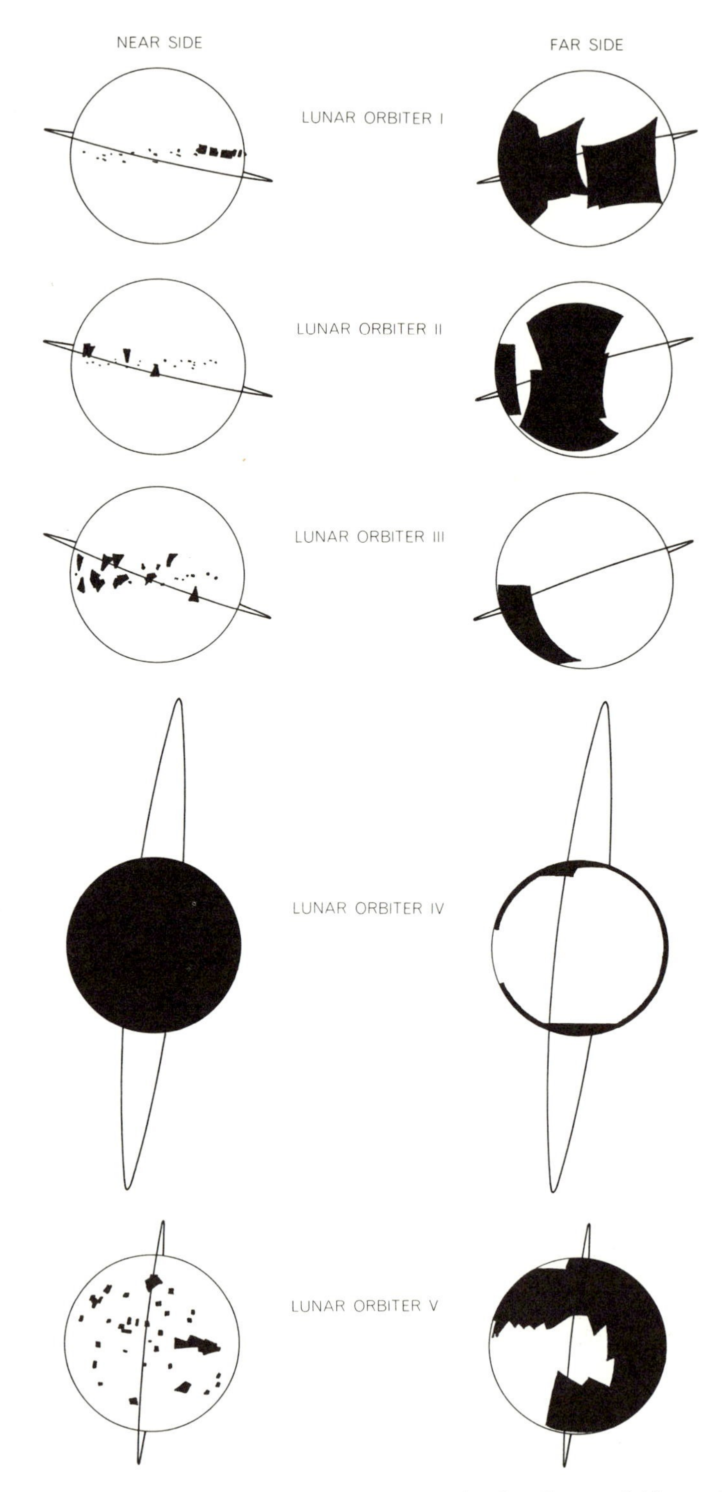

FIGURE III-6. Orbital paths and photographic coverage for the five Lunar Orbiter missions (from "The Lunar Orbiter Missions to the Moon" by Levin, Viele, and Eldrenkamp. Copyright © 1968 by Scientific American, Inc.).

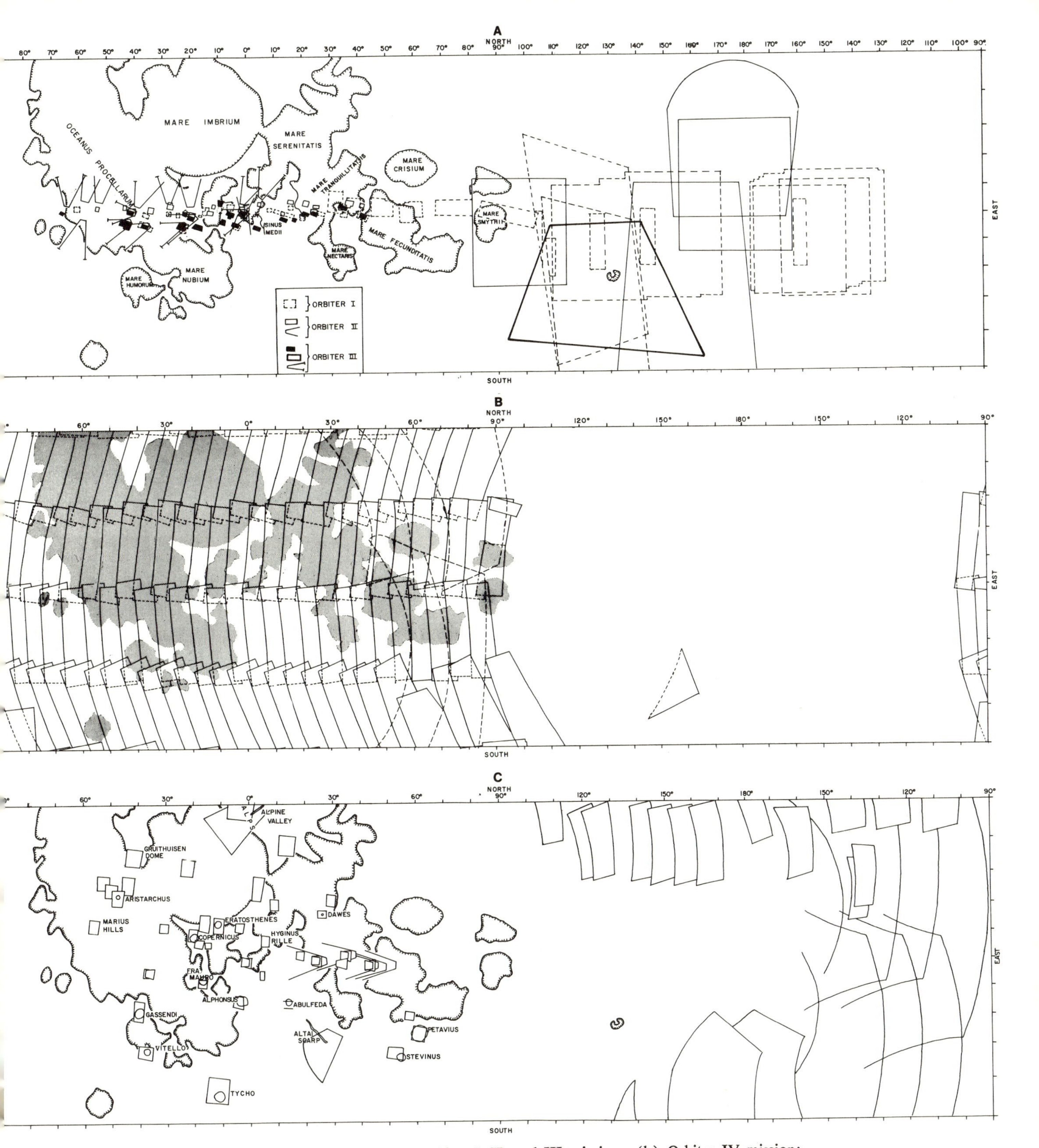

FIGURE III-7. Photographic coverage for (a) Orbiter I, II, and III missions; (b) Orbiter IV mission; and (c) Orbiter V mission. Coverage in the northern and southern polar regions is not shown. Major parts of these regions were photographed on the Orbiter IV mission.

have to make a comparative study of billions of craters. Like the proverbial needle in the haystack, any new crater would be almost impossible to find but—once discovered—would be unmistakable.

Because production of Orbiter photographs is quite unlike the operation of a conventional camera, it is worthwhile to spend several moments reviewing the system. Each spacecraft contained a camera with two lenses, one a wide-angle lens with a 80-mm focal length and the second a telephoto lens with a 610-mm focal length. The two were alined so that they could take simultaneous exposures. In each pair of photographs the wide-angle lens provided a "medium resolution" photograph with a field of view 20 times that of the "high resolution" photograph taken with the telephoto lens. The camera utilized a strip of 70-mm film, long enough to accommodate 211 photograph pairs. It was developed on board by the Bimat technique designed by Eastman Kodak Company. In this process the exposed film is pressed against a mat soaked with special developing and fixing solutions.

The developed film was "read" by a flying spot scanner housed in the spacecraft. This scanner consists essentially of a very small beam of light which is moved systematically across the film. A detector, located on the opposite side of the film, records the varying brightness values as the light is either transmitted through clear portions of the film or is absorbed by darker portions. The brightness values are then converted into video signals and transmitted back to Earth (Fig. III-8). Some idea of the sensitivity of this technique is indicated by the fact that there are 17,000 scan lines for each framelet.

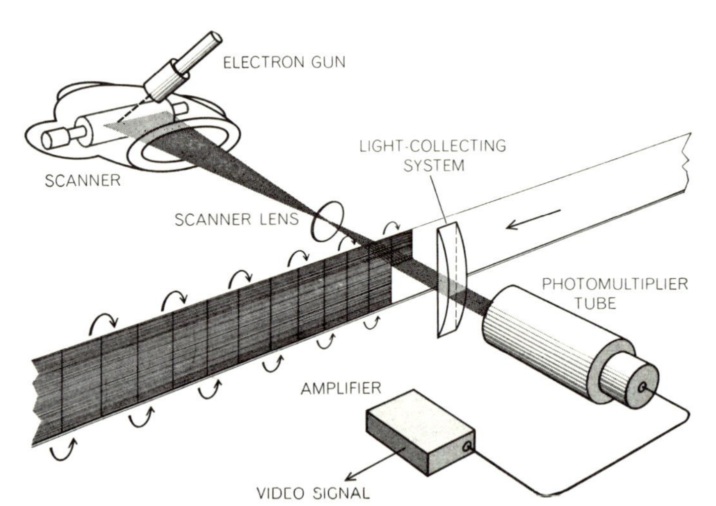

FIGURE III-8. The scanning process on Lunar Orbiter missions. A microscopic spot of light, generated by a beam of electrons aimed at a drum coated with phosphor, moved along a part of the film. The density of the image on the film governed the strength of the light reaching the photomultiplier tube, which generated electronic signals for transmission to Earth. One complete scan read out the information on a band of film that took up 2.54 mm along the length of the film and 60 mm of its width. These are the framelets that one sees in Orbiter photographs (from "The Lunar Orbiter Missions to the Moon" by Levin, Viele, and Eldrenkamp. Copyright © 1968 by Scientific American, Inc.).

The spacecraft scanner analyzes framelets 2.54 × 60 mm, with the long dimension perpendicular to the film strip edge. This information is returned to Earth where the video signal is recorded on magnetic tape and is regenerated much as any television image is produced. Synchronous with its regeneration it is transferred to 35-mm film. Successive framelets are mosaicked and printed on film. The composites are combined once more to reproduce the image originally recorded on the spacecraft film (Fig. III-9).

In photographs from the first four Orbiter missions the framelets extend approximately east-west, but in photographs taken by Lunar Orbiter V they extend approximately north-south. The precise orientation depends on the exact orbit of the spacecraft and the tilt of the camera. Framelet numbers, reference marks, and film density scales appear at the left end of each framelet.

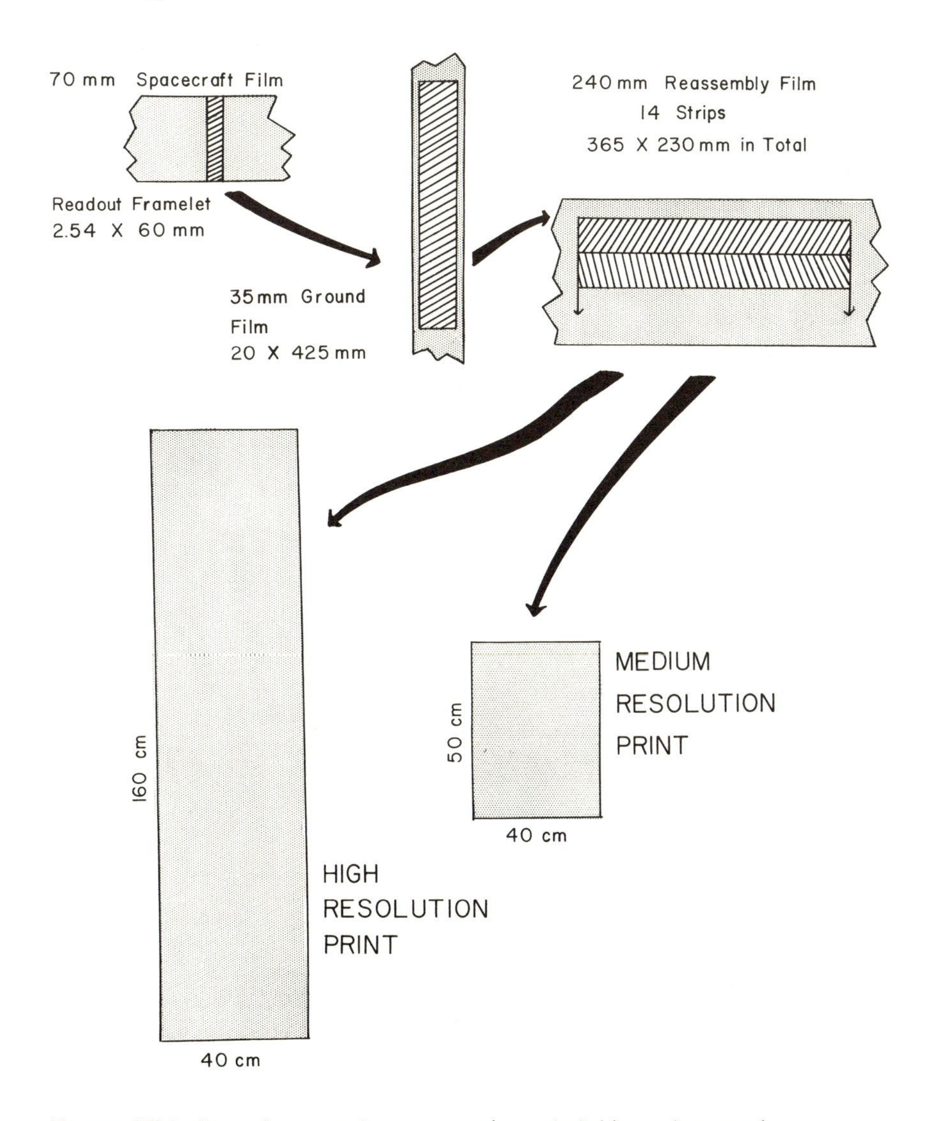

FIGURE III-9. Steps in ground reconstruction of Orbiter photographs.

The resolution limits of Orbiter photographs depend, in part, upon the distance of the camera above the surface of the Moon. All five spacecraft were inserted into elliptical orbits with periapsis (closest approach to the Moon) occurring on the nearside and apoapsis (farthest departure from the Moon) on the farside. For a given mission these distances changed between successive orbits due to the perturbing effects of the gravitational fields of the Moon, Earth, and Sun. For Orbiters I, II, and III periapsis fluctuated between 46 and 58 km. Pictures taken at these distances with a telephoto lens have a resolution limit of about 1 meter, and medium resolution photographs have a resolution limit of about 8 meters. Photographs on the farside were taken from altitudes in the vicinity of 1,000 km. The Orbiter IV spacecraft, designed as it was for a complete photographic survey of the nearside, had a periapsis of about 2,700 km, resulting in photographs with resolution limits almost always less than 100 meters and sometimes as small as 25 meters near the equator. Orbiter V photographs of the nearside, taken from an altitude of approximately 100 km, have a resolution limit of 2-3 meters near the equator.

Orbiter missions were designed so that photographs were taken when local Sun angles were between 10° and 30°, and the topographic details were accentuated by shadowing. The two possible times for photography, then, were just after sunrise or before sunset. Because the first three spacecraft were injected into orbits that carried them from west to east across the nearside of the Moon, the sunrise period was preferable. As soon as the photographs were taken, the spacecraft continued to travel in sunlight, thus drawing the solar power necessary for processing of the nearside photographs. As a result of the photographs being taken near lunar sunrise, shadows on almost all Orbiter photographs of the nearside are cast from east to west, or if the photograph is properly oriented, from right to left.

The quality of Orbiter photographs is extremely variable, and one should be cautious in comparative interpretations. Some variations result from different Sun angles and differences in surface slopes. Other variations result from the use of different exposures during film and print processing. One of the most critical processing steps is transferral of the image information from magnetic tape to 35-mm film. By varying the "gain" it is possible to enhance a feature of particular interest. If a frame contains features with a wide range of reflectivities, a single printing will be insufficient to display all the detail. Instead it is desirable to prepare a series of prints, each exposed to best reveal the density contrasts for a particular area or object.

Defects in the Bimat development process aboard the spacecraft produce clusters of circular and amoeba-like blemishes on many photographs. During regeneration, movement of 35-mm on-the-ground film synchronous with reception of the video signal generates delicate lineations parallel to the film edge. But the most characteristic features of Orbiter photographs are the sharp discontinuities along the join line between adjacent framelets. So common are linear artifacts parallel to framelet boundaries that any natural-appearing structural trend or geologic contact which trends in the same direction is automatically suspect. If one works regularly with Orbiter photographs he grows so accustomed to their various peculiarities that they begin to seem quite natural —all of which is preparation for Figure III-10.

FIGURE III-10. Drawing by J. W. Van Divier, U.S. Geological Survey.

Although Orbiter photographs contain a wealth of detail previously unobtainable, they certainly do not make further telescopic observations unnecessary. A great advantage in the use of the telescope is that one can select the desired illumination angle. At angles of very low illumination, gentle swells and depressions not visible on spacecraft photographs are clearly revealed (Fig. III-11).

FIGURE III-11. Comparison of (a) photograph of floor of Ptolemaeus taken by Ranger IX with (b) photograph of approximately the same area taken from a ground-based observatory. The terminator position is shown by a straight line. Arrows indicate two shallow depressions which are sharply shadowed by late evening sunlight (taken from Rackham, 1967).

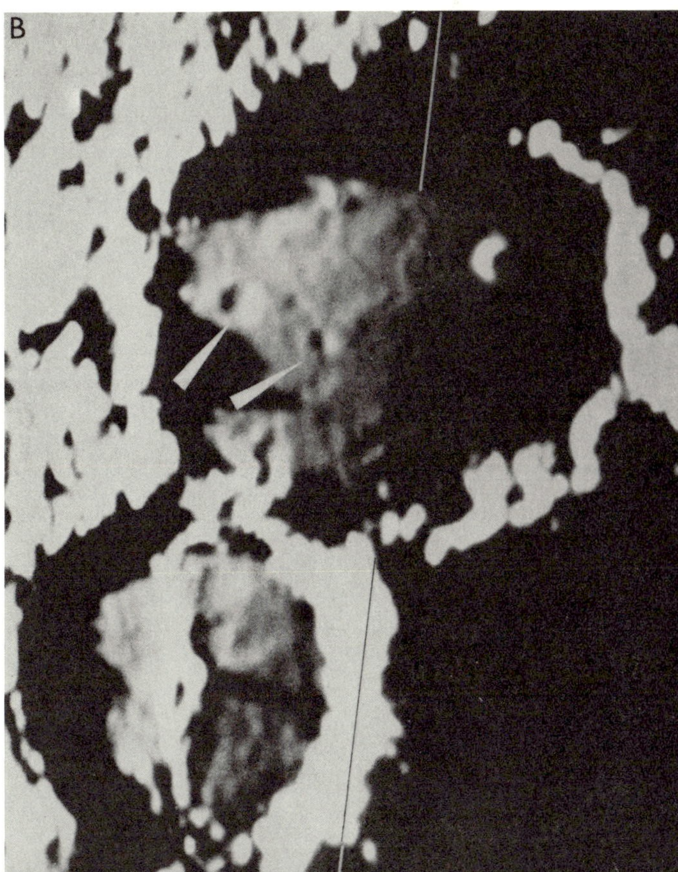

FIGURE III-12. Landing positions for Ranger, Surveyor, and Luna spacecraft. The Lunas were Russian soft-landing spacecraft.

RANGER AND SURVEYOR PHOTOGRAPHS

In addition to pictures from orbiting satellites, photographs of the Moon have been obtained from hard- and soft-landing unmanned spacecraft. There have been three successful hard-landing missions: Rangers VII, VIII, and IX; and five successful soft landings: Surveyors I, III, V, VI, and VII (Fig. III-12).

Obviously, the photographic constraints in the two types of mission are quite different. In the case of Ranger only several minutes were available for receipt and transmission of all photographs. Because of the limited time available and the rapidly changing base position, the spacecraft carried a battery of six television cameras. These had different focal lengths and preselected fields of view. The television images were transmitted directly to Earth without any intervening film development process, which, of course, would have been prohibitively time-consuming. Successive photographs do not overlap precisely, but, to a first approximation, the same area was photographed with successively increasing magnification and resolution. The last television pictures taken just before impact have a resolution of about 50 cm.

Unlike Ranger, Surveyor waited for a safe landing before doing its work. Photographs were obtained with a single television camera, which transmitted directly to Earth. The camera operated on command. Camera orientation could be specified, optical focal length adjusted, exposure time changed, focus adjusted from 5 feet to infinity, and various filters inserted in the optical system.

The unusual perspective of Surveyor photographs (unusual for Moon watchers although common enough for most people with their feet on the ground) presents special problems. Because of the inclined line of sight and the depth of the imaged field it is difficult to judge scales, distances, and real shapes. The chief value of the pictures resides in their clear delineation of surface texture. Grains as small as 1 mm can be individually counted, and delicate rock fabrics are revealed.

The Surveyor spacecraft were variously equipped to perform experiments on the lunar surface in addition to the photographic ones. The physical character of the top few inches of lunar "soil" was tested with scooping and impacting tools. Chemical analyses were obtained by an alpha-scattering technique that was elegantly simple. The findings suggest—by generous extrapolation—that large portions of the maria resemble basalt in composition (Turkevich and others, 1968*a*). Analyses obtained at the Surveyor VII site, just north of Tycho, reveal significant differences, suggesting rocks more anorthositic than basaltic (Turkevich and others, 1968*b*).

SPECIALIZED SENSING TECHNIQUES

Discussion to this point has dealt with photographs and visual observations of the Moon's surface. If the position of visible light in the electromagnetic spectrum is examined (Fig. III-13) it is obvious that only a small part of the spectrum has been discussed—and that in a rather specialized way.

The electromagnetic spectrum is divided according to wavelength. Wavelengths less than those for visible light include, successively, ultraviolet light, X-rays, and gamma waves. At wavelengths greater than visible light, one encounters, first, the infrared field, and beyond that, a broad region of micro and radio waves. Emission of the short wavelengths is largely controlled by atomic, molecular, and mineralogical configurations, and increasingly longer waves are influenced by progressively larger structural features. Reflection of radar waves, in particular, depends on the gross structure of the reflecting surface measured in centimeters or even meters. A second approximate rule of thumb is that the shorter wavelengths are absorbed more easily than the longer ones. For example, although the Moon is a weak X-ray emitter, atmospheric absorption prevents receipt of this energy at the Earth's surface.

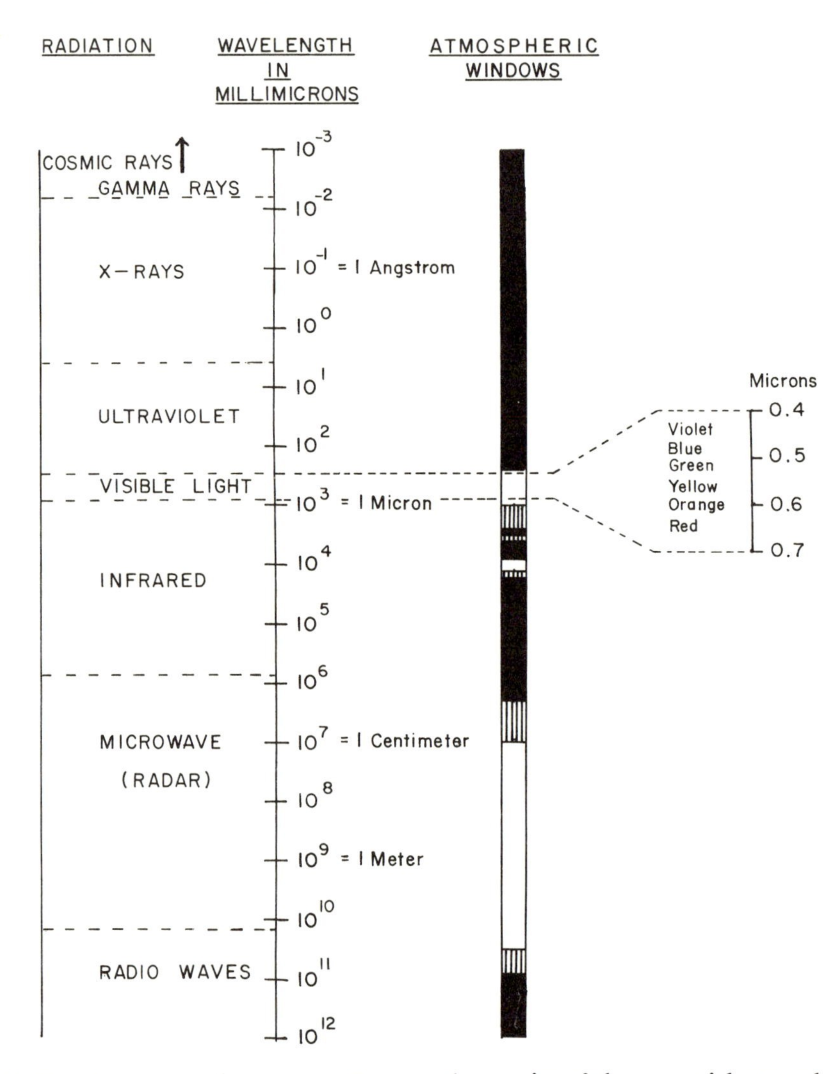

FIGURE III-13. The electromagnetic spectrum. Data on the opacity of the terrestrial atmosphere are taken from Arking (1967, p. 914). Dark areas represent regions where the atmosphere almost completely blocks electromagnetic radiation; cross-hatched regions represent zones of partial blocking; clear regions designate zones of very little attenuation.

It is clear that if one looks at the Moon with reference only to the gross properties of visible light, a great deal of pertinent information may be overlooked. A more comprehensive line of attack would be to examine the lunar surface in the entire electromagnetic spectrum (Fig. III-14). As more becomes known about the detailed behavior of lunar material, it should become possible to pinpoint its physical character. Dense basalt, vesicular basalt, rhyolite, iron-rich meteorites, volcanic ash, loosely packed dust—these are only a few of the possibilities. Each of these substances absorbs, reflects, and emits electromagnetic energy in a distinctive way.

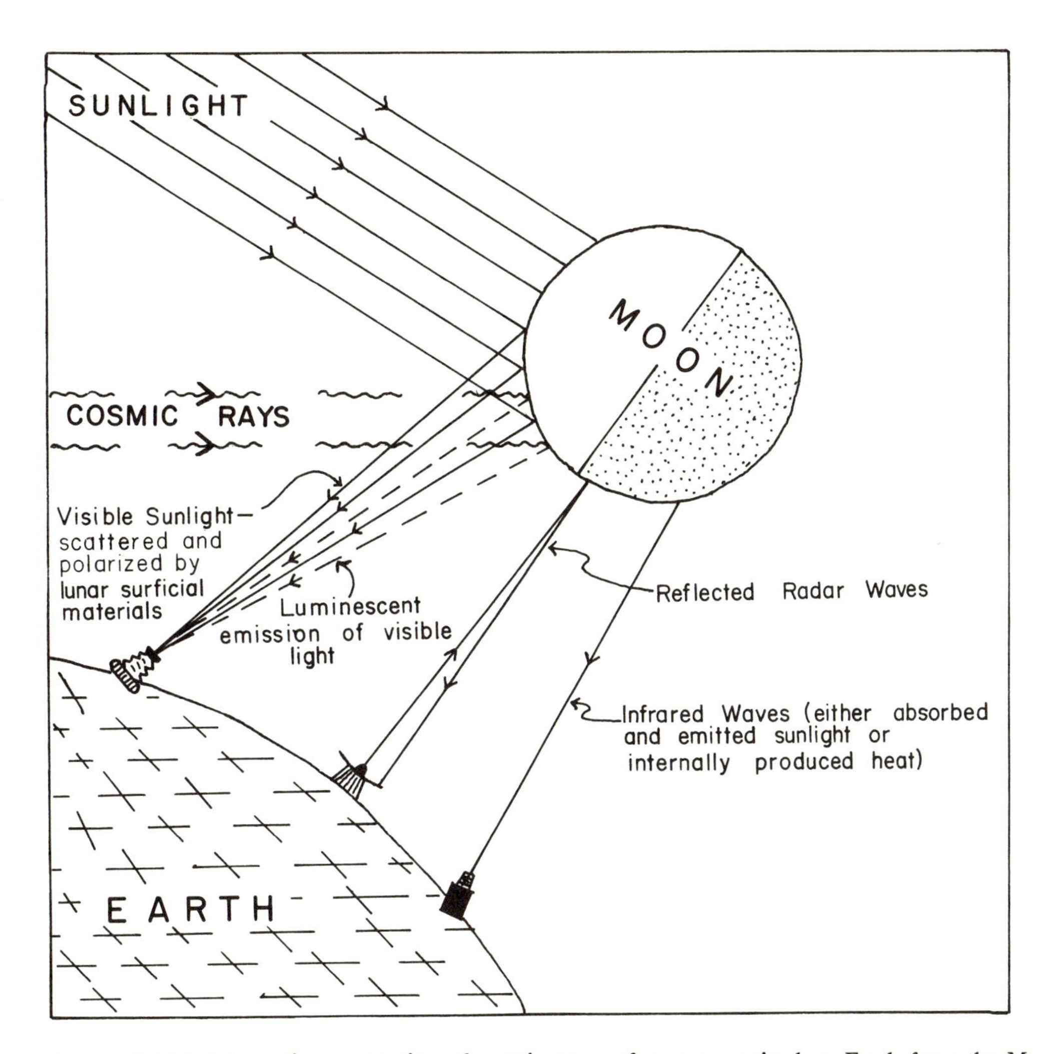

FIGURE III-14. Schematic presentation of certain types of energy received on Earth from the Moon. Records of the variation of energy across the lunar disc provide characteristic images of the Moon's surface.

There are a number of ways to augment the information present on a photograph and that obtained from visual observations. First, one can examine in more detail the visible light. Knowing the strength and spectral composition of sunlight, one can measure precisely the reflecting effectiveness or brightness of the Moon's surface. As we have already seen, different portions of the surface have quite different albedos

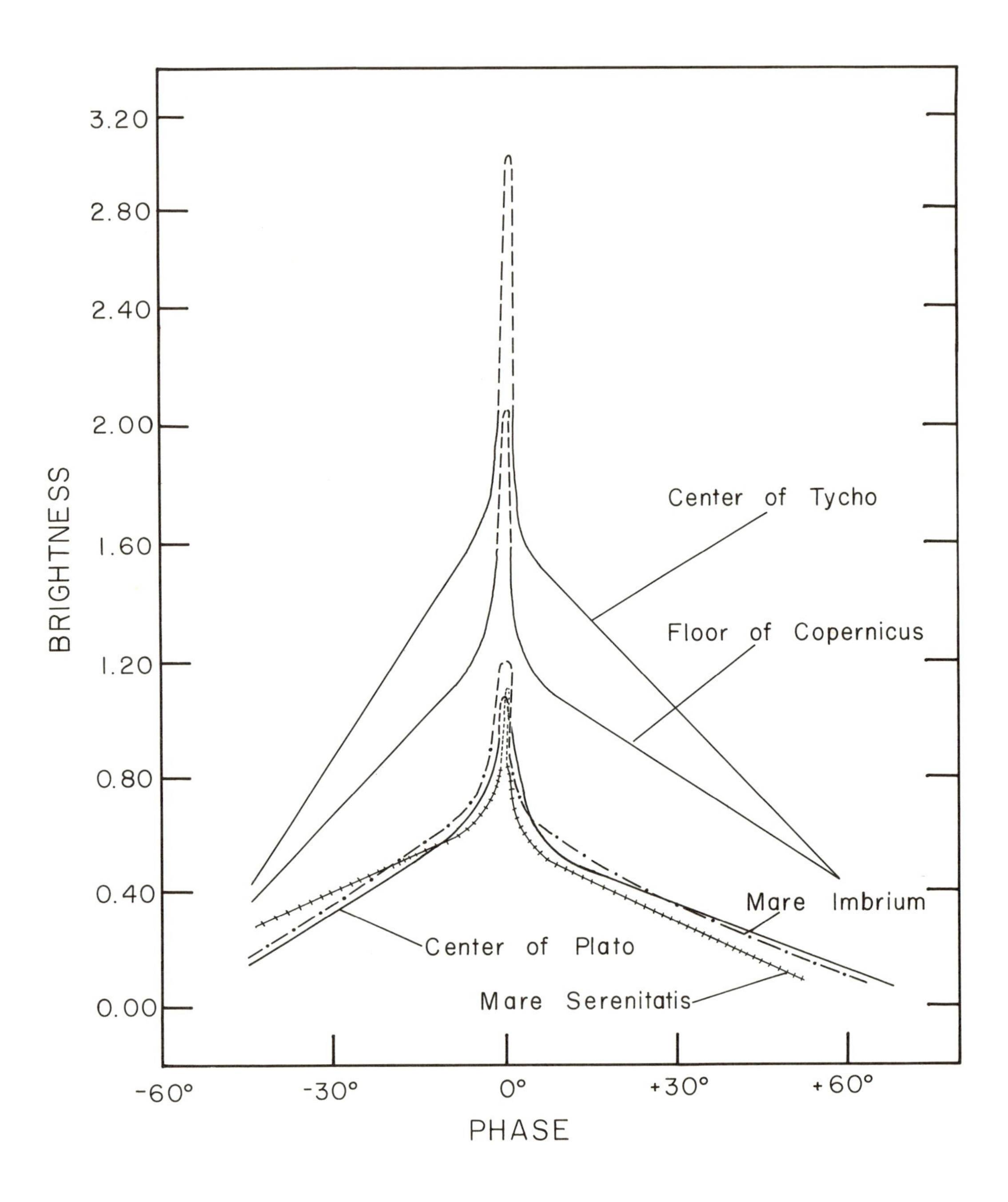

FIGURE III-15. Brightness of certain lunar features as a function of phase. Brightness units are arbitrarily normalized (after Gehrels, Coffeen, and Owings, 1964).

measured at full Moon. A possible way to differentiate particular areas further is to construct brightness curves as a function of lunar phase. Figure III-15 shows, however, that the differences between representative areas are not particularly striking.

Sunlight is both diffusely scattered from the Moon's surface and is also polarized. Instead of vibrating in all directions perpendicular to the direction of transmission, it has a tendency to vibrate in one or several directions. Geologists, of course, capitalize on this polarization phenomenon when examining minerals with a petrographic microscope. Unfortunately parts of the lunar surface are not so exotically different as are minerals viewed with the microscope. Following the same analytical format as for lunar brightness, percent of polarization can be plotted as a function of phase (Fig. III-16). Some differences are revealed, but a minimum of information additional to that from more conventional visual observations is gained.

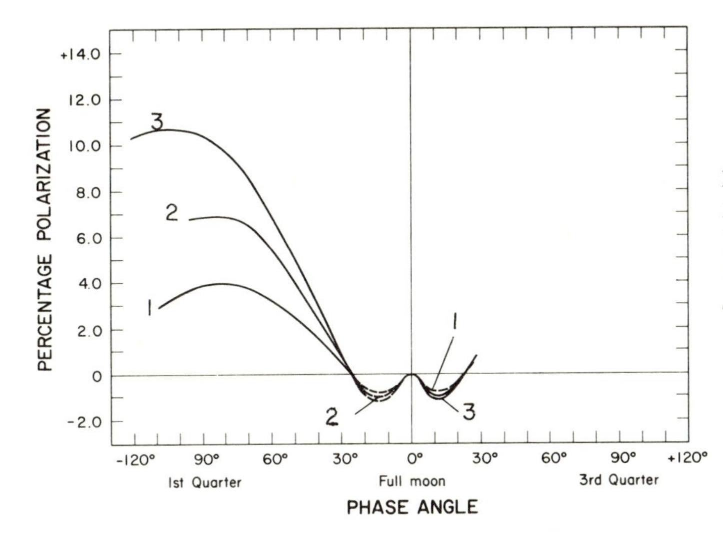

FIGURE III-16. Comparison of polarization-phase angle curves of three widely different units: Proclus ray material (1), Cayley Formation (2), and a subunit of the Procellarum Group (3) (from Wilhelms and Trask, 1965).

Still dealing with visible or near-visible light, one can subdivide that portion of the spectrum. Whitaker (1966) had considerable success in photographing the Moon with ultraviolet and infrared filters. By combining a positive transparency of a near infrared (7,800Å) photograph with an ultraviolet (3,500Å) negative, he obtained a composite which differentiated redder and bluer areas within the maria (Figure III-17). He interpreted these differences as an indication of volcanic flow materials of several types.

A number of transient color changes on the Moon have been reported over the past few hundred years (Middlehurst, 1967). In a program designed to quantitatively test the reality of such color variations, Roberts (1968) photometrically scanned selected areas of the Moon. By simultaneously measuring the light intensity in three narrow wavebands he was able to detect color variations in the range 5 percent to 15 percent with a fluctuation period of approximately one second. Although the reasons for these changes are still uncertain, he speculated that short-term variations in radiation from the Sun may change the luminescence of the lunar surface.

In reading about "reddish" and "bluish" regions of the Moon one should not be misled into thinking that those subtle, quantitatively measured variations are easily detected by the human eye. Colormetric observations of the lunar surface, designed to detect "visual" color, were carried out by Surveyor III (see Surveyor III Mission Report, 1967*a*). Television pictures were taken through three different filters, and the resulting images were standardized by reference to a control target mounted on the foot-

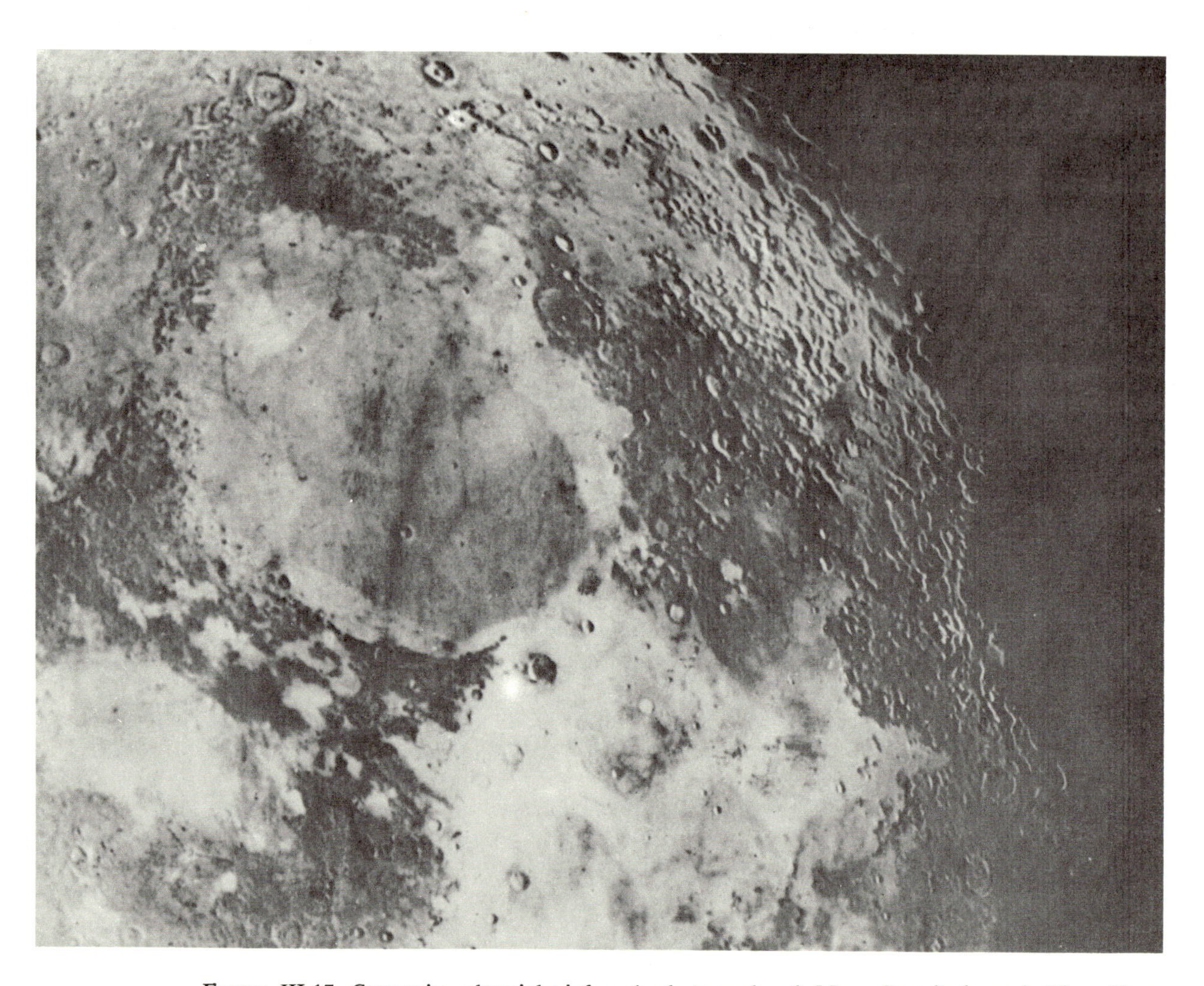

FIGURE III-17. Composite ultraviolet-infrared photograph of Mare Serenitatis and Mare Tranquillitatis region. Darker is redder (taken from Whitaker, 1966).

pad. The result? The disturbed surface material is dark gray. Several blocks are slightly lighter gray. It is a result sufficient to please the engineer but not the artist. The first persons to view the Moon at close range, the Apollo VIII astronauts, confirmed that the landscape is, indeed, colored a monotonously bleak gray.

At approximately the same time that Surveyor V made a chemical analysis of the lunar surface, Hapke (1968) made an independent confirmatory study based on telescopically determined properties of visible light. Considering albedo, intensity variation with phase, polarization, and the detailed configuration of the spectrum in the range 0.2 to 2 microns, he concluded that all these properties most nearly matched those of a powder made from basic rocks and irradiated by simulated solar wind.

Broad absorption bands occur at the boundary between visible light and infrared in the reflectance spectra of many iron-bearing minerals. Adams (1968) has interpreted a previously determined minimum at 1 micron for the spectral curve of Mare Tranquillitatis as evidence of a basic rock containing either olivene or pyroxene.

If we extend our consideration farther beyond the visible spectrum, atmospheric absorption effects rule out the use of the shorter wavelengths with Earth-based detectors. This leaves the categories far infrared and microwave. Photographic emulsions are insensitive to these energy forms, so conventional photographs are out of the question. However, photo-like images can be produced. Just as one can measure visual brightness at a single point with a light meter, so one can measure infrared radiation at a fixed point with a special thermal or photoconductive detector. If that detector is then systematically moved along a raster of closely spaced lines, a two-dimensional image is electronically produced, with the bright spots corresponding to areas of high infrared radiation. Were it not for the presence of scan lines, the resulting image would look very much like a conventional photograph.

Infrared is not confined to reflected energy. A major source is sunlight which is absorbed and later—sometimes much later—emitted by radiation.[1] For this reason one often encounters the term "thermal infrared" in reference to material which is heated and then emits in the infrared over a long cooling period. Such sources are easier to define during periods of total darkness when there is an absence of reflected energy. Accordingly the most informative infrared (10-12 microns) image of the entire nearside was obtained during a full-Moon eclipse (Saari and Shorthill, 1966). The eclipse provided the additional benefit of a standard energy background for the entire lunar disc. At one instant, just before the eclipse, the lunar surface was uniformly heated by sunlight. Then, in a matter of an hour, it was plunged into total darkness. The patches that were not cooling as rapidly as the surrounding terrain stand out as bright spots on the eclipsed Moon. While the reasons for this phenomenon are not known with certainty, there is a striking correlation between areas of high infrared radiation and young craters (Fig. III-18).

Infrared radiation can be produced not only from absorbed light but also from an internal heat source. This phenomenon is well known on Earth, where infrared aerial surveys have revealed many previously unknown hot spots in volcanic regions. Evidence for internally heated regions on the Moon is generally lacking, but Hunt and others (1968) have mapped one possibly internal anomaly, a linear thermal feature in the southwestern part of Mare Humorum. Its close relation to a fracture line and its anomalous cooling behavior relative to most infrared-enhanced craters suggests an internal origin.

There have been several attempts to analyze the detailed form of the spectrum in selected portions of the infrared. Goetz (1968), for example, took comparative measurements in the 8 to 13 micron band for twenty-two lunar points. Twenty of these points showed almost no differences, even though they included localities with major variations in topography and albedo. One would have expected that associated differences in grain size and chemical composition would produce some differences in the shapes

[1] A good discussion of the principles of infrared radiation is found in Simon (1966).

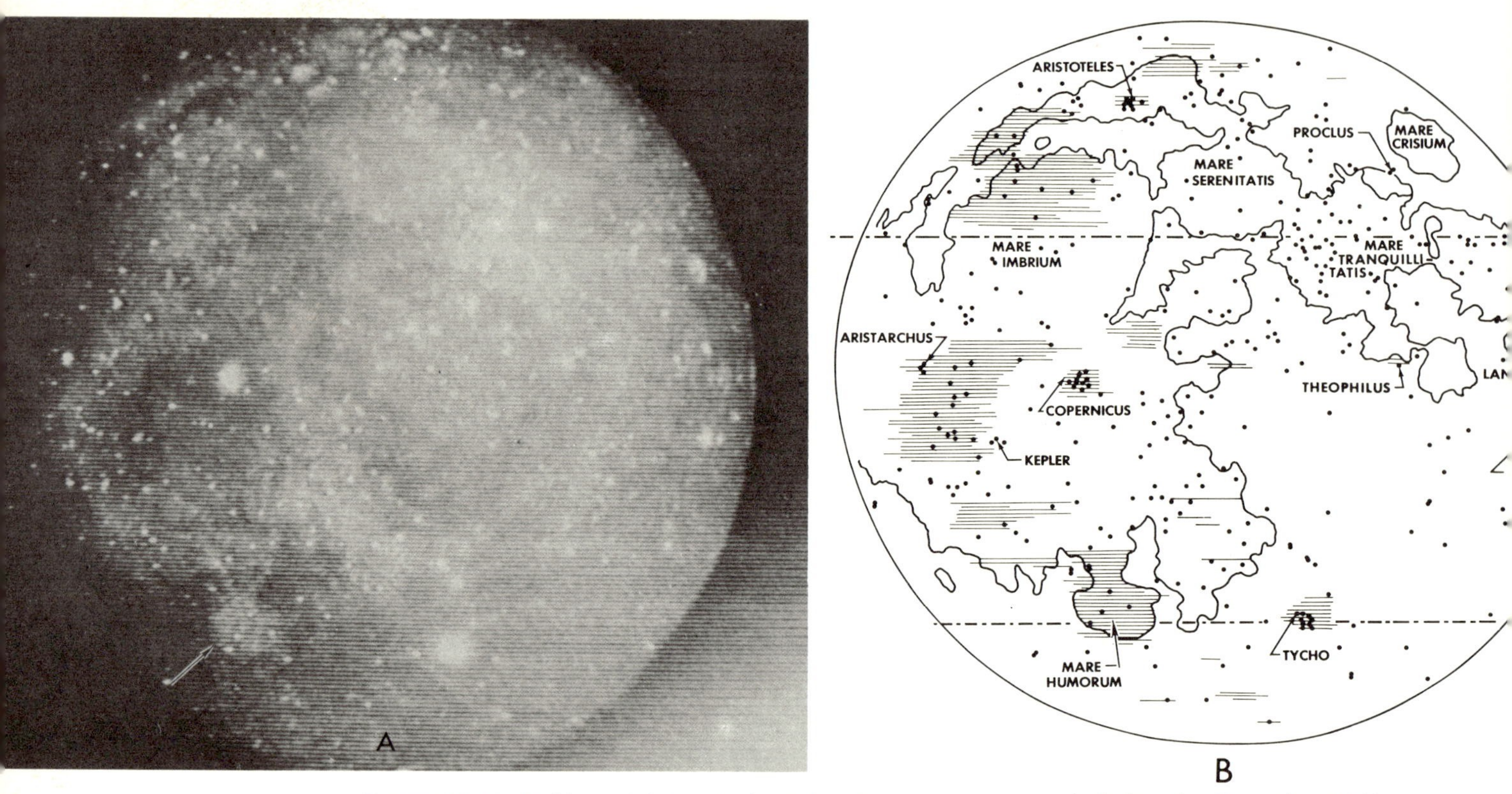

FIGURE III-18. Position of infrared thermal enhancements measured during the December 1964 lunar eclipse. (a) Reconstructed infrared image. Mare Humorum is shown by arrow. (b) Schematic position of hot spots (dots) and larger areas of thermal enhancement (parallel lines) (taken from Shorthill and Saari, 1966).

of emission spectra. Goetz's tentative conclusion is that either roughening of the surface by micrometeorite bombardment or unknown radiation effects are destroying the primary spectral contrast. The two remaining points, one on the crater Plato and a second in Mare Humorum, do show differences from the other twenty points. No singularly compelling explanation for these differences is available.

The final element of the electromagnetic spectrum to be considered is microwaves. The analytical tool is radar, a mechanism for generating a short pulse of microwaves and measuring the time and intensity of the echo return. A good statement of the fundamentals of radar astronomy as well as a comprehensive summary of radar studies of the Moon is found in Evans and Hagfors (1968).

There is a positive correlation between increase in wavelength and increase in penetration. Visible light and infrared provide information only about the upper one or two centimeters of the lunar surface, but radar penetrates to depths on the order of meters. Accordingly, the intensity of the radar return is governed not only by the composition and roughness of material directly on the surface but also by the behavior of material extending meters below the surface.

In early radar studies, beam resolution was poor, and observations for the entire disc were averaged together. Using this aggregate approach Evans and Hagfors as early as 1964 prophetically interpreted the back-scattering data to signify a surficial regolith, composed of a mixture of sand and broken rock occupying about 50 percent of the volume. Hagfors (1966) pointed out that a near-specular return observed at longer wavelengths but lacking for shorter wavelengths indicates that much of the surface is smooth (though not necessarily level) at the scale of meters, but rough at the millimeter scale.

In more recent studies of the Moon, radar beam resolution has been improved to encompass areas with diameters as small as one km. Using this approach Thompson and Dyce (1966) were able to determine that the back-scattered return from certain craters was a factor of 10 greater than the return from immediately neighboring regions. Many craters which showed radar enhancement were independently classified as youthful on the basis of sharp topography and bright halos or rays. Thompson and Dyce concluded that recent meteor impacts or internal explosions have locally disrupted a regional smooth, porous, fragmental layer and have produced a shattered, compressed rock surface with relatively high dielectric constant.

As in the case of infrared it is possible to scan the Moon systematically with a high-resolution radar beam, in this way producing a two-dimensional map of returned intensity values which bears a remarkable resemblance to a photograph. The dark areas on the radar map shown in Figure III-19 superficially resemble shadows cast by oblique

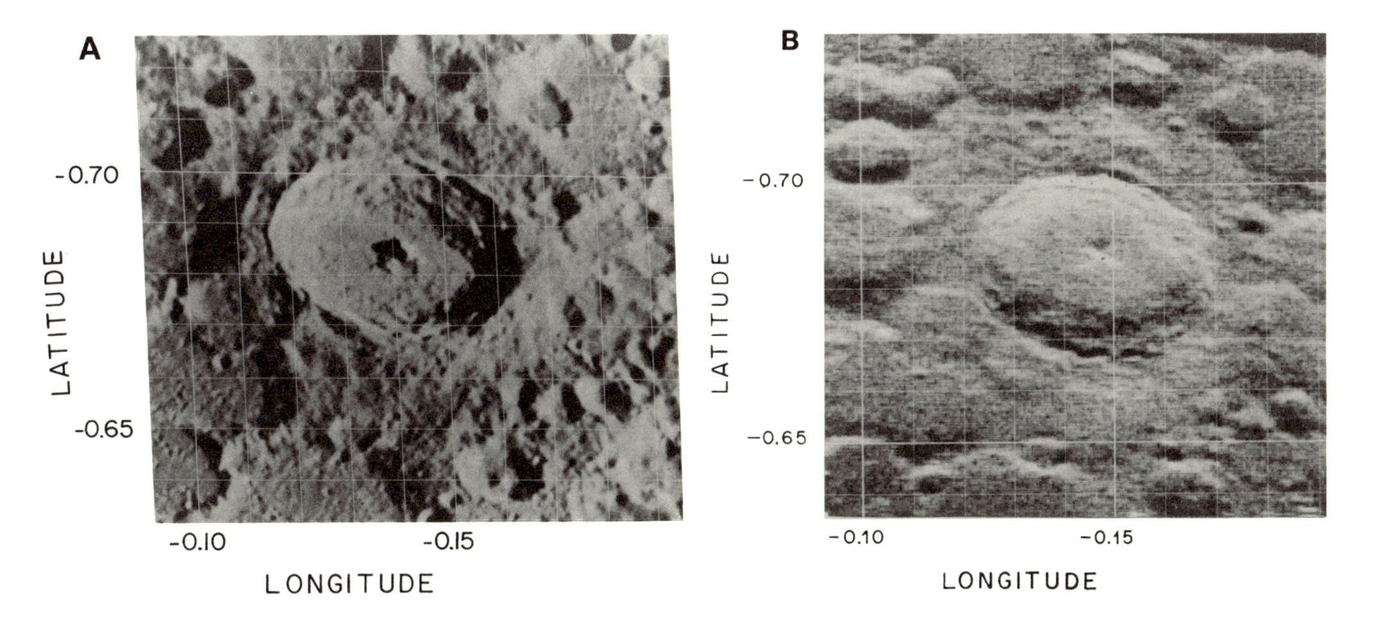

FIGURE III-19. Radar map of the crater Tycho compared with a telescopic photograph. (a) Telescopic photograph taken from *The Orthographic Atlas of the Moon*, G. P. Kuiper, ed., University of Arizona Press, 1960. (b) Radar map at 3.8–cm wavelength using circular polarization in the polarized mode (from Pettengill and Thompson, 1968).

sunlight. However, this comparison is without meaning since the radar illumination was of a full-Moon type. Apparently, variations in large-scale slopes with respect to the local horizontal control the intensity of the echo. Variations in the time required for the radar beam to reach the Moon and return to Earth, not shown in this display, permit one to construct a generalized topographic map for the lunar surface.

The theoretical potential of the various electromagnetic measurements as aids in regional mapping is obvious. Instead of subjectively describing the appearance of a formation one might, instead, indicate its albedo, color, cooling rate, radar roughness, etc. Although any one of these properties might prove an incomplete definition, the combination of numerous parameters should provide a unique description. However, practical success has been considerably less than theoretical considerations would suggest. The most important deficiency is that the reasons for different behavior of different materials is poorly known. One can assign a value for percent polarization, but its physical significance is uncertain. The same is true for infrared radiation, for radar return intensity, and for many other of the analytical values derived from the electromagnetic spectrum. Although one can assign a suite of numbers to a given geologic formation, it is not clear whether these numbers describe the same physical property, different physical properties, or indeed, irrelevant physical properties.

This problem of significant parameters and meaningful classification is a familiar one for petrologists and stratigraphers. The most easily quantified parameters are not always the best. For example, specific gravity of most sedimentary rocks reveals almost nothing about their major differences and origins. A Tertiary limestone and a Precambrian quartzite might have identical values, even though the two rocks are chemically and mineralogically dissimilar and were formed by completely different processes operating at widely separated times.

The entire field of "remote sensing" emerges as one in which instrumental elegance has far outstripped understanding of physical behavior. The potential remains exciting but unrealized.

Lunar Craters and Terrestrial Analogs

CHAPTER

IV

REASONING BY ANALOGY

THE principle of uniformitarianism is often condensed to an epigram: "The present is the key to the past." Phrasing this in language less striking but more accurate, one might say that Earth history can be interpreted in terms of natural forces observable today. For example, familiarity with present landscapes and the processes which modify them leads to detailed paleogeographic renderings of ancient scenes.

Many analogies involve comparison of distinctively shaped or patterned objects. However, in making these comparisons it is important to consider also the processes which formed the objects. Processes which are fundamentally similar may form products which look completely dissimilar. For example, although meandering and braided streams are distinctively different when viewed from the air, they are both the result of stream erosion and deposition. Relatively subtle variations in gradient and discharge account for the two patterns.

It is also possible that completely different processes may produce superficially similar products. A classic example is afforded by the study of extraterrestrial dust. Almost a hundred years ago spherical magnetic grains were recovered from deep-sea sediment samples. These were interpreted as iron-rich particles formed by ablation of larger meteorites falling through the Earth's atmosphere (Murray, 1876). In the following years numerous collections of magnetic spherules were made with surprising ease from the atmosphere, rain water, snow, and recent sediments. Detailed chemical analysis in the past decade has demonstrated that some of the spherules do have compositions consistent with a meteoritic origin (Schmidt and Keil, 1966). However, many turn out to be nothing more than a minor but common constituent of industrial smoke and ash (Handy and Davidson, 1953).

Since most interpretations of lunar geology depend on the use of terrestrial analogs, we should examine the assets and liabilities of such analysis in more detail. Consider the paleogeographic reconstruction shown in Figure IV-1. The ancient stream system is recognized as such by comparison of its deposits with those formed by recent streams. In both situations one can identify lenticular lenses of gravel, cross-stratified sand deposits of point bars, and sun-cracked clay formed on mud flats. More properly we should say that, in recent deposits, these sediment types are unmistakably related to a stream

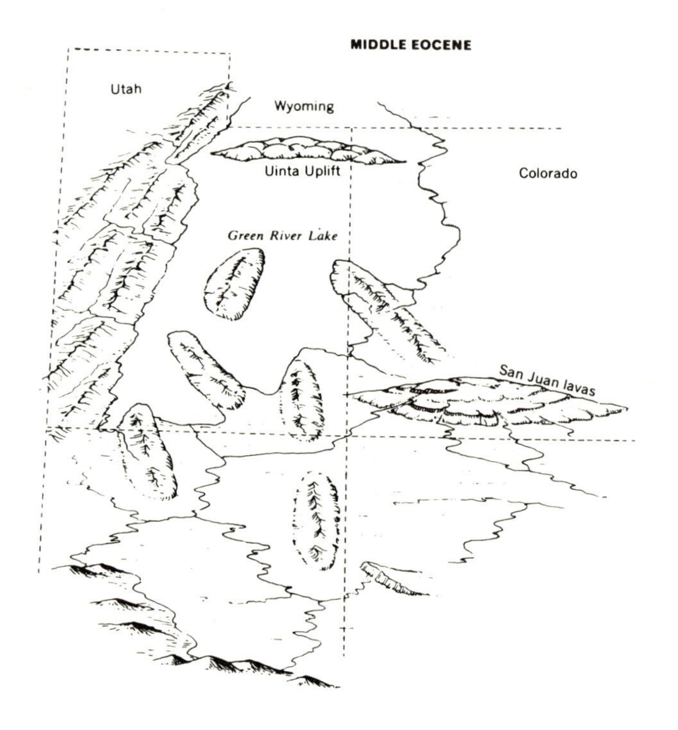

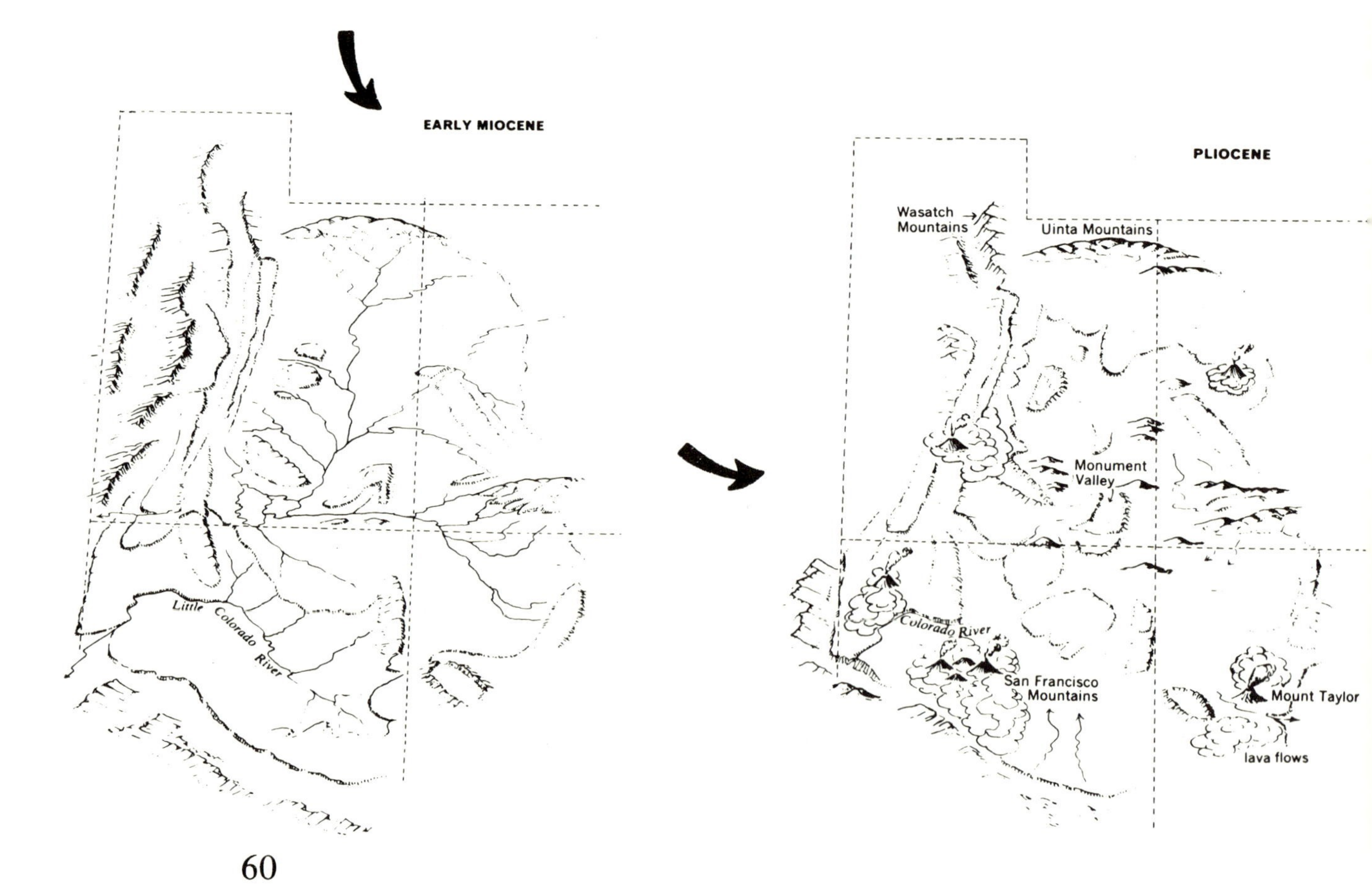

FIGURE IV-1. Paleogeographic maps showing development of Colorado Plateau during three epochs in the Cenozoic. The configurations and locations of lakes, rivers, mountains, and volcanic fields are based on an examination of the rock record coupled with an understanding of the processes which form present-day landscapes (adapted by Kay and Colbert, 1965, from Hunt, 1956).

which is even now flowing, and to bordering mud flats which are even now curling dry in the hot sun. When we look at the ancient scene we see only the rock record—and that confined to limited exposures (Fig. IV-2). But if gravel deposits are lenticular in the two dimensions of an outcrop, is it not reasonable to assume that they have a sinuous extension in the third dimension? If the sun-cracked shales occur several feet stratigraphically above the visible channel deposit, is it not reasonable that this particular mud flat existed when the channel had migrated laterally to some point beyond the present outcrop? The visible channel deposit, then, is interpreted to record a stage in the evolution of the stream older than that shown by the sun-cracked shale.

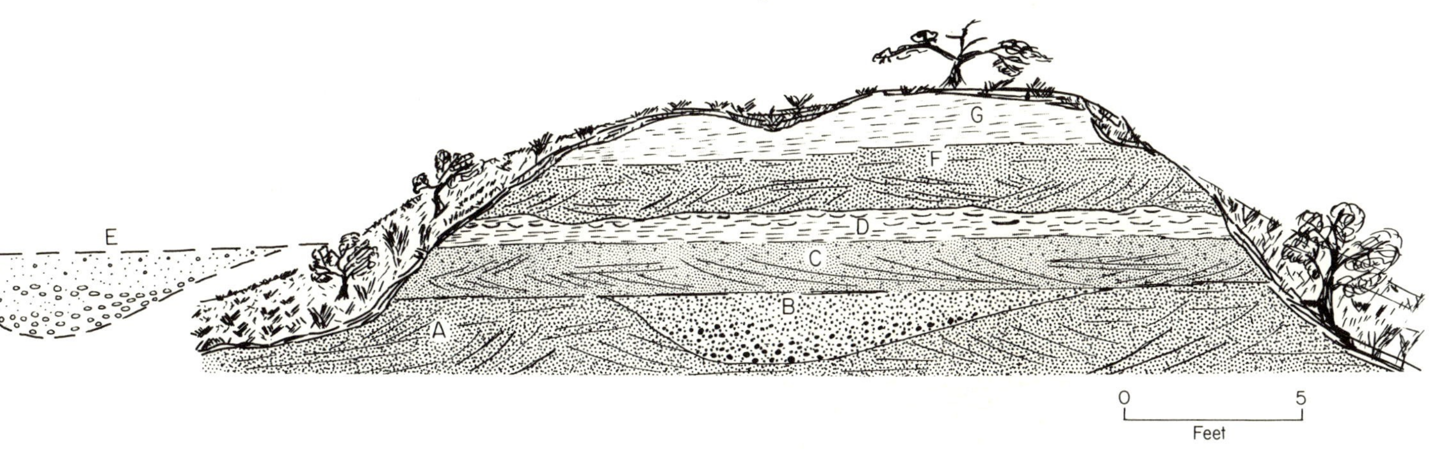

FIGURE IV-2. Schematic drawing of a road-side outcrop showing a vertical succession of fluvial sedimentary rocks. Lower sandstone bed "A" contains a coarse-grained channel deposit "B." Overlying sandstone bed "C" includes point bar and levee deposits. These grade upward into overbank mud deposits "D." As the muds were deposited, an associated stream deposited the sediments shown at "E." Recent erosion has destroyed this part of the rock record. The sand-shale sequence "F-G" documents another fluvial cycle.

Reviewing the details of this analogy we begin to see that it is more permissive than exclusive. The rock record contains a number of features which are similar to those seen in recent stream deposits. But lenticular gravel deposits occur on wave-swept ocean beaches; so, for that matter, do cross-stratified sands and sun-cracked shales. This being true, how can one be sure that the recent stream is the proper analog? The answer to this cannot take the form of mathematical certainty. The best one can do is say that in some situations there are so many lines of evidence leading from the recent analog to the ancient example that all—or almost all—Earth scientists are convinced. At that point the analog is accepted as "correct" or "true." In the case of the ancient stream deposit, there are many other features one might have mentioned: fresh-water fossils, plant debris, scour and fill structures, associated varved lake deposits, non-marine red beds—so many things that almost all scientists would agree that the analogy with recent streams is correct.

The use of analogs places a heavy premium on the geologist's awareness of the world around him. A student trained in the actively volcanic islands of Hawaii can recognize at a glance the special meanings of ancient lavas, but a student familiar only with New England metamorphic geology would be doing well to identify correctly the rocks underlying the White Mountains as lavas, let alone to interpret the significance of particular structures and textures. This fact of personal experience tends to make choice of analogs more fortuitous than consistently logical. If a geologist is unfamiliar with the features of recent deltas, he may be unable to make any sense of an ancient deltaic sequence. If he has studied the Mississippi delta and no other, he may note certain similarities and become convinced that he is standing on an ancient Mississippi-like deposit. However, if he has visited the mouth of the Niger River, he might realize that the rocks more closely resemble this deltaic example. Assuming the role of the complete sceptic, one can argue that since the major deltas in existence today throughout the world differ slightly, there is no reason to assume that any one of them will closely resemble those of the past.

One other characteristic of the analog argument should be noted. It is primarily descriptive, so that features can be correctly compared even though they are not completely understood from a dynamic or genetic point of view. For example, the fact that streams deposit particular suites of sediments has long been recognized, but application of fluid mechanics to interpretation of deposits and bed forms in streams has come only in the last decade.

Geologists—making frequent use of analogies as they do—have always been sensitive to the criticism that they describe and interpret that which they do not completely understand. Perhaps this is inevitable, stemming from the geologist's role as both scientist and historian. Many aspects of both natural and human history will always be clouded with uncertainty. In both fields the use of analogs is hazardous, but, employed intelligently, the gains far outweigh the losses.

With these cautionary introductory remarks in mind, we are now ready to observe some of the chief structures of the Moon and to examine possible counterparts here on Earth.

CRATERS ON THE MOON

The most striking features on the Moon are, of course, the craters. From Galileo on, observers have been fascinated with problems of crater form and origin to the virtual exclusion of other significant problems of lunar geology. It is surprising, therefore, that there are so few systematic and comprehensive discussions of crater morphology and classification. Instead, two particular parameters have been repeatedly studied in the hope that they will provide a numerical shortcut to the general problem of crater origin and significance. First, craters have been classified according to their dimensions. The most comprehensive attempt along these lines was made by Baldwin (1963) who ranked more than two hundred craters according to their depth-width ratios. The second statistic that has been calculated and recalculated is crater density, usually treated as a function of cumulative crater size (e.g., Kreiter, 1960; Opik, 1960; Dodd and others, 1963; Shoemaker and others, 1962; Fielder, 1965*a*, Hartmann, 1968).

A brief examination of representative photographs, especially those from the Orbiter and Surveyor series, suggests some of the problems in any comprehensive descriptive approach. Craters as small as 10 cm can be identified in Surveyor photographs. At the other end of the size spectrum, telescopic and Orbiter photographs reveal craters with diameters of more than 100 km, and circular basins in maria regions are more than 500 km across. To describe these end members within the same framework is something like comparing a small mud puddle with the Atlantic Ocean. It can be done, but not without some mental broad jumps.

Even considering craters of restricted size range, there are remarkable variations. It is tempting to single out those few craters whose geometry is most clearly defined and to treat the remaining ones as imperfect renditions of the "type" crater. So it is that proponents of impact craters point to Copernicus or Tycho and volcanic enthusiasts focus their attention on Alphonsus and Ptolemaeus. Finally, craters reveal a wide diversity of ages. Although it is clear that older craters will be more degraded and subdued than their younger counterparts, the exact features attributable to old age are debatable. For example, if a large crater has a small depth-width ratio relative to some arbitrarily chosen standard, is this evidence of degradation with time or is it indication of several different mechanisms of primary crater formation?

Perhaps the most instructive approach is to look first at a medium-sized, recently formed crater with simple morphology and then to consider more complex craters, both larger and smaller. Mösting C is 3.5 km in diameter, and in telescopic photographs appears as a perfect cup-shaped depression rimmed by a bright halo. Orbiter photographs show a wealth of unsuspected details (Figs. IV-3 and 4). The crater has a depth of approximately 330 meters, with rubble covered walls sloping inward at 10° and converging on a slightly flatter floor. Surrounding the central depression is a sharply raised rim mantled with large boulders which can be identified by the sharply bounded shadows they cast. Extending outward from the crater rim for a distance of approximately one crater diameter is a series of hummocky ridges concentrically disposed. These ridges and intervening troughs are less densely cratered than the surrounding mare on which Mösting C is situated, suggesting that the hummocky surface has had limited opportunity to undergo modification by small-scale cratering. The hummocky aspect of the ridges implies a depositional feature. They resemble somewhat terrestrial sand dunes, although there is not a systematic arrangement of steep lee slopes and gentle stoss slopes. The concentric arrangement implies a central source, although the situation is complicated by the fact that dune orientations have slight maxima at N60°E and N30°W, matching the orientation of linear grooves which are interpreted as structural lineaments (T. W. Offield, personal communication). The disposition of surficial materials, then, may reflect both depositional and structural controls.

Traced outward, the hummocky terrain gives way to a braided surface. Rill-like patterns of gouges and indistinct crater-like depressions form rough "V" shapes pointing back to the center of Mösting C. Local relief is less than in the hummocky terrain. Apparently the cover of ejected debris is thin or discontinuous, and the surface has been

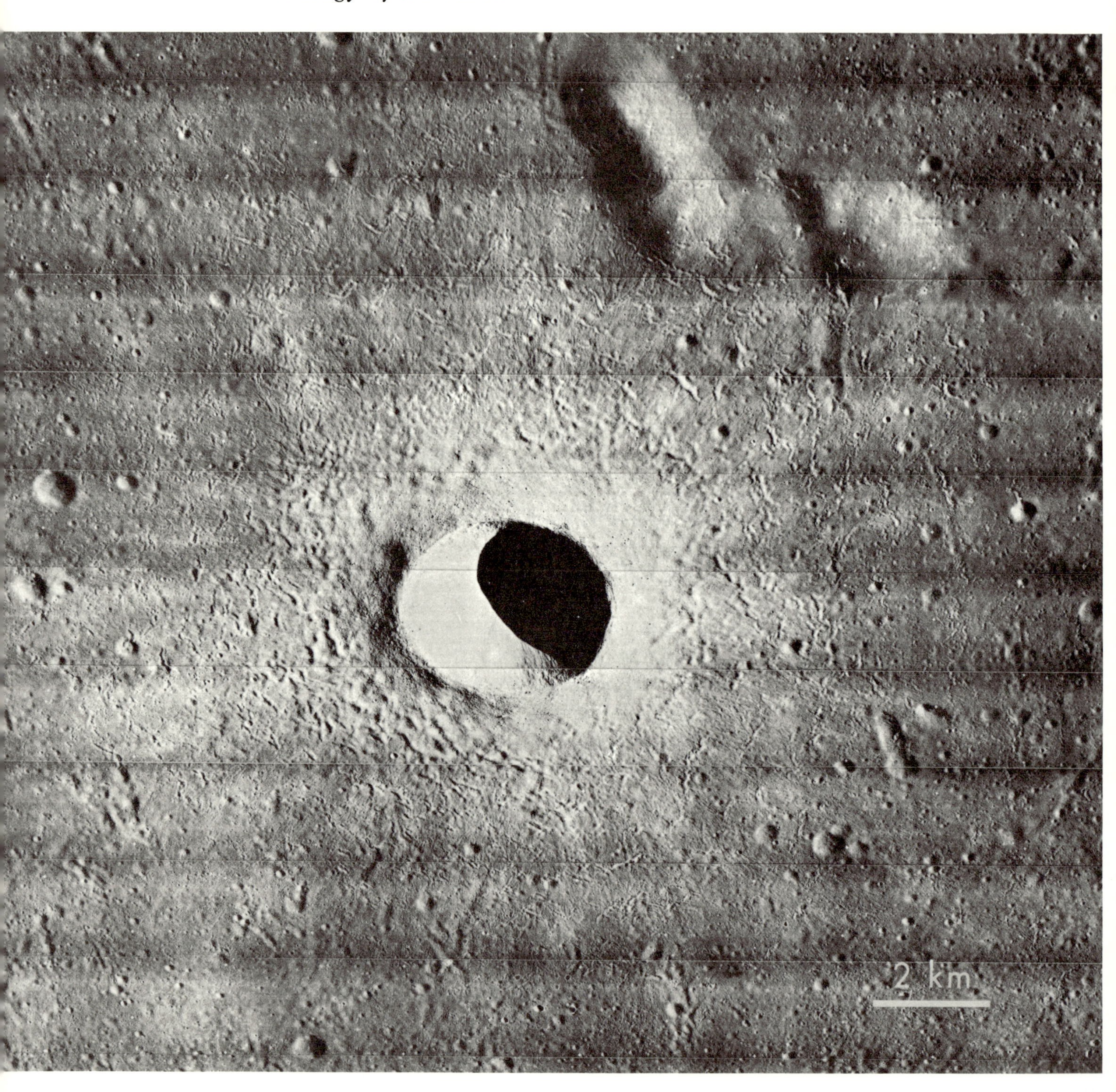

FIGURE IV-3. Crater Mösting C (Lunar Orbiter Photograph III 113M). The locations of this crater and all other lunar structures referred to throughout the text are shown on an index map of the Moon, Appendix B.

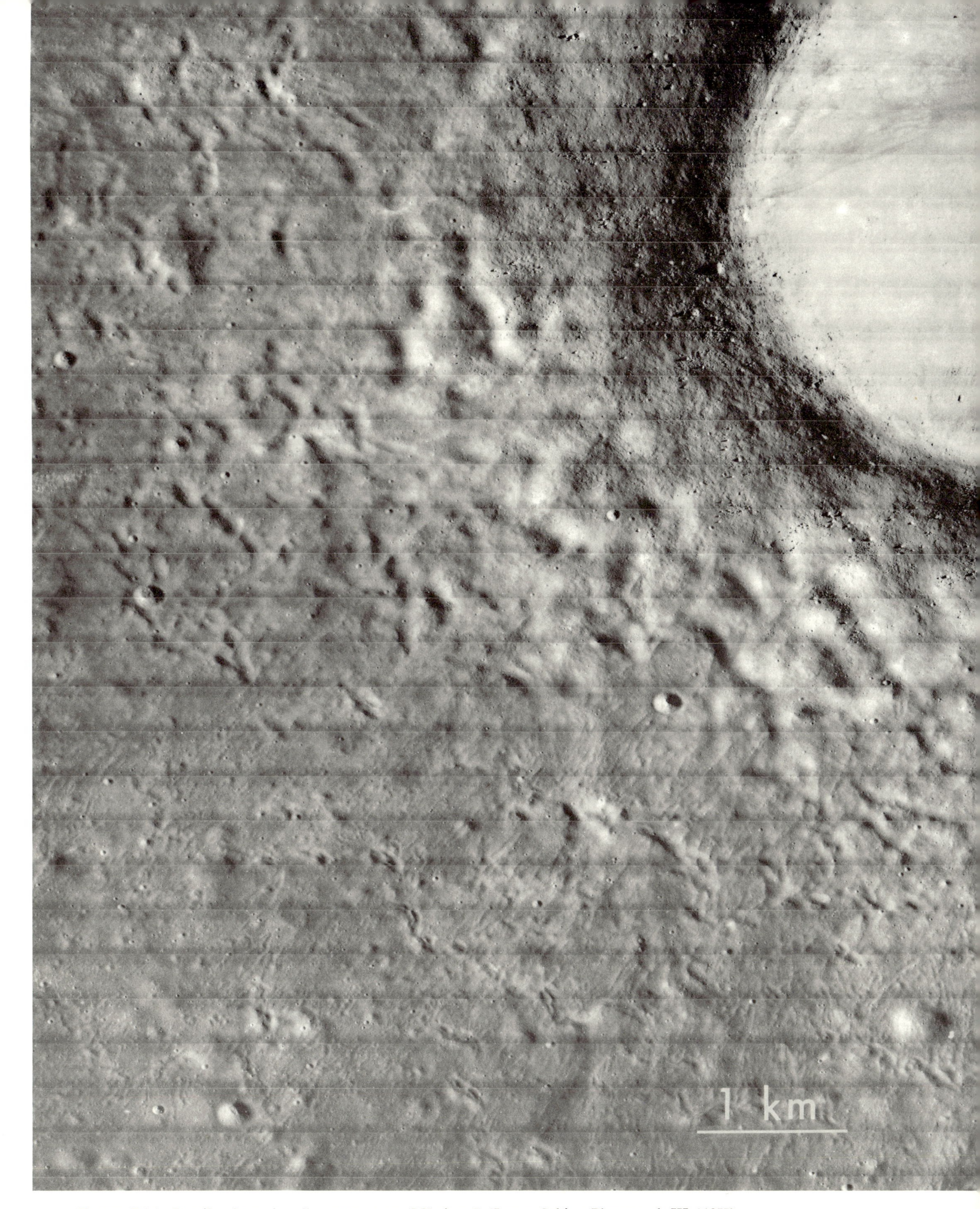

FIGURE IV-4. Details of terrain adjacent to crater Mösting C (Lunar Orbiter Photograph III 112H).

modified as much by erosional scouring from ejected blocks as by constructional piling up of ejecta. Some secondary craters are radially arranged; others inscribe looping arcs which appear to mark the intersection of a curved shell of ejected material with the ground.

Many fresh craters up to 20 km in diameter have a shape similar to that of Mösting C. Although the available photographs often do not show the fine details visible in Mösting C, craters in this group can be characterized by circular plan views, simple cup-like profiles, sharply raised rims, and prominent surrounding ejecta blankets, frequently with a hummocky inner facies and a radial outer facies. Euclides and Censorinus (Figs. IV-5 and 6) are typical.

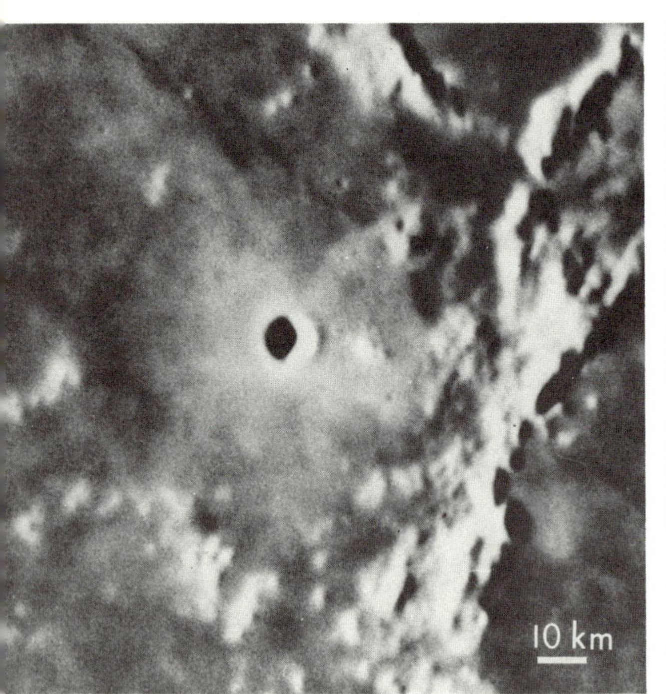

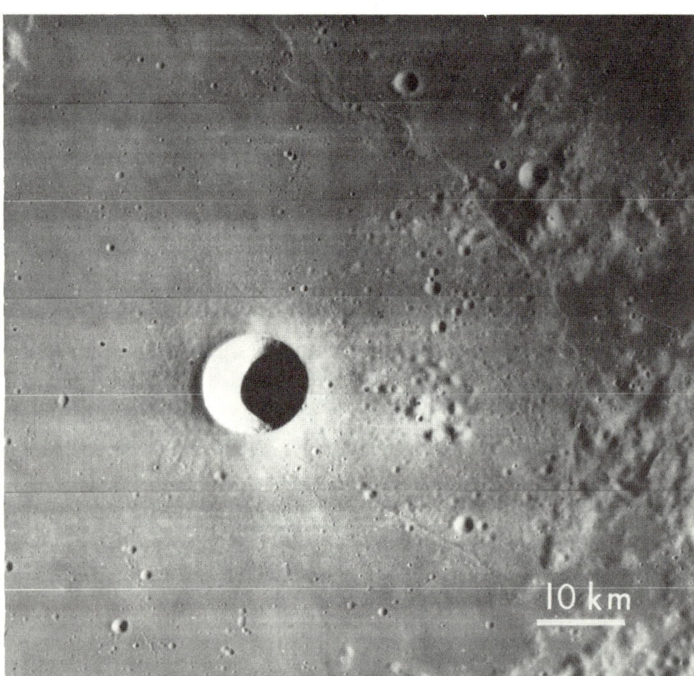

FIGURE IV-5. The crater Euclides. The left plate is a telescopic view showing a bright halo surrounding the crater (McDonald Observatory, Plate 191). The right plate reveals a hummocky terrain characteristically present around many young lunar craters (Lunar Orbiter Photograph IV 132H).

Craters larger than 20 km differ in having much more complicated internal structures, commonly including terraced walls and central peaks. One of the best examples is Aristarchus (Fig. IV-7), a crater in the northwest quadrant, which at full Moon is one of the brightest objects on the entire nearside. The crater is 40 km in diameter,

Opposite page:
FIGURE IV-6. The crater Censorinus. The upper plate shows bright streaks radiating from the crater, especially prominent where they are superposed on dark mare material in the upper left of the plate (Lunar Orbiter Photograph V 63M). The lower plate, a detailed view of Censorinus, shows that these radial features have topographic expression. Note that Censorinus appears younger than the slightly larger crater on the right edge of the plate. Numerous large blocks are situated on the rim of Censorinus but are relatively rare around the older crater. In addition, the linear ridges and channels radiating from Censorinus are superposed on the rim of the older crater (Lunar Orbiter Photograph V 63H).

10 km

1 km

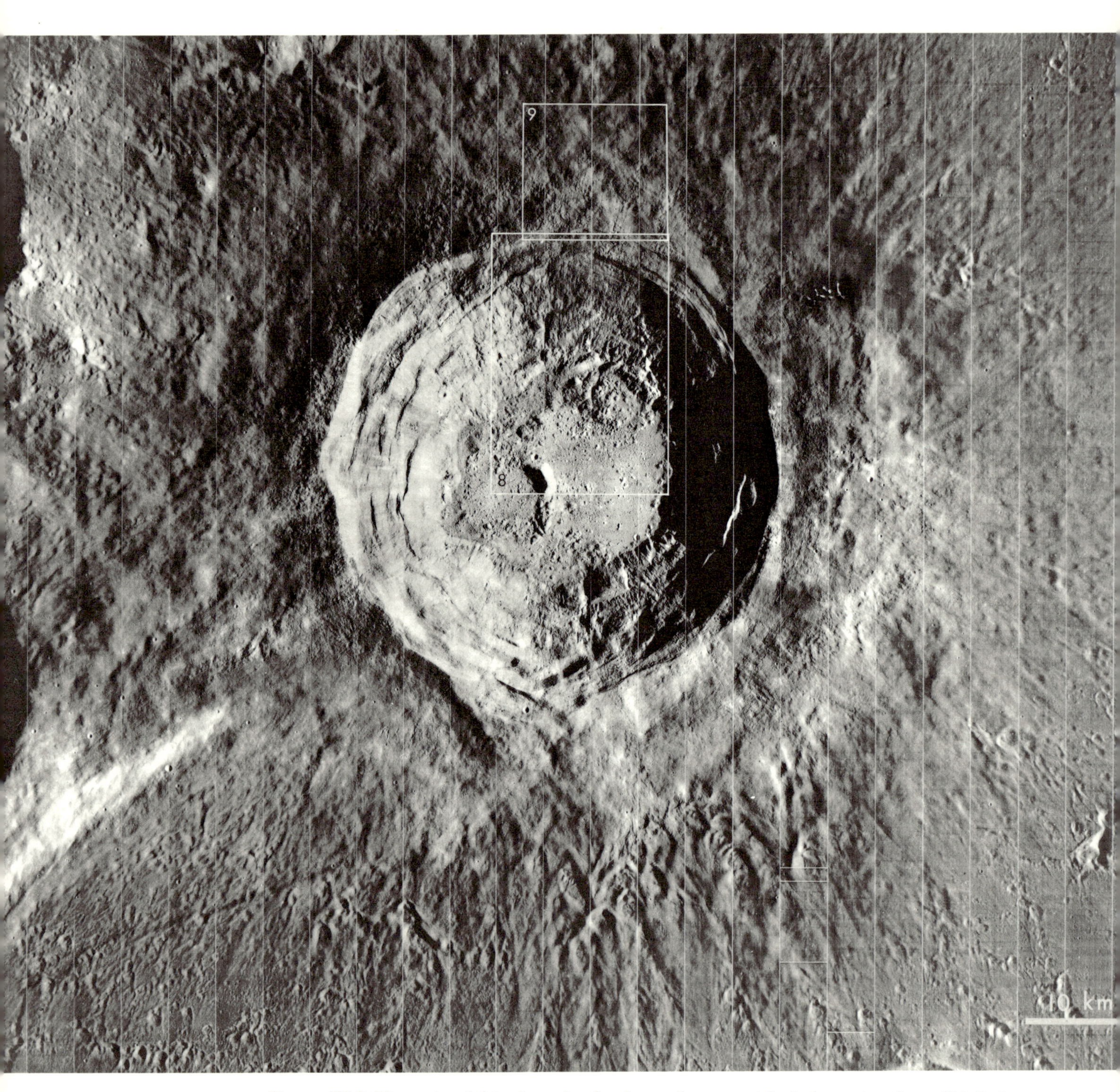

FIGURE IV-7. The crater Aristarchus, showing inner rim concentric facies, outer rim radial facies, and secondary craters. Locations of detailed views shown in Figures IV-8 and 9 are indicated by white outlines (Lunar Orbiter Photograph V 198M).

and its marginal ramparts rise approximately 2,000 meters above the floor. Near the center an elongate peak projects more than 300 meters above the floor. Discontinuous terraces rim the interior walls, formed by rough promontories of bedrock partly buried in smooth featureless slopes of fine sediment or fluidly emplaced material. The floor has a tortuously crenulated appearance, the sort of wormy pattern commonly seen in viscous flowing materials (Fig. IV-8). In contrast, the central peak has relatively smooth slopes littered with large boulders.

The rim features of Aristarchus are considerably more complicated than those displayed by Mösting C. There is an inner collar, approximately 5 km in width, characterized by rough, elevated terrain, concentric lineations, and superposed radially arranged channels apparently scoured into pre-existing deposits (Fig. IV-9). Surprisingly, the distinction between wall and rim, so clear at small scales, is not striking on the high-resolution Orbiter pictures. Both areas contain the same concentric structures and radial scour channels.

Beyond the inner facies and extending a maximum distance of about 40 km from the rim, there is a moderately hilly terrain with shallow radial furrows. (A rule of thumb is that the hilly terrain adjacent to an unmodified crater extends outward one crater diameter.) Still farther from the crater are swarms, lines, and loops of secondary craters that pockmark the dark, flat surface of Oceanus Procellarum.

Examples of craters with many of the same features as those shown in Aristarchus include Tycho (Figs. IV-10 through 15) and Copernicus (Figs. VII-12 and 13). Fields of secondary craters are particularly well developed around Copernicus. Some of these secondaries can be distinguished by their shallow, irregularly elongate shapes, paucity of rim materials, steep continuous walls toward the primary crater, breached walls in the opposite direction, and a herringbone texture extending outward from the breached end (Fig. IV-16). Such features are most common close to the primary crater, where ejected fragments presumably travelled relatively slowly in low trajectories. At greater distances fragments had higher velocities and steeper impact angles. Resulting craters are round and indistinguishable from small primary ones.

Some craters have a fundamentally different appearance from those in the Mösting C or Aristarchus groups. One type is exemplified by a crater in the floor of the Orientale basin on the western limb of the Moon (Fig. IV-17). Its interior is flooded, and its rim is atypically smooth and slopes gently outward. Although the general appearance is youthful, no radial rim facies or secondary crater chains are present (McCauley, 1968*a*). Schiller, a crater in the southwestern highlands south of Mare Humorum, demonstrates still another type (Fig. IV-18). The most obvious feature is the marked departure from circularity. Rim deposits are confined to a narrow band, and again radial deposits and secondary craters are lacking.

So far we have considered only craters large enough to be easily resolvable with telescopic observations. Figure IV-19 illustrates that they make up a very small percentage of the total. For example, less than three hundred craters with diameters greater

FIGURE IV-8. Detailed view of the floor and wall of Aristarchus (Lunar Orbiter Photographs V 199H and 200H).

than 25 km are visible on the Earthside hemisphere. By contrast, extrapolation of the cumulative curve to a lower limit of 1 cm indicates approximately one crater center per ten square centimeters of surface.

Some idea of the character of very small craters is gained from Surveyor photographs. Starting at the lower size limit the craters are merely shallow depressions with no raised rims or other distinguishing features. Above 3 meters most of the recently formed craters have sharp rims and steep inner walls. In the range of 10 to 20 meters many craters are surrounded by fields of ejected blocks. For craters in this size range the blocks seldom exceed 1 meter. Of course, this summary is valid only for the small areas examined by Surveyor spacecraft, and it is likely that morphologies of small craters will vary slightly elsewhere due to different composition and strength of the surface layer.

FIGURE IV-9. Detailed view of rim material adjacent to Aristarchus. Note the sinuous channels cut into an older concentrically patterned terrain. Topographic gradient is from lower left to upper right (Lunar Orbiter Photograph V 201H).

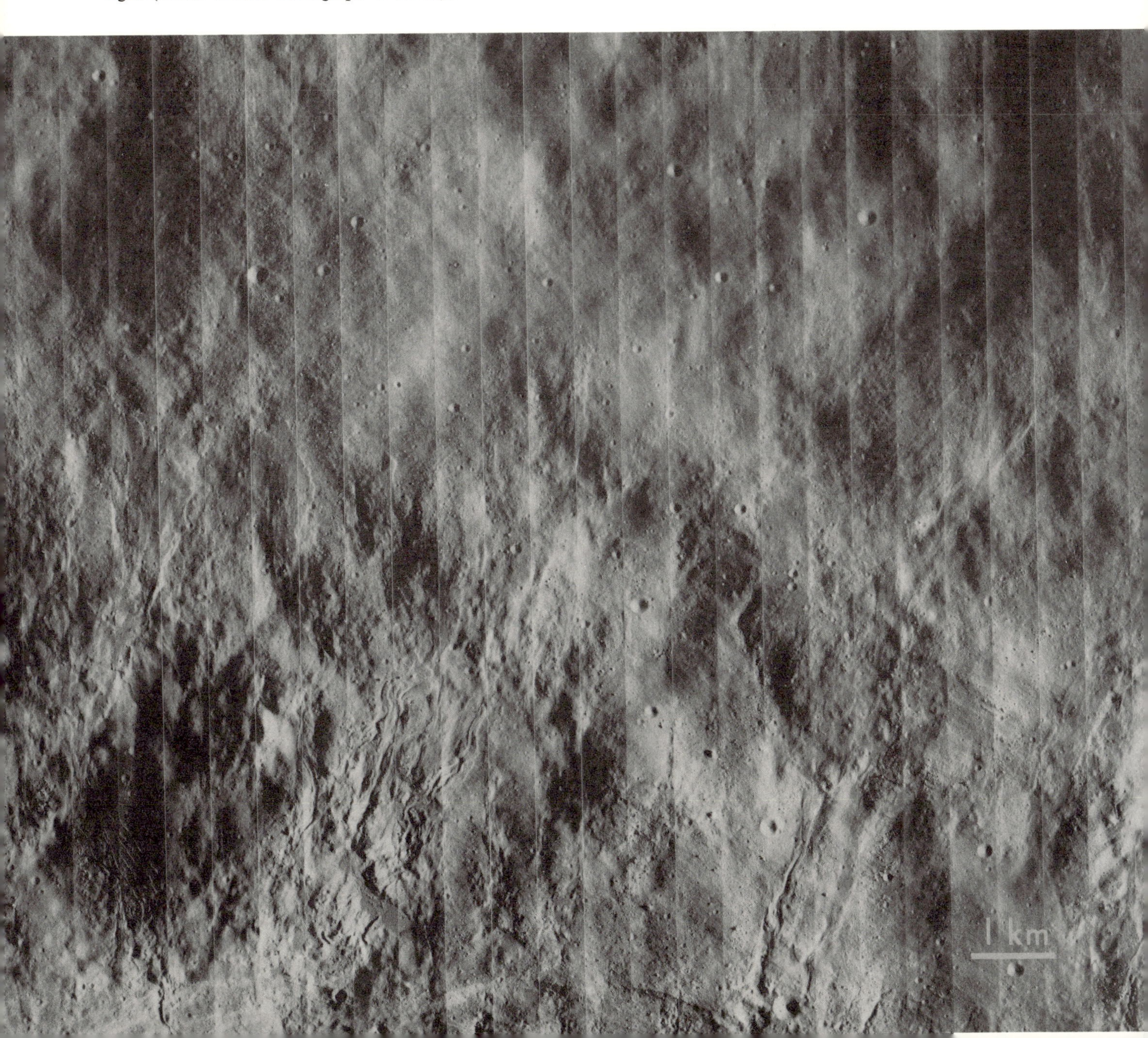

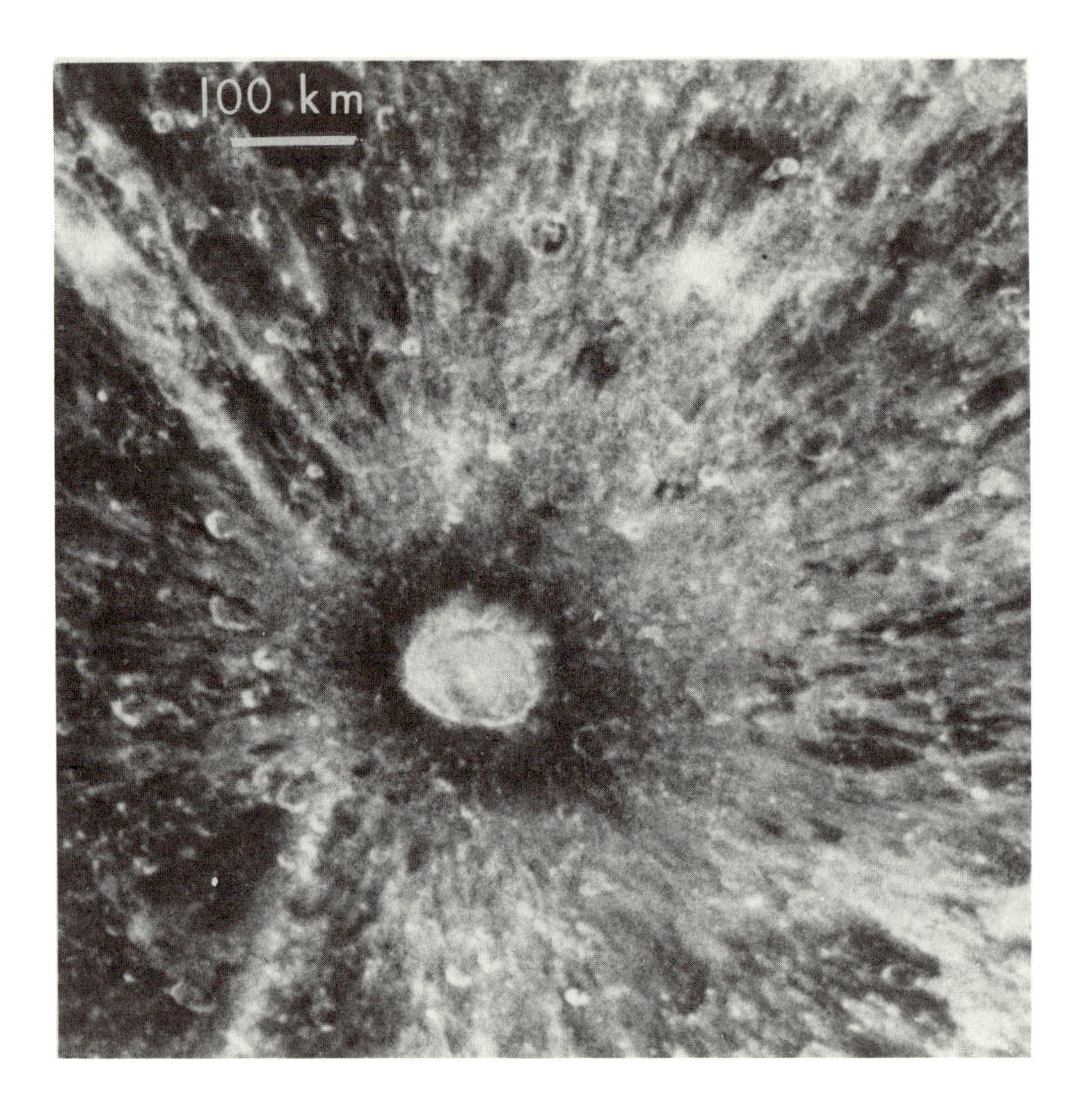

FIGURE IV-10. Full Moon telescopic photograph of crater Tycho, showing dark inner halo, bright outer halo, and rays. Interior crater walls and a central peak are especially bright (U.S. Naval Observatory Photograph).

FIGURE IV-11. The crater Tycho. Locations of more detailed views contained in Figures IV-12, 13, and 14, are indicated by white outlines. The orientation of the Surveyor VII photograph shown in Figure IV-15 is indicated by the arrow (Lunar Orbiter Photograph V 125M).

Figure IV-12. Detailed view of crater floor of Tycho showing crackled and crenulated surface (Lunar Orbiter Photograph V 125H).

FIGURE IV-13. Rim of Tycho showing concentric ridges of inner rim and irregularly braided outer rim. Note that pools of dark, smooth material overlie the concentric ridges. Materials with wormy texture in the middle left may have been emplaced by viscous extrusive flow (Lunar Orbiter Photographs V 127H and 128H).

FIGURE IV-14. Rim of Tycho showing lobe of dark crenulated material (arrow). The texture of this material resembles that displayed by terrestrial viscous igneous extrusive rocks (Lunar Orbiter Photograph V 127H).

FIGURE IV-15. Mosaic of Surveyor VII pictures showing panoramic view to the north and northeast. The large boulder in the foreground is a little less than 1 meter in diameter. The most distant ridges are about 20 km from the camera. The position of Surveyor VII relative to the crater Tycho is shown in Figure IV-11.

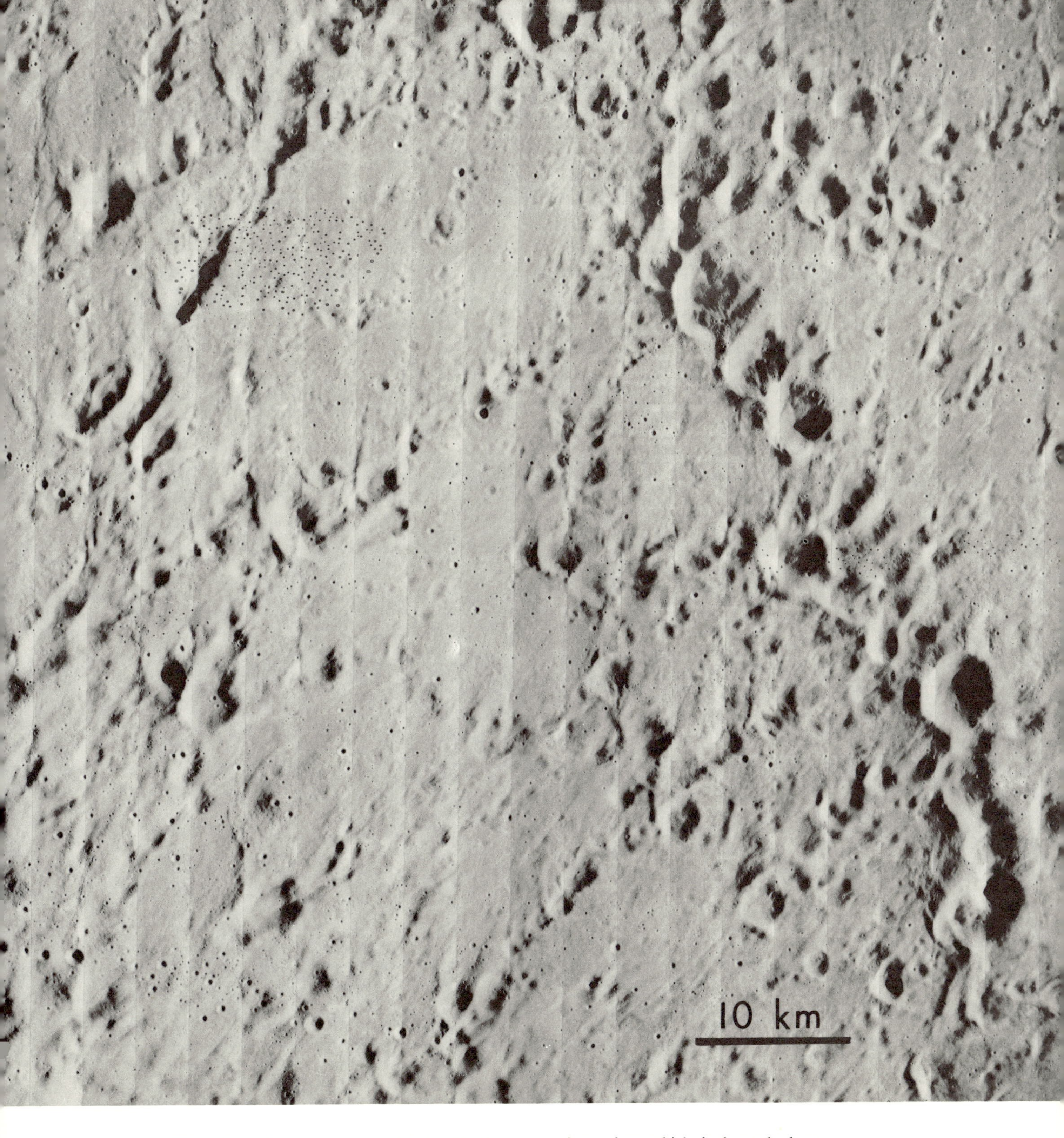

FIGURE IV-16. Secondary craters associated with the crater Copernicus which is located about 100 km to the southwest. Many of these secondary craters are characteristically irregular in shape with steep inner walls on the side adjacent to the primary crater and breached walls on the opposite side (Lunar Orbiter Photograph V 142M).

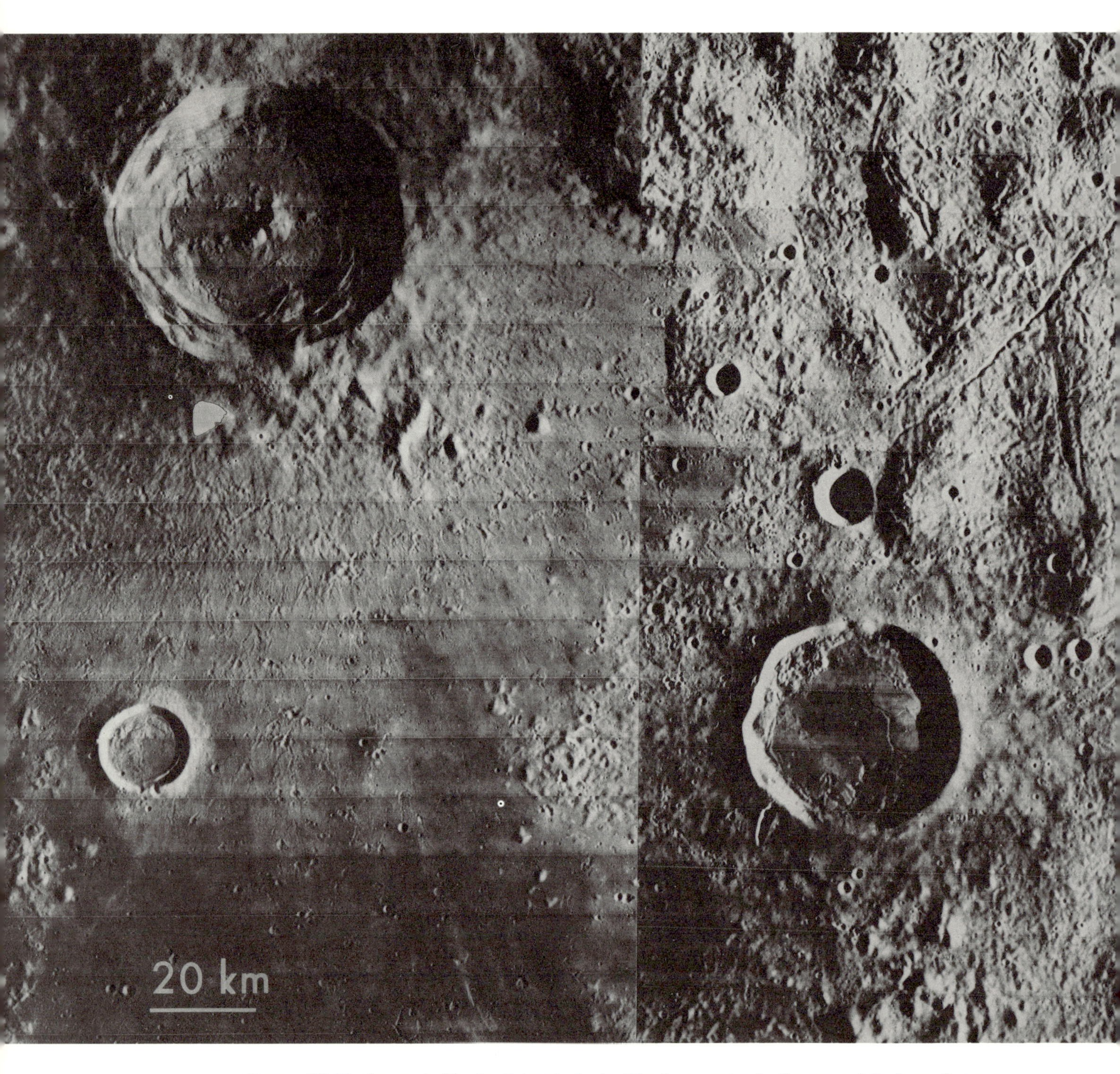

FIGURE IV-17. Craters inside the Orientale basin. The large crater in the upper left shows features indicating an impact origin. These include secondary craters, hummocky and radial rim material, terraced interior walls, and a central peak. The large crater in the lower right shows none of these features. Secondary craters are lacking; rim deposits are smooth; the inner wall is relatively featureless; and lobes of dark mare-like material fill the crater interior. Cracks and rilles are present within the crater. The combination of features present—and absent—suggests a volcanic origin for the crater (Lunar Orbiter Photographs IV 187H and 195H).

FIGURE IV-18. The crater Schiller. The elliptical composite crater form and the absence of prominent rim facies or secondary craters suggest a volcanic rather than an impact origin (Lunar Orbiter Photograph IV 155H).

FIGURE IV-19. Size-frequency distribution of craters measured on Ranger, Surveyor, Orbiter, and telescopic photographs. Generalized from several sources, including Surveyor Program (1967*a*).

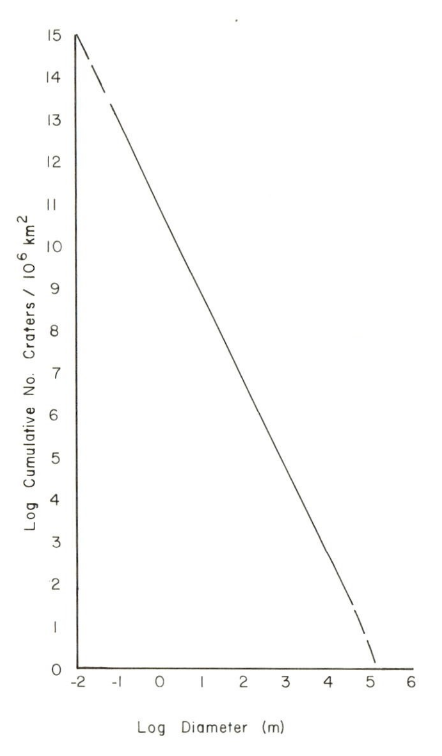

If we broaden our view and consider the many fresh craters from ten to several hundred meters in size, and visible on Orbiter photographs, distinctions can be made on the basis of interior shape and structure. Some craters have a perfect cup shape; others are flat-bottomed or possess a central mound; still others display a series of concentric ledges on the inner wall.

Maximum sizes of ejected blocks are roughly related to crater diameter, both of which are indirect measures of total cratering energy. Around craters of several hundred meters, blocks are commonly as large as 10 or 20 meters. Some blocks show a penetration scar where they have struck the surface along the trace of a ballistic trajectory and skidded to a stop at the end of a scoured furrow. However, for the great majority of blocks genetic association with a particular crater is assumed simply because of proximity.

All descriptions to this point have been for presumably young craters. As mentioned at the outset, craters are modified with age. The principal processes are erosional degradation by small-scale cratering, filling of the crater interior with erosional debris and slumped blocks from the rim and wall, flooding or blanketing from an external source, and internal flooding or isostatic rising of the crater floor. These effects are illustrated in Figures IV-20 through 24. Additional discussion of this important but complicated subject is saved for later sections on stratigraphy and age relations.

FIGURE IV-20. Craters with a spectrum of morphologies believed to be related to relative age. Recently formed craters have sharp rims and are surrounded by blocky debris. Old craters have subdued rims and shallow pan shapes. Recognizable rim deposits are lacking around old craters (Lunar Orbiter Photograph III 143H).

FIGURE IV-21. A large crater, Hommel, with its original shape greatly modified by more recent cratering events and by general down-slope erosion. Arrows indicate position of crater rim (Lunar Orbiter Photograph IV 82H).

FIGURE IV-22. Debris blanket associated with evacuation of nearby Orientale basin mantles preexisting craters. The vaguely defined outline of one such mantled crater is indicated by arrows (Lunar Orbiter Photograph IV 173H).

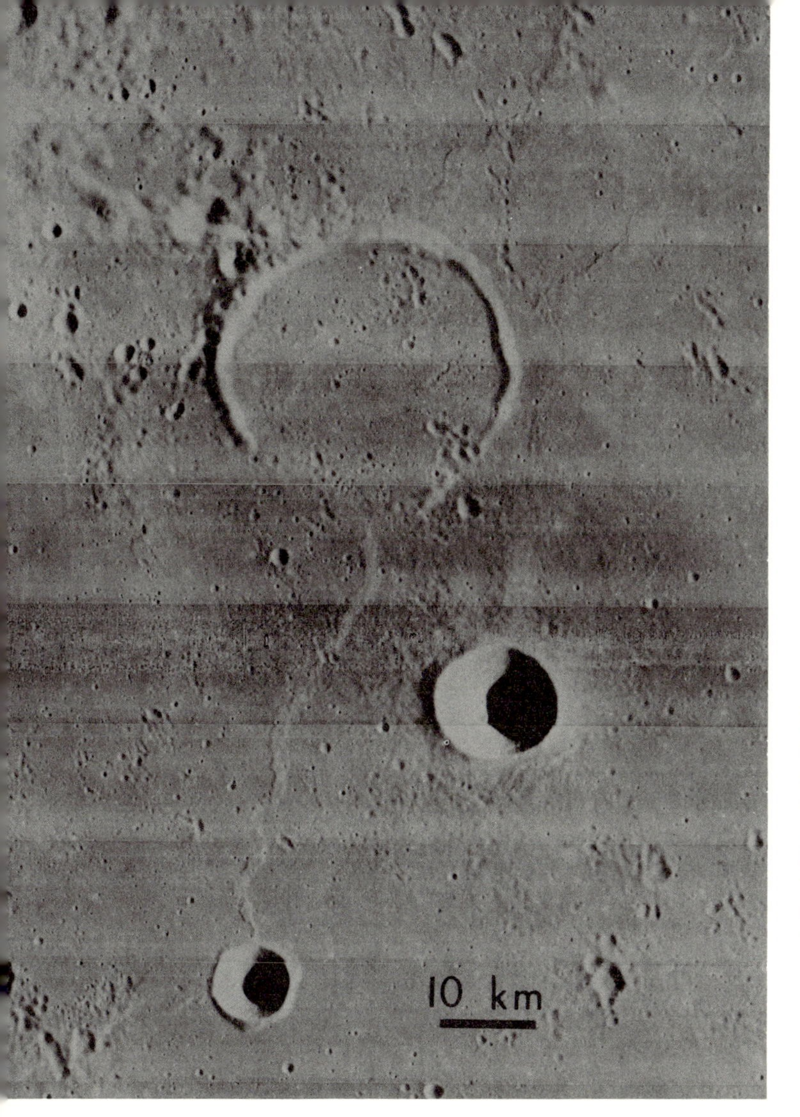

FIGURE IV-23. Craters showing several different relationships to adjacent dark mare material. Large crater near the center has rim deposits and secondary craters superposed on mare materials. Large crater in lower left has a steep outer rim sharply terminated by mare material, suggesting an overlap relationship. The large crater at top has been breached and flooded so that only an incomplete narrow rim is visible (Lunar Orbiter Photograph IV 175H).

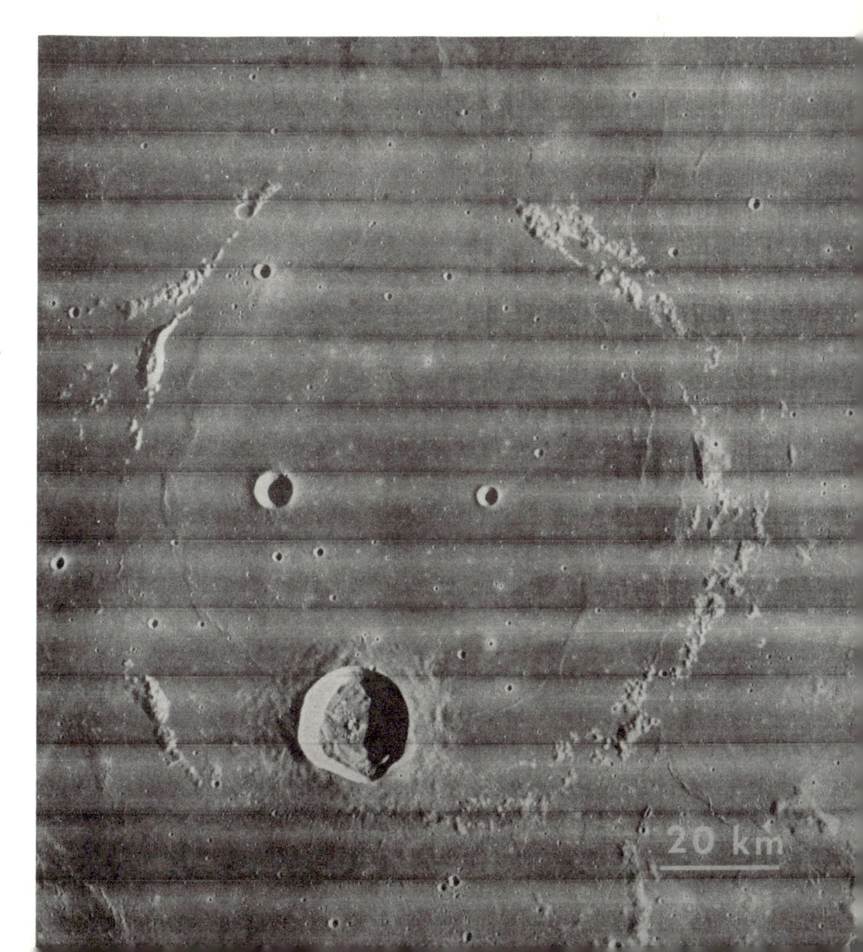

FIGURE IV-24. An ancient crater so completely flooded that only small remnants of the originally continuous rim crest are visible. This discontinuous circle of hills is known as the Flamsteed Ring (Lunar Orbiter Photograph IV 143H).

POSSIBLE METEORITE CRATERS ON EARTH

Now that we have some idea of the features present in lunar craters we can return to Earth to search for some convincing analogs. There are two likely possibilities: craters formed by impact—real or simulated—and craters formed by some volcanic process.

Impact structures are known in greater detail, chiefly because they can be simulated in model studies. In addition to study of natural impact structures, information can be gained from observation of fixed explosions of nuclear devices or chemical charges, armed and inert missile impacts, and pellet impacts from high-velocity guns. A recent comprehensive review of shock-related features is contained in French and Short (1968).

Circular depressions which may be natural impact structures are being continually discovered. In 1930 less than 10 such structures were known. By 1966, 33 probable impact structures had been described, and many more possible candidates were known. A tentative upper limit following a systematic search throughout the Earth is 250 (Short, 1966).

Before considering general criteria for identifying impact structures it would be well to study one undoubted example, undoubted because of the occurrence of abundant meteoritic fragments in and around the crater. This is Meteor Crater, located in Arizona in a region of flat-lying Permian and Triassic sedimentary rocks. Although the structure stands out as a striking anomaly on a flat plateau, had the impact occurred 30 km to the northwest it would have been nestled inconspicuously among Quaternary volcanic craters.

The crater and its geology are shown in Figure IV-25. Several features are noteworthy. Below the floor there is a breccia lens containing fragments from all formations intersected by the crater. Sedimentary rocks are bent upward beneath a raised rim. Detached, overturned flaps of these rocks locally rest on Quaternary sediment along the rim crest. A number of small vertical faults are visible in the crater wall (Shoemaker, 1960). Many of these are parallel to regional joints trending northeast and northwest; the gross plan view of the crater itself resembles a square with its diagonals oriented along these directions. For some distance away from the crater, debris from excavated Permian and Triassic rocks rests on Quaternary alluvium.

Many large fragments from a nickel-iron meteorite have been recovered from the surface around the crater as well as from depth in the breccia lens. This latter occurrence suggested to D. M. Barringer, a prominent industrialist and public figure in the early part of the century, the possibility of an economically important ore body within the crater. Though he showed prophetic insight in championing the meteoritic origin of the crater, Barringer did not know then that the meteorite was substantially fragmented and dissipated as it hit. His exploration of the property and his efforts to gain public support for developmental costs were both colorful and controversial (Nininger, 1956). After an extensive drilling program the search for the main ore body (which, following a traditional prospecting axiom, was always "a little deeper") was abandoned.

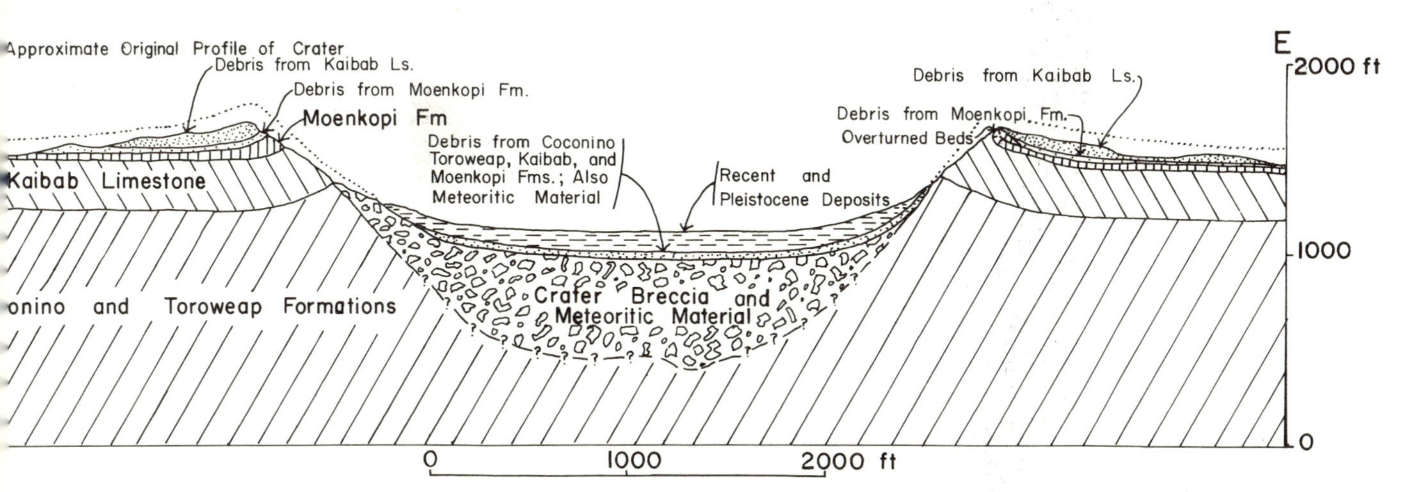

FIGURE IV-25. Meteor Crater, Arizona. In the oblique aerial view, north is at the right (courtesy of United States Air Force). The geologic cross section is adapted from Shoemaker (1960).

The reasonably complete preservation of Meteor Crater permits a detailed reconstruction of the original event. When a meteorite strikes the ground a shock wave is generated. This is a pulse of pressure which moves radially outward from the impact point at supersonic velocity. Rocks experiencing this transient pressure are severely compressed, often beyond their elastic limit. When this occurs, the rocks are plastically distorted, crushed, and faulted. At higher pressures certain minerals are converted to new phases through solid-state transformation or are altered to glass-like substances either through short-range molecular disordering or through true melting (Fig. IV-26). In Meteor Crater there is abundant evidence of disturbance and shock metamorphism: fracturing, brecciation, presence of glass, and occurrence of two high-pressure forms of silica—coesite (Chao and others, 1960) and stishovite (Chao and others, 1962).

The discovery of these two minerals was particularly exciting, since it simply and compellingly indicated the occurrence of pressures which could only be shock-induced. Figure IV-27 shows the pressure-temperature fields for various silica phases. Prior to their discovery at Meteor Crater, coesite and stishovite were known only from laboratory synthesis. Naturally induced high pressures are generally associated with great depths. The only natural near-surface phenomenon which couples high pressure with high transient temperature is a very-high-energy shock wave.

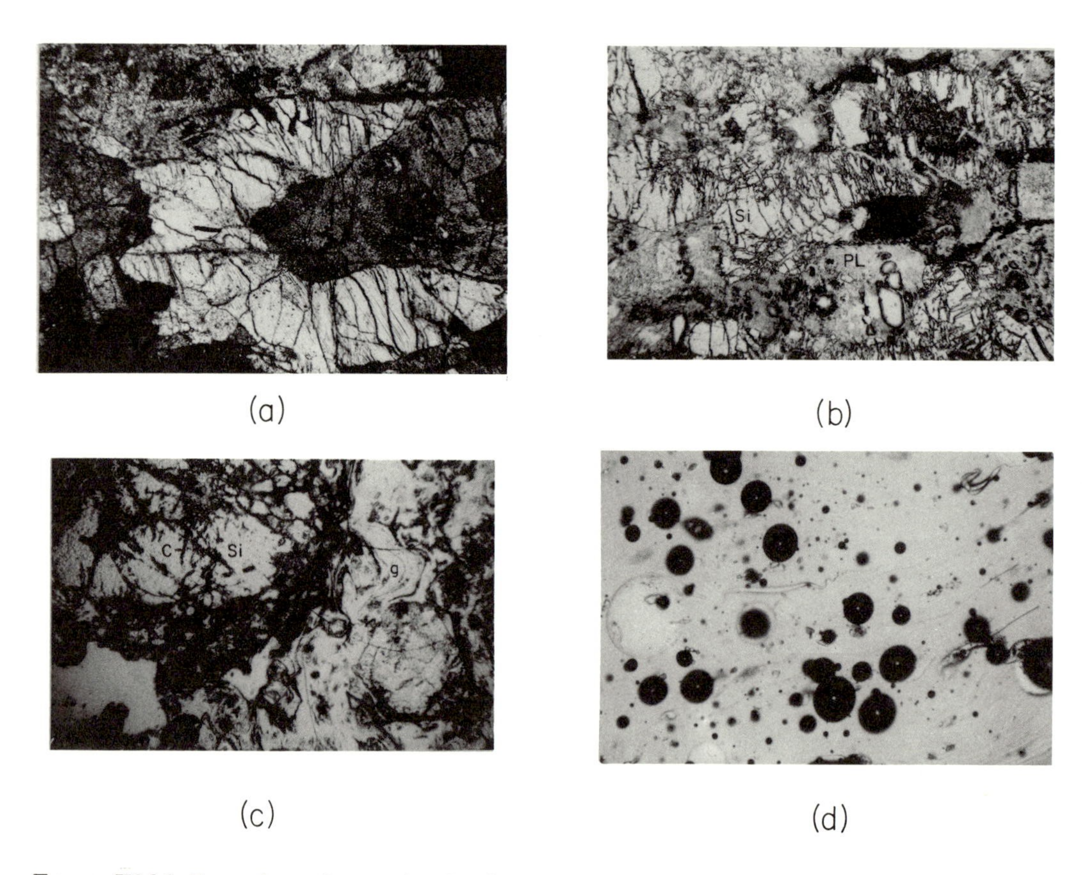

FIGURE IV-26. Four photomicrographs showing progressive shock metamorphism of granite gneiss in the Ries Crater, Germany. All three samples are from the same quarry. (*a*) weakly shocked rock, quartz shows closely spaced fractures; (*b*) moderately shocked rock, some silica glass (si) with stringers of recrystallized quartz and turbid plagioclase glass (pl); (*c*) strongly shocked rock, schlieren (g) and relict inclusions of silica glass (si) with stringers of coesite (c); (*d*) intensely shocked rock, glass with sparse mineral fragments (from Chao, 1967).

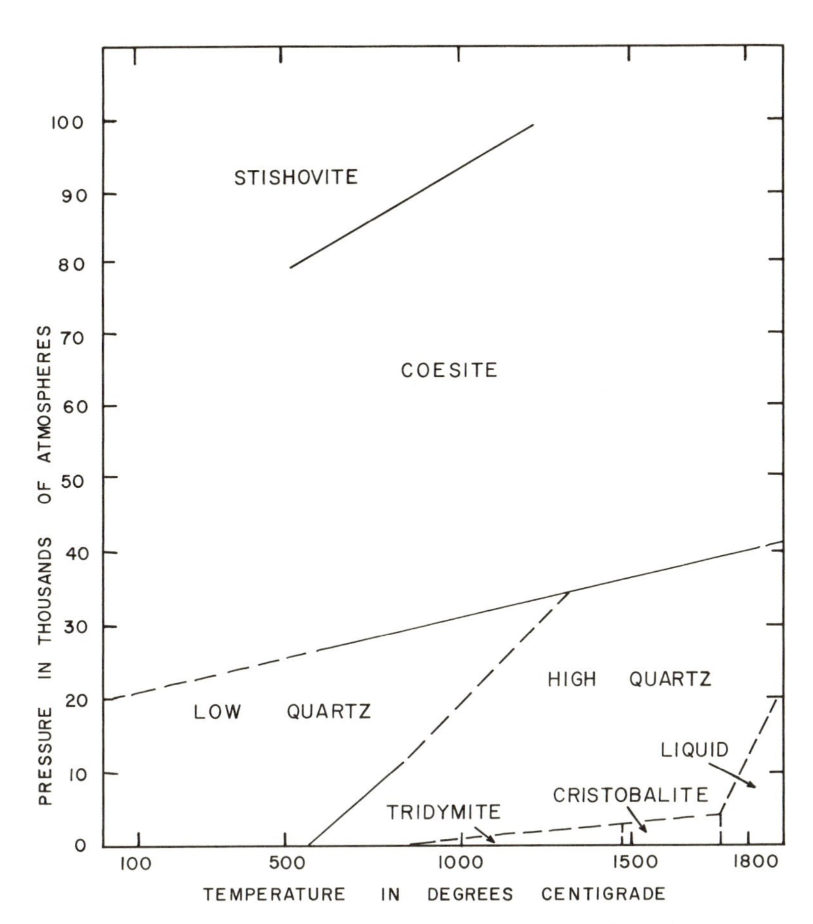

FIGURE IV-27. Pressure-temperature curves for several inversions between forms of silica. Modified from Boyd and England (1959). Boundary curve between coesite and stishovite is from Akimoto and Yasuhiko (1969).

When compressional shock waves encounter the ground surface, a reflected rarefaction wave is produced. As this wave propagates in an approximately reversed direction, material formerly compressed is now placed in tension. It is the returning rarefaction wave and the attendant tension which are largely responsible for ejection of material from the crater. It is important to remember that the crater will always be much greater in size than the projectile responsible for its formation. Meteor Crater is about 1,200 meters in diameter, but the original meteorite probably was only 25 meters across (Shoemaker, 1962).

Not all the suspected impact craters are of such recent origin. The Flynn Creek structure in Tennessee is a partly exhumed crater 3.6 km wide and about 150 meters deep, its interior partly filled with Upper Devonian and Lower Mississippian undeformed marine sedimentary rocks (Figs. IV-28 and 29). The crater intersects Middle and Upper Ordovician strata, so its time of formation must fall either in the Silurian or Devonian; Roddy (1968) presents evidence for an Upper Devonian age. Despite its great age the crater retains many of its original features. Apparently it formed either in a shallow sea or on a lowland plain close to sea level, and was buried by marine sediments almost immediately (geologically speaking). As at Meteor Crater, there is a zone of complex structure in the crater rim. Parts of the rim are uplifted, tilted, and folded; in other parts of the rim large sections have been moved outward along thrust faults. The ejecta blanket outside the crater is absent, presumably eroded, except for one area on the rim which contains a down-dropped fault block overlain by breccia arranged in a crudely inverted sequence, suggesting ejection from the crater. Unlike Meteor Crater but similar to many lunar craters, a central uplift is present at Flynn Creek. This uplift is structural rather than depositional or erosional in origin. The Lower and Middle Ordovician limestones which comprise the central uplift have been raised as much as 350 meters above their original stratigraphic position.

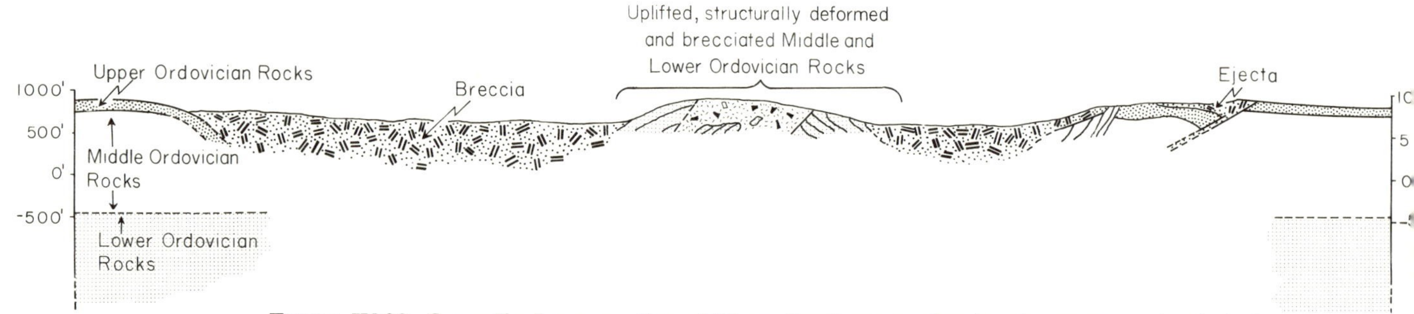

FIGURE IV-28. Generalized cross section of Flynn Creek crater, showing the structure shortly before deposition of the Chattanooga Shale in Late Devonian time (after Roddy, 1968).

The only two possible mechanisms of formation for the Flynn Creek structure are impact and volcanism. Bucher (1963) pictured a violent gas explosion, perhaps related to increasing pressures associated with a rapidly crystallizing water-rich magma at depth. In a critique of this idea, Roddy (1968) has pointed out that there is no evidence of volcanism either in the form of volcanic ash, igneous rock fragments, or any sort of hydrothermal mineralization and alteration. Further negative evidence is found in intense rim structure and presence of a central uplift, neither of which are characteristic of undoubted volcanic gas explosion craters. The positive evidence favoring impact is suggestive but not conclusive. Most of the structural features match similar features found in Meteor Crater, in nuclear cavities, and in craters formed by chemical explosion. Two important items of evidence, however, are missing. First, high-pressure mineral phases have not been found. Probably these are not to be expected since all pre-impact country rocks in the vicinity of Flynn Creek are silica-poor limestones. Diagnostic shock effects have been found at other structures almost exclusively in silicate minerals. Unlike most silicates, calcite is easily deformed under conditions of moderate pressure and temperature so that it is difficult to differentiate shock-induced twinning uniquely. In addition, the plastic calcite forms a sheath about the few grains of quartz that are present, protecting them from pressure effects.

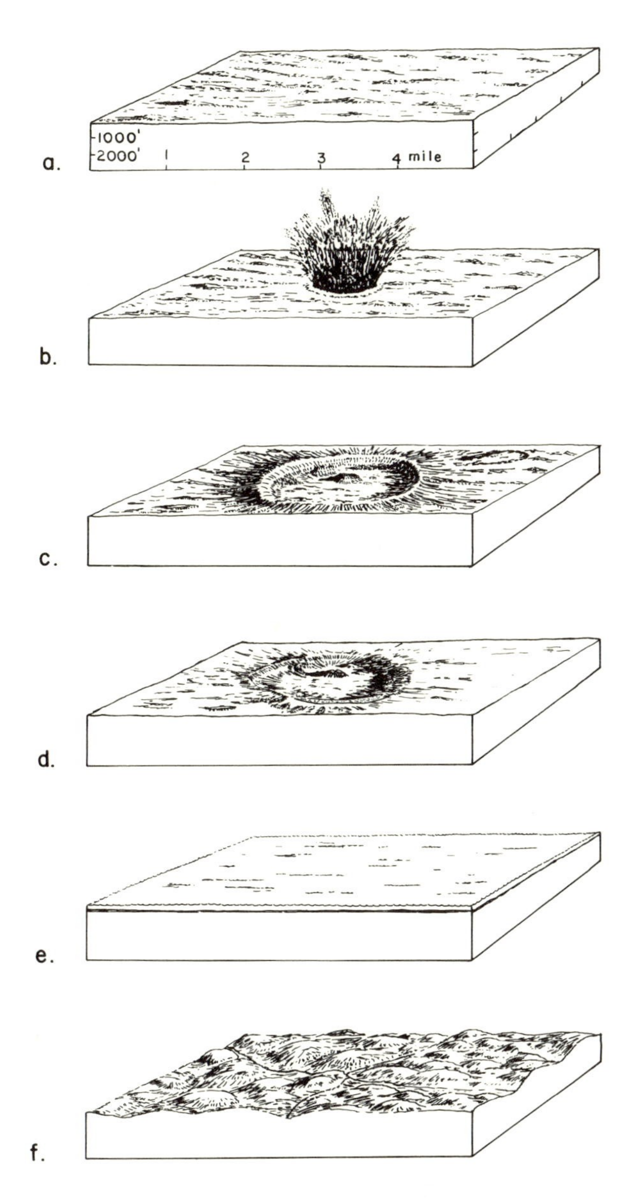

FIGURE IV-29. Diagrammatic sequence of events during formation and subsequent history of the Flynn Creek crater (from Roddy, 1968). (a) The Flynn Creek area, shortly before formation of the crater, about 360 million years ago, was apparently a rolling lowland or coastal plain exposing rocks of Upper Ordovician age. (b) Formation of the Flynn Creek crater by comet impact about 360 million years ago. The area may have been inundated by the first shallow waters of the Chattanooga Sea. (c) The final crater immediately after formation was 3.6 km in diameter and about 150 meters in depth; it contained a central hill that stood about 120 meters above the crater floor. An ejecta blanket surrounded the crater and overlay the highly deformed strata in the crater rim. (d) A period of erosion followed formation of the crater; many small valleys were formed on the crater rim. The ejecta blanket was removed, much of the ejecta being washed back into the crater. Subaqueous erosion and marine deposition of bedded breccias and bedded dolomite occurred in the crater during early Late Devonian time in shallow waters of the Chattanooga Sea. (e) The Chattanooga Sea filled the crater with black muds in early Late Devonian time. (f) Uplift and erosion during Quaternary (and probably Tertiary) time has produced a highly dissected region of steep-sided hills and valleys with an average relief of 150 meters. The structure of the Flynn Creek crater is now well-exposed along the valley floors and walls.

The second important item of evidence missing is meteorite fragments. However, the general shape of the crater and the presence of a central uplift, coupled with analog studies of certain man-made explosion craters, suggest that the impacting body may have been a surface-exploding comet, not a deeply penetrating meteorite (Roddy, 1968). There is an important distinction between these two kinds of impacting bodies. Meteorite compositions though complicated are well known. Free iron, nickel and ferrosilicates are the chief components. The substance of comets can be studied only at a distance by spectrographic techniques. Results are inconclusive, but the presence of frozen gases binding less volatile dust particles is suggested, the sort of material which would dissipate on impact leaving little trace of its former presence. Cometary explosion, of course, would satisfactorily explain the absence of meteorite fragments at Flynn Creek.

The Flynn Creek crater is but one of a large group of circular structures characterized by a central uplift of brecciated rocks. The detailed geometry of a central dome is well displayed at the Sierra Madera structure in Texas (Fig. IV-30). Permian and Cretaceous sedimentary rocks have moved both upward and inward (Wilshire and Howard, 1968). The upward movement is demonstrated by the fact that the lowest strata exposed in the uplift are about 1,200 meters above their normal position. The inward movement is demonstrated by thickening of the stratigraphic section due to compressional folding and faulting. Drill hole information shows that there is almost no deformation at depths below 2 km under the central uplift.

The inward movement of the rocks at Sierra Madera and the lack of deformation at depth disprove any hypothesis of origin involving basement movement or volcanic doming. Deformation by uplift would cause extension, not compression, of the overlying beds. There is a superficial resemblance between the uplift and salt diapirs, but there is no evidence that the central dome has actually intruded the surrounding rocks as diapirs commonly do. Further, there are no density contrasts at Sierra Madera to indicate the upward movement of relatively light rocks.

Central uplifts do occur in craters produced by chemical explosives detonated on the surface (Diehl and Jones, 1967 *a*, *b*) (Fig. IV-33), and an analogous surface-explosion impact origin for Sierra Madera seems reasonable. Additional evidence at Sierra Madera includes the presence of shatter cones. These are distinctive conical rock fractures which are formed under shock pressures and are oriented so that they point towards the energy source. If the beds at Sierra Madera are returned to horizontal position the cones point upward and inward, reflecting the geometry of the shock wave at the moment of impact (Howard and Offield, 1968).

Discussion of Meteor Crater, Flynn Creek, and Sierra Madera illustrates that the appearance of an impact structure will vary considerably according to the depth of erosion and the subsequent history of sedimentation (Fig. IV-31). The central uplifts preserved following relatively deep erosion indicate craters which were significantly larger then the remnants now preserved.

The largest suspected terrestrial craters are usually the poorest documented, their origin suggested only by a roughly circular perimeter with a depressed and water-filled

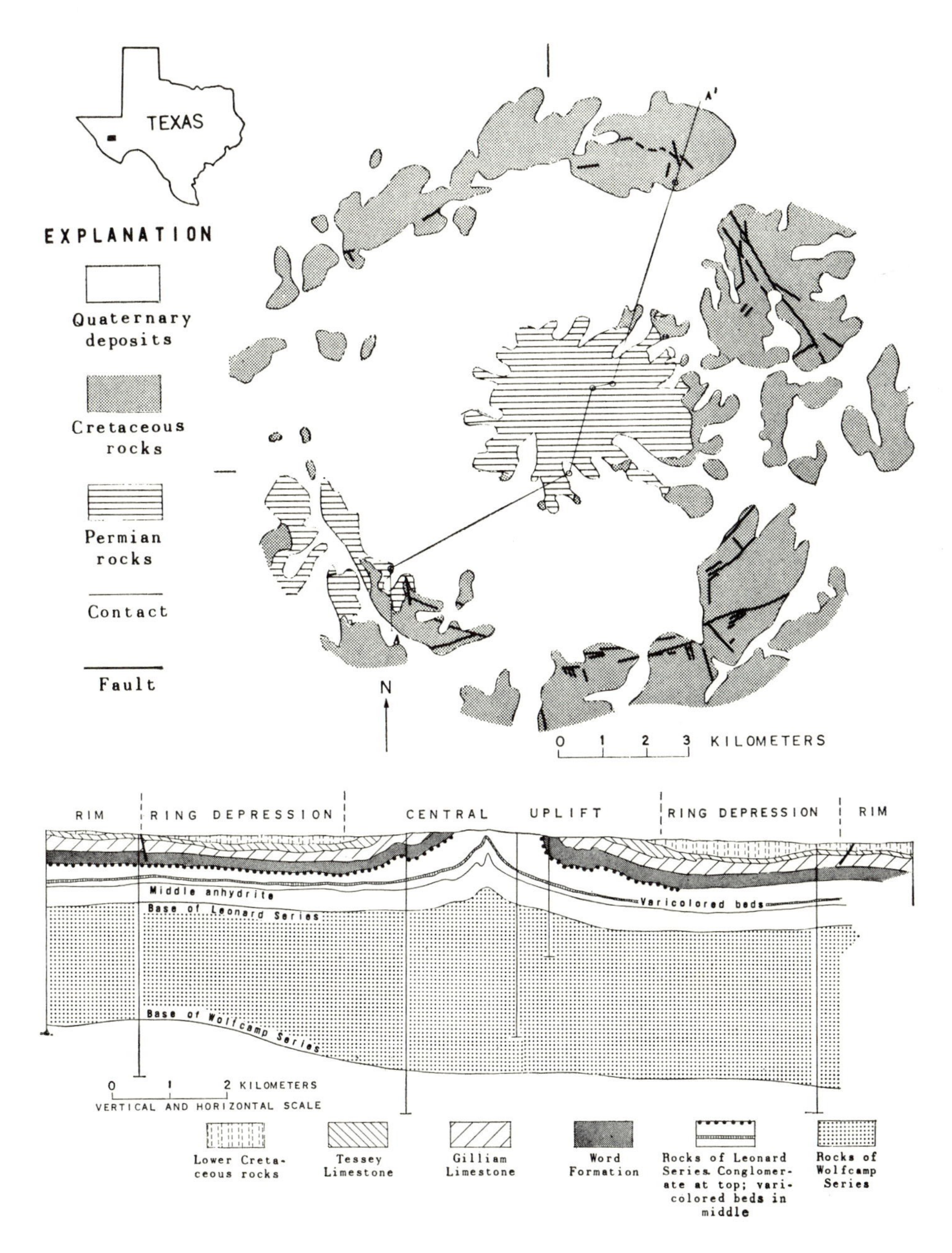

FIGURE IV-30. Generalized geologic map and cross section through the Sierra Madera structure. Line of section is shown on the map. Vertical lines on the cross section are drill holes (taken from Wilshire and Howard, copyright 1968 by the American Association for the Advancement of Science).

interior. An unusually large example of a questionable impact structure is an arc-shaped segment of shoreline on the east side of Hudson Bay, which, when extended, forms a circle with a diameter of 350 km (Fig. IV-32) (Beals, 1968). The Belcher Islands may indicate a central uplift, but so far no evidence has been assembled, structural or otherwise, to lend much weight to an impact hypothesis.

FIGURE IV-31. Schematic cross section showing loss of information with progressive erosion of an impact crater. Following minor erosion (level 1), ejecta deposits are completely removed, but deformed rocks in crater walls are still visible. When erosion has proceeded to level 2, an annular exposure of breccia remains, surrounding a central uplift. At level 3 only sub-crater structural displacements are visible. Post-crater fill may cover crater breccia at erosion level 1. This simple model fails to consider modification of the primary crater shape and redeposition of crater materials which will accompany erosion.

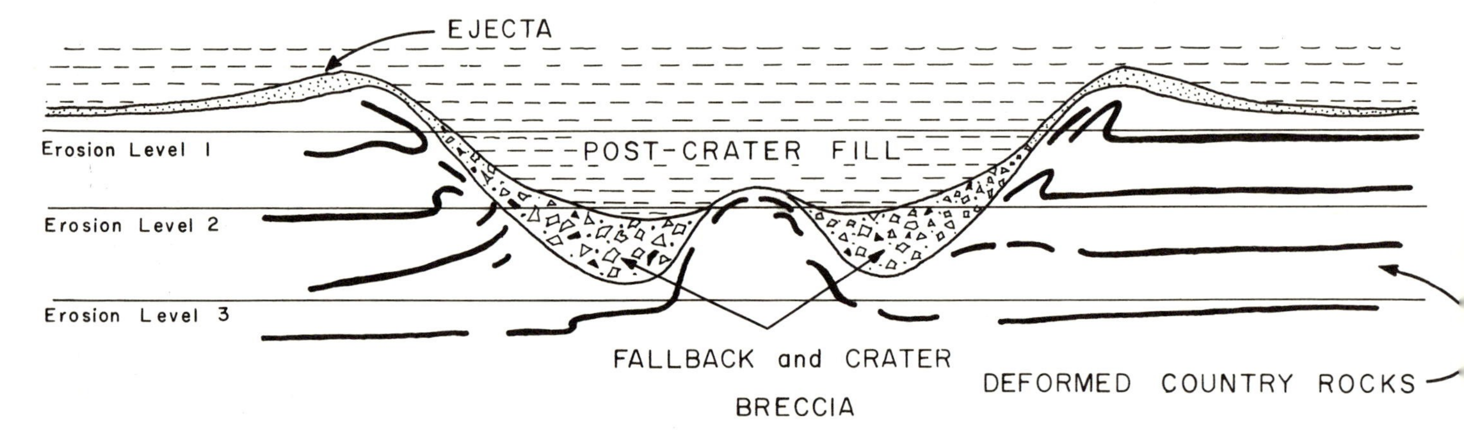

On an even larger scale, Cohenour and Sharp (1968) suggest that roughly arcuate continental margins—the Gulf of Mexico, for example—are asteroidal impact scars. They propose that a swarm of asteroidal impacts in late Paleozoic time may have initiated continental drift. If the idea is outrageous, still its proof—or disproof—will involve some refreshingly new analysis of the rock and fossil record.

From a stratigraphic point of view it is the extremely large craters that are the most important. The most extensive stratigraphic unit associated with craters is the ejecta blanket. For craters the size of Meteor Crater or Flynn Creek the ejecta have such limited extent, even if restored to their original boundaries, that they would comprise insignificant local formations on the Moon. The possibility of ever observing a fresh ejecta blanket here on Earth from a natural crater as large as that indicated by the Hudson Bay arc is remote, and on general humanitarian grounds, not to be hoped for. But we can turn our attention to smaller man-made craters and observe not only the detritus but also the very processes and patterns of deposition.

EXPLOSION CRATERS

Viewed at the simplest level, craters are little more than holes blasted in a planetary crust. The biggest man-made blasts are, of course, nuclear explosions. Several nuclear devices have been exploded underground, producing craters grossly similar to lunar craters. This was a purely coincidental result of the nuclear testing program. Scientists involved in this effort expected to learn a great deal about the design of nuclear devices and the behavior of the Earth's crust under shock stress, but they did not set out with the goal of digging lunar-like holes.

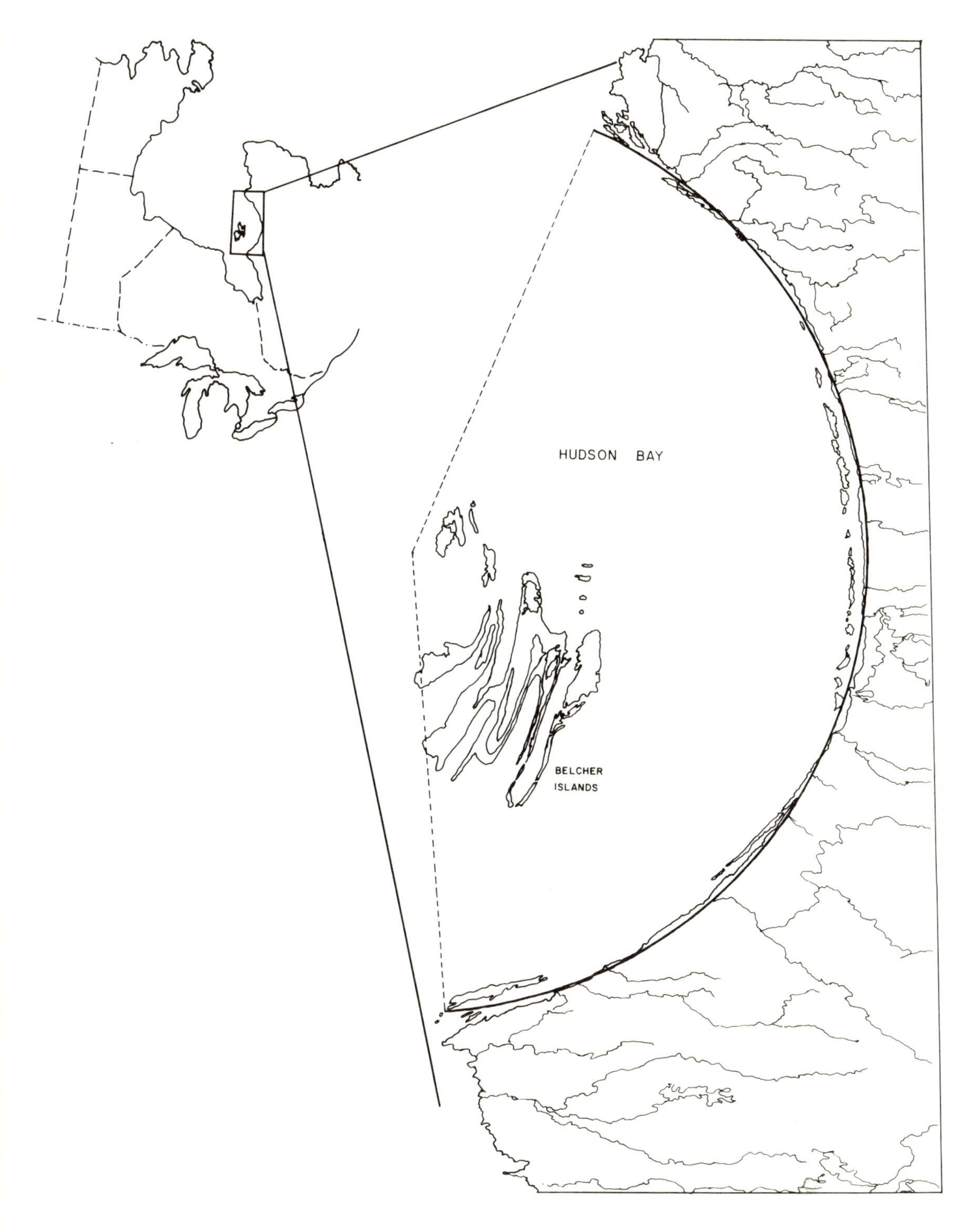

FIGURE IV-32. Arc-shaped segment of shoreline on the east side of Hudson Bay which may mark the margin of an ancient impact crater (after Beals, 1968).

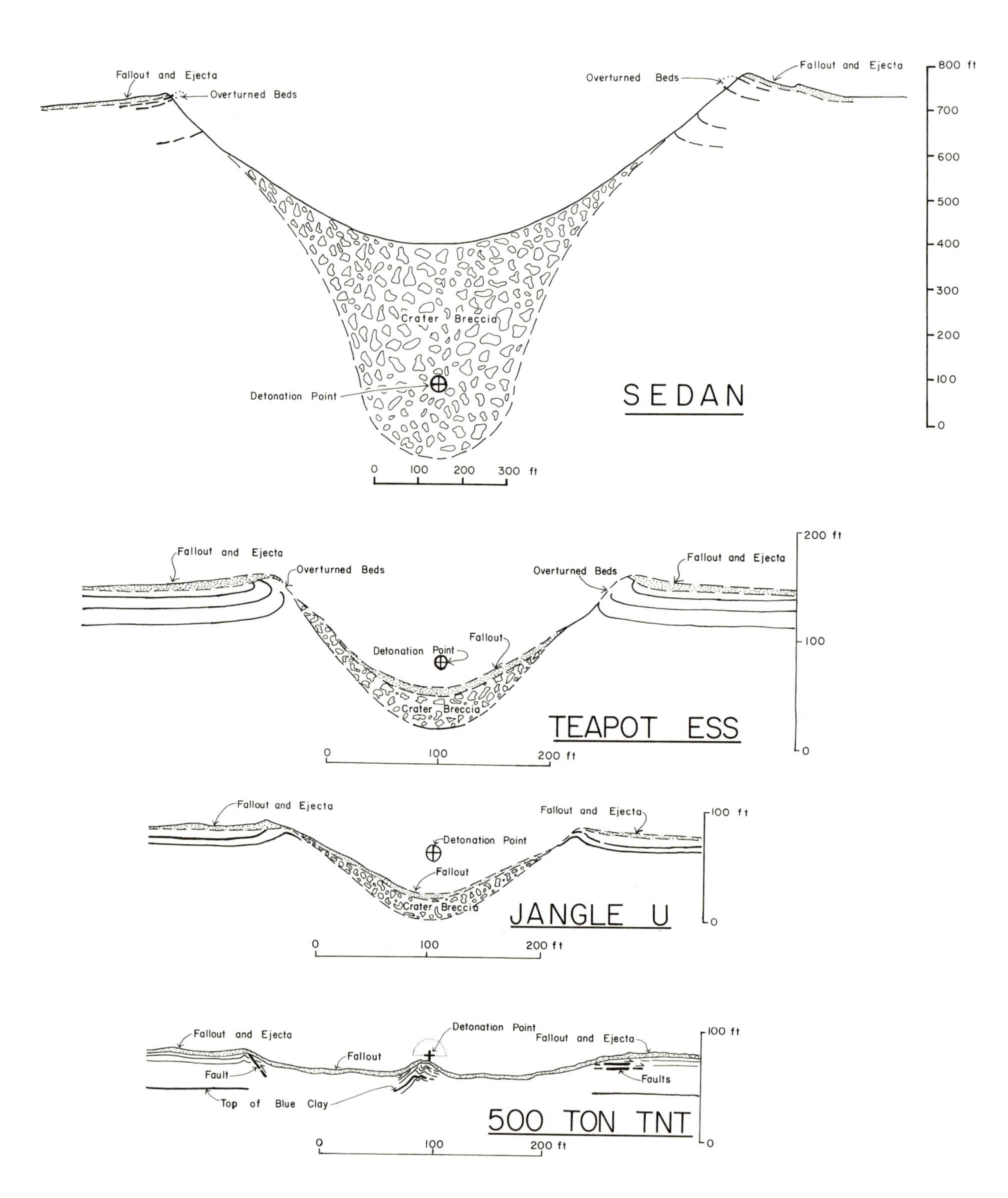

FIGURE IV-33. Cross sections of experimentally produced craters. (a) Cross section of the Sedan nuclear explosion crater, Nevada (modified from Short, 1965). (b) Cross section of the Teapot Ess nuclear explosion crater, Nevada (modified from Shoemaker, 1960, 1962). (c) Cross section of the Jangle U nuclear explosion crater, Nevada (modified from Shoemaker, 1960). (d) Diagrammatic cross section of 500-ton TNT explosion crater, Suffield Experimental Station, Canada. The dotted half-circle indicates the original configuration of the charge (taken from Roddy, 1968).

Details of three crater-producing nuclear explosions and one chemical explosion are shown in Figures IV-33 and 34. The absolute size of the craters is obviously a function of the magnitude of the explosion, but the depth-width ratios seem to be related to the depth of the burst. Shallow bursts have correspondingly shallower craters than deep bursts. All four craters shown in Figure IV-33 have central cores of breccia and structurally raised rims topped by ejected debris. Two of the craters, Teapot Ess and Sedan, have overturned flaps displayed in the upraised rims. For surface or very-near-surface

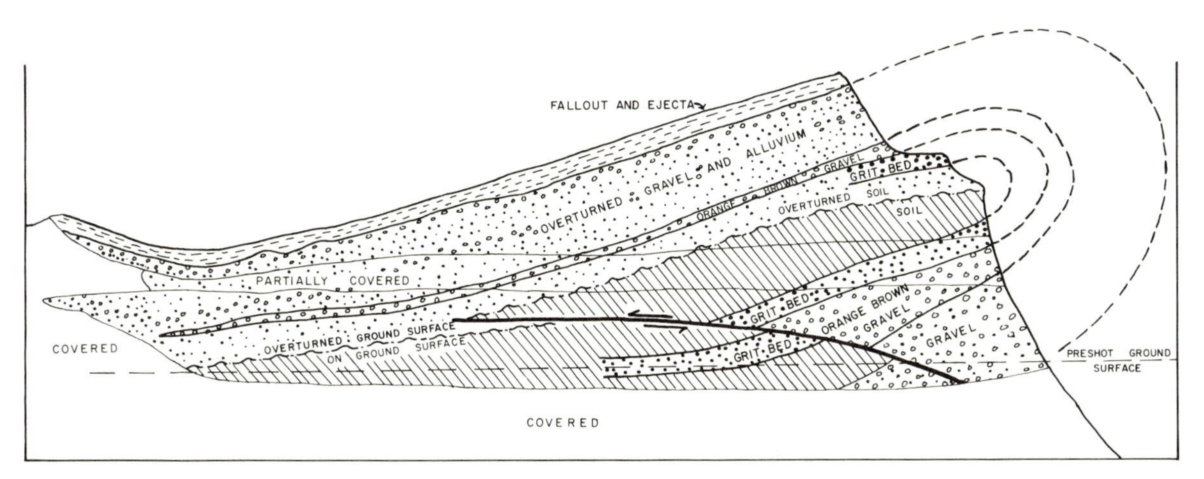

FIGURE IV-34. Geologic cross section through the lip of the Sedan nuclear crater (adapted from Richards, 1964).

explosions, craters are relatively small, chiefly because a great deal of the energy is dissipated both through the atmosphere and through the surficial rocks, which are compressed downward rather than being displaced upward or laterally. A central peak was formed in several surface-detonated TNT craters but is not present in any of the below-surface nuclear explosions. For relatively deep explosions the major portion of the energy is expended in breaking up of rock which falls back to its approximate original position. Laterally displaced and ejected material is of minor importance.

MISSILE AND PELLET IMPACT CRATERS

Missile impact scars show many of the features displayed in explosion and presumed meteorite craters. Several craters produced at the White Sands Missile Range, New Mexico, have been studied in considerable detail (Moore, 1966). The target materials were unconsolidated alluvium and gypsiferous sediments. Craters formed by the impact of inert missile warheads have cup-shaped interiors and raised rims (Fig. IV-35). There is a noticeable asymmetry both in crater shape and in ejecta distribution. The wall nearer the projectile source is steeper; ejecta is sprayed away from the projectile source. Crater walls have slopes of as much as 30° where they are underlain by unconsolidated debris displaced and dispersed by impact. Steeper slopes, some overhanging, occur where deformed and shattered target material occurs more or less in

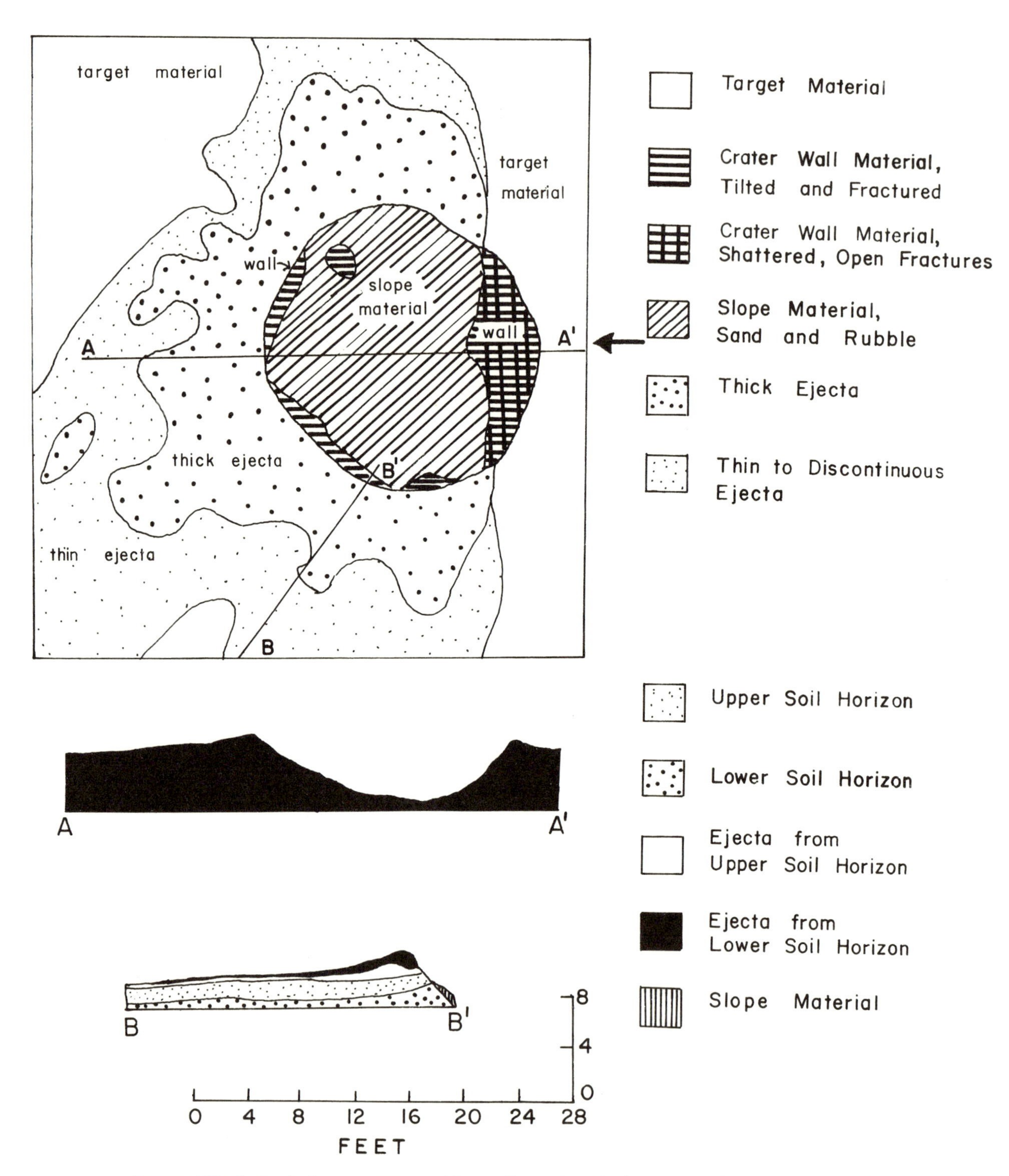

FIGURE IV-35. Geologic map, topographic profile, and geologic cross section for crater produced in gypsum by missile impact (after Moore, 1966).

place. The local steep slopes are a consequence of the cohesion of target material caused by high shock pressures. Much of the ejected debris is composed of sheared and compressed target material with densities exceeding 2.2 g/cm^3. The density of uncompressed target material a short distance from the crater sites is only 1.4 g/cm^3.

As unconsolidated as the target alluvium is, it still appears to maintain a rudimentary large-scale cohesiveness during ejection so that the ejecta deposit comprises an overturned flap of the lower soil horizon overlying the upper soil horizon. The total amount of material displaced during cratering by impact appears to be approximately the same as that displaced during cratering by a chemical explosion with TNT energy equal to the kinetic energy of the missile (Moore, 1966).

Gault and others (1968) have studied the structures formed by firing small cylindrical pellets a fraction of an inch in diameter at high velocities on the order of 1 km per sec into loose stratified layers. Generalized particle movement is shown in Figure IV-36. Beneath the crater the movement is almost vertically downward; around the margins it is nearly tangential to the ground surface.

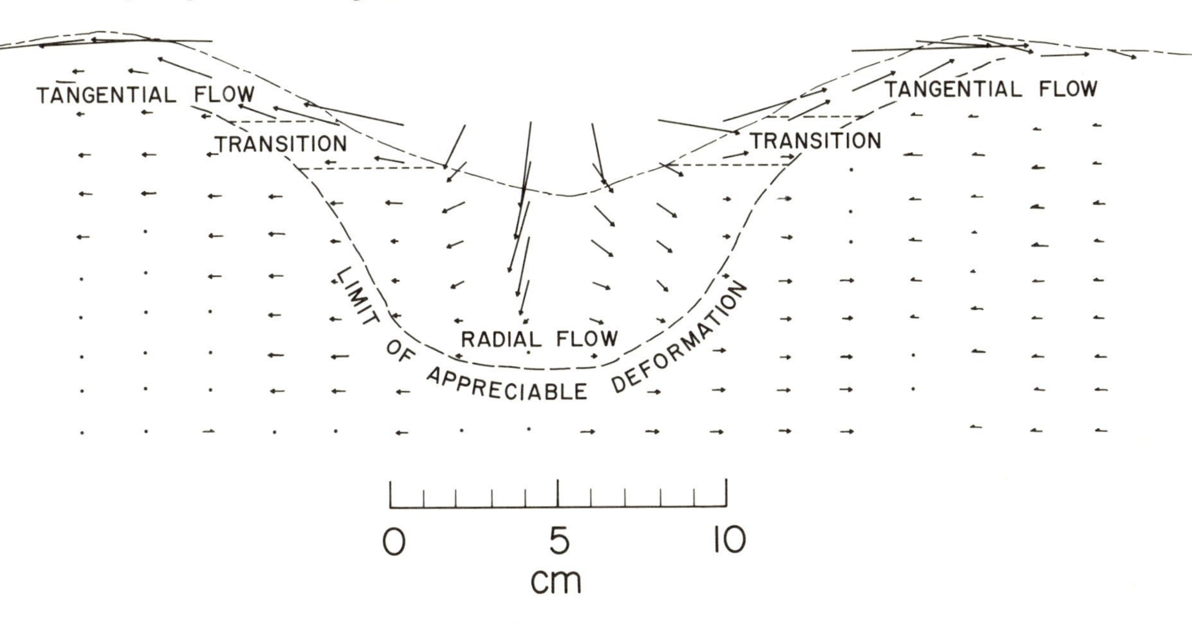

FIGURE IV-36. Flow pattern of granular materials near an impact crater formed by pyrex projectile striking vertically with velocity of 0.69 km per sec. Flow lines are determined from positions of known points within the target materials before and after the crater formed (from Gault and others, 1968).

The results of several experiments in which a color-keyed, loosely consolidated sequence of variable thickness overlies a strong substrate are shown in Figure IV-37. If the upper sequence is thick, a relatively simple crater is formed with particle displacement following the theoretically predicted pattern. As the upper sequence is made thinner, the central part of the crater assumes a series of different shapes and structures which reflect the fact that the shock wave transmitted to the strong substrate has an energy less than the dynamic yield strength of the substrate. In those situations where the crater does not intersect the substrate surface, a structural dome or central mound is formed. For thinner surficial sequences the upper layer is completely evacuated, producing a crater with a flat floor.

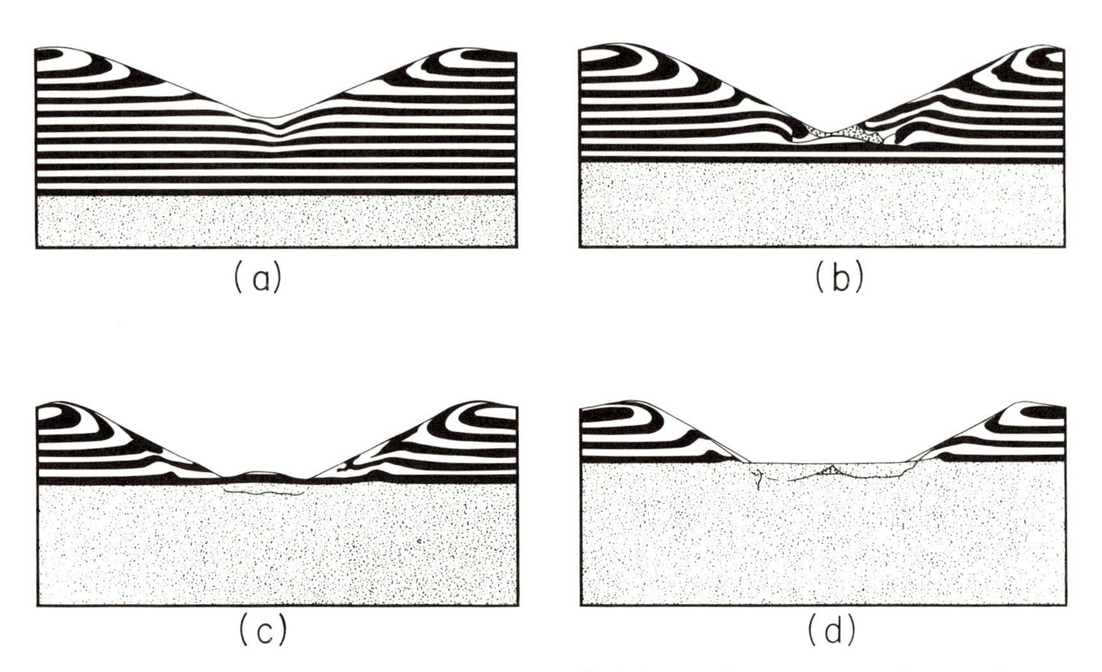

FIGURE IV-37. Near crater structure of loose stratified layers above stronger substrate. Craters produced by 1 km per sec normal impact of small pyrex projectile. (a) normal crater in a thick surficial layer; (b) normal crater in a thin surficial layer; (c) central mound crater; (d) flat-bottomed crater (from Quaide and Oberbeck, 1968).

VOLCANIC CRATERS

Volcanic landforms visible on the Earth are extremely varied, including a myriad of both positive constructional forms and negative forms resulting from evacuation of material. Before Orbiter photographs became available, those who argued for volcanic origin of structures on the Moon concentrated their attention on large lunar craters which they compared to scaled-up examples of calderas and maars as they are known on Earth. With the arrival of Orbiter pictures, opportunity for comparison was greatly improved. Many relatively small features were seen which resemble cones, domes, plugs, and flows formed on Earth.

The large number and diversity of lunar features which are now considered possibly volcanic suggest the superficiality of that standard question: "Do you think the surface features of the Moon are formed by impact or volcanic activity?" Few scientists familiar with the photographic record would respond with unqualified support for one or the other process. Virtually all geologists agree that certain features such as breached domes, level plains, or lobate scarps provide evidence of volcanic activity on the Moon. Moreover, most geologists believe that emplacement of volcanic flows and ash deposits has been widespread and volumetrically important. The origin of lunar craters is more controversial, but even here one should make distinctions. Many geologists subscribe to a volcanic origin for certain varieties of small craters—for example, some dark halo craters or sets of alined craters. Origin of large craters such as Alphonsus or Ptolemaeus is much more uncertain, and many who reject the volcanic hypothesis at this point nonetheless accept it for other crater types, such as the two just mentioned.

CALDERAS

A useful general definition of a caldera is that supplied by Williams (1941, p. 242): "Calderas are large volcanic depressions, more or less circular or cirque-like in form, the diameters of which are many times greater than those of included vent or vents, no matter what the steepness of the walls or form of the floor . . . calderas are almost invariably large volcanic basins produced by engulfment."

Subdivision is possible according to the character of the eruption and the composition of the magma. One type of caldera characteristically occurs on shield volcanoes built up by the eruption of basaltic magmas (Fig. IV-38). The calderas may form by vertical rise of magma into near-surface chambers followed by withdrawal into flank rifts and

FIGURE IV-38. The summit area of Mauna Loa, Hawaii, displaying four collapse calderas (courtesy of United States Air Force).

attendant sinking of the summit cap (Williams, 1941). Recent geophysical studies of the Hawaiian shield volcanoes demonstrate that the entire shield periodically swells in response to pressure of the underlying magma. Collapse often occurs following this doming, even though little lava breaks through to the surface in the area of collapse (MacDonald and Eaton, 1964). In any event caldera subsidence in the Hawaiian Islands is unaccompanied by violent or explosive eruptions. Shortly after their initial formation, calderas display steep walls with inward facing scarps interrupted by step faults. Caldera floors are the sites of lava lakes, whirl-pool drainage fissures, and secondary eruptive cones. Eventually the depression may be completely filled by successive lava eruptions.

A second, more common caldera type is formed by collapse due to spectacularly rapid explosion of large volumes of ash. Such calderas commonly form over andesitic or rhyolitic magmas, these being more viscous and susceptible to violent particulate explosion than basaltic magmas (Fig. IV-39). Some of the expelled ash rises high in the air and falls back as sediment accreted in the cold state. Much material, however, falls back on the volcano and combines with volcanic gases and heated air to form fluidized ash

FIGURE IV-39. Crater Lake, Oregon, a caldera whose formation followed rapid explosion of lava (courtesy of Washington National Guard). Generalized cross section through the crater, disregarding the water level, is drawn with the same horizontal and vertical scale.

flows (Smith, 1966). These can cascade down outer volcanic slopes much like a stream of water. Streams of ash can flow great distances along gentle gradients; those associated with the Aso caldera of Japan have travelled more than 50 km (Fig. IV-40). Continuing volcanic activity frequently results in formation of small secondary cones and domes within the caldera.

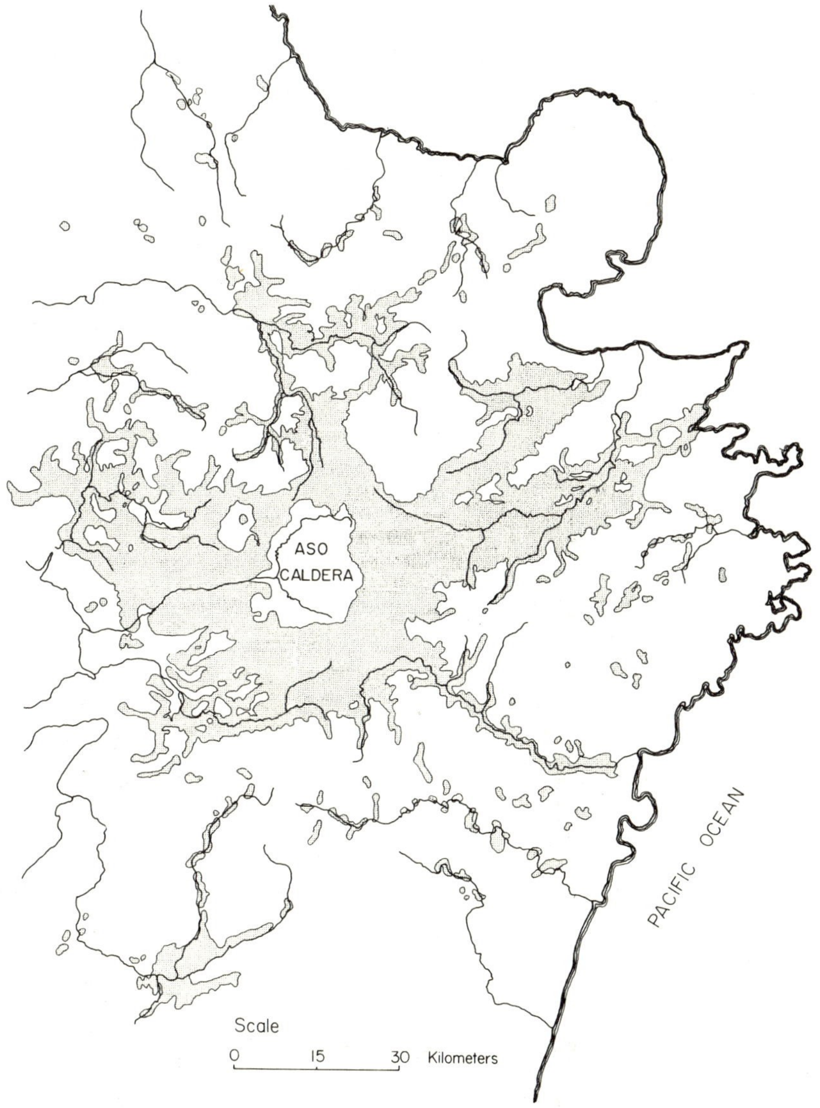

FIGURE IV-40. The Aso caldera, Japan, showing the present extent of ash flows (stippled pattern) and their correlation with stream valleys (after Matumoto, 1943).

The Krakatoa eruption of 1883 led to the formation of a caldera of this second type. A series of eruptions lasted over a period of three months, culminating in catastrophic explosions of two days' duration. The effects of this awesome event, which took place on a little-visited island between Java and Sumatra, were worldwide. Air waves, sea waves, and high-altitude clouds of dark dust travelled for thousands of miles. The largest tidal wave engulged the coasts of Java and Sumatra, killing more than 36,000 people. The climactic explosions at Krakatoa completely destroyed a pre-existing cone more than a kilometer high. The magmatic eruptions discharged a tremendous volume of pumice and ash. The exact volume cannot be calculated, since most of the deposits rest on the floor of the Sunda Straits.

A third type of caldera resembles Krakatoa in that tremendous volumes of pumice and ash are produced by eruption of andesitic magmas. However, these calderas generally are not associated with a single pre-existing cone but, instead, are controlled by fissures which lie athwart previously formed structures.

Circular craters representative of the last two classes attain diameters of at least 25 km. Elongate forms are as large as 100 km (Smith, 1966). Caldera floors are commonly uplifted following their initial collapse; the secondary dome so formed may be modified by later collapse. The final structure, then, has a composite origin (Fig. IV-41).

There is still another structure which, though not itself a caldera, may indicate the former presence of a caldera now eroded. These are ring structures or complexes of ring dikes. They are characterized by concentric rings of intrusive dikes which surround

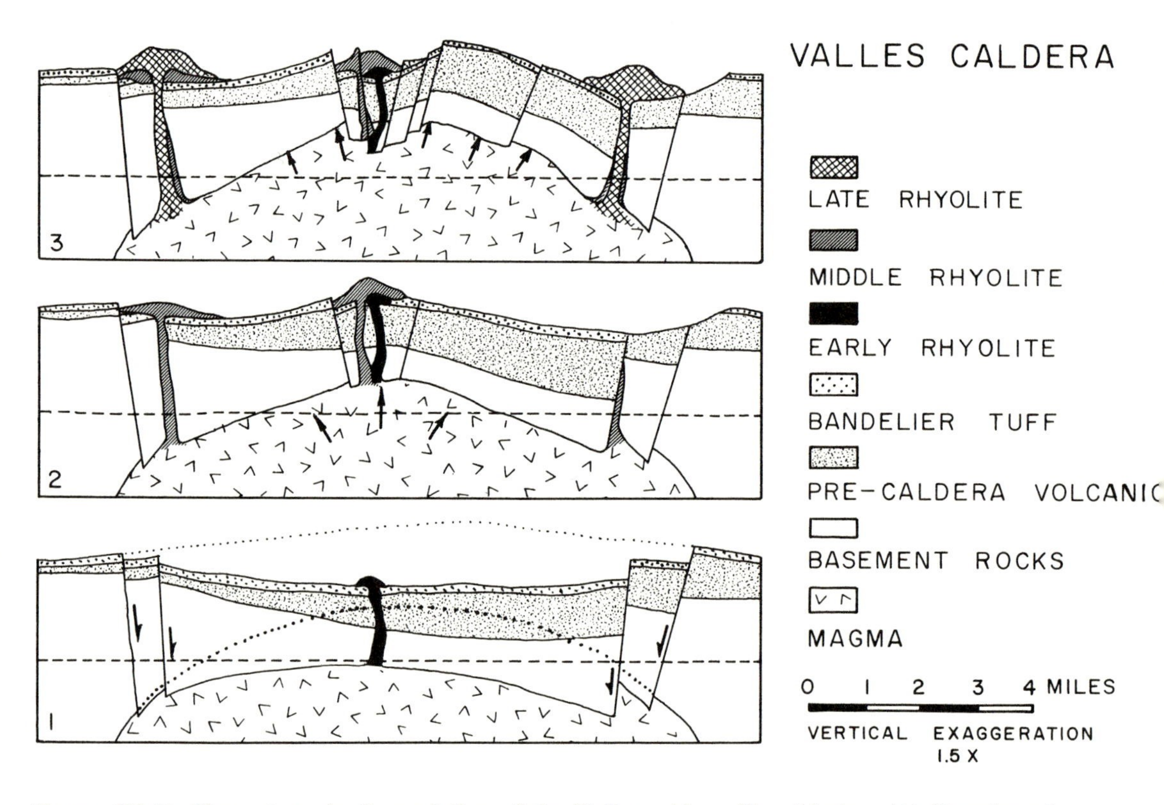

FIGURE IV-41. Three stages in the evolution of the Valles caldera, New Mexico. (1) Eruption of Bandelier ash flows followed by subsidence of cylindrical block 10 miles in diameter to depth of several thousand feet; partial burial of subsided block with ash, lava, and erosional detritus. (2) Uplift and doming of central block; emplacement of ring volcanoes. (3) Continued uplift and volcanic activity; formation of interior graben over central part of uplift (taken from Smith, 1966).

a central core of plutonic and volcanic rocks. Some ring structures show a subdued topographic depression over the central core. The precise configuration of rocks exposed at the surface will vary greatly according to the depth of erosion (Fig. IV-42).

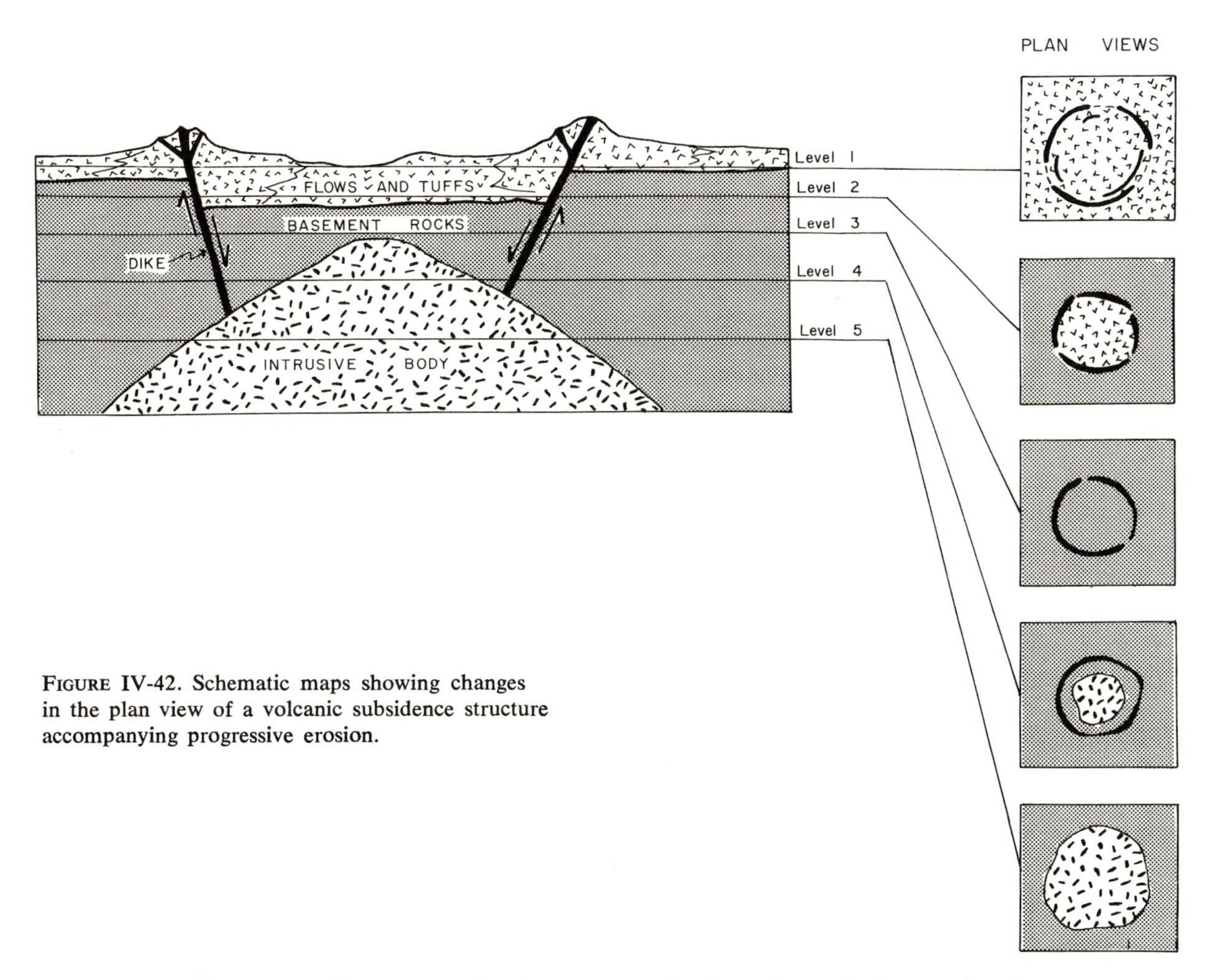

FIGURE IV-42. Schematic maps showing changes in the plan view of a volcanic subsidence structure accompanying progressive erosion.

Consideration of ring dikes opens the door to examination of certain large volcanic complexes which may have formed by some subsidence mechanism. Elston (1965) reviews this problem, focusing his attention on the Mogollon plateau of southwestern New Mexico. This is a geologic province roughly circular and 140 km in diameter, termed by Elston a "volcano-tectonic" basin. It is ringed by faults and has a structurally raised rim. The central depression is filled with thousands of cubic kilometers of Tertiary volcanic rocks, chiefly rhyolite ash-flow tuffs. Elston speculates that if the rocks were eroded down to the Precambrian basement, a ring-dike complex would be revealed.

The Mogollon Plateau is an order of magnitude larger than those structures usually classified as calderas. Even though the geology and topography do not at first glance suggest a simple concentric configuration, it is possible that the landscape existing at the end of the last eruptive phase was more clearly caldera-like. Following this line of reasoning, Elston points out certain similarities of size and shape with the lunar crater Theophilus (though other workers regard Theophilus as one of the better examples of an impact crater).

MAARS

Maars are vents formed by the explosive erosion and enlargement of an initially small opening by gas-charged volcanic eruptions. The name "maar" is taken from examples in Germany, which are circular funnels approximately half a kilometer in diameter, now occupied by lakes. They occur in a nonvolcanic terrain, and the explosions which formed the depressions apparently expelled only small amounts of volcanic debris. In the formation of other maars large volumes of ash may have been expelled.

Typically the ejected materials, either fragmented surficial rocks or volcanic lava and ash, build up a rim around the central vent of a maar (Fig. IV-43). The rim is generally a very subtle feature with gentle outward-facing slopes. In some situations, however, the rims rise more than 100 meters above the surrounding countryside so that the maar resembles a conventional volcanic cinder cone.

The precise mechanics of a maar-producing explosion are complicated and conclusions about them are partly based on inference. However, the underlying cause of the explosion is generation and sudden release of gas, either from heating of ground water or from boiling of water-rich magma intruded close to the surface where confining rock pressures are low. Observation of an actively erupting maar, the Nilahue maar of southern Chile, demonstrates that the eruptive history may include multiple explosions (Müller and Veyl, 1957). Alternating periods of violent gaseous discharge and quiescence extended over more than three months.

Probably a disproportionate amount of attention has been directed toward maars as the chief candidate for volcanic analogs of lunar craters. In this role they are little more than straw men, chiefly because they are so small, less than several kilometers in diameter. However, if any maars do exist on the Moon, the feeder vents may have brought fragments of deep crustal rocks to the surface—as have several vents, or "diatremes," in the Colorado Plateau (McGetchin, 1968). In this event the lunar maars will be ideal localities for easily obtaining information concerning the vertical profile of rock types in the lunar crust.

OTHER CIRCULAR DEPRESSIONS

This discussion of Earth analogs for lunar craters has been far from exhaustive. Whitaker (1966) and Kuiper (1966), among others, have pointed out certain geometric similarities between some lunar depressions and sinkholes on Earth formed by either

Opposite page:
FIGURE IV-43. Hole-in-the-Ground, a late Pleistocene maar in Oregon. The crater is 1.6 km in diameter and the highest rim point is 153 meters above the crater floor. Basalt flows exposed in the far wall underlie the explosion tuff breccias which were emplaced during maar formation. (a) Oblique aerial view (courtesy of Oregon Dept. of Geology and Mineral Industries). (b) Topographic profile with the same horizontal and vertical scales. (c) Generalized geologic cross section (from Peterson and Groh, eds., 1965).

(A)

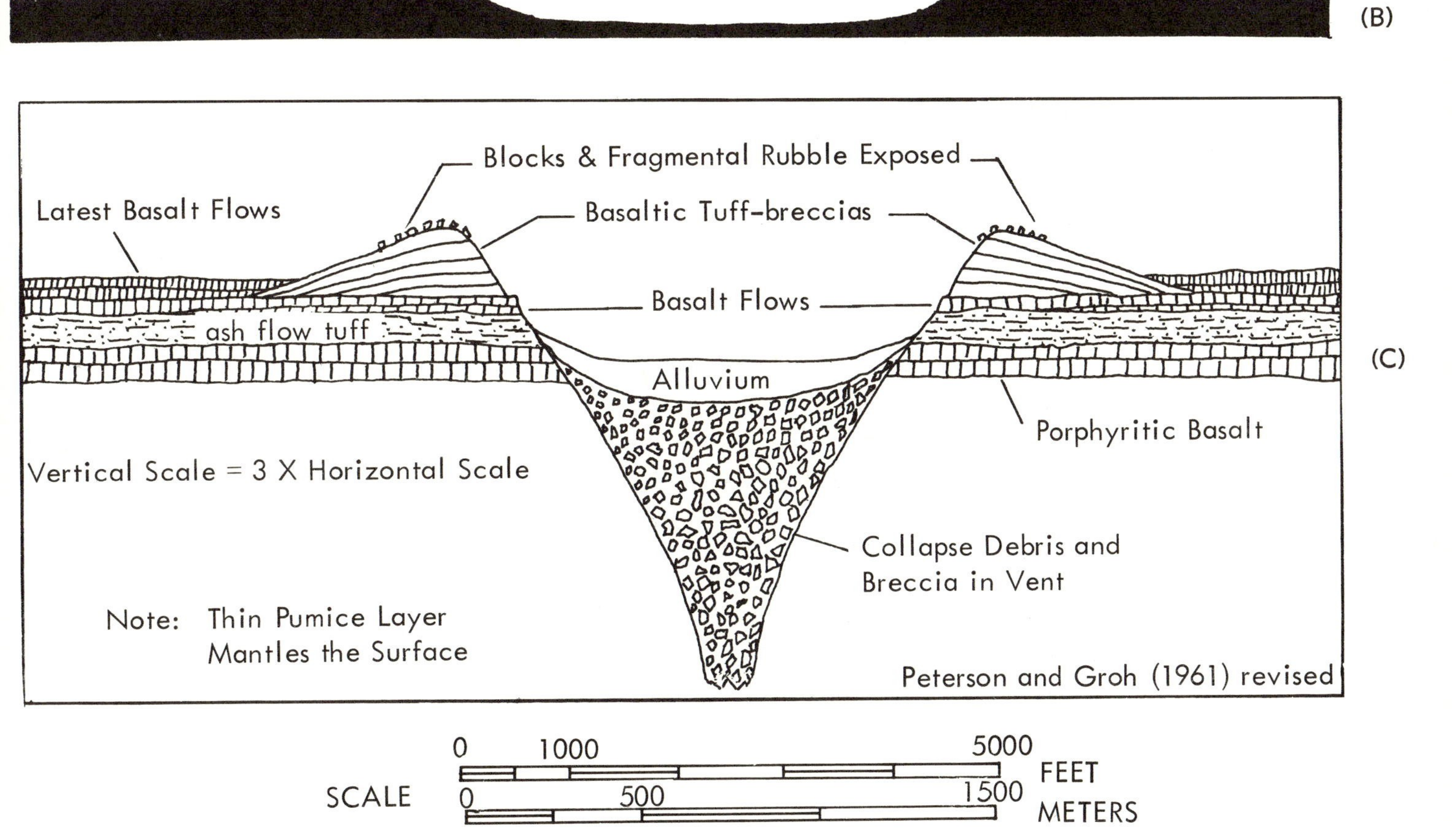

(B)
Blocks & Fragmental Rubble Exposed
Latest Basalt Flows
Basaltic Tuff-breccias
Basalt Flows
ash flow tuff
Alluvium
Porphyritic Basalt
Vertical Scale = 3 X Horizontal Scale
Collapse Debris and
Breccia in Vent
Note: Thin Pumice Layer
Mantles the Surface
Peterson and Groh (1961) revised
(C)
0
1000
5000
FEET
SCALE
0
500
1500
METERS

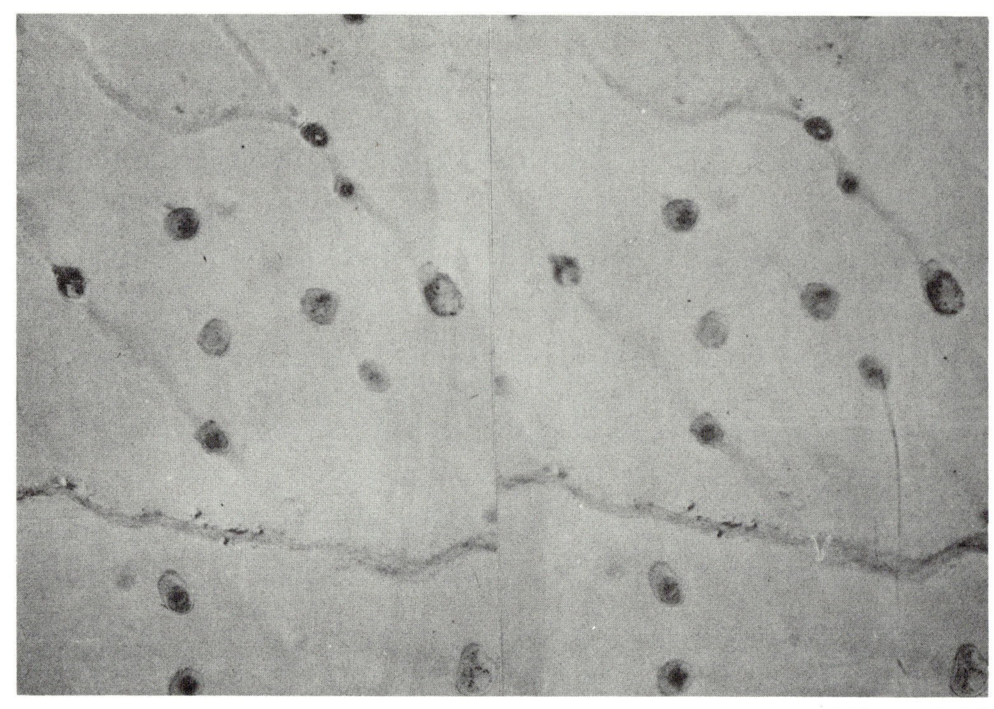

FIGURE IV-44. Stereographic pair of photographs showing terrain in North Africa underlain by horizontal limestone beds. Vegetation-filled sinkholes are alined along parallel joints (from von Bandat, 1962).

physical or chemical removal of material beneath the surface (Fig. IV-44). In the case of karst topographies the analogy is not to be taken literally, since the process of ground-water solution almost certainly does not operate on the Moon. However, the general process of downward drainage of material is perhaps important. For example, Shoemaker has commented on the alinement of small rimless craters along lineaments in the vicinity of the Surveyor III site (Surveyor III Mission Report, 1967*a*). These depressions may have been formed by downward drainage of sediment at separated points along a fissure.

CRATER EJECTA STRATIGRAPHY

Ejection and deposition of material from impact craters and explosive craters is one of those esoteric subjects that has become a matter of general interest only in the past decade. Recent studies indicate that there are at least three different types of deposits: bulk deposits, missile ejecta, and base surge sediments. Each type is characterized by a different transportation mechanism. The resulting sediments display distinctive petrographic features, primary sedimentary structures, and morphologic expression.

As a result of detailed field studies, laboratory simulation, and theoretical modelling, the dynamics of ballistic (missile) ejection are now fairly well understood. Gault and others (1968) have pointed out the necessity to distinguish between impact events and underground explosive events. In the latter case a spherically expanding shock wave is reflected from the ground above. The resulting tension imparts a near-vertical motion to the surface material over the explosion (Fig. IV-45). In addition, the explosion cavity is filled with high-pressure gases which vent vertically through tears in the surface layer carrying sediment with them. Ejection of material is thus accomplished both by spallation and by aerodynamic processes, neither of which plays a significant role in impact ejection.

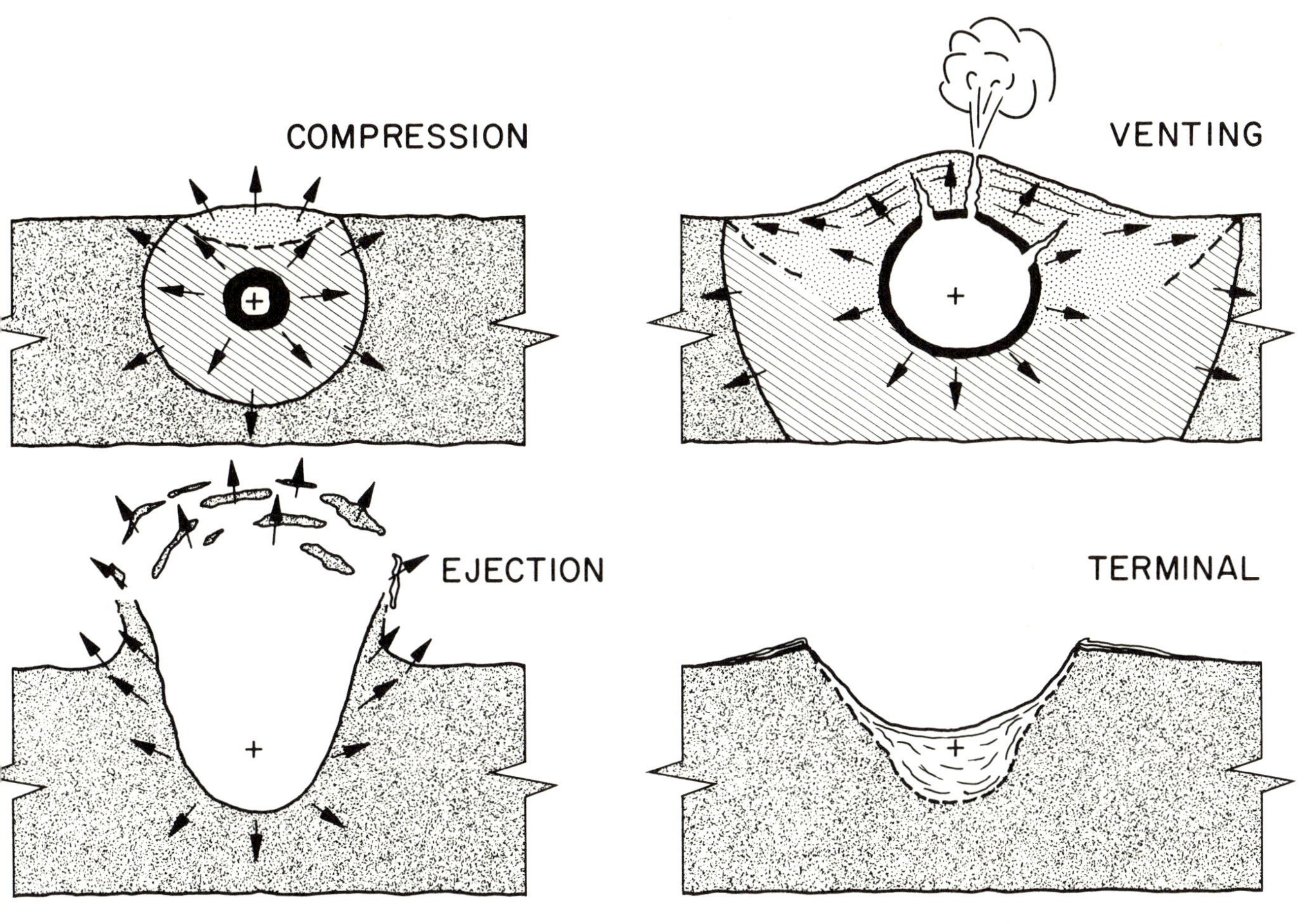

FIGURE IV-45. Schematic representation of the formation of a crater by a deeply buried explosive (from Gault and others, 1968).

In an impact event two systems of shock waves are generated, one in the projectile and a second in the target material. In the early stage of crater formation these shock waves race outward across the target face and upward along the side of the projectile. Behind them there develops a system of rarefaction waves. In this rapidly decompressed envelope situated between the target and the projectile, ejected material squirts out at very high velocities, five or more times greater than impact velocities. This early phase of ejection is termed "jetting." The material in the jet is subjected to such high pressures and attendant high transient temperatures that it will commonly be in a liquid or vapor state (Gault and others, 1968).

The jetting phase is terminated when the shock wave reflects from the back side of the projectile, thereby completing its destruction. Subsequently the shock wave continues to travel through the target with accompanying rarefaction waves generated along the expanding circle of intersection with the ground. The interaction of these two wave systems causes material to be thrust outward and upward along the margins of the expanding crater. The major amount of crater evacuation takes place during this stage, while material is ejected at modest velocities. A cone-shaped curtain of debris moves up along the crater wall and away from the crater, and tears apart to form delicate filaments (Fig. IV-46). The resulting depositional pattern is a thick blanket close to the crater gradually changing to discontinuous patches and rays away from the crater.

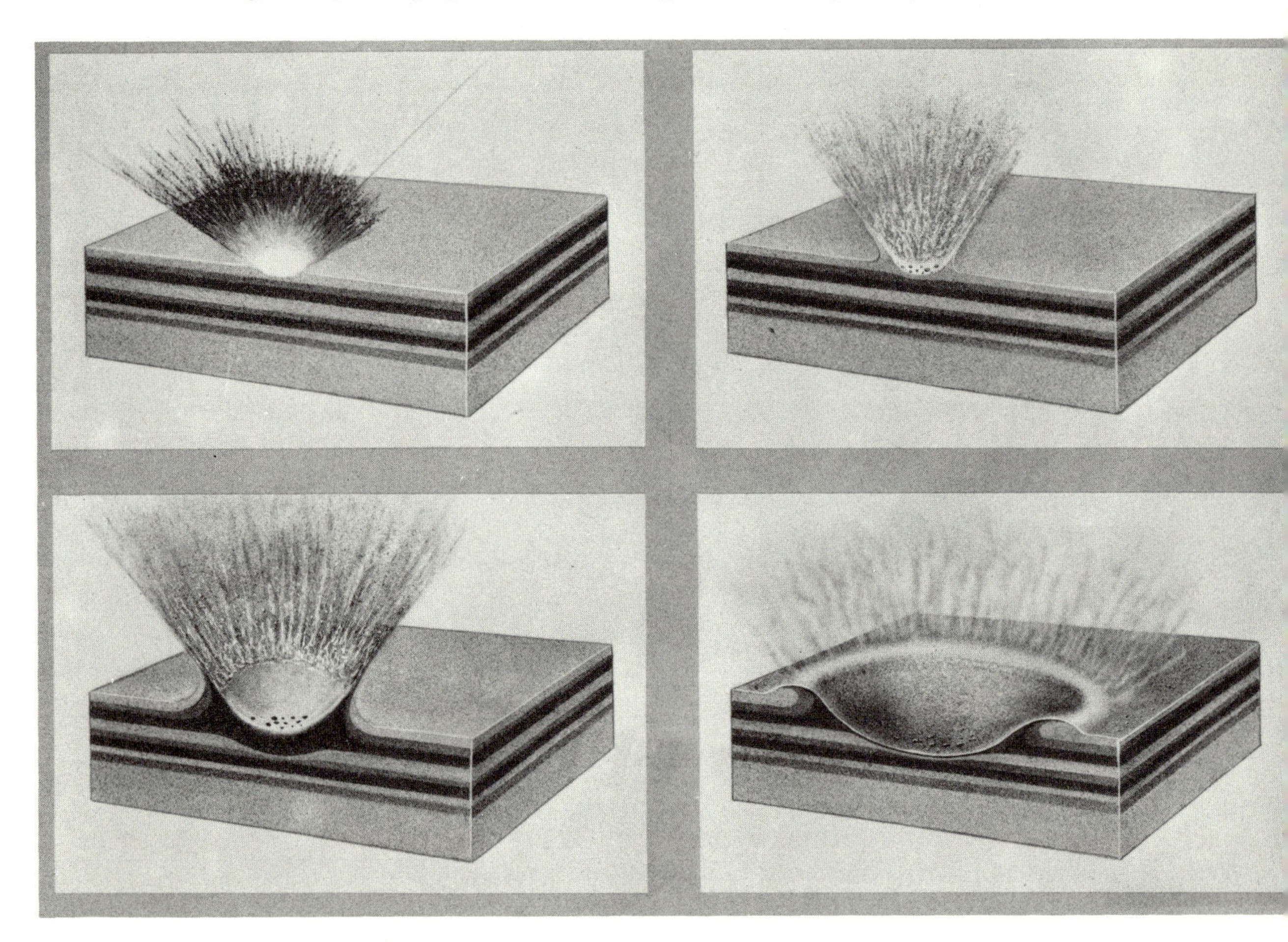

FIGURE IV-46. Schematic representation of the ejecta patterns and crater development for an impact in a homogeneous noncohesive medium (from Gault and others, 1968).

The situation just described is an idealized generalization that assumes a homogeneous target. Obviously, physical variations in the target material will alter the perfect symmetry of the debris cone. In addition, an inclined projectile trajectory will produce an ejecta deposit with only bilateral symmetry (Fig. IV-47).

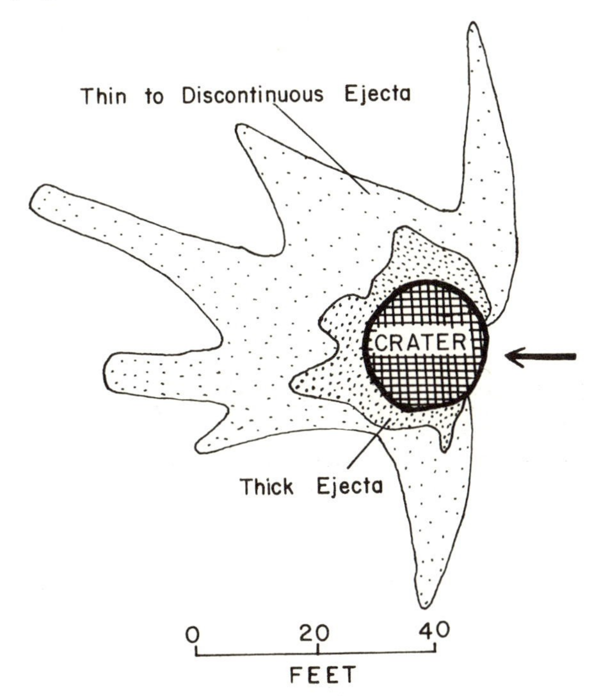

FIGURE IV-47. Ejecta distribution around a crater formed by a missile warhead impacting well-cemented gypsum sand. Trace of missile path is shown by arrow. Thick ejecta ranges from 2 feet to a fraction of a foot in thickness and contains relatively undeformed blocks of target material up to 1 foot long. Note that the thin ejecta distribution is markedly asymmetric with prominent tongues extending forward of and at right angles to the missile path (adapted from Moore, 1966).

A second type of transportation mechanism, less well understood than ballistic transportation, has recently come to light. This is base surge deposition. The term was coined by Glasstone (1950) in reference to clouds moving horizontally away from nuclear explosion sites. Subsequently the term has been more generally applied to all "ring-shaped basal cloud(s) which sweep outward as a density flow from the base of a vertical explosion column" (Young, 1965) (Fig. IV-48). Base surges can be formed by volcanic explosions and perhaps by those supersonic impact events which, either by virtue of the involved materials or because of the high shock pressures and temperatures, are accompanied by production of gases or gas-like media.

The first detailed study of the base surge phenomenon was made by Young (1965) following the 1948 Bikini underwater explosion. He concluded that the surge formed along the rim of a crater-like depression in the water, fed by jets of spray released from the wave which constituted the crater rim. The surge formed before the main vertical column of water vapor started to fall and so could not be a result of bulk subsidence. Close to the explosion site the base surge moved with a velocity better than 150 km an hour. It continued outward for a radial distance of 4 km.

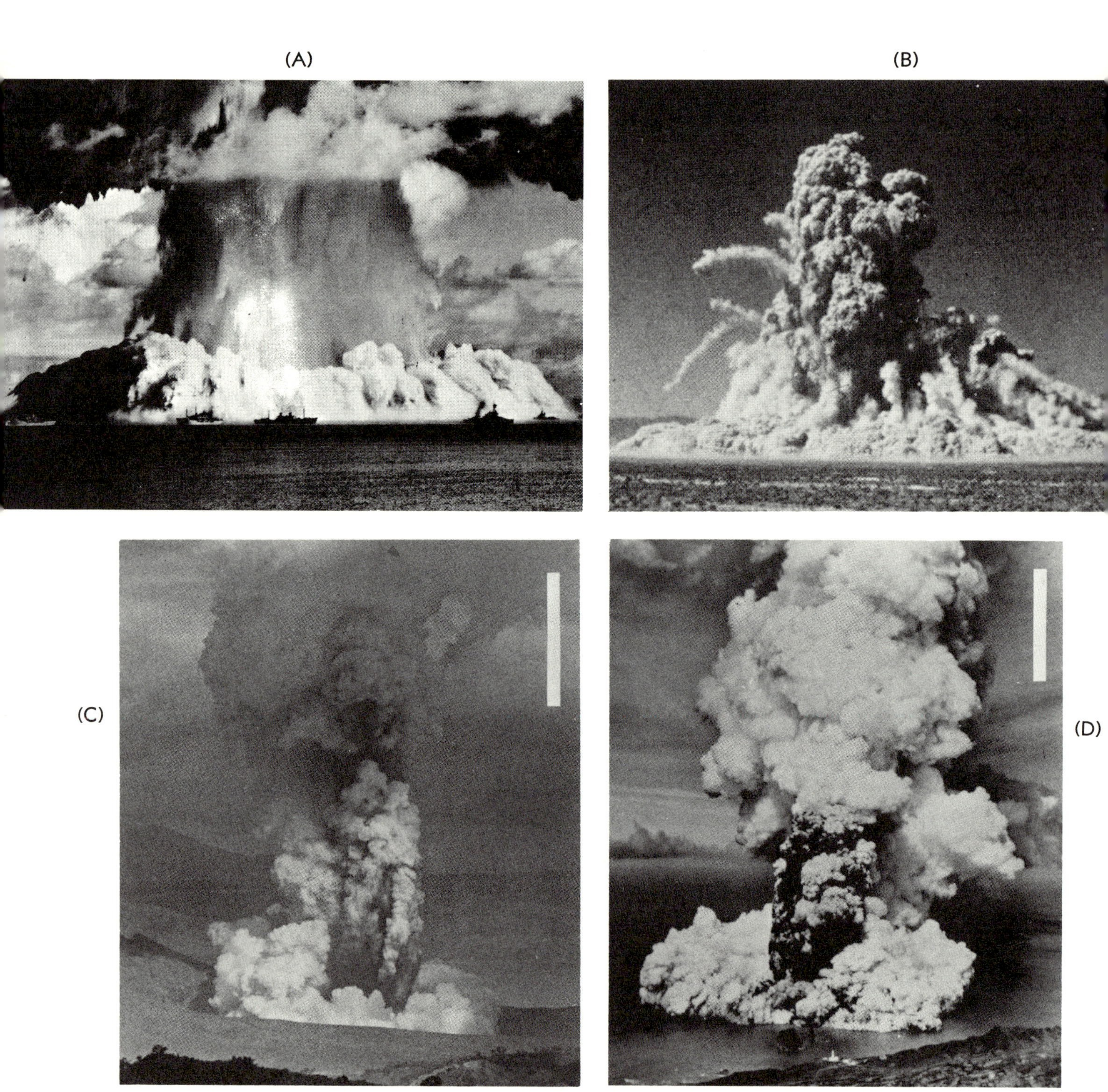

FIGURE IV-48. Several examples of base surge formation. (a) 1946 Bikini underwater test explosion 17 seconds after burst (from Young, 1965). (b) 1962 Sedan nuclear test explosion 39 seconds after burst (from Nordyke and Williamson, 1965). (c) 1965 eruption of Taal Volcano, Philippines (from Moore, 1967). The white bar is approximately 500 meters long. (d) 1957 eruption of Capelinhos Volcano, Azores (courtesy of United States Air Force). The white bar is approximately 500 meters long.

The Sedan thermonuclear explosion was accompanied by a prominent base surge. Following the explosion both the base surge and missile deposits were mapped. The surge sediments, characterized by the presence of concentric dunes, extend beyond the missile deposits (Fig. IV-49).

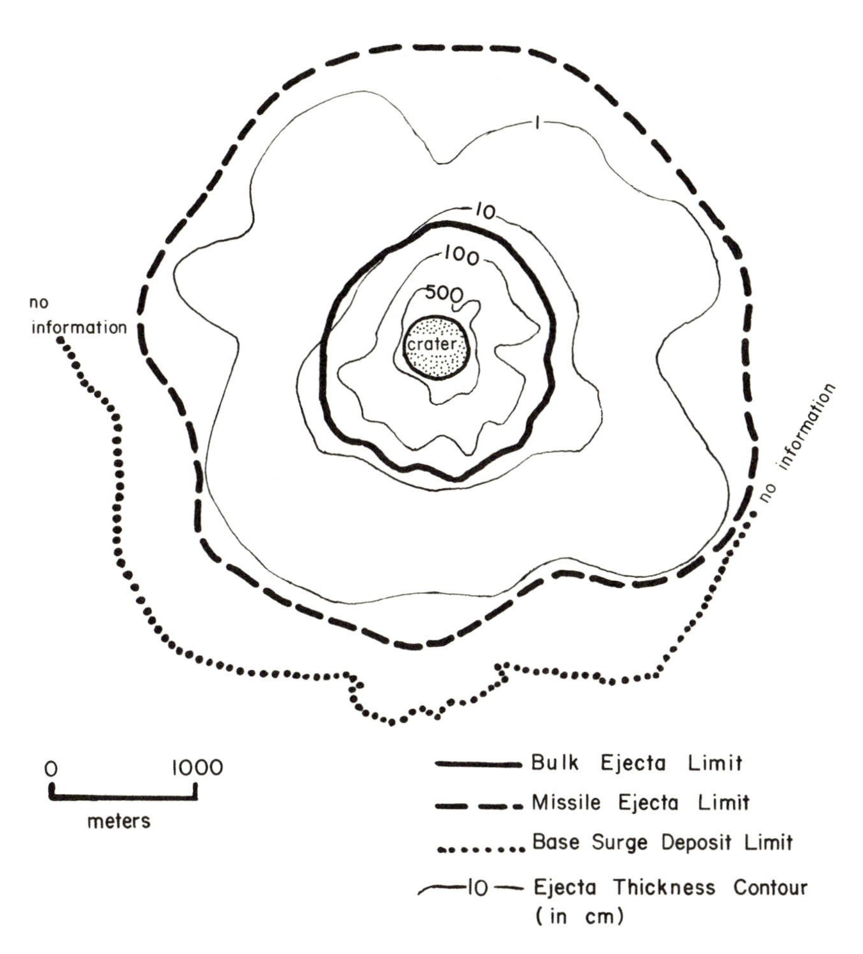

FIGURE IV-49. Ejecta distribution around the Sedan crater (after Nordyke and Williamson, 1965, and Carlson and Roberts, 1963).

A particularly instructive example of base surge deposition is provided by the multiple eruptions of the Taal Volcano in the Philippines over a period of several days in 1965 (Moore, 1967). As in the case of artificial explosions, it is easy to distinguish between a high pillar of clouds and a horizontally spreading basal cloud. Horizontal movement of material is uniquely documented by mud coatings on the windward side of tree trunks. Coatings comprise multiple layers, each grading outward from silt to sand. Thickness of the coatings gives a measure of the amount of base surge materials (Fig. IV-50), although there is no reason to think that a vertical succession through ground deposits at the same points would yield identical figures.

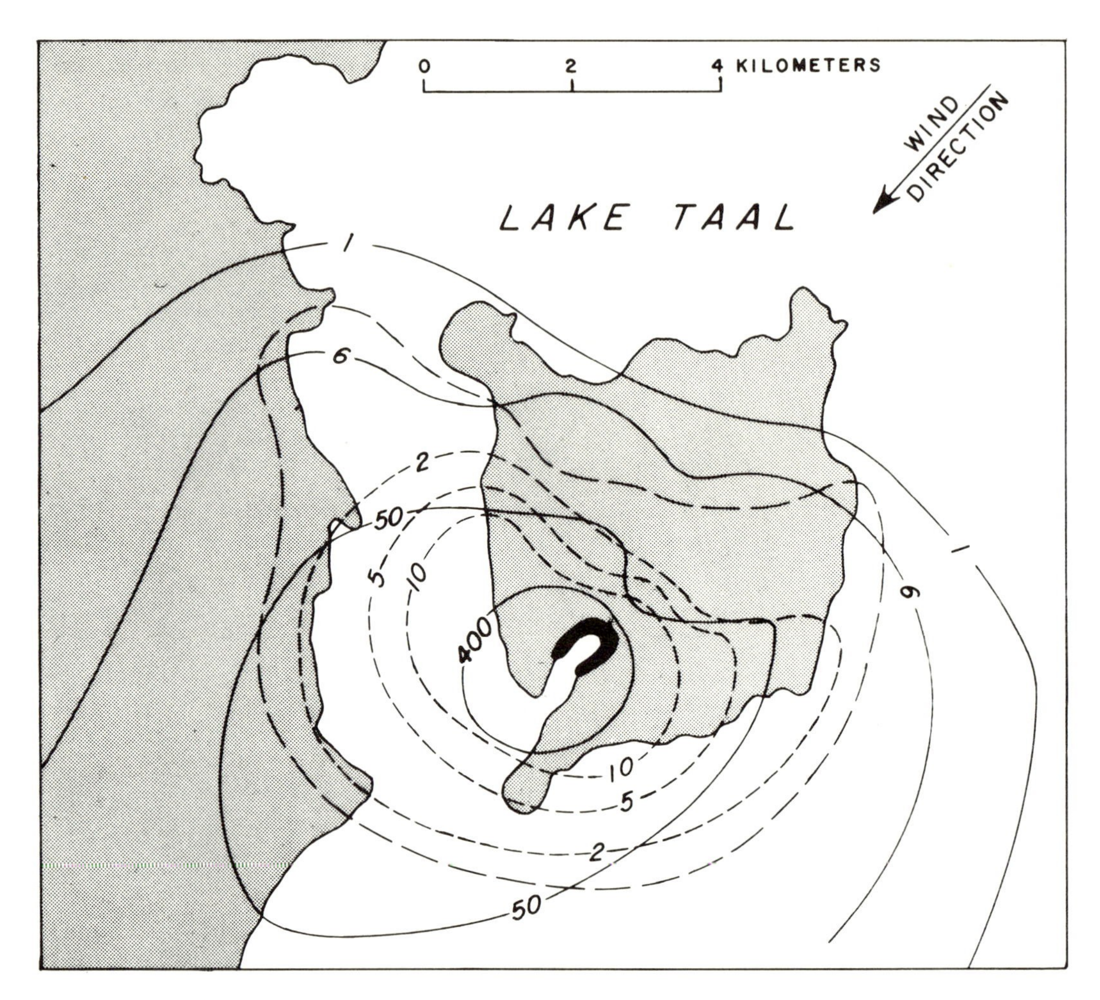

FIGURE IV-50. Distribution of ejecta around Taal Volcano, Luzon, Philippines, following September 1965 eruption. Solid lines show thickness of total ejecta deposits measured in centimeters. Short dashed lines show thickness of base surge deposits. Long dashed line indicates outer limit of base surge as determined by faint sandblasting (adapted from Moore, 1967).

The base surge deposits around Taal Volcano show prominent concentric dunes with wavelengths up to 19 meters close to the explosion center, decreasing to 4 meters 2.5 km away from the crater (Fig. IV-51). At Bárcena Volcano in Mexico an interesting relationship between furrowed and dune-covered terrains is shown close to the main crater (Fig. IV-52). Apparently the furrows formed on the steep outer crater slopes where the high-velocity surge clouds were sufficiently competent to transport contained sediment and to erode the underlying substrate. Within a short distance, however, the surge slowed to a point where erosion ceased, to be replaced by deposition and dune formation.

The vertical zonation of bulk, missile, and base surge deposits is presently more in the realm of hypothesis than in fact. Roberts (1964) has shown a schematic arrange-

FIGURE IV-51. Aerial photograph showing explosion crater of Taal Volcano on the extreme right, outlined in white. Trends of ash dunes concentric to explosion crater are indicated by white lines (adapted from Moore, 1967).

ment for explosion craters (Fig. IV-53) which may be essentially correct but is still considerably simpler than the natural situation. It is reasonable to expect that bulk ejecta will be concentrated close to the crater and at the base of the section. These deposits include the large blocks and overturned flaps so closely related to the main crater that they are as much structural elements as sedimentary accumulations. For underground blasts spallation ejection will slightly precede the breaking apart of the roof and release of gases. To that extent one might argue that missile ejecta will reach the ground before the base surge passes over a particular point. But certainly one would expect some projectiles with high-angle trajectories to fall in or behind the expanding base surge cloud. The presence of a sharp interface between the two types of deposits seems unlikely.

Not shown on Figure IV-53 is an ash-fall deposit resulting from fine sediment carried to great heights and then falling back to the ground. Relatively speaking this is a very slow process, and resultant deposits will be sharply differentiated from other ejecta by grain size, stratification, and position at the top of the section.

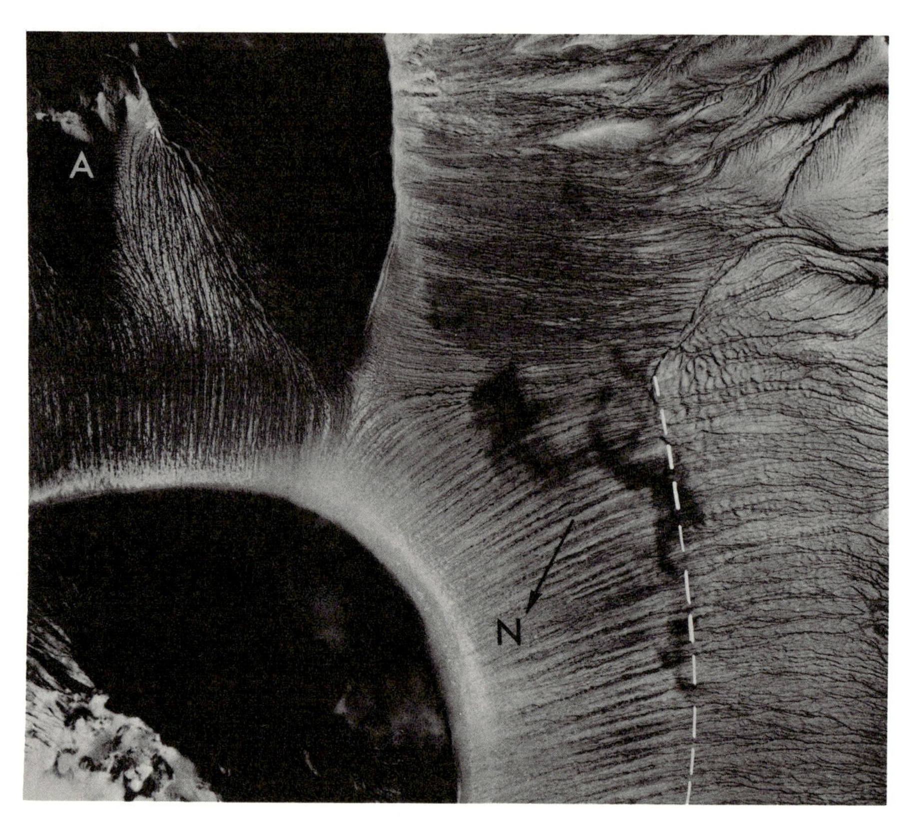

FIGURE IV-52. Aerial photograph of Bárcena Volcano, Islas Revillagigedo, Mexico. Main crater is about 750 meters diameter. Furrows eroded by surge clouds radiate from the crater, changing downslope to dunes. The transition zone, where surge clouds lost their erosive capability and began deposition, is shown by a dashed white line. Small dunes are well displayed on ridge crest at locality "A." The original topography adjacent to the crater has been slightly modified by subsequent rainfall and associated runoff erosion (adapted from Moore, 1967).

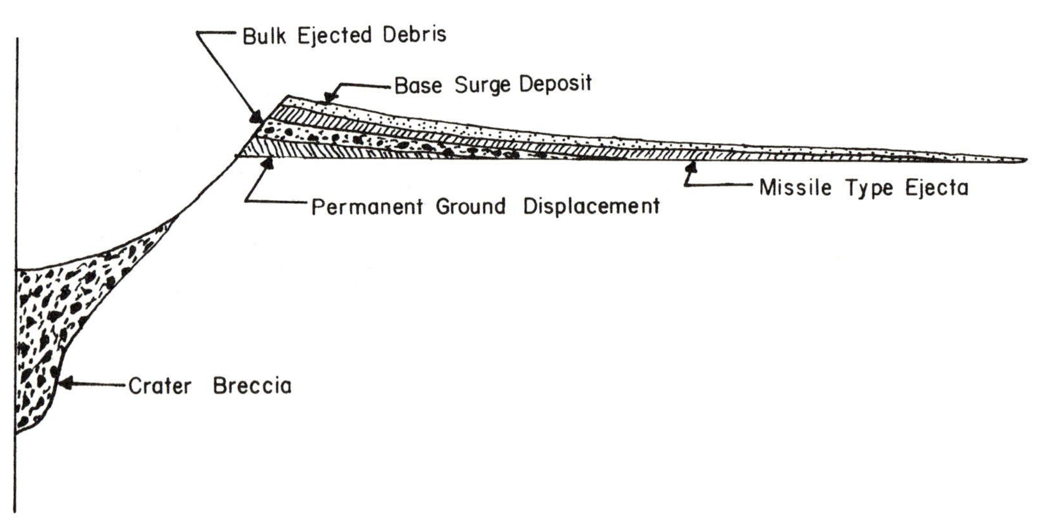

FIGURE IV-53. Schematic diagram showing stratification of ejected materials adjacent to an explosion crater (adapted from Roberts, 1964).

Moore (1967) distinguishes materials around the Taal Volcano similar to those around artificial explosion craters: (1) direct throwout, (2) base surge deposits, and (3) material blasted to high elevation and carried back to Earth by mud rains. The first two sediment types are, in practice, difficult to differentiate. Five columnar sections described by Moore show random interfingering of coarse and fine material which cannot be explained by a simple model involving one or several fining or coarsening upward cycles. As mentioned above, the tree-trunk coatings do show regular progression from fine to coarse material within an individual layer. Some layers rest unconformably on partially eroded previous deposits. Apparently each base surge cloud was graded so that fine silt was first to reach a given point, this to be followed by coarser sand which in some cases was sufficiently abrasive to erode previously deposited material. Note that this arrangement is opposed to the general case of turbidity current deposition in water where initial strong currents erode older sediments and deposit coarse material. As the current wanes, successively finer material overlies the coarse basal deposits.

In impact events there is reason to expect an even more intimate mixture of missile and base surge deposits than in underground explosions. Since generation and dissipation of gases is contemporaneous with the first stages of evacuation, there is no time lag between spallation and surficial discharge of gases as there is in both artificial and volcanic underground explosions. In this connection McCauley and Masursky (1968) have drawn attention to bedded sands adjacent to Meteor Crater which were apparently deposited contemporaneously with crater evacuation. The bedded sequence contains sharply delimited layers of crushed angular quartz alternating with layers of rounded quartz grains typical of the parent Coconino Formation. The authors tentatively ascribe the bedded sands to "shock disaggregation and fragmentation of the Coconino sandstone" which, during crater evacuation, behaved like a "dense fluid suspension." The precise way in which sedimentary sorting and bedding were achieved by processes operating for microseconds is unknown.

The relevance of studies on the sedimentology of crater evacuation to the study of lunar stratigraphy is obvious. Some features present in the terrestrial situation—dunes, for example—find a direct equivalent in the dunes around Mösting C. Quantitative study of ejecta velocities and blanket dimensions supplies constraints for sedimentary and stratigraphic models of the lunar surface. Documentation of the base surge phenomenon lends some authority to speculation that hummocky terrains surrounding large lunar basins may be massive turbidity-current deposits transported by gas or gas-like clouds spreading horizontally in the wake of cometary or meteoritic collisions.

ANALOGS RECONSIDERED

If we were more secure in our present understanding of crater formation and attendant sedimentary processes this section might more happily be entitled *Summary*. As it is, the reader can best construct his own argumentative summary. At the very least, we have gained some appreciation of the great diversity of features visible in or around lunar craters. Many of these features can be duplicated on Earth in natural impact or man-made explosive craters. Comparisons suggest that lunar craters such as Mösting C and Tycho—as well as the many others showing similar features—were formed by impact. Other craters, such as the one inside the Orientale basin and shown in Figure IV-17, most closely resemble volcanic structures. For the many craters which have been modified and subdued subsequent to their initial formation, it is impossible to make a meaningful choice between the two competing origins, impact and volcanism.

For the stratigrapher and regional geologist the processes of ejecta generation and distribution are as important as formation of the actual crater. Although the latter subject has so far received more attention than the former, first-hand examination of lunar rocks will certainly stimulate interest in crater-related stratigraphy. Emphasis will then shift to the comparative petrography of lunar ballistic deposits, base-surge deposits, ash flows, ash falls, extrusive lavas, and regolith (soil) materials. Correspondingly there will be increased interest in defining the physical and chemical environments of formation for this spectrum of rock types.

Imbrium Basin Stratigraphy

CHAPTER

FOR many years observers of the Moon have studied and interpreted the surface features. Until a few years ago, however, interest centered on particular structures and random oddities. Craters were described with little reference to surrounding relationships. Evidence for displacement along prominent lineaments was argued apart from any considerations of regional structure. Most of the observations were topographic, not geologic, so that particular craters commanded attention more for their deep precipices and striking perfection than for any geologic reasons. Evidence for the absence of any stratigraphic concern appears in the common offhand reference to craters as "formations." It is obvious that the word is being used in its general sense, a procedure that no stratigrapher would follow.

Several observers have tried to classify features by relative age according to their geomorphic characteristics. The most inclusive attempt was that of Khabakov (1960), who mapped the entire nearside of the Moon, distinguishing between "very ancient," "ancient," and "young" "mountain rings" and "linear features" (Fig. V-1). The distinction between geomorphology and stratigraphy can be a subtle one, especially if stratigraphic units are defined chiefly on their geomorphic expression. Because of potential confusion between the two approaches it is worthwhile to review the principles of stratigraphy as they apply to terrestrial situations.

STRATIGRAPHIC NOMENCLATURE

Over the years that terrestrial rocks have been studied, certain stratigraphic concepts have gained general acceptance. Accompanying this acceptance, standard procedures for nomenclature have evolved. One result, just mentioned, is that stratigraphers employ general words for specific purposes. A formation, for example, is defined as "the fundamental unit in rock stratigraphic classification . . . a body of rock characterized by lithologic homogeneity; it is prevailingly but not necessarily tabular and is mappable at the earth's surface or traceable in the subsurface" (American Commission on Stratigraphic Nomenclature, 1961, p. 650). Unless the reader recognizes these special definitions, he can read a stratigraphic article with superficial understanding but still lack appreciation for the rigorous development of the subject.

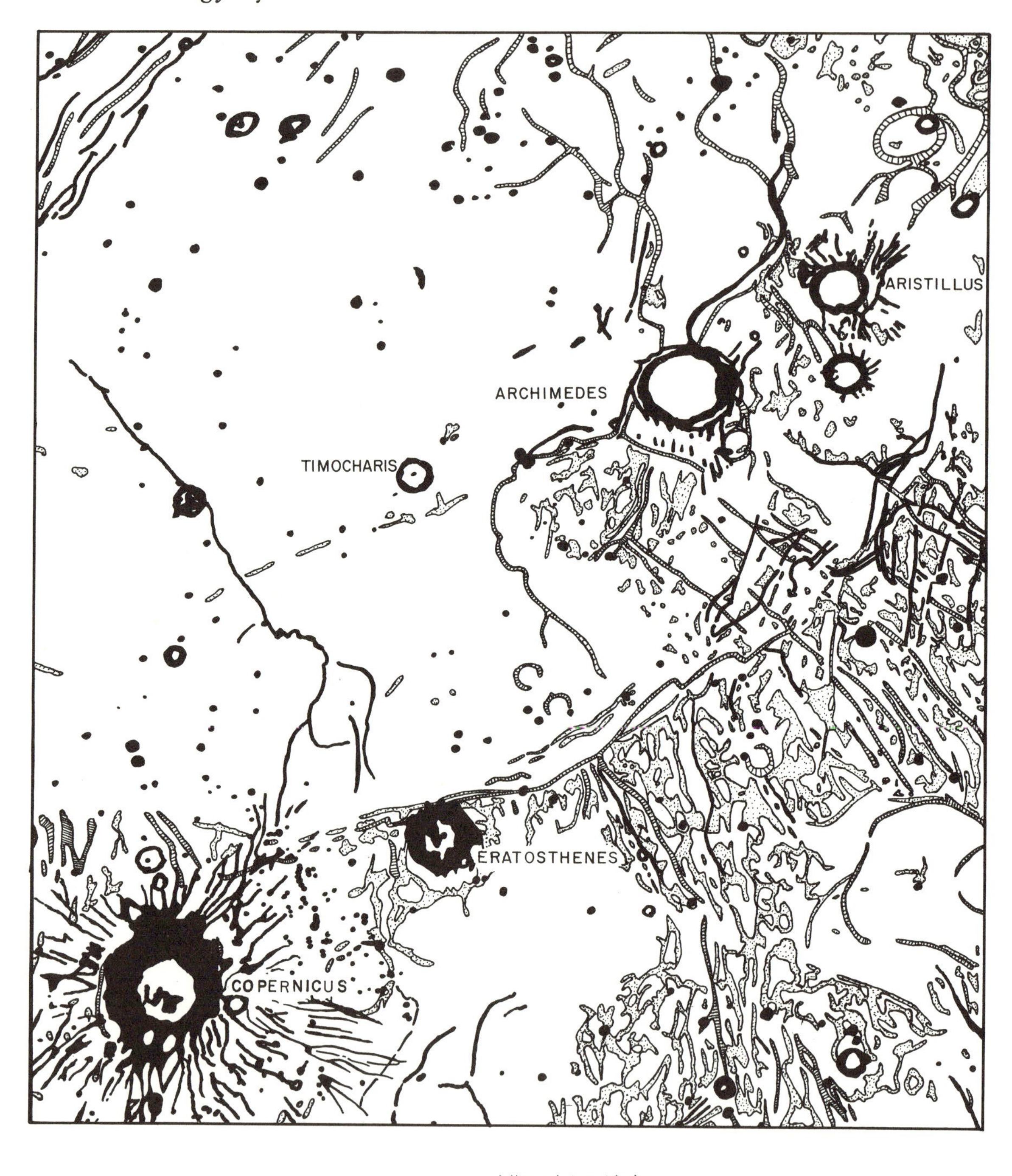

FIGURE V-1. A part of a map of the Earthside hemisphere of the Moon, prepared by Khabakov (1960). Features are ranked in age as young, ancient, and very ancient.

The conventions employed by most American stratigraphers follow the "Code of Stratigraphic Nomenclature," a report published in 1961 following fifteen years of discussion and publication of six interim reports. The final report reads very much like a legal document, which is hardly surprising in light of the authors' intention to compile a "systematic collection of rules of formal stratigraphic classification and nomenclature" (ibid., p. 649).

Stratigraphic units are divided into four categories: *rock-stratigraphic, soil-stratigraphic, biostratigraphic*, and *time-stratigraphic*. Closely related to time-stratigraphic units are *geologic-time* units.

Rock-stratigraphic units are subdivisions of the crust "distinguished and delimited on the basis of lithologic characteristics" (ibid., p. 649). Several critical points are implied by this seemingly innocuous definition. First, units are defined by observable physical features, not by inferred geologic history. As an example let us assume that an area is being mapped wherein marine rocks overlie nonmarine rocks. Between the two sequences there is an unfossiliferous sandstone which is lithologically uniform throughout (Fig. V-2). There appears to be no break in sedimentation at either the upper or

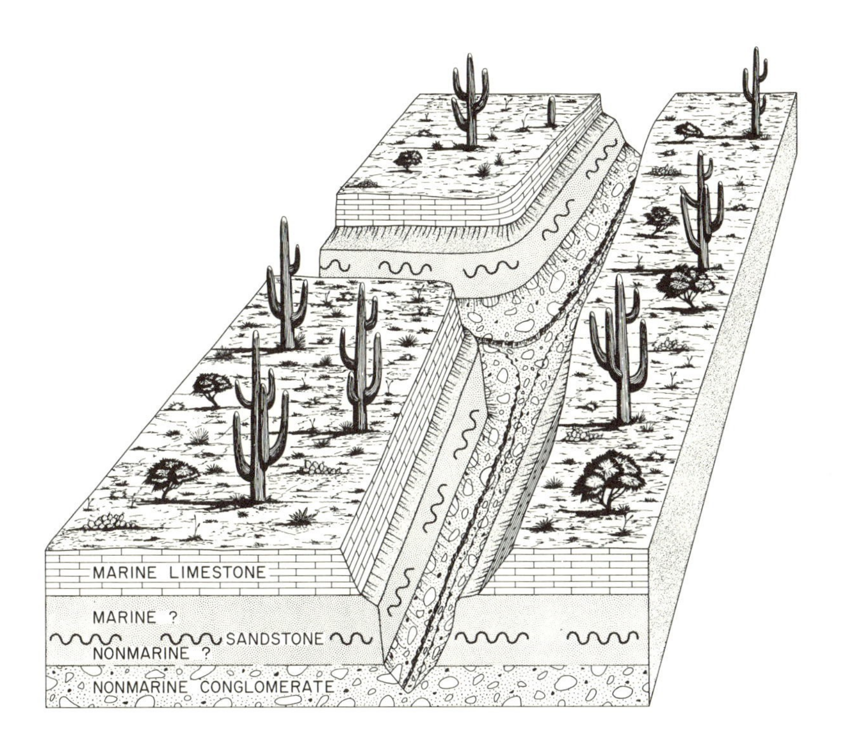

FIGURE V-2. A schematic drawing showing hypothetical relationships in a sequence of sedimentary rocks. An obscure unconformity occurs in the middle of an unfossiliferous sandstone unit. Because the underlying conglomerate is nonmarine and the overlying limestone marine, a reasonable inference is that a marine-nonmarine transition occurs in the sandstone. However, in the absence of fossils or other physical criteria, this cannot be confirmed (drawing by J. W. Van Divier, U.S. Geological Survey).

lower contact, but there is an obscure, discontinuous unconformity near the middle. A reasonable inference is that the lower part is a continuation of nonmarine deposition and that the unconformity marks the base of a similar-appearing sandstone deposited under marine conditions. However, mapping of two rock-stratigraphic units and characterizing one as marine and the other as nonmarine in the absence of physical proof is unacceptable. In so doing, one would be mapping an inference, not a physical observation. Instead the homogeneous sandstone should be mapped as a single rock-stratigraphic unit.

A second critical point in defining rock-stratigraphic units is that "concepts of time-spans, however measured, properly play no part in differentiating or determining the boundaries of any rock-stratigraphic unit" (ibid., p. 649). Accordingly, a homogeneous limestone with Silurian fossils in its lower part and Devonian fossils in its upper part is mapped as a single rock-stratigraphic unit. The fact that the unit straddles a time boundary is not relevant to its definition. Historically this has been a difficult proposition for many geologists to accept. Some subscribe to a catastrophic—or at least compartmentalized or noncontinuous—view of Earth history. Because the Silurian and Devonian are fundamental divisions of time, these geologists maintain that the crossover point should be recorded in the rocks. The expectation is not completely without reason, since many time-stratigraphic boundaries were first defined on breaks in the rock record: changes in lithology, erosional unconformities, or structural unconformities. For this reason boundaries between rock-stratigraphic units in some areas are also time boundaries, though the latter play no part in definition of the former.

There are other ways in which rock-stratigraphic units are independent of time. One unit may be deposited in a matter of days; the second represents slow accumulation over millions of years. Both are defined on observable physical features and may well have the same rank as rock-stratigraphic units. A laterally continuous homogeneous sand body may be deposited in a transgressive sea so that its age varies from point to point (Fig. V-3). Even though contained fossils unequivocally demonstrate a difference in age, the unit is not subdivided if it is otherwise lithologically homogeneous.

The fundamental rock-stratigraphic unit is the *formation*. It may be subdivided into *members*, and several formations may be included in a *group*. Formal names of formations are binomial, "consisting of a geographic name combined with a descriptive lithologic term or with the appropriate rank term alone" (ibid., p. 652). Examples would be Chattanooga Shale, Waynesboro Formation, Newark Group. This system of nomenclature can easily be transferred to the Moon with its profusion of crater names. But it should be remembered that a name like the "Cayley Formation" does not necessarily refer to rocks geographically confined to or even genetically associated with the crater Cayley. All that can be inferred is that the formation is typically exposed in the vicinity of the crater Cayley.

Another term commonly used in stratigraphic description is *facies*. The meaning and proper use of this word have been vigorously debated (e.g., Longwell, 1949), but most geologists subscribe to the original definition that a facies is an aspect of a formation, either paleontologic or petrographic, which serves to distinguish it from adjacent rocks of different aspect (Gressly, 1838). Following this definition, geologists talk of the sandy facies of the Martinsburg Formation or, on the Moon, of the hummocky facies of the Fra Mauro Formation. Although facies is commonly employed as a rock-stratigraphic term, there is no authority for its use within the code. A formation or member designation is used instead and, by definition, precludes any necessity for units called facies.

FIGURE V-3. A schematic drawing showing hypothetical relationships in a sequence of rocks formed in late Paleozoic time. Gravel, sand, and coal are deposited in nonmarine stream and swamp environments. Relatively homogeneous sand bodies accumulate close to the shoreline. Further offshore, finer silt and clay are deposited. Because of a progressive rise in sea level the shoreline moves from left to right. Simultaneously there is shift in facies distribution. At time "A," littoral sands are being deposited at the far left, but by time "C," similar sands are accumulating in the middle right (drawing by J. W. Van Divier, U.S. Geological Survey).

Description of a single typical exposure plays an important role in the definition of rock-stratigraphic units. The *type section* is a specific location at which the vertical extent of the unit is exposed and its characteristic features exhibited.

One of the most confusing elements in stratigraphic nomenclature is the establishment, abandonment, and redefinition of particular rock-stratigraphic names. Stratigraphers are heirs to a massive legacy of formation names, most of them established in the literature well before any thought had been given to a standard code. Many of these formations have never been adequately defined and mapped. Modern workers sometimes redefine the formations but retain the old names, sometimes introduce new names for previously defined units, and sometimes change both the names and the physical divisions. Add to this the fact that geologists are rarely unanimous in acceptance of the revisions, and you begin to get some idea of the complexity. Further compounding the confusion is the inclination that some stratigraphers have to sidestep the imposing requirements for naming new units by employing an informal terminology. Simply by virtue of repetition, the informal nomenclature often becomes the established standard.

A biostratigraphic unit is "a body of rock strata characterized by its content of fossils contemporaneous with the deposition of the strata" (American Commission on Stratigraphic Nomenclature, 1961, p. 655). These units are of great importance in determining time-stratigraphic boundaries, but criteria for biostratigraphic and time-stratigraphic units differ fundamentally. For example, fossil assemblages could vary laterally due to ecologic controls as well as vertically due to evolutionary changes. Of course, the lack of fossils in lunar rocks (at least at the present resolution limits) precludes the use of biostratigraphic units, and they will not be further mentioned here.

A time-stratigraphic unit is a "subdivision of rocks considered solely as the record of a specific interval of geologic time" (ibid., p. 657). Units are defined on the basis of actual sections of rocks and, consequently, are "material units." Each is "the record of an interval of time that extended from the beginning to the ending of its deposition or intrusion" (ibid., p. 657). The fundamental division is the *system*; smaller subdivisions are termed *series* and *stages*.

Units appearing on geologic maps are frequently identified by several letters which stand as abbreviations for rock-stratigraphic and time-stratigraphic designations. The time-stratigraphic label appears first as an upper-case letter and the rock-stratigraphic label follows. For example, the Martinsburg Formation of Ordovician age would be designated "Om."

Geologic-time units comprise the final major category. Distinction between these and time-stratigraphic units is subtle but important. Geologic-time units are "divisions of time distinguished on the basis of the rock record, particularly as expressed by time-stratigraphic units. They are not material units" (ibid., p. 659). Further the authors note that

> Historically the definition of a period as a unit of geologic time depended on chosen sections in the type area of the system which is the corresponding time-stratigraphic unit. The period comprised an interval of time defined by the beginning and ending of the deposition of the system. To define periods rigorously in this manner is to create unnamed time units between periods, in other words, gaps in formal geologic time. By later work supplementary sections largely or wholly filling the hiatuses have been found elsewhere in the world and their rocks, by common con-

sent, have been assigned to one or another of the contiguous systems. Many of the gaps have thereby been essentially filled. . . . The units of geologic time are no more valid than the time-stratigraphic units on which they are based (ibid., pp. 659–660).

This argument is illustrated in Figure V-4. In the first measured rock section at the far left, the Permian and Triassic Systems were deposited in time intervals separated by an interval of nondeposition. To equate the two systems with time periods is to leave the intervening period of nondeposition unaccounted for. Subsequent discoveries of additional rock sections make possible increasingly more precise delineation of geologic-time units.

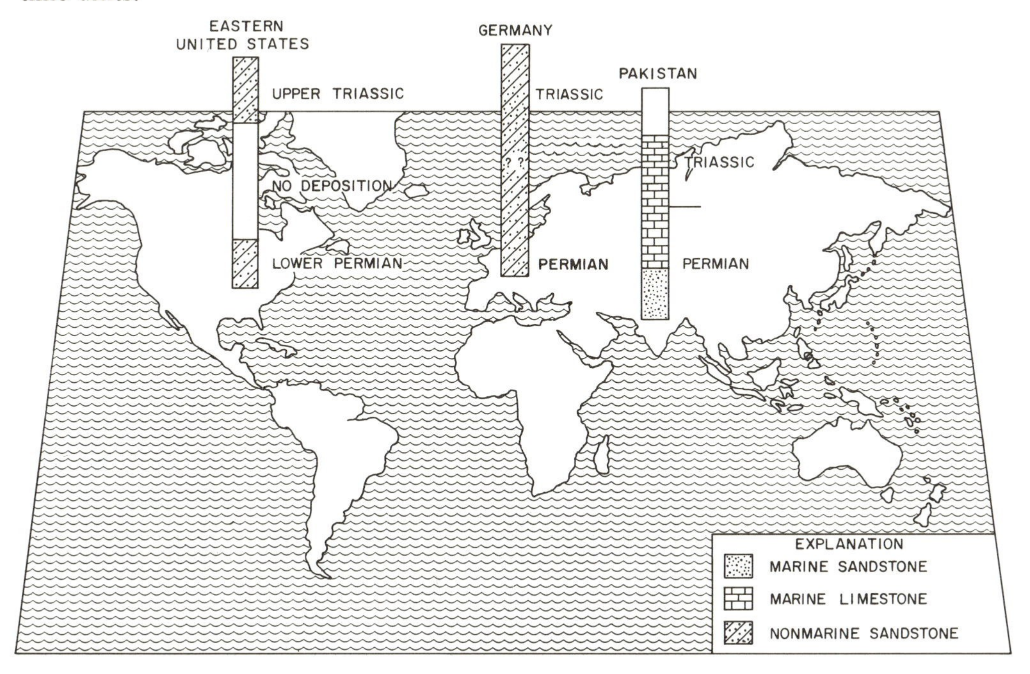

FIGURE V-4. Schematic illustration of the fact that recognition of a continuous rock record often involves integration of numerous sections. In the eastern United States, lower Permian and upper Triassic depositional episodes are separated by a period of nondeposition. The Permian-Triassic section in Germany possibly represents continuous deposition, but it is difficult to establish precise ages for the nonmarine strata. In Pakistan the point in geologic time signalling the crossover from the Permian to Triassic Period is more clearly recorded in fossiliferous marine strata (drawing by J. W. Van Divier, U.S. Geological Survey).

Ranks of geologic-time units in order of decreasing magnitude are *era, period, epoch,* and *age*. The last three correspond respectively to the time-stratigraphic divisions *system, series,* and *stage* (Fig. V-5).

The criteria chosen to measure geologic time are admittedly meagre, rising from necessity not from choice. By analogy with human history it would seem more convenient to talk about the year 3,000,000 B.C. or the minus ninetieth century. But no way is known to measure the absolute ages for the deposition of the majority of sedimentary rocks. Even though the times of formation of certain minerals and rocks can be determined by radioactive methods, this technique is not widely applicable to sedimentary events. Discovery of a generally applicable technique to measure sedimentation ages on an absolute scale is one of the most exciting advances imaginable in the field of geology.

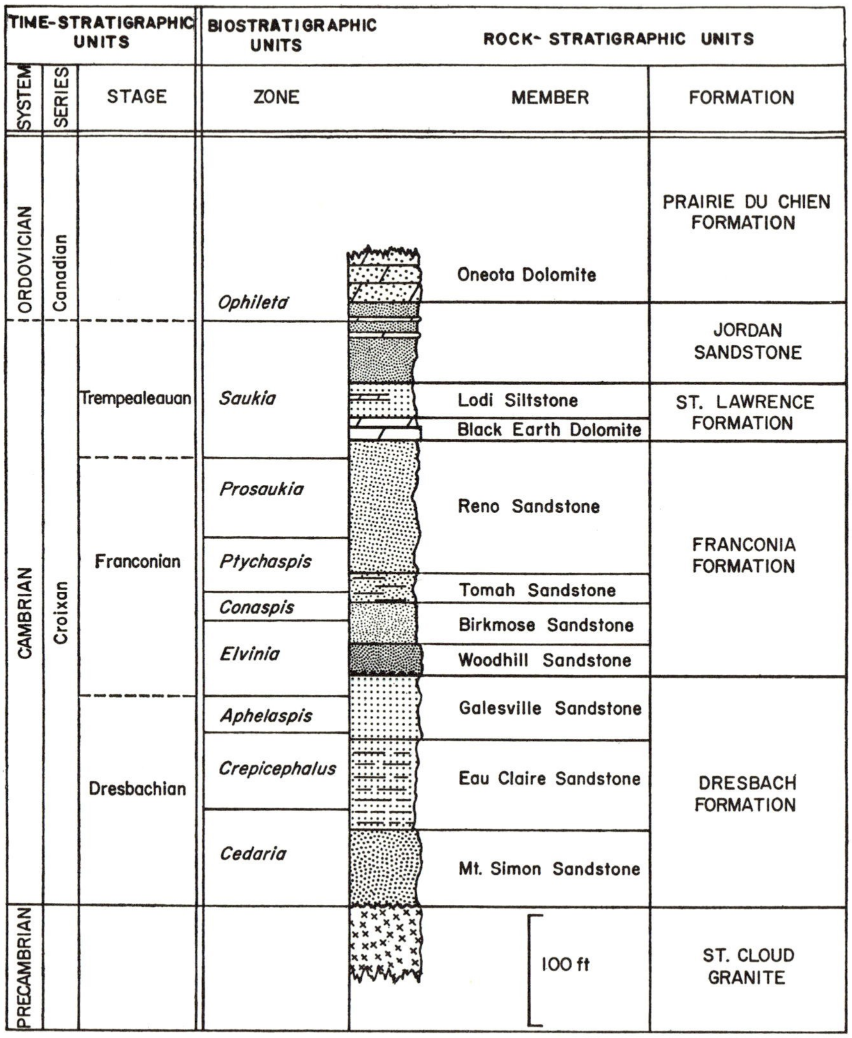

TIME-STRATIGRAPHIC UNITS			BIOSTRATIGRAPHIC UNITS	ROCK-STRATIGRAPHIC UNITS	
SYSTEM	SERIES	STAGE	ZONE	MEMBER	FORMATION
ORDOVICIAN	Canadian		*Ophileta*	Oneota Dolomite	PRAIRIE DU CHIEN FORMATION
CAMBRIAN	Croixan	Trempealeauan	*Saukia*		JORDAN SANDSTONE
				Lodi Siltstone	ST. LAWRENCE FORMATION
				Black Earth Dolomite	
		Franconian	*Prosaukia*	Reno Sandstone	FRANCONIA FORMATION
			Ptychaspis		
			Conaspis	Tomah Sandstone	
				Birkmose Sandstone	
			Elvinia	Woodhill Sandstone	
		Dresbachian	*Aphelaspis*	Galesville Sandstone	DRESBACH FORMATION
			Crepicephalus	Eau Claire Sandstone	
			Cedaria	Mt. Simon Sandstone	
PRECAMBRIAN				100 ft	ST. CLOUD GRANITE

FIGURE V-5. Lower Paleozoic sedimentary rocks from the Upper Mississippi Valley. Note that the boundaries for rock-stratigraphic, biostratigraphic, and time-stratigraphic units are generally not coincident. For example, deposition of the Jordan Sandstone continued from Cambrian time into Ordovician (from *Stratigraphy and Sedimentation* by W. C. Krumbein and L. L. Sloss. W. H. Freeman and Company. Copyright © 1963).

The preceding discussion is biased in favor of the majority view of American stratigraphers regarding stratigraphic classification. Their opinions are by no means universally accepted outside—or even inside—this country. Disagreement exists particularly about the definition and long-range correlation of time-stratigraphic units. Some geologists (e.g., Miller, 1965; Newell, 1966) maintain that strata are arranged temporally chiefly by their contained fossils. To this extent it is argued that separation of biostratigraphic units from time-stratigraphic units is more illusionary than real. Not unexpectedly, many Russian geologists (e.g., Ovechkin and others, 1961) disagree with the American code. Their chief complaint is that establishing separate time-stratigraphic units is incorrect in principle. According to them, stratigraphic subdivisions should represent natural stages of geological change for large regions, based on analysis of all organic and inorganic features. They consider each of these subdivisions to be simultaneously a time-stratigraphic unit.

It might seem that we have spent an inordinate amount of time considering stratigraphic nomenclature, splitting rocks—and perhaps hairs. But each one of the points we have considered is fundamentally relevant to a stratigraphic understanding of the Moon. There—as here on Earth—periods, systems, and formations must be differentiated. There—as here—informal nomenclature, imprecise definitions, and unjustified correlations will lead to confusion.

THE FIRST LUNAR STRATIGRAPHY

As mentioned in Chapter I, the first stratigraphic discussion of the Moon was published in 1962 by Shoemaker and Hackman. Because of its importance we should review their article in some detail. The basic principles remain unchallenged, but many of the stratigraphic units have been reinterpreted and redefined in the course of additional mapping.

Shoemaker and Hackman concentrated their attention on the area around Copernicus. The keystone of their entire argument is that "the geological law of superposition is as valid on the Moon as it is on Earth," and that "the lunar surface is locally built up of an intricate and complexly overlapping set of layers of ejecta from craters, material underlying the crater walls and floors, and material that occupies the maria" (p. 289). They singled out ejecta from craters for special attention and demonstrated that Copernicus was surrounded successively outward by a hummocky facies, a radial facies, and ray streaks with secondary crater scars. These ejecta facies were shown to overlie certain craters such as Eratosthenes and Reinhold which, accordingly, must be older than Copernicus. These two craters are superposed on a mare surface which must be still older (Figs. V-6 and 7).

In all, they recognized five overlapping sets of deposits which, except for the first, they called systems. From oldest to youngest these were pre-Imbrian, Imbrian, Procellarian, Eratosthenian, and Copernican. The pre-Imbrian was not definitely recognized in the Copernicus region but was presumed to be present on ridges and hills with great relief. The origin of the rocks was deemed "diverse" and "complex," and the authors thought it likely that much of the terrain was composed of multiple overlapping rim deposits. Hope was expressed that detailed mapping in the southern hemisphere in areas of more extensive pre-Imbrian exposure would reveal "many separate mappable

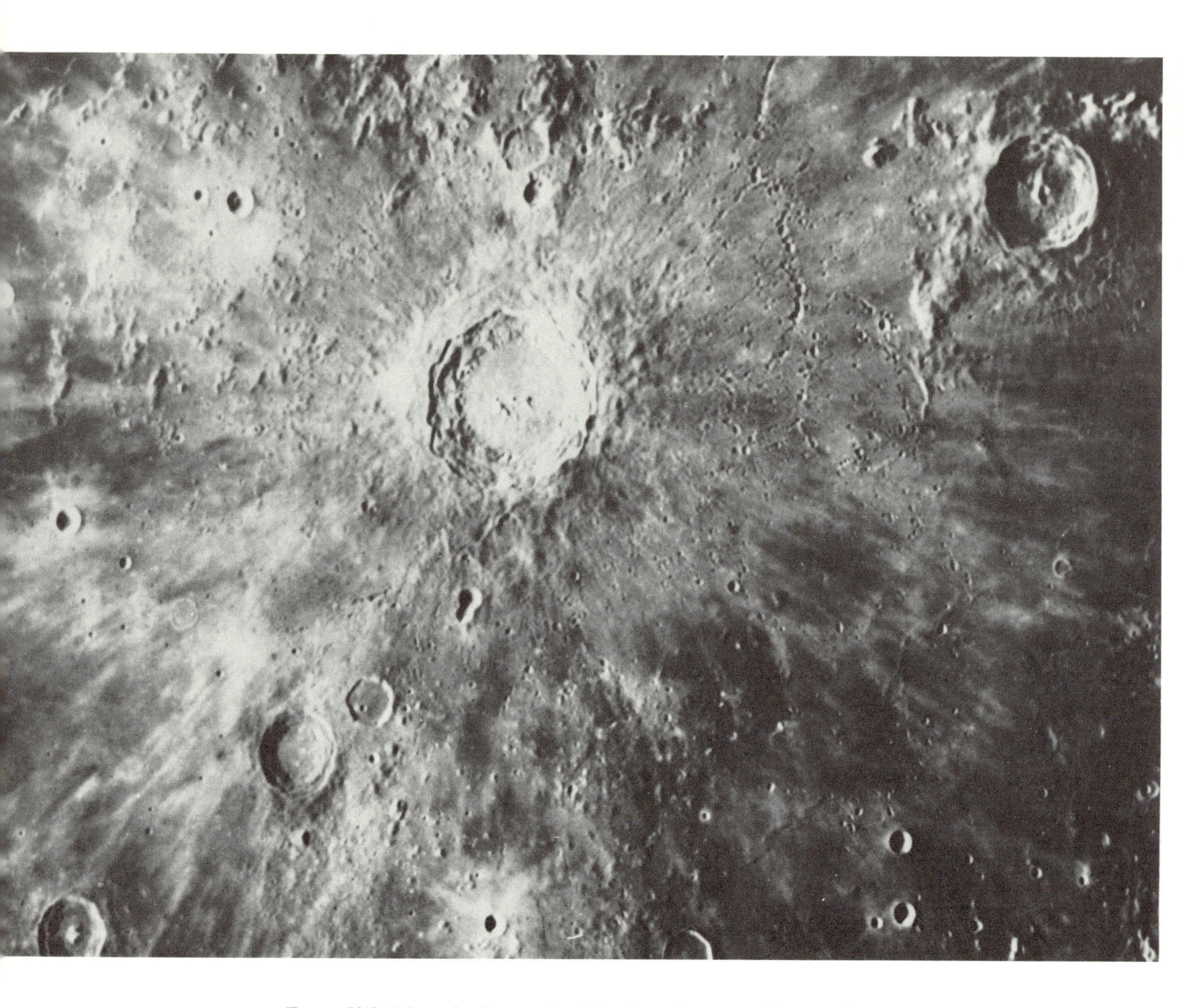

FIGURE V-6. Telescopic photograph of the Copernicus area (Mount Wilson Observatory).

Opposite page:
FIGURE V-7. Geologic map of the Copernicus area. Both the map and the legend follow the presentation of Shoemaker and Hackman (1962). Several minor units appearing in the original map have been omitted. Figure V-6 shows a telescopic photograph of the same area taken from a slightly different perspective.

STRATIGRAPHIC UNITS

COPERNICAN

EJECTA BLANKET
Probably chiefly crushed rock with large blocks.
Forms hummocky layers ranging from a few feet to about 3,000 feet thick around small and large craters. Initial surface probably rough at the scale of feet and inches.

Cb

BRECCIA
Probably chiefly crushed rock with large blocks.
Probably forms deep lenses inside of small and large craters. Topography is irregular to smooth. Initial surface probably rough at the scale of feet and inches.

Ct

TALUS
Probably partially sorted accumulation of fragments ranging in size from dust to large blocks.
Generally forms sheets mantling smooth slopes of about 30°. Initial surface probably rough at the scale of feet and inches.

SURFACE CHARACTERISTICS
Topography at the scale of miles varies from pitted to hummocky to smooth. Probably rough to very rough at the scale of feet and inches; initial surface characteristics probably fresh to partially modified.
High to very high reflectivity.

ERATOSTHENIAN

Ee

EJECTA BLANKET
Probably chiefly crushed rock with large blocks.
Forms hummocky layers ranging from a few feet to about 2,000 feet thick around small and large craters. Initial surface probably rough at the scale of feet and inches.

Eb

BRECCIA
Probably chiefly crushed rock with large blocks.
Probably forms deep lenses inside of small and large craters. Topography is irregular to smooth. Initial surface probably rough at the scale of feet and inches.

Topography at the scale of miles varies from pitted to hummocky to smooth. Probably smooth to slightly rough at the scale of feet and inches; initial small scale relief probably much reduced by small meteorite bombardment, insolation and mass movement.
Low to moderate reflectivity.

PROCELLARIAN

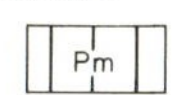

MARE MATERIAL
Probably volcanic flows. Great extent and generally smooth topography suggest thick sheets of basalt or ignimbrite. Forms layers ranging from a feather edge to several thousand feet thick. Initial surface may have been rough or relatively smooth at the scale of feet and inches.

Smooth plain broken by small meteorite and secondary impact craters and rounded ridges of several hundred feet relief. Probably smooth at the scale of feet and inches; initial small scale relief probably largely reduced by small meteorite bombardment. Low reflectivity.

IMBRIAN

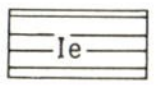

EJECTA BLANKET
Probably chiefly crushed rock and great blocks derived mainly from the region of Mare Imbrium. Forms a layer ranging from a few hundred to a few thousand feet thick. Layer is probably heterogeneous in composition either because of heterogeneity of source material or because of local alteration.

Hilly to locally smooth at the scale of miles. Topography characterized by numerous hills and depressions 1 to 2 miles across; locally controlled by relief on contact with pre-Imbrium rocks. Probably smooth at the scale of feet and inches; initial small-scale relief largely reduced by small meteorite bombardment, insolation, and mass movement. Low to moderate reflectivity.

PRE-IMBRIAN

PRE-IMBRIAN ROCKS
(undifferentiated)
Probably includes breccia, layers of ejecta, possibly volcanic or other igneous rocks. May in some places be covered with a thin mantle Imbrium ejecta.

Not well exposed.

stratigraphic units" (p. 292). Students of the Precambrian on Earth will recognize a familiar tone to these observations—dimly perceived relationships but optimistic hopes for clarification in the wake of more detailed mapping.

The Imbrian System is exposed north of Copernicus around the Carpathian Mountains. According to Shoemaker and Hackman the topography developed on the Imbrian is "unique . . . a gently rolling surface studded in most places with closely spaced low hills and intervening depressions . . . the numerous small-scale topographic features impart a shagreen appearance to the Imbrian" (p. 293) (Fig. V-8). (Any reader might be excused if he were unfamiliar with shagreen: "an untanned leather covered with small round granulations and dyed green" according to Webster's *New International Dictionary*, 1966.) McCauley (1967*a*, p. 433) states that Imbrian topography is "generally described as hummocky." These several quotations illustrate a problem inherent in all lunar mapping—describing appearances that are distinctive but difficult to envision except through firsthand experience, either telescopic observations or study of photographs.

The Imbrian is widely exposed around the southern margins of Mare Imbrium. As it is traced away from this mare, it becomes less hummocky and finally grades into a relatively smooth deposit. These relationships suggested to Shoemaker and Hackman that the deposit represents a vast ejecta blanket formed as the result of a cratering event which excavated Mare Imbrium.

The Procellarian System forms the smooth dark floors of major lunar basins. Named for extensive exposures within Oceanus Procellarum, it fills all other mare regions as well. It was tentatively interpreted to have a volcanic origin, in part because of the presence of volcano-like domical hills, sinuous scarps resembling lava flow fronts, and discontinuous ridges possibly caused by shallow intrusions. The Procellarian floods Imbrian rocks and is therefore younger. Its upper contact is a matter of more uncertainty. Preliminary crater counts for areas within eight different maria suggested a generally similar age for the upper surface (Shoemaker, Hackman, and Eggleton, 1962), and it was presumed that the time period associated with emplacement of the entire sequence was relatively short.

Crater deposits overlying the Procellarian were divided into two systems. The older is characterized by an absence of rays and by relatively low albedo. The system was named Eratosthenian after the crater Eratosthenes. The authors' belief that most craters were of impact origin necessarily led to the conclusion that the many craters included within the Eratosthenian were not of identical ages. Instead a wide range in age for the numerous individual crater deposits is implied.

The youngest crater system was named Copernican. Included craters possess bright rays and relatively unsubdued rim deposits and interior structure. Rays and ejecta blankets are superposed on all previously described units, indicating that the Copernican is the youngest system of all.

FIGURE V-8. The Apennine Mountains, bounding the Imbrium basin to the southeast. The linearly arranged hills probably reflect a radial depositional pattern for material ejected from the Imbrium basin at the time of its formation (Lunar Orbiter Photograph IV 109H).

It is instructive to review Shoemaker's and Hackman's use of stratigraphic nomenclature, keeping in mind the discussion of the preceding section. Although rock-stratigraphic units are the most fundamental stratigraphic divisions, they do not figure prominently in descriptions made by Shoemaker and Hackman. No formations are rigorously defined. Precisely located type sections are lacking, as are detailed descriptions and discussions of mappability. Instead those five sets of deposits just described are defined as systems, and these correspond to "five intervals of time which we will call periods" (p. 292). Nowhere is there any mention of the stratigraphic constraints for the naming of a system or a period. Indeed, the general reader would be completely unaware of the fact that the use of the terms follows conventions of Earth stratigraphy.

At the outset of the book, we mentioned that this first stratigraphic analysis of the Moon represents a flash of insight comparable to that of early nineteenth-century British geologists classifying the Earth's stratigraphic record. Where before there had been only a profusion of craters and assorted structures on the Moon, now there is an orderly progression of events recorded in systems and their equivalent periods. But the liabilities of this stratigraphic model are also obvious. Lack of detail leaves the door open for confusion of stratigraphic terms—and the ideas which the terms represent. Perhaps the most speculative observation of Shoemaker and Hackman's article is that "reconnaissance studies indicate that the stratigraphic systems recognized in the Copernicus region can be correlated and mapped over most of the visible hemisphere of the Moon" (p. 298). What if detailed studies prove otherwise? Should existing systems be redefined, should new systems be added, or should we turn to a more cautious approach and speak only of rock-stratigraphic units?

SYSTEMATIC MAPPING PROGRAM

Following the pioneering reconnaissance work of Shoemaker and Hackman, systematic mapping of the lunar surface has been undertaken by the U.S. Geological Survey under the auspices of the National Aeronautics and Space Administration (NASA). These investigations are compiled on lunar topographic charts prepared by the U.S. Air Force Aeronautical Chart and Information Center (ACIC) at a scale of 1:1,000,000. The Earthside hemisphere has been divided into forty-four quadrangles, sometimes called regions. Geologic maps for most of these quadrangles are available, either published as *U.S. Geological Survey Miscellaneous Geologic Investigations* or in preliminary form as U.S. Geological Survey open-file reports (Appendix A).

Following standard geologic practice, the maps contain a text and explanation. For some of the earlier maps the texts are brief and general. Indeed, for the first two maps they are identical. Texts for the first six maps all begin with the sentence: "Material exposed on the surface of the Moon is heterogeneous." Inevitably, the sentence has been modified and then dropped in subsequent maps. At the start of the mapping program, a claim for heterogeneity was necessary in the same way that it would be important for an eighteenth-century geologist to stress the fact that rocks on the Earth's surface are of different ages. But no twentieth-century geologist would feel it necessary to preface an article with the assertion that "materials on the surface of the Earth are heterogeneous." Now that we are becoming more familiar with the Moon, the simple statement that there are differences in surficial rocks is becoming similarly superfluous.

Unit descriptions, appearing within the explanations of lunar maps, are divided into two parts: *characteristics* and *interpretation.* As the titles suggest, characteristics include objectively determinable features, and interpretation contains a speculative discussion of the significance of these features. The intent of the mapper is to clearly separate observations from geologic inferences. Units are mapped according to their characteristics, not according to their inferred origin.

From the start of the systematic mapping program in 1962 until 1966, publication of maps was preceded by preliminary geologic summaries which appeared in the *Astrogeologic Studies Annual Progress Reports.* These reports, which appeared in the U.S. Geological Survey open files, were prepared annually for the National Aeronautics and Space Administration. In many instances the geological discussion in the preliminary report is more detailed than that contained in the text and explanation of the final map. This unusual division of information is especially troublesome for the early maps. In later publications most of the information contained in the annual report also appears in the text next to the map. In this way one can easily consider both the objective distribution of map units and the author's interpretation of the geologic history for the area.

NOMENCLATURE FOR THE IMBRIAN SYSTEM

Stratigraphic nomenclature for the Imbrian System has had a checkered history. Originally defined to include only the presumed ejecta material around the Imbrium basin (Shoemaker and Hackman, 1962), the system was almost immediately enlarged. The lower part (the original Imbrian System) was named the Apenninian Series after typical exposures in the Apennine Mountains and comprised the "regional material of the Imbrian System" (Shoemaker and others, 1962). The upper unit, the Archimedian Series, included crater-rim material superposed on the Apenninian and overlapped by the Procellarian System (Figs. V-9 and 10).

Next, the Procellarian System (time-stratigraphic) was changed to Procellarum Group (rock-stratigraphic) (Hackman, 1964). No reason was given in print for the change, although most stratigraphers would agree that it is best to start with rock-stratigraphic, not time-stratigraphic units. McCauley (1967*a*) states that the system designation was abandoned because it included rocks of uncertain and widely ranging age, mare filling materials which comprise an integral rock-stratigraphic unit but not necessarily a single time-stratigraphic unit. Despite this possibility the Procellarum Group has been pigeon-holed in the upper part of the Archimedian Series (Hackman, 1964). It therefore also defines the top of the Imbrian System. In this position it is as restricted in time as was the Procellarian System.

FIGURE V-9. Telescopic photograph of southeastern part of Imbrium basin. Largest crater is Archimedes, 80 km in diameter. Rugged terrain in lower right corner is Apennine Mountains. Light terrain south of Archimedes is the Apennine bench. Archimedes is younger than the bench but older than the mare material (Catalina Observatory Photograph).

At the same time that the Procellarian was reduced to rock-stratigraphic status, the regional material of the Apenninian Series was subdivided to include the Fra Mauro Formation (Eggleton, 1964) and the overlying Apennine Bench Formation (Hackman, 1964). Although the establishment of rock-stratigraphic units was a necessary clarification, the subdivision of a time-stratigraphic unit (series) into rock-stratigraphic units (formations) carries with it the potential for confusion between observational and inferential categories. The Fra Mauro corresponds to the inferred ejecta blanket surrounding Mare Imbrium, the same unit which was called the Imbrian System by Shoemaker and Hackman (1962).

The Apennine Bench Formation is exposed just to the north of the steep basin-facing scarps of the Apennine Mountains. It has less local relief than the Fra Mauro and a

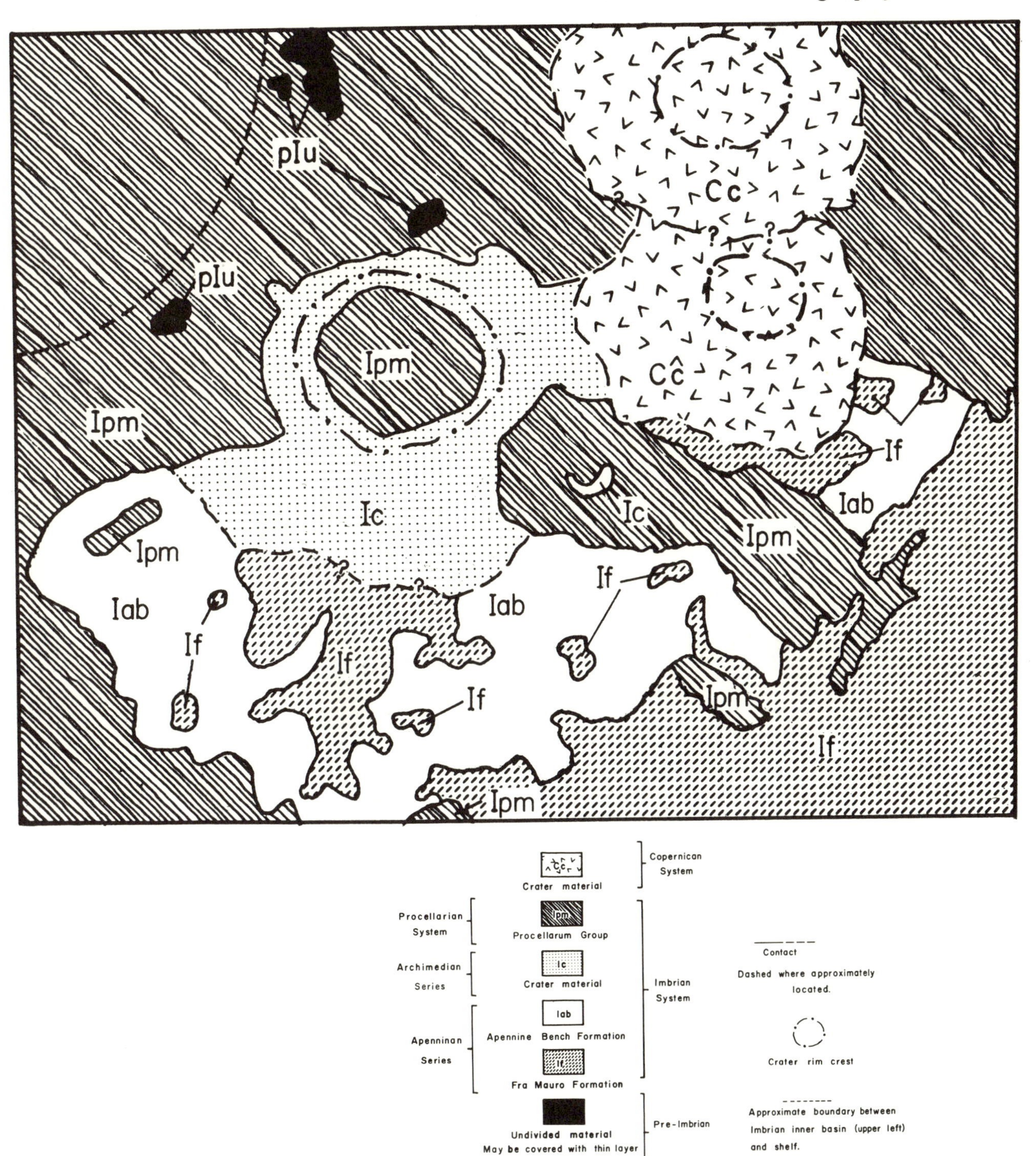

FIGURE V-10. Geologic sketch map of area in Figure V-9 (after Hackman, 1966). Time-stratigraphic units on the left side of the legend represent subdivisions according to Shoemaker and others (1962).

more uniform albedo. The genetic significance of the unit is not clearly stated, but its deposition is presumed to be related to formation of the Imbrium basin. Included materials are thought to be volcanic flows and ash beds; this contrasts with ejected debris included within the Fra Mauro Formation (Hackman, 1964).

As mapping proceeded, it became apparent that the Archimedian and Apenninian Series could be differentiated only in the areas where they were first defined. For this reason their further use was discouraged (Wilhelms, 1970). Recently published maps show the Imbrian System with no internal time-stratigraphic divisions.

The Apennine Bench Formation has not been recognized outside its type area. The Fra Mauro Formation, however, has been recognized and mapped extensively south and east of Mare Imbrium.

This complex development of Imbrian stratigraphy is perhaps difficult to follow. Some of the discarded names and ideas already constitute ancient history. Nonetheless, it is a history well worth telling since it illustrates the problems which attend casual use of stratigraphic terminology.

FRA MAURO FORMATION

This formation, presently defined as the basal unit of the Imbrian System, includes characteristically hummocky material exposed in a belt around the Imbrium basin and becoming smoother away from the basin. Type areas for two facies, mapped as members, have been defined in the vicinity of the crater Fra Mauro (Eggleton, 1964, p. 52). The hummocky facies is characterized by an "irregularly patchy areal distribution and by abundant close-spaced, low, rounded, subequidimensional hills and intervening depressions generally 2 to 4 kilometers across." The smooth facies has "fairly smooth local topographic texture with generally rounded but sometimes angular, intervening forms." Although delineation of individual facies is difficult, the composite formation has been mapped extensively in the approximate central portion of the lunar disc.

The genetic interpretation which suggests the scheme for stratigraphic subdivision of the Fra Mauro Formation is readily apparent. As already mentioned, many geologists believe that the unit is an ejecta deposit from the Imbrium basin. A further assumption is that different types of material should be revealed at different distances from the basin center, as can be demonstrated for terrestrial explosion and volcanic craters as well as for smaller lunar craters. Accordingly, Wilhelms (1965) points out that the facies of the Fra Mauro closely resemble facies of rim material around lunar craters, especially the crater Aristoteles.

A critical question in interpretation of the Fra Mauro is: How much of the visible relief is produced by constructional deposition, and how much is a reflection of localized bedrock movement? (Fig. V-11). This is a continuing problem in interpretation of almost all lunar formations. A moment's reflection will reveal that we are really worrying about an unknown third dimension, that of depth.

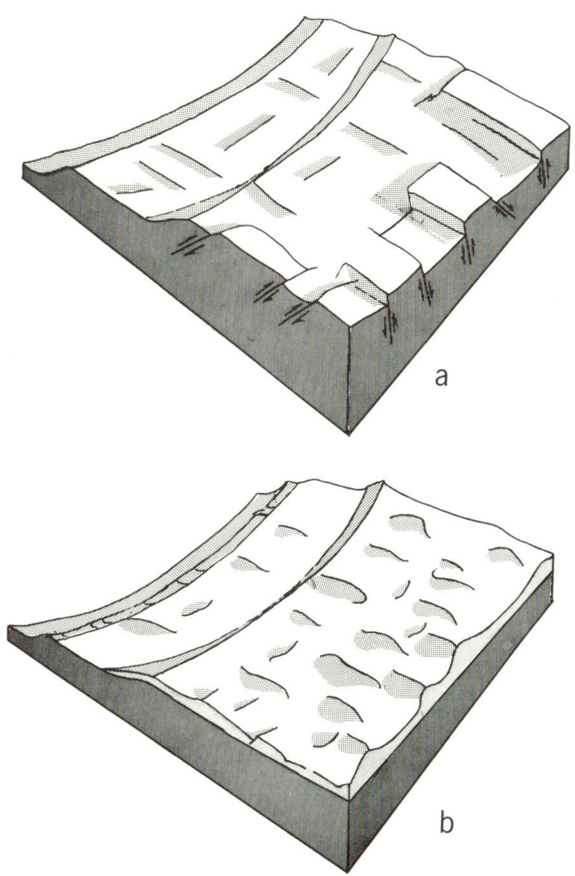

FIGURE V-11. Schematic block diagrams showing two possible mechanisms for production of a strongly lineated topography: (a) by structural faulting, or (b) by depositional relief.

There are two ways in which the reality of the Fra Mauro as a deposit with significant thickness has been demonstrated. The formation mantles pre-existing craters and partially obscures them. Since crater depth is proportional to width, the lower size limit for craters visible beneath the formation provides an indication of its thickness. Thickness calculated according to this technique is 900 meters for the area just south of Mare Imbrium. Traced 900 km to the south, the thickness progressively decreases to about 550 meters (Eggleton, 1963).

A second indication of the depositional character of the formation is provided by relationships around the crater Julius Caesar 600 km southeast of Mare Imbrium. The blanket appears to be plastered against the northwestern rim and the southeastern wall of this crater, possibly as a result of emplacement along ballistic trajectories originating within Mare Imbrium. The northern inner part of the crater, protected by the "ballistic shadow" of the rim, was subsequently flooded by Procellarum mare material (Morris and Wilhelms, 1967).

FIGURE V-12. Telescopic photograph of the western margin of Mare Tranquillitatis, showing dark mare materials in right side of the photograph and lighter plains materials in left side (Catalina Observatory Photograph).

In summary it should be emphasized that the Fra Mauro is one of the most problematical formations mapped on the Moon. Early workers, influenced by the hope that an Imbrium basin ejecta blanket may have been deposited over much of the nearside, were extremely liberal in identification of "characteristic" Fra Mauro topography at considerable distances from the Imbrium basin. More recently the pendulum has swung far in the opposite direction (e.g., Pohn and Offield, 1969). However, the existence of a thick ejecta deposit close to the Imbrium basin remains entirely reasonable on both observational and inferential grounds.

CAYLEY FORMATION

As previously mentioned, crater ejecta in general—and the Fra Mauro Formation in particular—become smoother as they are traced away from the crater of origin. The smooth facies of the Fra Mauro Formation is characterized by three properties: it mantles subjacent terrain, it is contiguous with the hummocky facies of the Fra Mauro, and it displays a subdued braid-like texture. Some early mappers enlarged this definition to include materials of flat plains within the smooth facies of the Fra Mauro. This is an approach which inevitably leads one deeper and deeper into a blind alley. As Wilhelms

FIGURE V-13. Geologically annotated photograph of western margin of Mare Tranquillitatis. Approximately the same region as Figure V-12, showing relationships of the Procellarum Group and Cayley Formation (Lunar Orbiter Photograph IV 90H; geology after Morris and Wilhelms, 1967).

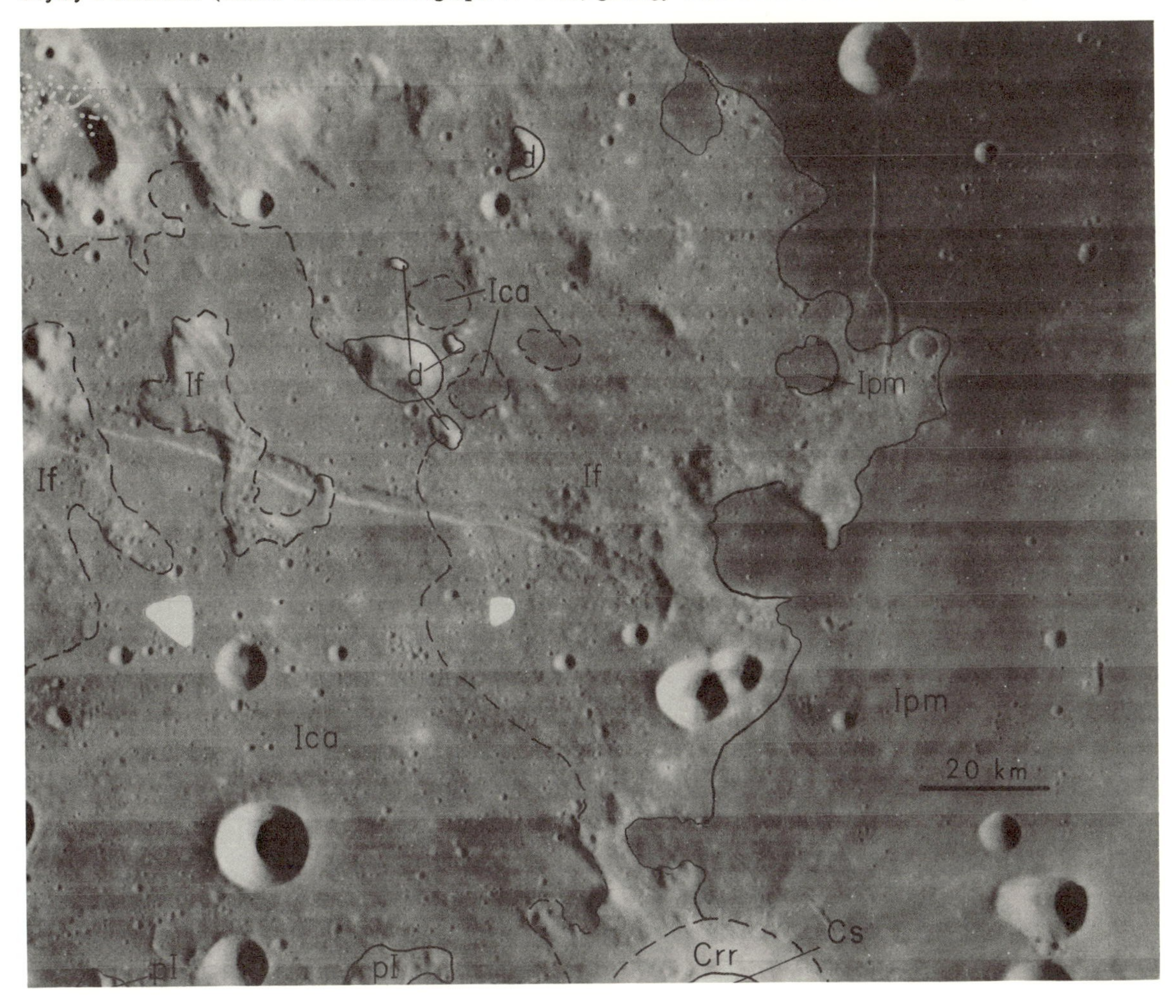

SLOPE MATERIAL Cs

CRATER RIM MATERIAL Crr

PROCELLARUM GROUP Ipm

CHARACTERISTICS: Flat mare material. Low albedo.

INTERPRETATION: Volcanic materials: flows or ash beds, or both. Uppermost layer may be fragmental debris.

CAYLEY FORMATION Ica

CHARACTERISTICS: Surface generally flat and smooth; appears to mantle subjacent terrain. Albedo intermediate, about the same as the Fra Mauro Formation, higher than Procellarum Group. Crater density greater than on Procellarum Group.

INTERPRETATION: Volcanic materials, may be ash flows.

FRA MAURO FORMATION If

CHARACTERISTICS: Forms rolling or undulating surface, in places studded with low subequidimensional hummocks several km across. Albedo intermediate.

INTERPRETATION: Thick blanket of debris ejected from the Imbrium basin by the impact of a large solid body.

TERRA DOME MATERIAL d

CHARACTERISTICS: Round or elliptical domes with convex profile.

INTERPRETATION: Volcanoes composed of relatively viscous volcanic deposits. Imbrian or younger.

(1965) pointed out, rock-stratigraphic units can only be mapped on observable physical properties, uniquely recognizable. "Extension of the definition of the Fra Mauro to include flat and smooth materials makes this unique recognition impossible, because smooth plains-forming materials occur over much of the surface and not all can be Fra Mauro" (p. 25).

In order to avoid this dilemma, smooth plains-forming materials, lighter than the Procellarum Group and without any braid-like texture, were included within a new unit, the Cayley Formation. In the area of its definition near the crater Cayley, the formation is unaffected by Imbrian sculpture but is embayed by the Procellarum Group and also has a higher crater density than the Procellarum (Figs. V-12 and 13). These relationships established an age for Cayley younger than Fra Mauro but older than Procellarum. The materials are tentatively interpreted as a combination of flows and free-fall tuff (Wilhelms, 1970).

The Cayley Formation has been mapped extensively in the central part of the Moon. Redefining as Cayley the flat, relatively featureless plains has achieved the desired end of divorcing the rocks from an Imbrium ejecta origin. However, it remains likely that rocks of diverse age and possibly diverse origin are presently included within this formation.

PROCELLARUM GROUP

At telescopic resolution the maria are strikingly homogeneous, especially in comparison with the cratered and corrugated appearance of surrounding terra. Mare surfaces have a small range in albedo, generally appearing dark. Superimposed craters are rare, so that large areas within the maria appear smooth and featureless. However, the rays from craters such as Tycho, Copernicus, and Kepler strike across the dark plains, indicating that the mare fill must be older than the rayed craters. For these several reasons the basin-filling rocks are included in a single rock-stratigraphic unit, the Procellarum Group. The areal extent of this group is great, covering about one-third of the Earthside hemisphere. Equally impressive is its thickness, which can be calculated by assuming probable depth-to-width ratios for the filled basins. The initial depths of the basins have almost certainly been modified by isostatic rebound, but, even after taking this into account, the average thickness of the Procellarum Group, calculated from assumptions of isostatic equilibrium, may be in excess of 7 km (Baldwin, 1970).

Wilhelms (1970) has recently proposed that the name Procellarum Group be discontinued since it has retained all the undesirable time-stratigraphic connotations of the earlier designated Procellarian System. For example, nearly all maps designate the Procellarum as Imbrian (Ipm). Many authors are aware that some mare materials are almost certainly younger than Imbrian, possibly as young as Copernican (e.g., Moore, 1967). Where they have chosen to acknowledge this inference, they have dropped the name Procellarum Group and have identified the unit more noncommittally as mare materials (m) of appropriate age (e.g., Carr, 1966). Rather than once again redefine the Procellarum Group, this time to include basin-filling dark materials without respect to age, Wilhelms suggests calling this large group of rocks "mare materials." Where local relationships permit, the unit may be further identified according to its observable features or inferred age.

Further discussion of the Procellarum Group—or mare materials—appears in Chapter VIII, Volcanic Stratigraphy.

Other Basins—Other Stratigraphies

CHAPTER

THE SINGLE BASIN MODEL

THE stratigraphic model of Shoemaker and Hackman placed special emphasis on the formation of the Imbrium basin and the deposition of an ejecta blanket for great distances away from that basin. The authors realized that the Imbrium basin was only one of several comparable features on the Moon; they did not, however, specifically propose that the other basins might have ejecta deposits which, in other portions of the Moon, could serve as time planes of fundamental importance. By default, then, the Imbrium event assumed unique stratigraphic significance.

HUMORUM BASIN

The first quadrangles completed in the systematic 1:1,000,000 geologic mapping program were those adjacent to the region of Copernicus. The apparent intent was to extend the stratigraphy away from the region in which it was first defined. As mapping proceeded toward the southwest, the unrefined model became more and more difficult to apply.

One of the first major problems occurred in the Letronne quadrangle. In the southern part of this area there is an extensive rough terrain, the northern edge of which is flooded by mare material. This rugged highland was mapped as Apenninian Series and was interpreted as "probably chiefly crushed rock and great blocks derived mainly from the region of Mare Imbrium . . . [forming] a layer probably ranging from a few meters to about 500 m in thickness" (Marshall, 1963). Here, then, is a presumed massive thickness of Imbrium debris occurring about 1,000 km from the edge of the Imbrium basin. If deposits of this magnitude are present at these great distances, it is reasonable to suppose that recognizable Imbrium ejecta are present across most of the Earthside hemisphere.

But it is quickly seen that the rough topography in the Letronne quadrangle is not randomly disposed. Instead it is located on the northern rim of Mare Humorum, a well-developed circular basin (Fig. VI-1). Similar hummocky terrain discontinuously encircles Mare Humorum on all sides. This being the case, one possible interpretation is that the Humorum basin is a slightly smaller analog of the Imbrium basin. Both,

FIGURE VI-1. The positions of several major basins in the western half of the Earthside hemisphere of the Moon.

then, are giant impact scars surrounded by their individual ejecta deposits. This is the interpretation advocated by Titley (1967) on a geologic map of the Mare Humorum quadrangle. Rough, hummocky deposits in the northern part of the quadrangle, contiguous with rocks of the Letronne quadrangle that had been mapped as Imbrium ejecta four years earlier, are now included in the Vitello Formation and interpreted as "ejecta from the Humorum basin" (Titley, 1967). Still another possibility is that some of the hummocky materials are volcanic in origin and unrelated to the event which excavated the Humorum basin. In any case, emplacement of the materials was apparently controlled by some local event and not by formation of the distant Imbrium basin.

A second rock-stratigraphic unit was introduced in the Humorum quadrangle to include crater materials superimposed on Humorum rim material but overlapped by rocks of the Procellarum Group (mare material) (Fig. VI-2). These crater materials, comprising the Gassendi Group, have a local stratigraphic position directly analogous to crater materials of the Archimedes type which overlie Imbrium ejecta but are flooded by mare material. The crater Letronne, visible as an incomplete arc flooded with mare material along the southern margin of Oceanus Procellarum, should probably be assigned to the Gassendi Group. When Marshall mapped the Letronne quadrangle, he failed to make this distinction and instead included the hummocky materials directly adjacent to Letronne within the Apenninian Series.

FIGURE VI-2. The crater Gassendi, located astride the northern rim of the Humorum basin. The southernmost part of the crater interior is flooded with mare material similar to that immediately south of the crater (Lunar Orbiter Photograph IV 143H).

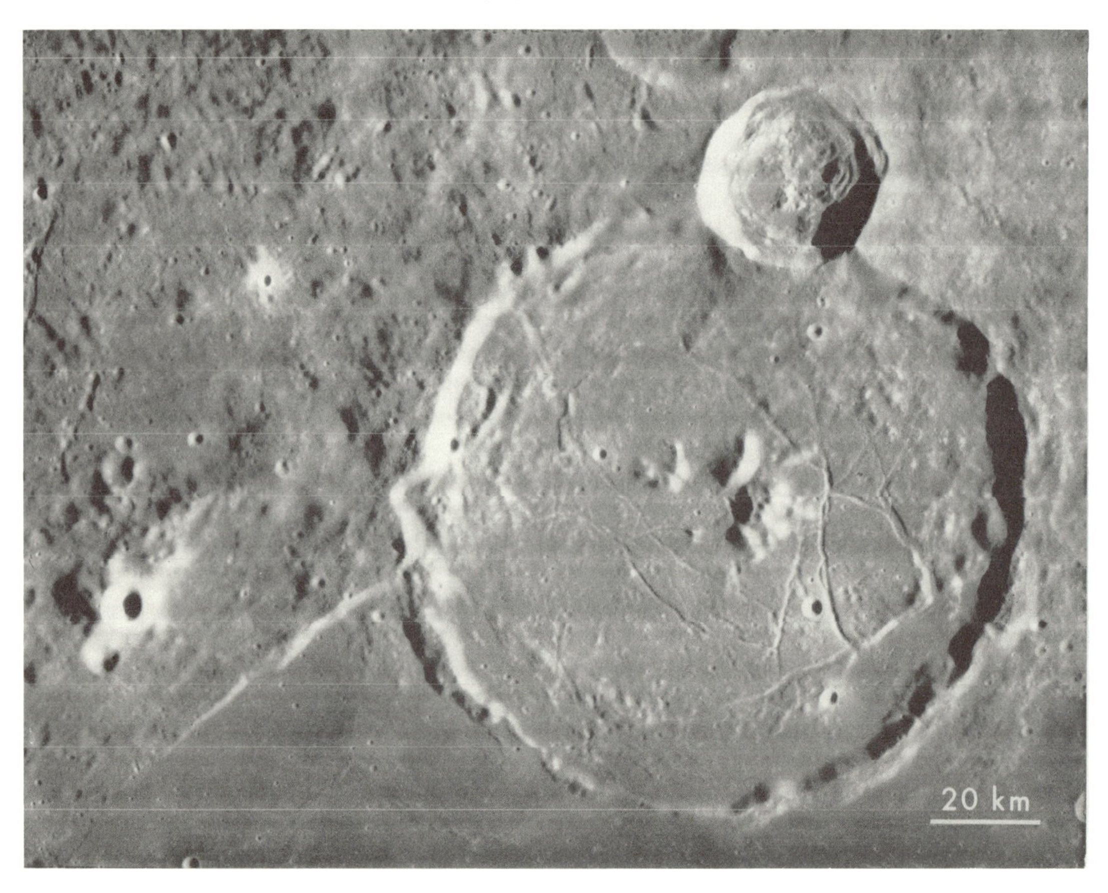

The next question is obvious. What are the age relationships of the Vitello Formation (possible Humorum ejecta) and Fra Mauro Formation (probable Imbrium ejecta)? The clearest criterion would be an overlap relationship between the two deposits. Although such a relation, with Imbrium material overlapping Humorum material, was once thought to be revealed in the Riphaeus Mountains region (Titley and Eggleton, 1964) more recent photographic data offer no support for this interpretation.

Another possible criterion of age is the density of superposed craters. The Humorum rim has more craters than the Imbrium rim, suggesting an older age for the Humorum material (McCauley, 1967*a*). The diverging slopes of the two crater density curves indicate that the difference in abundance of large craters is great—a factor of 10 or more—but that cumulative numbers to a lower limit of 7 km are essentially identical (Fig. VI-3). The paucity of small craters on the Humorum rim may indicate burial of small craters by volcanic deposits.

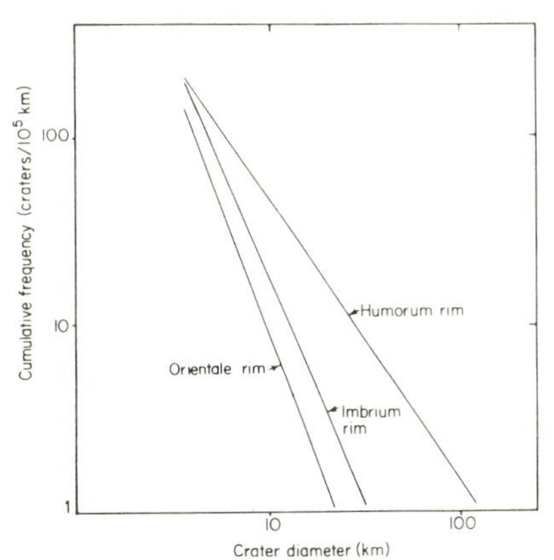

FIGURE VI-3. Comparison of cumulative crater frequencies on the rims of the Orientale, Imbrium, and Humorum basins (from McCauley, 1967*a*).

A final feature suggesting that the Humorum rim is older than the Imbrium is the relatively more subdued and weathered appearance of the concentric structural blocks bounded by the scarps which ring the Humorum basin.

ORIENTALE BASIN

West of Copernicus and Kepler, close to the western limb of the Moon, the original stratigraphic model again proved inadequate to explain all the relationships. This first became apparent during the course of mapping within the Hevelius and Grimaldi quadrangles (McCauley, 1964).

These two quadrangles contain the western margin of Oceanus Procellarum and the bounding western highlands. The highlands have a rough, hummocky appearance similar to that of the Fra Mauro Formation. The chief problem in correlating the two is again one of distance. The center of the Imbrium basin is located some 1,500 km to the northeast. Between these two highland terrains most of the region is covered with the more recent mare material which fills Oceanus Procellarum. Correlation of rocks in this situation is just as hazardous as matching rock types from one side of the Atlantic Ocean to the other.

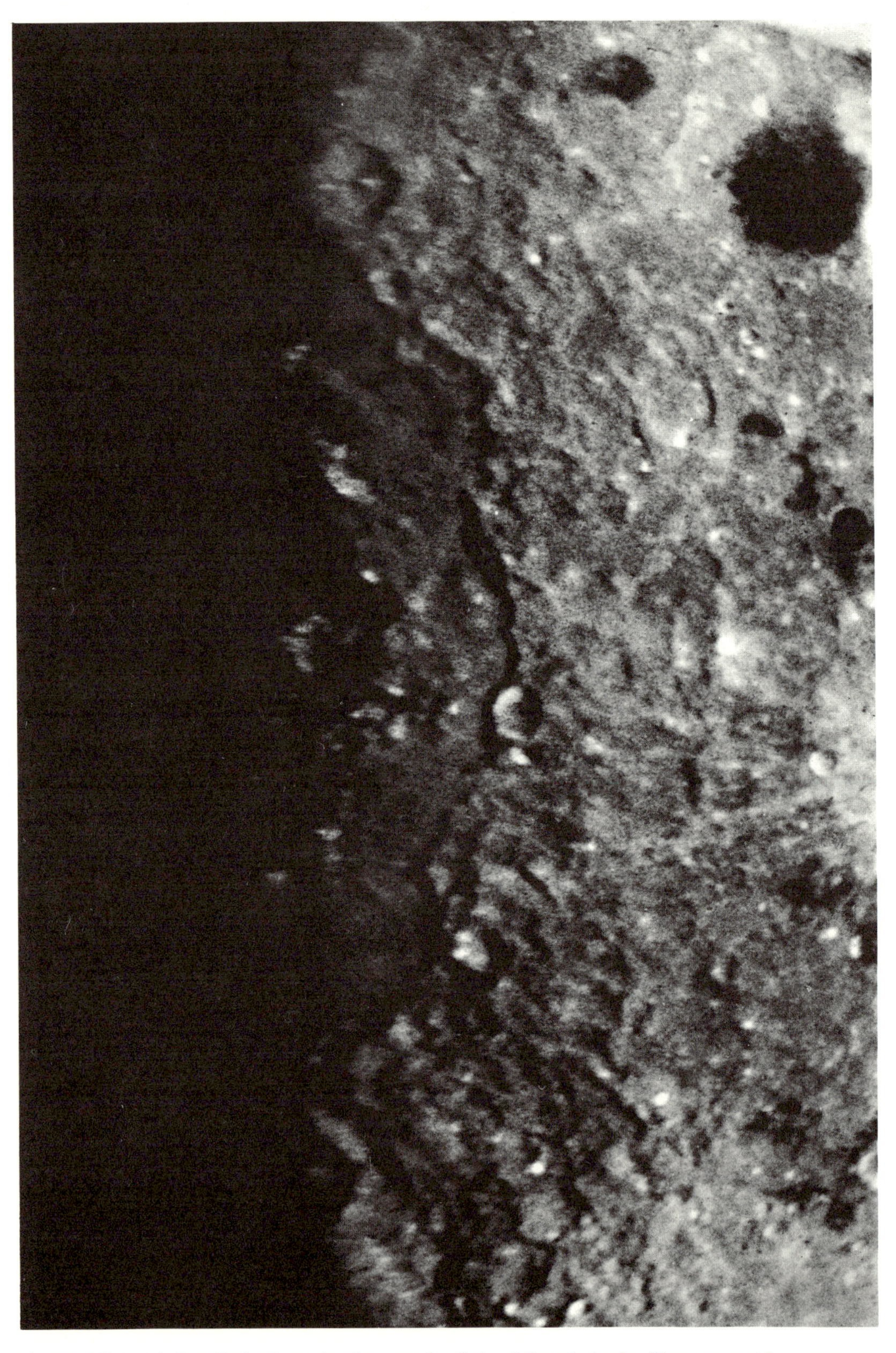

FIGURE VI-4. (a) Rectified telescopic photograph of the Orientale basin. Two concentric scarps are displayed. The outer ring is called the Cordillera Mountains. Radial lineaments are visible at considerable distances from the basin, particularly in the area to the southeast of the Cordillera Mountains (taken from Hartmann and Kuiper, 1962).

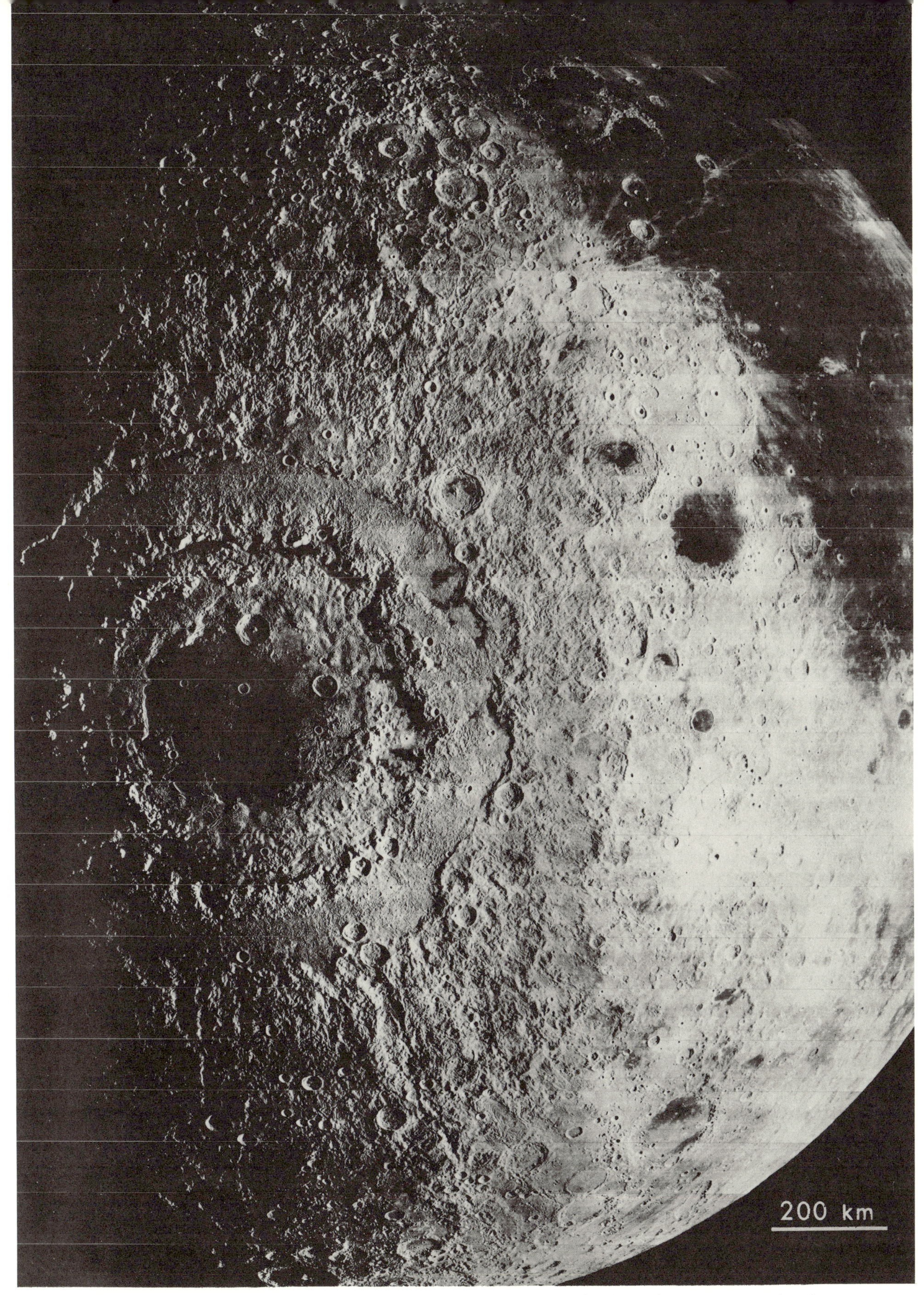

FIGURE VI-4. (b) The Orientale basin photographed by Lunar Orbiter IV. The western margin of Oceanus Procellarum can be seen in the upper right (Lunar Orbiter Photograph IV 187M).

Searching for a way out of this dilemma, McCauley looked still farther west, a region so close to the limb that it was visible only under rare conditions of favorable libration. His attention centered on Mare Orientale. Here Hartmann and Kuiper (1962) had previously described a series of concentric scarps surrounding a centrally located, flat dark plain about 320 km in diameter (Fig. VI-4a). The outermost scarp is particularly prominent, rising more than 6,000 meters above the adjacent plains. This ring of high peaks is called the Cordillera Mountains. These several features are clearly shown in Figure VI-4b, an Orbiter photograph taken when the spacecraft was directly over the Orientale basin. However, it should be remembered that the early interpretations of Hartmann and of McCauley were based exclusively on telescopic observations which provided only an incomplete and distorted view of the basin.

McCauley noted that the region outside the Cordillera Mountains is extremely rough and hummocky. Many old craters are partially obscured by a hummocky blanket which covers crater rims, drapes over walls, and spreads across floors. The degree of crater infilling decreases to the east, away from Mare Orientale. At a distance of about 1,000 km, older craters are covered by only a thin veneer of blanketing material. The thickness of the blanket can be estimated by determining the maximum size of filled craters, the same technique which Eggleton (1963) used for analysis of the Apenninian Series (Fra Mauro Formation). A maximum thickness is 4,000 meters about 300 km from the Orientale basin center. At a distance of 800 km this has decreased to 1,000 meters. In the floor of Hevelius, some 1,200 km from the basin center, the thickness of the blanket is less than 100 meters.

Just as in the region south of the Imbrium basin, the terrain adjacent to the Orientale basin is cut by radial lineaments which impart a "horst and graben" appearance to the surface (Hartmann, 1964). This raises the possibility that some of the irregular topography is structurally as well as depositionally induced.

The features just described lead to an interpretation so obvious that most readers have probably anticipated it. Again we are looking at a vast deposit of ejected debris, this one thrown from the Orientale basin at the time of its origin and blanketing as much as three million square kilometers of the adjacent countryside.

McCauley first included all blanketing deposits adjacent to the Orientale basin in an informally defined rock-stratigraphic unit, the Cordillera Group. Hummocky surface, radial structure, and albedo intermediate between that of dark maria and light terrae in the south-central part of the visible disc were cited as the characteristic features of this group. Subsequently he identified a relatively smooth subunit along the outer margin of the ejecta blanket; this was termed the Hevelius Formation (1967*a*, *b*).

The Orientale basin and its associated deposits are particularly interesting because they appear to be relatively recent. The evidence for this is manifold. The basin itself has sharp concentric scarps with relief of thousands of meters. In the Imbrium basin multiple ring-like features are present, but only the outer scarp—the rim of Carpathian and Apennine Mountains—has major relief. Secondly, the Orientale basin has only a

FIGURE VI-5. The crater Crüger, situated on the Orientale ejecta blanket and flooded with more recent dark mare material (Lunar Orbiter Photograph IV 168H).

small central pool of dark mare material. In contrast, the Imbrium basin has been flooded so deeply that mare units lap onto the outermost structural ring. An additional test of age is provided by densities of superposed craters. Such an analysis shows the Orientale rim material to be slightly younger than the Imbrium rim (Fig. VI-3).

Similar to both the Imbrium and Humorum basins there are several craters superposed on the Orientale ejecta and flooded with dark mare materials. The crater Crüger is a good example (Fig. VI-5). Its narrow rim, smooth interior walls, and deeply flooded interior suggest that the entire structure may have a volcanic origin.

A CLOSE LOOK AT ORIENTALE

All the preceding results were obtained from telescopic data. Photographs of the extreme margins of the visible disc taken from orbiting satellites provided a uniquely exciting test for previous interpretations.

In the center of the Earthside hemisphere, telescopic photographs record crater geometry to a lower limit of about 1 km, so major topographic elements were clearly delineated before Orbiter missions. On the farside of the Moon nothing of the topography was known prior to Orbiter missions. In this sense, then, there was very little to test. The margins of the Earthside hemisphere occupied a position between these two extremes. Telescopic photographs showed craters and maria, but the features were foreshortened and distorted. Crater interiors were hidden by surrounding ramparts. In short, one could see enough to make some guesses but not enough to make the guesses very convincing.

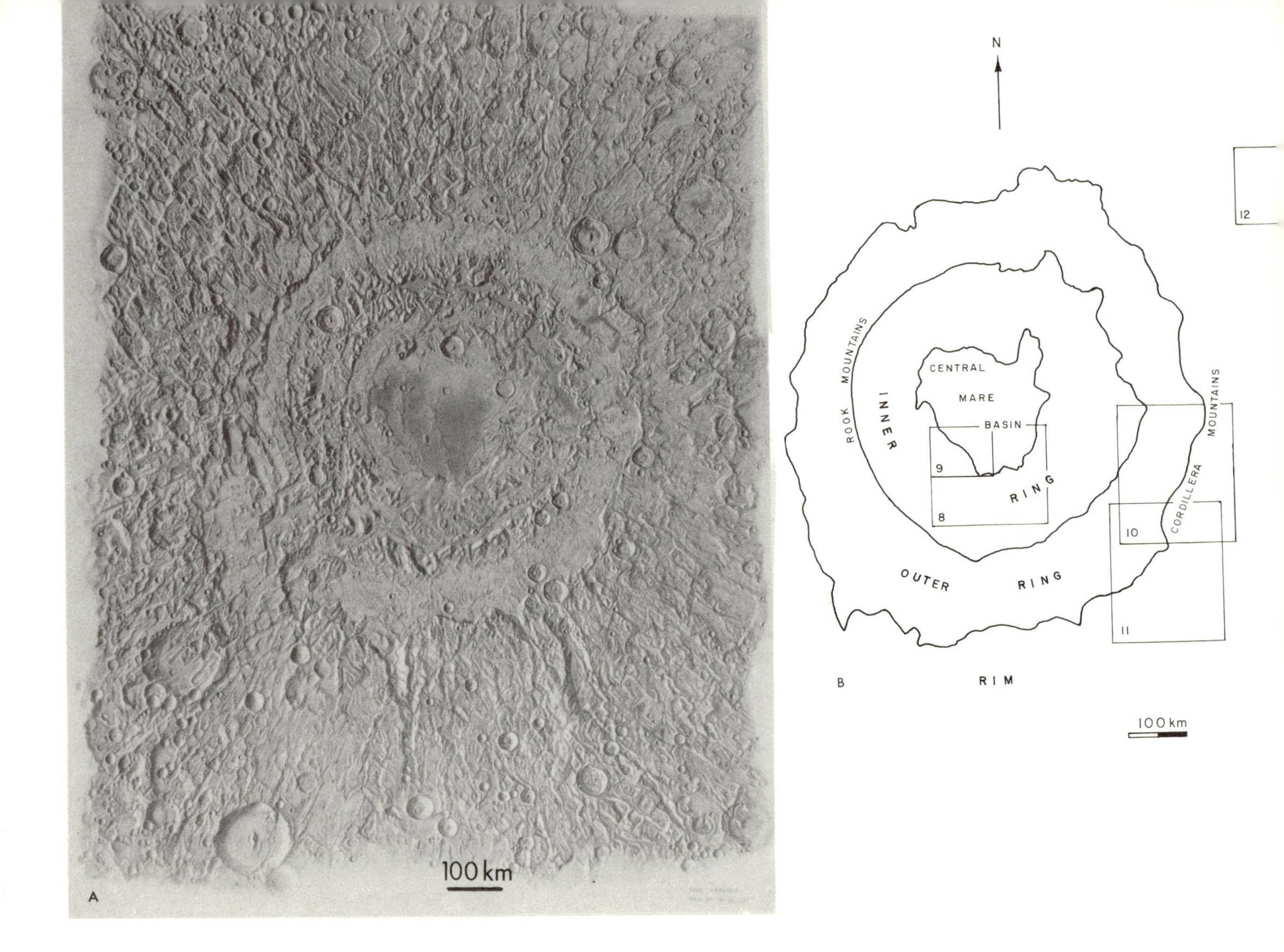

FIGURE VI-6. (a) Generalized topographic relief map of the Orientale basin (drawn by Patricia Bridges, U.S. Air Force, Aeronautical Chart and Information Center). (b) Location map showing principal features in Orientale basin and also location of Figures VI-8-13. This map is drafted at the same scale as the topographic map (a).

It was in this context that the spacecraft photographs of the Orientale basin were examined. The first photograph was taken from the Soviet *Zond III* spacecraft in 1965. Despite poor resolution the photograph does show a circular basin with a dark mare-like pool in the inner basin. Similar dark patches occur scattered across the benches and outer rim (McCauley, 1967*a*).

The Zond photograph, although interesting, failed to suggest the spectacular detail apparent in Orbiter IV photographs. A brief examination of these pictures would convince almost anyone that the Orientale basin is one of the younger—if not the youngest—lunar basins, and that it is surrounded by a flood of ejected debris.

The basin itself is divided by five concentrically arranged scarps at radial distances of 180 km, 240 km, 300 km (Rook Mountains), 465 km (Cordillera Mountains), and 730 km (Figs. VI-4, 6, and 7). The outermost ring is partly made up of extensively

FIGURE VI-7. (a) Photomosaic of the Orientale basin compiled from Lunar Orbiter IV high-resolution photographs. (b) Geologic map of the Orientale basin drafted to the same scale as the photomosaic (a). Both map and legend are adapted from McCauley (1968*b*).

CRATER MATERIALS, UNDIVIDED

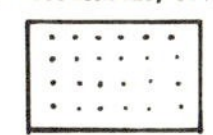

MARE MATERIAL

Dark plains material, which fills local depressions in center of basin, in Mare Veris at the base of the Rook scarp, in Mare Autumni at the base of the Cordillera scarp, in Mare Aestatis, and in the central parts of the craters Grimaldi, Crüger, Schlüter, and Riccioli.

CENTRAL BASIN PLAINS MATERIAL

CHARACTERISTICS:
Includes smooth light plains material and material of closely spaced, rolling, highly fractured hills. Appears to mantle pre-Orientale terra material. Is restricted to Orientale basin within Rook Mountain scarp. Contains several large collapse depressions. Surface texture is grossly similar to that on floors of young craters such as Tycho and Aristarchus.
INTERPRETATION:
May consist of impact-melted materials or volcanic materials which filled the lowest part of the basin shortly after its formation.

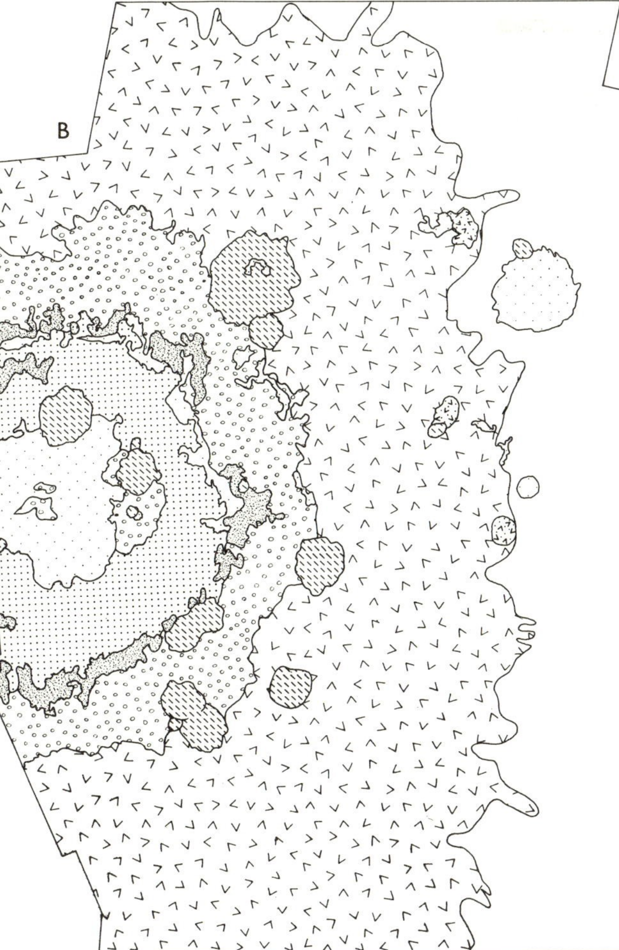

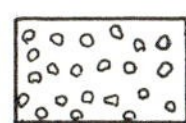

MONTES ROOK FORMATION

CHARACTERISTICS:
Forms small closely spaced smooth hills. Weakly to moderately lineated but not coarsely braided like the Cordillera Formation. Contains local patches of moderately smooth terrain. Contact with Cordillera Formation is locally gradational. Occurs mostly between Cordillera and Rook Mountain scarps but northwest of Schlüter occurs outside depressed part of Cordillera scarp. Also occurs in patches in center of basin.

INTERPRETATION:
Origin uncertain. May consist mostly of fallback from the base-surge column. Alternatively, may consist of intensely fractured rim material which was originally similar to the Cordillera Formation but slumped inward during crater filling and scarp formation, thereby modifying the primary surface texture.

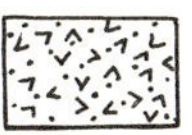

CORDILLERA FORMATION

Radially braided material.
CHARACTERISTICS:
Characterized by coarse to fine subradial ridges and grooves with a swirly to braided texture. Occurs mostly outside Cordillera scarp and completely surrounds Orientale basin. Extends farthest from basin towards the north and south. Overlies complexly cratered older terra surface. Braided texture becomes progressively less distinct with increasing distance from the basin, as thickness apparently decreases. Becomes indistinguishable from cratered terra materials in vicinity of the craters Grimaldi, Darwin, and Byrgius.
INTERPRETATION:
Consists of ballistically deposited ejecta overlain by base-surge deposits of unknown thickness; both materials produced by the impact of an asteroidal body near the center of Mare Orientale.
Transverse ridge material.
CHARACTERISTICS:
Characterized by fine dune-like structures oriented circumferentially to the basin and at right angles to the lineation of the radially braided material. Occurs on distal walls of pre-Orientale craters. This unit is not present within 300 km of the Cordillera scarp.
INTERPRETATION:
Composed of base-surge materials whose radial momentum was dissipated at the base of obstacles such as crater walls.

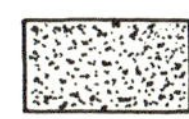

PRE-ORIENTALE TERRA MATERIAL

CHARACTERISTICS:
Mapped only within Orientale basin. Forms smooth-textured large blocks often with rectilinear outlines. Occurs mostly in Rook Mountains but also present in the inner ring.
INTERPRETATION:
Probably consists of highly fractured pre-Orientale bedrock of diverse origin. Present surface expression suggests that these blocks have been structurally depressed less than the materials of the adjacent crudely concentric troughs. Fallback and rim deposits thin to absent at surface.

FIGURE VI-8. Interior of the Orientale basin showing light, hilly, fractured materials of the inner ring and dark mare materials at the center of the basin. Location of photograph is shown in Figure VI-6b (Lunar Orbiter Photograph IV 195H).

deformed pre-basin crater walls. These concentric structures are believed to be caused by the impact of an asteroidal-size body at the center of the inner basin (McCauley, 1968*a*). McCauley proposed that during the early stages of shock-wave propagation, there was compressive failure and uplift of large structural blocks. Some uplift may have occurred along concentric inward dipping thrust faults. Much of the circularly disposed topography may therefore represent a "frozen" shock wave, frozen in the sense that segments of the crust were vertically displaced beyond the point of elastic recovery. An alternate proposal is that the surficial crust, perhaps weakened by fracture during the impact event, collapsed along concentric rings a short time later as a result of large-scale extrusion of subsurface magma (Hartmann and Yale, 1968). A variation on this idea, presented by Mackin (1969) in a general explanation for the origin of maria, is that concentric structural rims form only in large craters which penetrate to depths where temperatures are high enough to permit the country rocks to deform rapidly and to be expelled explosively by shock-assisted melting of material already close to its melting

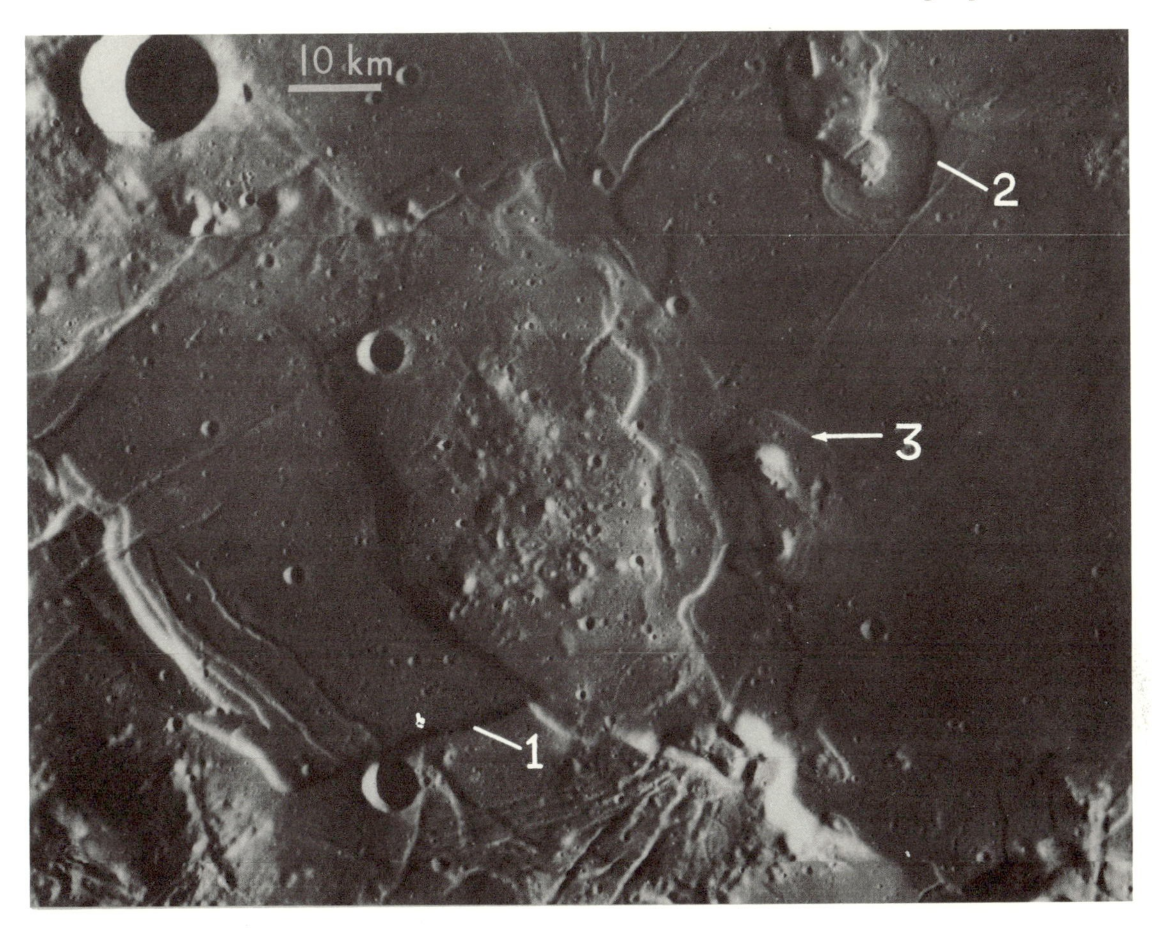

FIGURE VI-9. Details of the Orientale central mare basin showing slump scarps (1), collapse depression (2), and intermediate-level plateaus around projecting "basement" spines (3). This is a detailed view of part of the area shown in Figure VI-8. Exact location of photographs is shown in Figure VI-6b (Lunar Orbiter Photograph IV 195H).

temperature. Under these conditions lateral support for the crater walls would be removed, leading to the formation of huge concentrically arranged slump blocks.

Orbiter IV photographs showed details within the Orientale basin that permitted a much more elaborate subdivision of surficial units than was possible with telescopic data. McCauley (1968*b*) recognized three major units, concentrically arranged (Fig. VI-7). The outermost, situated between the Rook and Cordillera Mountains, was informally called the Montes Rook Formation. It is characterized by small closely spaced smooth hills. Inside this unit are lighter materials which form closely spaced and highly fractured hills. At the very center of the basin—and also in scattered pools at the foot of the Rook and Cordillera Mountains—dark smooth mare material can be seen (Fig. VI-8).

These several materials doubtless have been formed by some combination of brecciation, ejection, fall-back, shock melting, and volcanic activity. Sorting out these

FIGURE VI-10. Orientale basin, showing light plains materials of the inner ring, hummocky materials of the outer ring, and the radial rim facies. The first two provinces are separated by the Rook Mountains, the last two by the Cordillera Mountains. Location of photograph is shown in Figure VI-6b (Lunar Orbiter Photograph IV 181H).

FIGURE VI-11. Orientale basin showing walls of the Cordillera Mountains separating the inner hummocky materials from radially disposed rim materials. Location of photograph is shown in Figure VI-6b (Lunar Orbiter Photograph IV 180H).

FIGURE VI-12. Rim facies of the Orientale basin blanketing the crater Riccioli. Linear hills of the radial facies trend from lower left to upper right. In the lee of the northeast (upper right) wall of Riccioli, ejected material has piled up in a series of transverse dunes. Subsequent to ejecta deposition the interior of Riccioli has been flooded with dark mare materials. Location of photograph is shown in Figure VI-6b (Lunar Orbiter Photograph IV 173H).

Opposite page:
FIGURE VI-13. Outer rim materials approximately 700 km southeast from the center of the Orientale basin. Note that the linear pattern from upper left to lower right is in places disrupted by curved sets of transverse hills. These may be depositional dunes formed in places where the prevailing base surge currents flowing towards the southeast (lower right) broke up into turbulent eddies due to influence of surface topography. Approximate location of photograph is shown in Figure VI-6b (Lunar Orbiter Photograph IV 167H).

30 km

processes and identifying their resulting rock record is difficult. McCauley (1968*b*) suggests that the Montes Rook Formation may be either fall-back from a base-surge column or fractured rim deposits which slumped back into the basin shortly after its formation. The lighter plains could be either impact-melted or volcanic materials. The dark central pool almost certainly has a volcanic origin. Saunders (1968) has argued that many of the features within the Orientale basin can be explained by assuming that the basin was volcanically flooded several times after its initial formation. In certain instances a consolidated crust formed on the magma pool. Subsequent drainage of the magma either into the Moon's interior or into a flank chamber caused the crust to collapse and drape across the submagma topography. An abundance of apparent slump scarps, collapse depressions, and protruding ridges with surrounding plateaus of intermediate elevation lend weight to Saunder's interpretation (Fig. VI-9). The widespread fractures in the light plains material and prominent ridges in the central dark mare pool may be the consequence of local extension and compression respectively in the collapsing magma crust.

In addition to revealing details of basin structure, Orbiter IV photographs showed spectacular sedimentary features within the ejecta blanket. The two main facies are (1) braided, and (2) dune-like (Figs. VI-11, 12, and 13). Unlike many smaller craters where dunes are displayed adjacent to the crater and braided deposits occur at greater distances, the two facies are intermixed around the Orientale basin. Dunes appear to pile up in front of major topographic obstacles, especially on the distal sides of crater floors. These have been termed "deceleration dunes" by McCauley (1969). Large braids are best developed on level upland surfaces. Some braids are deflected around protruding peaks. Commonly, sets of braids with different orientations are vertically overlaid. In depressed, protected regions braids and dunes are either subdued or absent.

These several sedimentary features provide powerful evidence for deposition of ejecta from a laterally moving base surge. The resemblance of the features to smaller scale sedimentary current structures on Earth (Harold Masursky, personal communication) suggests the presence of trapped gases which imparted a hydrodynamic or aerodynamic-like behavior to the transporting medium and its sediment load. Hartmann and Yale (1968) have made some preliminary calculations which indicate that gases released from the impacted rocks might be sufficiently dense to entrain large volumes of solid particles in a rapidly moving cloud. Mackin (1969) favors formation of true *nuées ardentes* and attendent emplacement of ignimbrites as a volcanic phase integrally related to the impact event.

AGE RELATIONS

Although we have established relative ages for the Imbrium, Humorum, and Orientale basins, the relative ages for many of the formations associated with these basins remain uncertain. A highly schematic and interpretive summary is shown in Figure VI-14. This is a simplified paleogeographic rendering, not a geologic map of relationships now

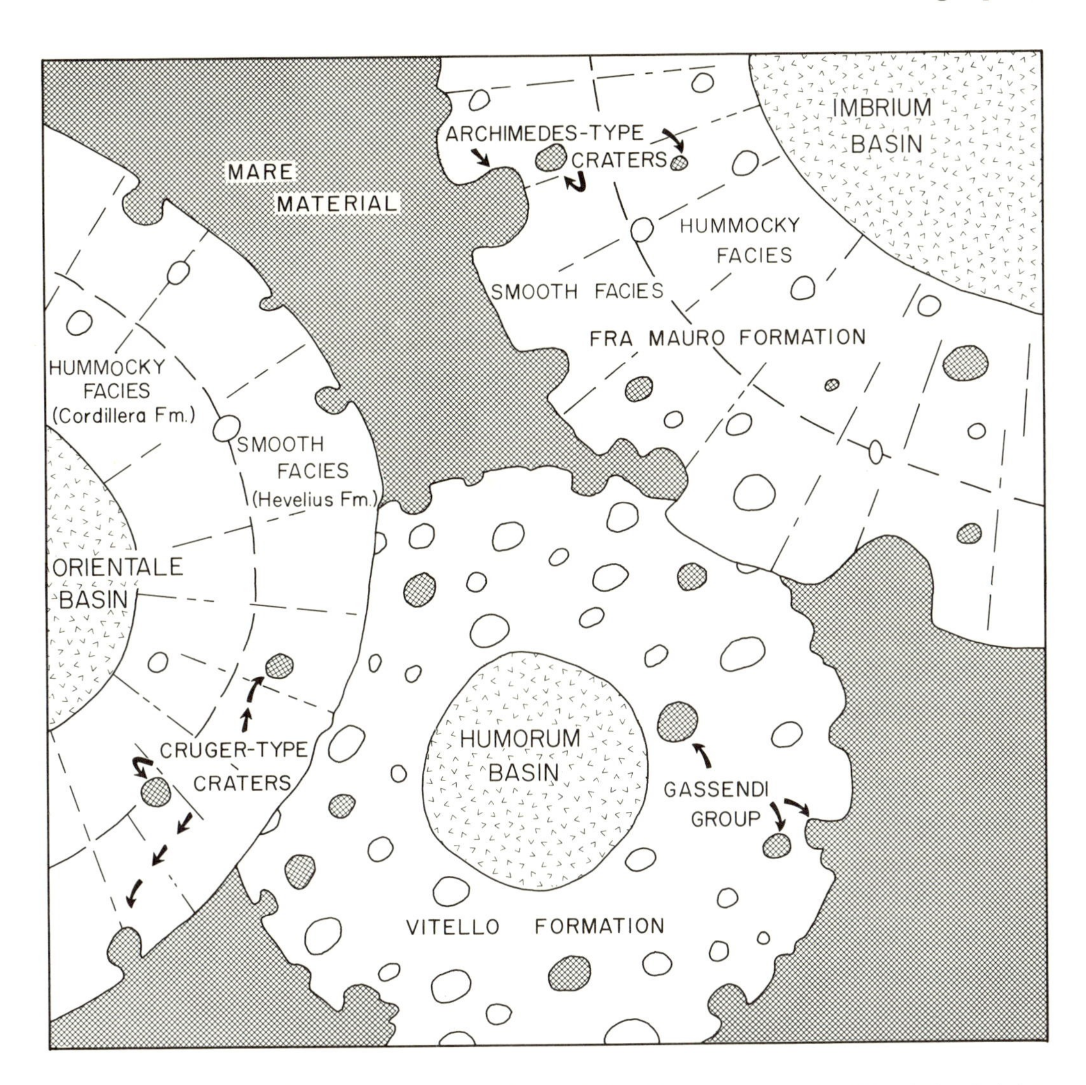

FIGURE VI-14. Schematic map showing idealized relationships between materials of the Imbrium, Humorum, and Orientale basins, assuming emplacement of no post-basin regional units other than dark mare materials. In fact, the rim facies of all three basins have been so obscured by subsequent events that it is impossible to map their original limits.

visible on the Moon's surface. In particular, the extent and overlapping relations of the three ejecta blankets cannot be observationally confirmed. The difficulty in identifying smooth distal parts of the Fra Mauro Formation and differentiating them from later light plains-forming materials has already been mentioned. A similar problem attends identification of the smooth facies of the Orientale basin ejecta. Even though McCauley (1967*b*) recognized a "relatively smooth material with a fine hummocky to braided texture" in the western part of the Hevelius quadrangle, Trask (1965), mapping directly south in Byrgius quadrangle, chose to define contiguous rocks more noncommittally as Imbrian or pre-Imbrian plains and terra materials. Since the smooth distal parts of the ejected blankets are so equivocally recognized, it is obvious that overlapping relations between adjacent ejecta blankets cannot be directly observed.

If one accepts the relationships shown in Figure VI-14, there remain a number of acceptable correlation schemes. Perhaps the simplest is shown in Figure VI-15. Emplacement of ejecta blankets associated with each of the three basins is staggered in time; the depositional events are essentially instantaneous. Once formed, the basin rim deposits are overlaid by later crater deposits formed by a presumably steady influx of impacting bodies. These craters are subsequently "tagged" by emplacement of the mare material, which is shown here as occurring over a short period of time and everywhere at the same time. Emplacement of the light plains-forming materials is interpreted as a

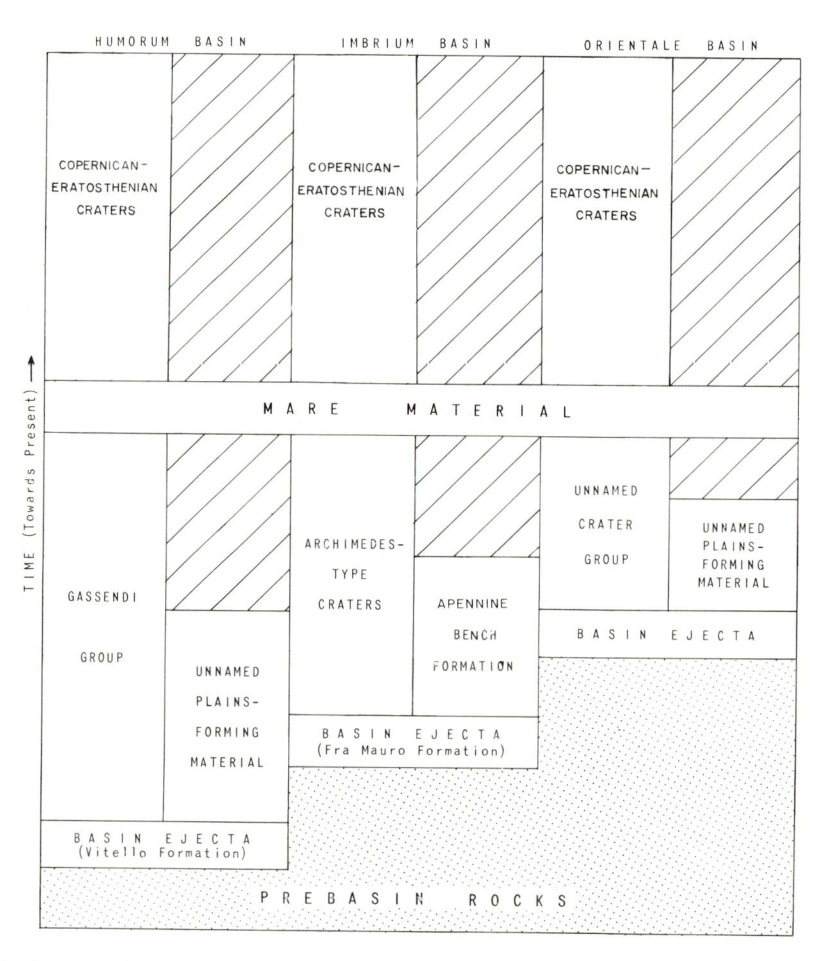

FIGURE VI-15. Hypothetical stratigraphic columns for units of the Humorum, Imbrium, and Orientale basins constructed on the assumption that mare materials are emplaced over a short period of time and everywhere at the same time. As a variation on this model emplacement of mare materials may have been staggered in time, closely following the formation of individual basins.

sequence of volcanic events directly following basin formation but separated in time from the later emplacement of mare material.

A second model (Fig. VI-16) postulates that the mare material is deposited everywhere at the same time, and that emplacement occurs over a long period of time. Considering the estimated thickness of the mare material, the latter assumption is entirely reasonable. In this model the post-basin cratering episode is not sharply terminated by emplacement of the mare material. Any crater formed before or *during* mare emplacement may show identical evidence of flooding. Undeniably, the probability of flooding is greater for a crater formed before the start of mare emplacement than it is for one formed near the top of this time interval.

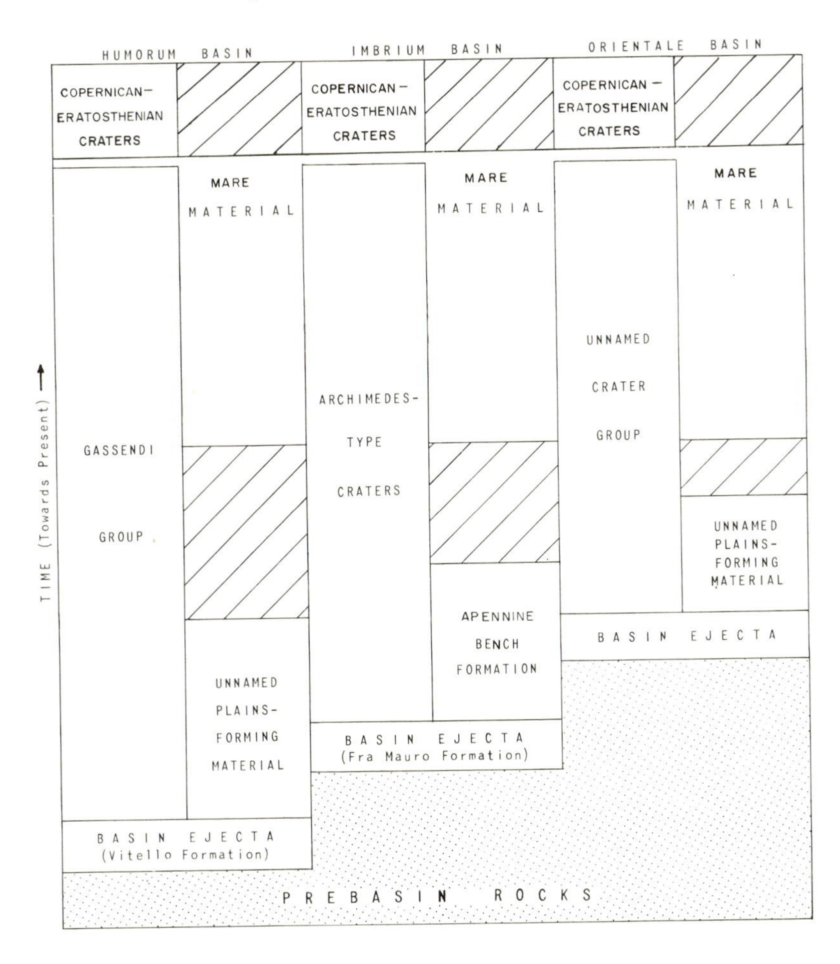

FIGURE VI-16. Hypothetical stratigraphic columns for units of the Humorum, Imbrium, and Orientale basins constructed on the assumption that mare materials are emplaced over a long period of time; the period starts and ends everywhere at the same time.

A third model (Fig. VI-17) introduces two additional complexities. It assumes that deposition of the mare material starts at different times in different places, and also that it extends to the present. Under these circumstances relatively recent craters of Copernican and Eratosthenian age might be flooded with mare material along with craters of pre-Imbrian age. Establishment of relative ages for these craters becomes extremely uncertain, so much so that one would be hard pressed even to rank the numerous probabilities in any order of priority.

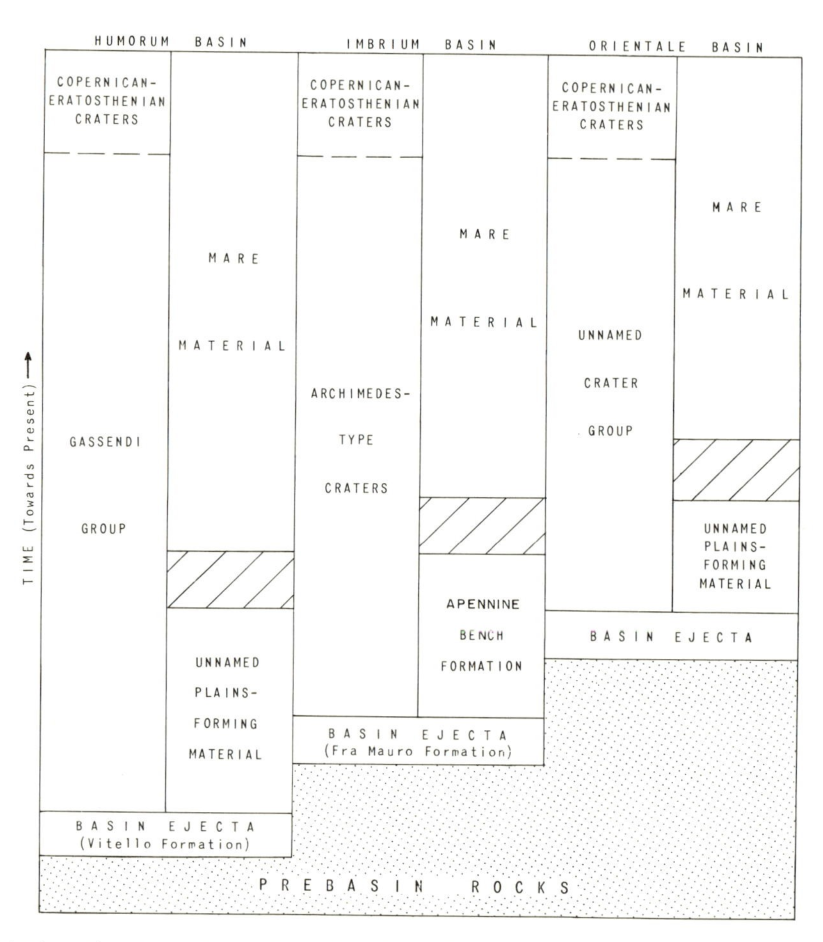

FIGURE VI-17. Hypothetical stratigraphic columns for units of the Humorum, Imbrium, and Orientale basins constructed on the assumption that mare materials start to form locally at different times and that deposition continues to the present.

At present there is no information to indicate which one of these models is most probable. Occam's razor—the maxim that assumptions introduced to explain a situation must not be multiplied beyond necessity—would dictate a choice of the first model. However, a more realistic understanding of the complexity of the natural world would suggest one of the latter two models. In any event, it is clear that the key to the solution is establishment of the depositional age—or ages—of the mare material.

Crater Stratigraphy

CHAPTER

VII

CRATERS VERSUS BASINS

IN the two preceding chapters we have devoted attention to stratigraphic units which have significant lateral extent at regional scales. Notable examples are the Procellarum Group which underlies most of the mare regions and the Fra Mauro Formation which mantles the central part of the Earthside hemisphere south of Mare Imbrium.

Arguing from a presumed similar impact origin for large basins and for many of the small craters, one would expect a similarity of stratigraphic features between the two groups. As previously pointed out, similarities do, in fact, exist. Both basins and craters are surrounded by ejecta deposits with inner concentric and outer radial facies. However, if there are similarities in kind, nonetheless there are differences in degree. Multiple concentric structural rims are especially prominent around basins. Extensive ejecta deposits are present around some basins and are assumed to be present around others. Their unique identification is frequently difficult. Basin interiors are characteristically flooded with dark mare materials. Basins displaying these several features include Orientale, Imbrium, Crisium, Humorum, Nectaris, Serenitatis, and Fecunditatis.

Craters are characterized by a single steep-walled cavity and by ejecta deposits with an intricate but clearly decipherable pattern of interfingering facies: hummocks, radial braids, rays, and secondary crater clusters. Mare-like flooding confined to individual craters is relatively rare. Many craters, notably some in the Southern Highlands, appear to be internally flooded, but with lighter Cayley-like materials.

Large craters such as Copernicus, Tycho, and Aristarchus show complexities intermediate between those of small craters and large basins. In Tycho, for example, the floor is covered with wormy, crackled material which resembles "volcanic" or "shock-melted" material revealed in the Orientale basin. The walls and inner rim of Tycho are dissected by concentric lineaments which imply a structurally complex terrain. Just as is the case around most basins, rim materials around Tycho contain a diverse assemblage of probable missile ejection, base-surge, and volcanic deposits. Identification and mapping of these subunits proves much more difficult than for smaller craters.

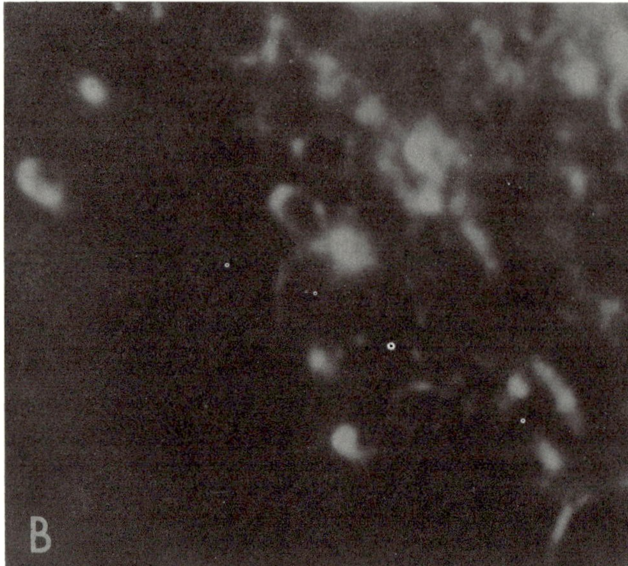

FIGURE VII-1. Two telescopic photographs of the eastern edge of the Humorum basin. (a) Low Sun-angle accentuates topographic relief (Catalina Observatory). (b) Full Moon photograph showing regions with high albedo in and around topographically sharp craters identifiable in (a) (U.S. Naval Observatory).

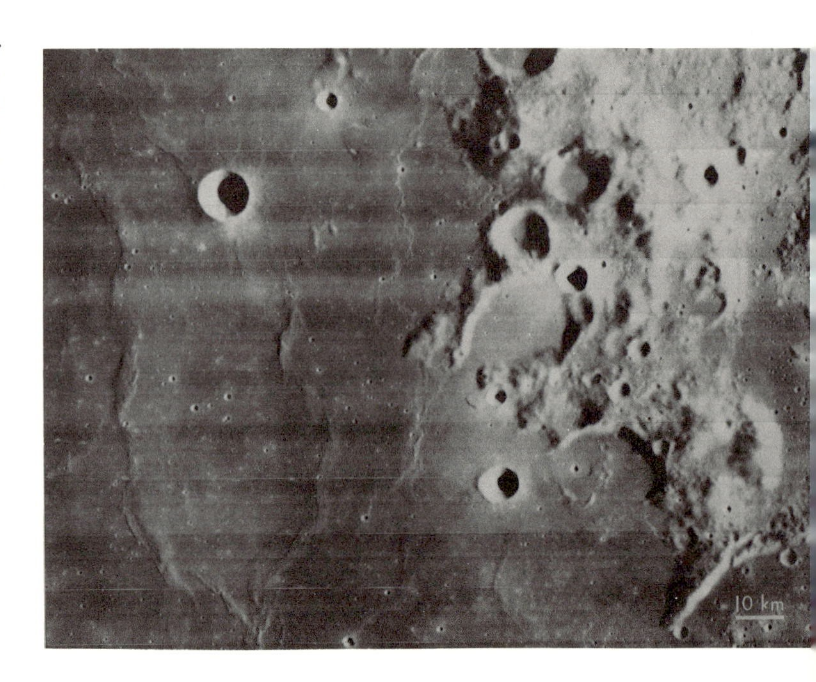

FIGURE VII-2. Orbiter photograph of same approximate region shown in Figure VII-1 (Lunar Orbiter Photograph IV 137H).

CRATER DEPOSIT NOMENCLATURE

The stratigraphic convention in mapping craters has been to consider all craters formed in a limited interval of time to constitute either a system or a group. Deposits associated with a particular crater constitute an unnamed formation. Figure VII-3 illustrates these conventions. Four large craters are included within the Gassendi Group. As previously discussed, the associated time interval is limited by emplacement of the Vitello Formation and the Procellarum Group. A second suite of craters is mapped as formations within the Copernican System. The age assignment here is made on the basis of crater morphology, particularly the presence of bright halos around craters (Figs. VII-1, 2, and 3).

FIGURE VII-3. Geologic map of eastern edge of the Humorum basin. Photographs of the same approximate region are shown in FIGURES VII-1 and 2. Note the temporal classification of crater materials according to superposition, topography, and albedo (after Titley, 1967).

COPERNICAN — Cc — **CRATER MATERIALS**

CHARACTERISTICS:
Bright halo or rayed craters with high albedo. Crater rims are sharp. Rim materials are smooth-appearing around smaller craters and hummocky with subconcentric and subradial ridges around larger craters.

RIM CRESTS

IMBRIAN

MARE MATERIAL

IMBRIAN AND PRE-IMBRIAN — Iplg — **GASSENDI GROUP**

CHARACTERISTICS:
Crater material that is partly or wholly covered by mare material. Craters have irregular and discontinuous rim crests.

PRE-IMBRIAN

UNDIFFERENTIATED MATERIALS

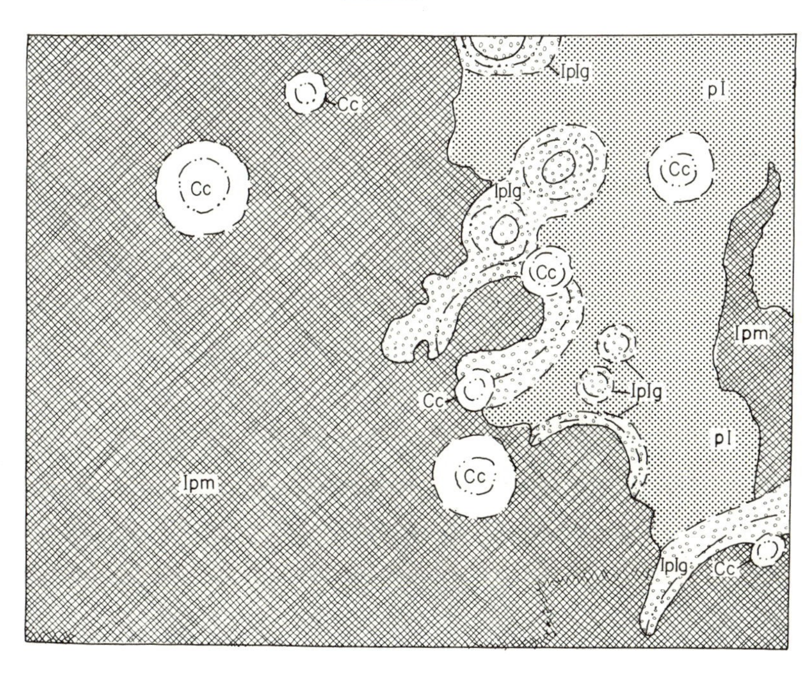

It would be inaccurate to justify these conventions in terms of accepted stratigraphic procedures. The principal difficulty lies in the fact that stratigraphy normally deals with vertically stacked rocks. More often than not terrestrial sedimentary rocks are underlain and overlain by other rocks which define a significantly small time interval for deposition of the enclosed strata. In contrast, lunar craters of closely related age are randomly disposed on units of different age and lithology. They commonly are not overlapped by younger deposits. For these reasons it is a hopeless task to determine age by superposition for *each* crater formation.

A second novelty in crater stratigraphy concerns the depositional mechanism. On Earth most sedimentary units are deposited by extensive, laterally continuous current systems, either water or air. But craters are essentially "point" deposits associated with a laterally discontinuous energy input. Only in exceptional situations do crater deposits overlap each other to give some semblance of lateral continuity. Earth analogs for this type of stratigraphy are rare. There are many examples of irregular and discontinuous distribution of facies, as in volcanic terrains and in reefs, but here the discontinuous facies are usually enclosed in other sediments which provide a rudimentary temporal control.

Put very simply, the problem is that most recognizable lunar crater deposits are bounded by free surfaces except on their lower side (Fig. VII-4). Conventional stratigraphic units are bounded on all sides by other rock units (Fig. VII-5). Exceptions, such as recent sediments which are not yet buried, are easily identified and interpreted.

There is, however, a valid basis for considering crater deposits stratigraphically, especially if we assume that most craters form by one process, such as impact. In that case deposits for all craters will have essentially identical properties at the time of their

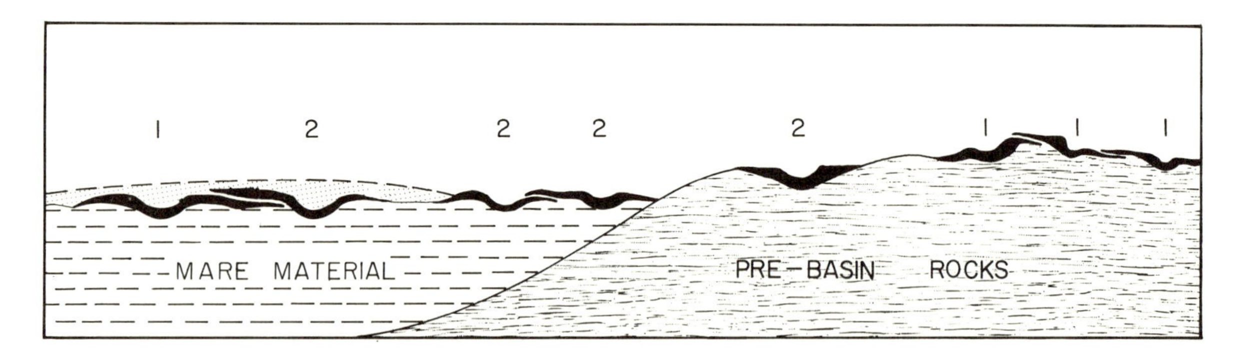

FIGURE VII-4. Schematic stratigraphic relationships for crater deposits in a hypothetical situation where, by independent criteria, craters numbered 1 are known to have formed in a brief interval of time and craters numbered 2 are known to have formed much more recently and also in a brief interval of time. Using this independently determined information it is possible to state that both cratering episodes follow emplacement of mare material and precede deposition of the regional ejecta blanket shown by stippled pattern. Superposition relations are consistent with this situation, but they certainly do not suggest it as a unique solution.

FIGURE VII-5. Schematic stratigraphic relationships for terrestrial sedimentary rocks. The present distribution of rocks is discontinuous because of erosion, and deposition of patch reefs was discontinuous in space and time; nonetheless correlation is possible because of easy reconstruction of the enveloping sandstone-limestone-shale sequence.

formation in the sense that they have responded to similar depositional mechanisms. As the craters age, they will be subjected to relatively uniform degrading processes: slumping, smaller-scale impact, and radiation sputtering. Consequently, the morphologic appearance of crater deposits will systematically vary with age, just as fossil assemblages vary in sedimentary rocks or radioisotopes vary in plutonic rocks as a function of age. The obvious conclusion is that relative crater morphology provides the most informative technique for stratigraphic classification of craters. This argument is developed further in Chapter X. Superposition relations are extremely valuable but are so imprecisely and infrequently displayed that their chief value is in providing critical temporal tie lines between two pre-existing stratigraphic columns, one for blanket deposits and a second for crater deposits.

SUBDIVISION OF CRATER DEPOSITS

Deposits associated with a single crater are commonly subdivided. These subunits are members within a single formation. The members are informally designated "materials," and the same informal member occurs in many separate formations. The identification of members depends greatly on evidence obtained from study of craters on Earth. As more has become known about craters close at hand, there has been a corresponding development of stratigraphic terminology for lunar craters.

On early maps crater deposits were separated into two subdivisions: crater rim material and slope material. Rim deposits were characteristically mapped over the rim crest and partially down the inner crater walls (Fig. VII-6). This was in accordance with relationships observed and inferred in Meteor Crater and nuclear craters (Shoemaker, 1962).

The inner crater is distinguished by an unusually high albedo (Fig. VII-7). Bright rays and halos seen *outside* craters are best explained as disruption of old surface material and exposure of lighter soil or bedrock. The same explanation was extended to deposits *inside* craters, these being interpreted as freshly exposed bedrock and recently accumulated talus deposits. This interpretation accords with the observation that the brightest areas commonly occur along steeper portions of the crater wall. Orbiter and Surveyor photographs provide vivid confirmation of the features just described. Cliffs of bare rock, faulted and slumped blocks, and talus aprons all can be identified along those crater walls which are unusually steep.

FIGURE VII-6. Geologic map of the crater Kepler. Both map and legend are adapted from Hackman (1962). The crater is assigned a Copernican age on the basis of its fresh topography and bright rays. Topography and albedo for the same approximate region are shown in FIGURE VII-7.

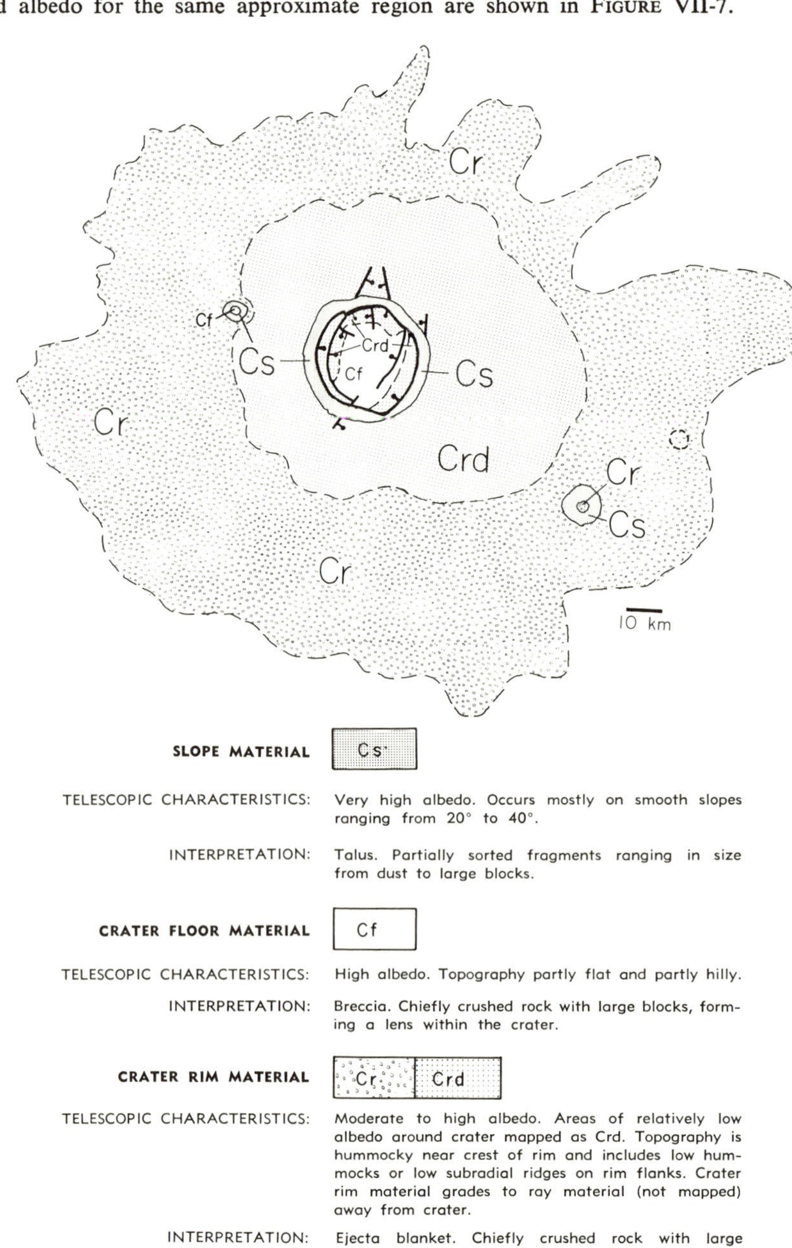

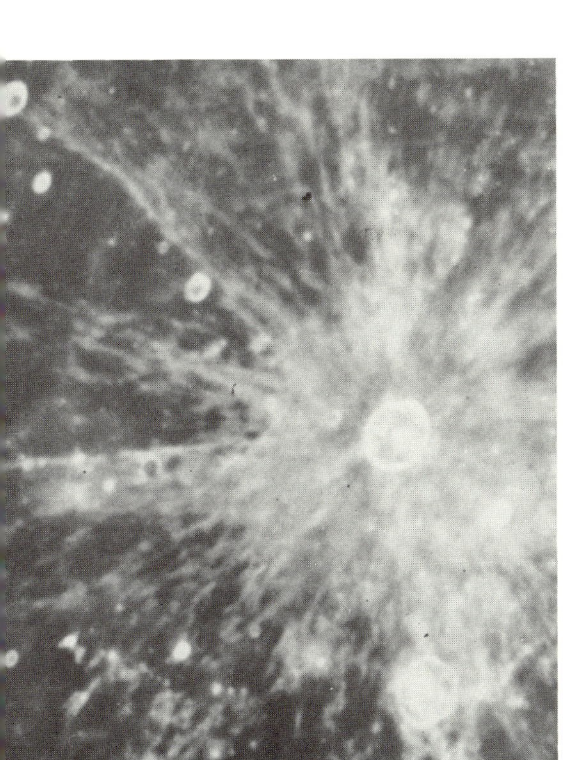

FIGURE VII-7. (a) Full Moon telescopic photograph (U.S. Naval Observatory) and (b) Orbiter photograph (Lunar Orbiter Photograph IV 138H) of the crater Kepler.

The areal distribution of slope materials is slightly anomalous. Their source is included within the slope unit and may, indeed, constitute the major mapped portion of the unit. For example, steep cliffs from which talus has been stripped are part of the slope member. Paradoxically, high-albedo materials typically do not spread across the crater floors as one might expect from knowledge of terrestrial sedimentary aprons at the base of steep scarps.

Because slope modification and associated exposure of light materials are thought to be recently active processes, slope materials are typically mapped as Copernican in age. This accounts for what at first glance appears to be a stratigraphic inconsistency on the geologic maps. Many craters, regardless of their primary age, have Copernican-age slope materials as one of their stratigraphic units.

The initial two-fold division of crater materials into slope and rim material eventually proved to be artificially restrictive. The contact between the two subunits midway down the inner wall was more often inferred than observed. Furthermore, many crater walls, unrelated in appearance to rims and rim deposits, were not sufficiently bright to map as Copernican. Finally, continuing examination of fresh missile craters demonstrated that crater walls were covered with a heterogeneous mixture of finely and coarsely comminuted rock, more or less in place. For these several reasons a third stratigraphic subdivision was introduced, crater wall materials. The contact between rim and wall deposits generally is placed along the rim crest, an easily identifiable morphologic feature (Figs. VII-8 and 9). Wall materials cover all parts of the wall not bright enough to be mapped as Copernican slope materials.

FIGURE VII-8. (a) Full Moon telescopic photograph (U.S. Naval Observatory) and (b) Orbiter photograph (Lunar Orbiter Photograph IV 85H) of the crater Plinius.

(A)

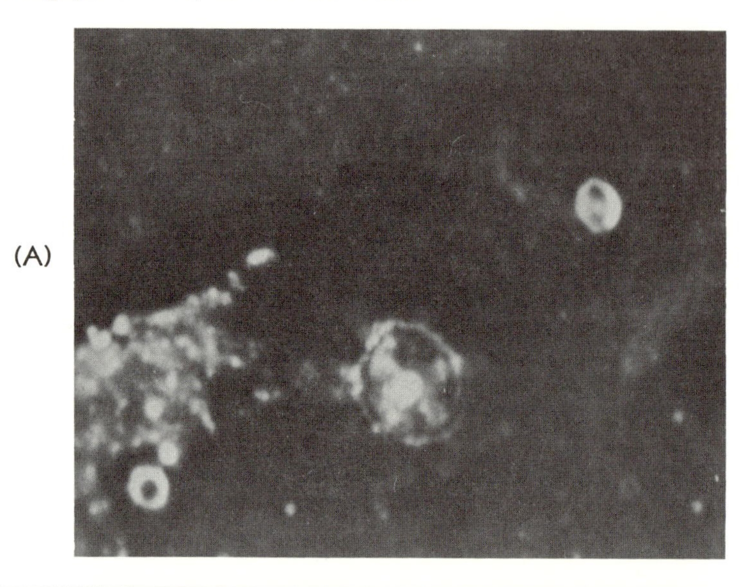

(B)

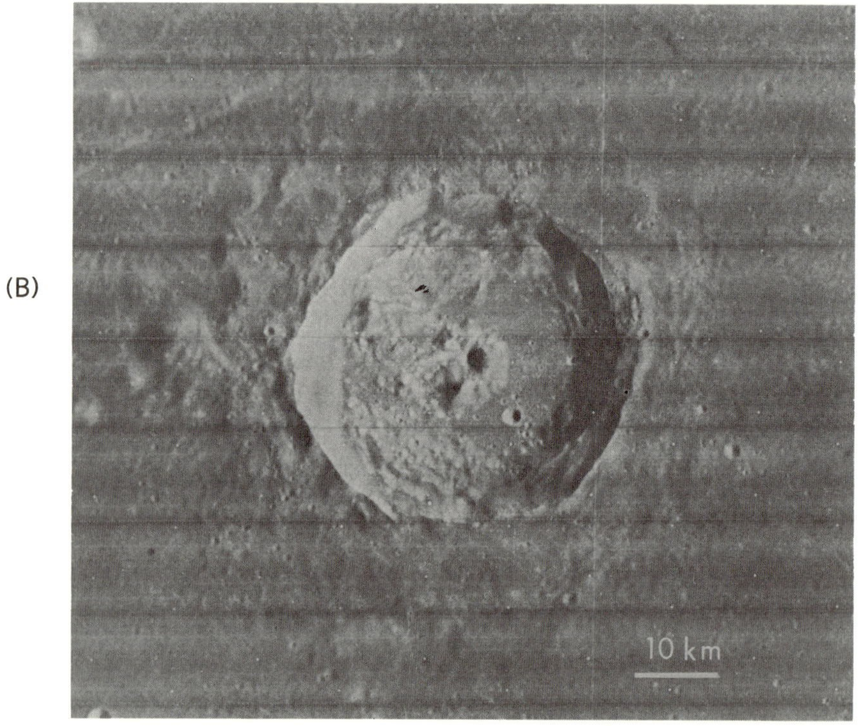

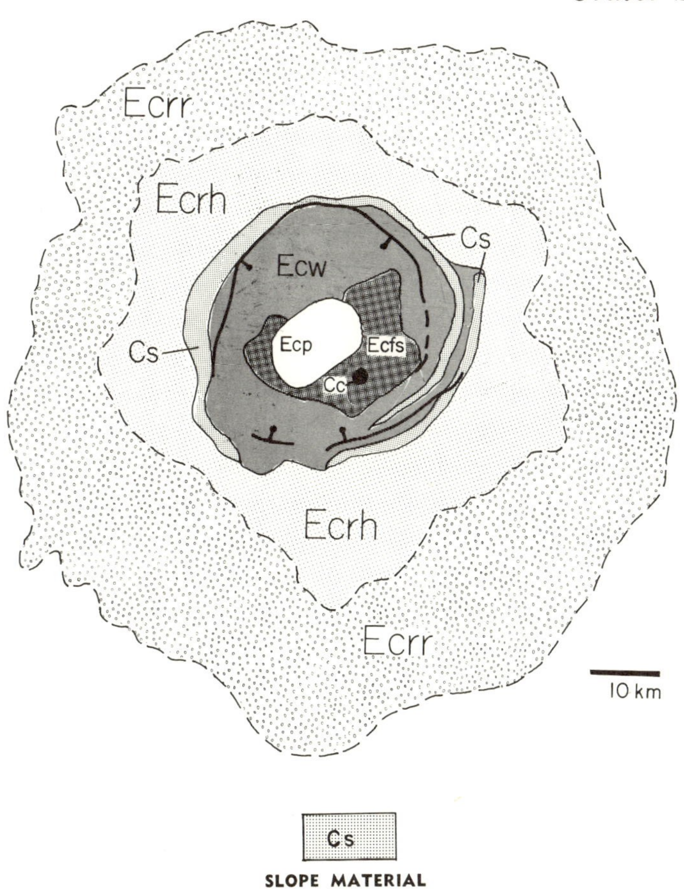

Cs

SLOPE MATERIAL

CHARACTERISTICS:
Very high albedo. Occurs on steep slopes, mostly on the inner walls of craters.

INTERPRETATION:
Freshly exposed rock, probably both talus and bedrock uncovered during talus formation. Process of talus formation may still be continuing.

MATERIALS OF RAYLESS CRATERS

CHARACTERISTICS:

Ecrr, rim, radial. Surface has moderate local relief with low ridges subradial to the crater. Albedo intermediate and patchy.

Ecrh, rim, hummocky. Surface has high local relief; discontinuous hills and valleys roughly concentric with the crater in the inner part grading outward to a branching pattern. Albedo intermediate.

Ecw, wall. Hummocky topography on steep to moderate slopes inside crater; locally crossed by narrow benches roughly concentric to the crater rim. Albedo intermediate to low.

Ecfs, floor, smooth. Albedo intermediate to low.

Ecp, peak. Forms high ground at or near center of crater. Albedo intermediate.

INTERPRETATION:

Ecrr, stringers of impact-produced ejecta.

Ecrh, Impact-produced ejecta of variable thickness and fragment size at the surface grading downward to overturned and fractured bedrock.

Ecw, darkened talus and slumped rim material.

Ecfs, Either a later volcanic cover or a smoothed and eroded part of an impact-produced breccia lens.

Ecp, Lens of intensely brecciated bedrock uplifted just subsequent to impact.

FIGURE VII-9. Geologic map of the crater Plinius. Both map and legend are adapted from Morris and Wilhelms (1967). Because of the absence of bright rays and the presence of slightly subdued rim topography, Plinius is assigned an Eratosthenian age. Topography and albedo for the same approximate region are shown in FIGURE VII-8.

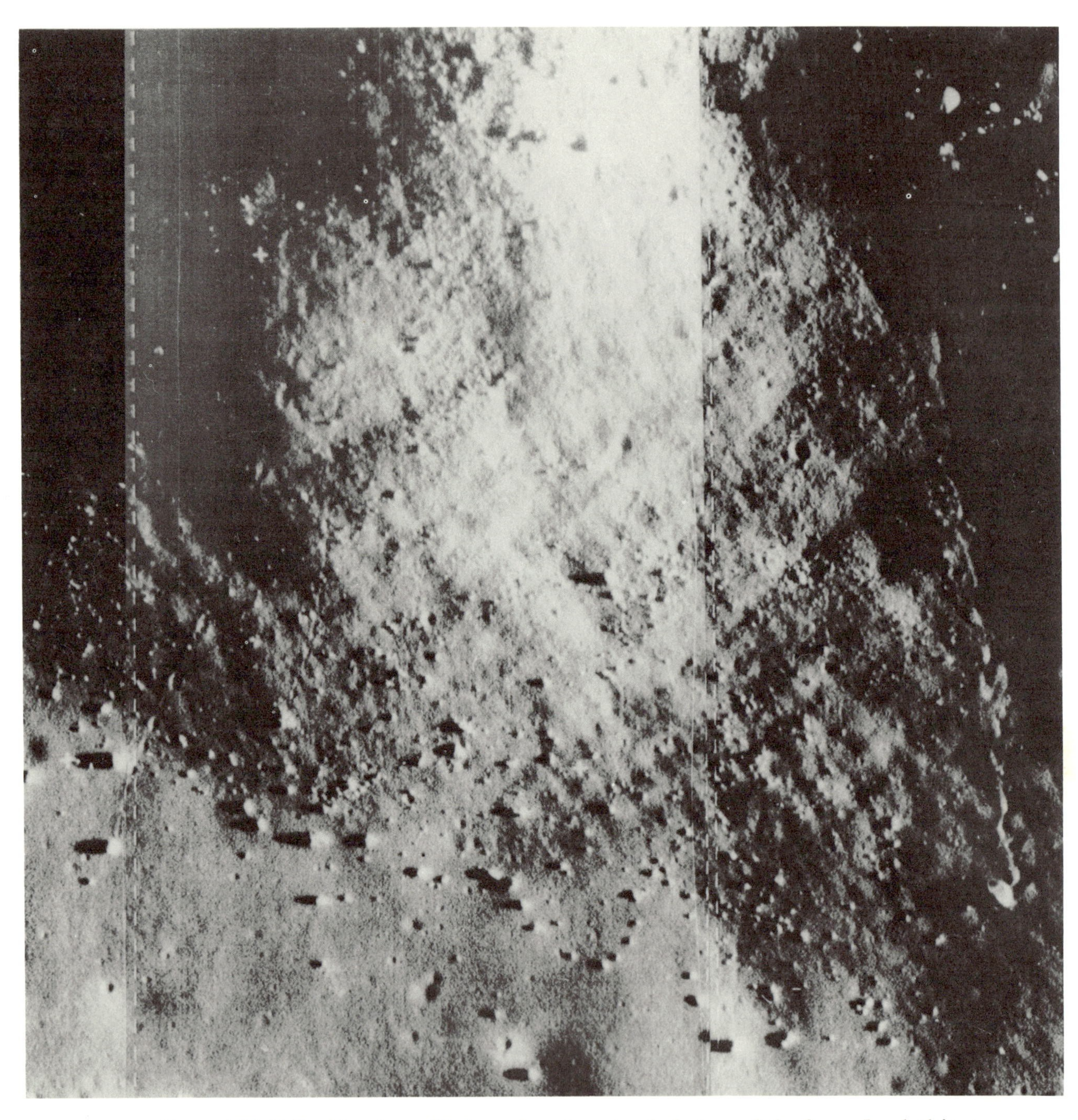

FIGURE VII-10. Boulders have slid down slope from top to bottom of the figure. Level plain appears in the lower part of the figure. Two boulders on the left and one on the far right are trailed by distinctive skid marks. Some boulders have sharply rectangular outlines, suggesting the presence of orthogonal joints or fractures in the rocks from which they were derived (Lunar Orbiter Photograph V 204H).

Many large craters have extensive, relatively flat interiors, clearly distinct from bordering walls. Where these interiors are hummocky terrains they are mapped as crater floor materials. By analogy with impact craters on Earth they are interpreted to comprise a central lens of breccia. Hummocky floor materials are commonly flooded with smooth, level plains-forming units. They are most logically interpreted as volcanic flows issuing from a vent inside the crater.

Some workers have interpreted the plains materials as an accumulation of debris from adjacent walls, but the precise importance of down-slope movement of sediment in alteration of crater morphology remain uncertain. The simple fact that particles on the lunar surface do move downhill is eloquently documented by photographs showing skid marks left by blocks which have skipped down sediment-veneered slopes (Fig. VII-10). However, Orbiter photographs contain very few examples of alluvial fans with morphology similar to fans which accumulate on Earth at the base of steep mountain fronts, especially in arid regions. Some craters do have curvilinear accumulations of materials, but these lack the details of terrestrial alluvial fans: apex at the point of origin and convex pattern away from the point of origin. Instead, the crescents have atypically large radii and intersect each other in complicated overlapping forms suggestive of internal origin (Fig. VII-11).

FIGURE VII-11. Interior of the crater Dawes showing overlapping swirls of material in the floor, possibly formed by erosion of materials from the walls, but more probably formed by extrusion of viscous lava from below (Lunar Orbiter Photograph V 70M).

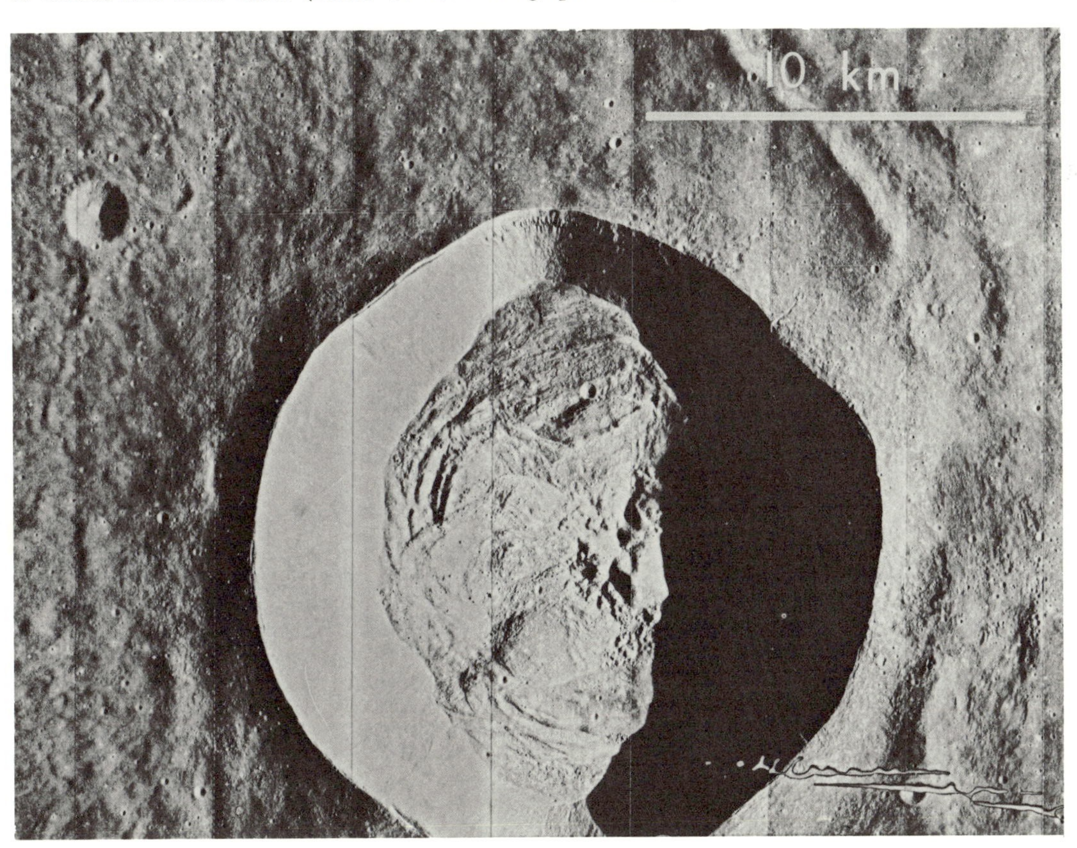

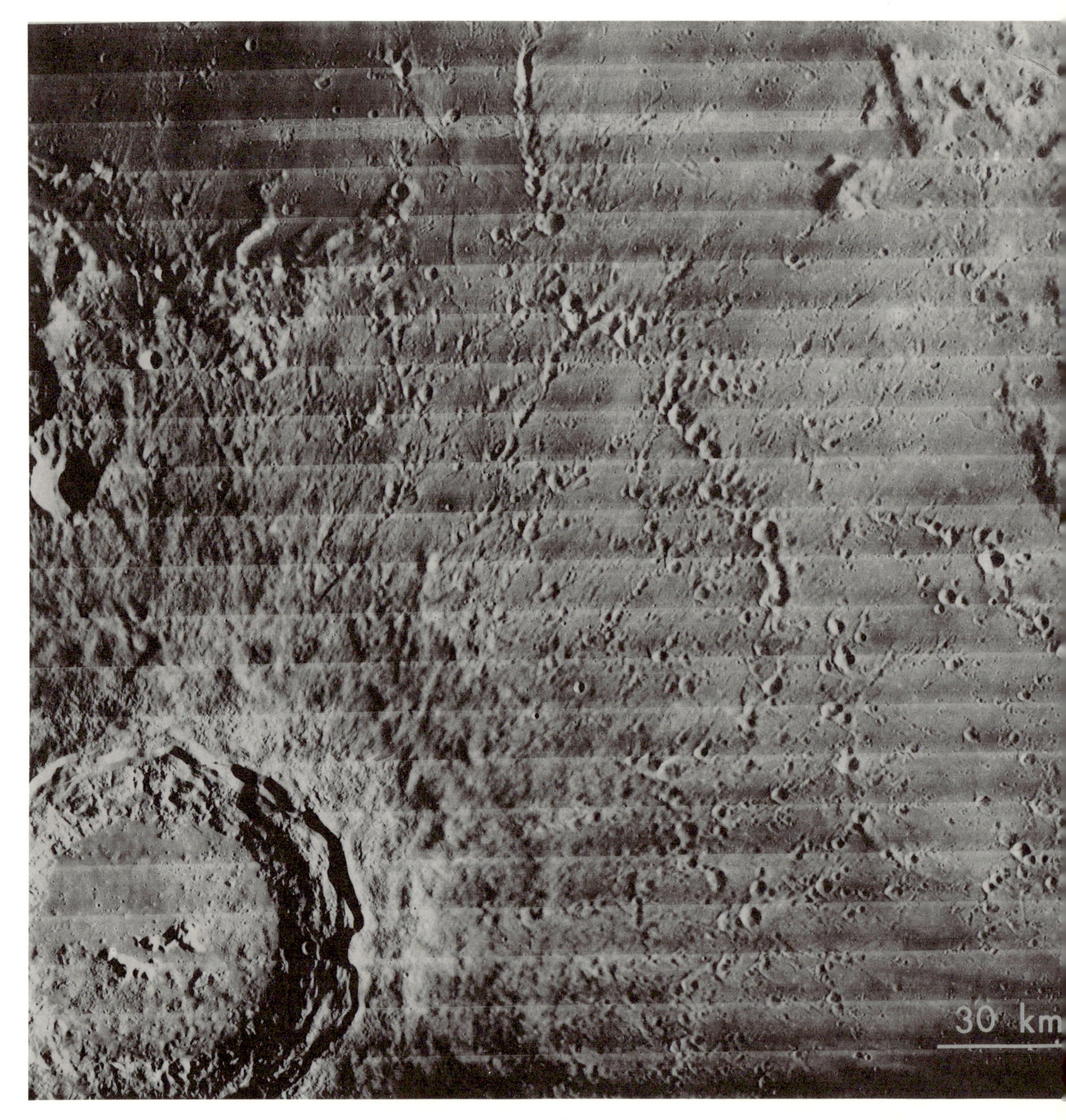

FIGURE VII-12. Rim topography northeast of Copernicus (Lunar Orbiter Photograph IV 121H).

Opposite page:
FIGURE VII-13. Geologic map of the several rim facies around Copernicus. Both map and legend are adapted from Schmitt, Trask, and Shoemaker (1967).

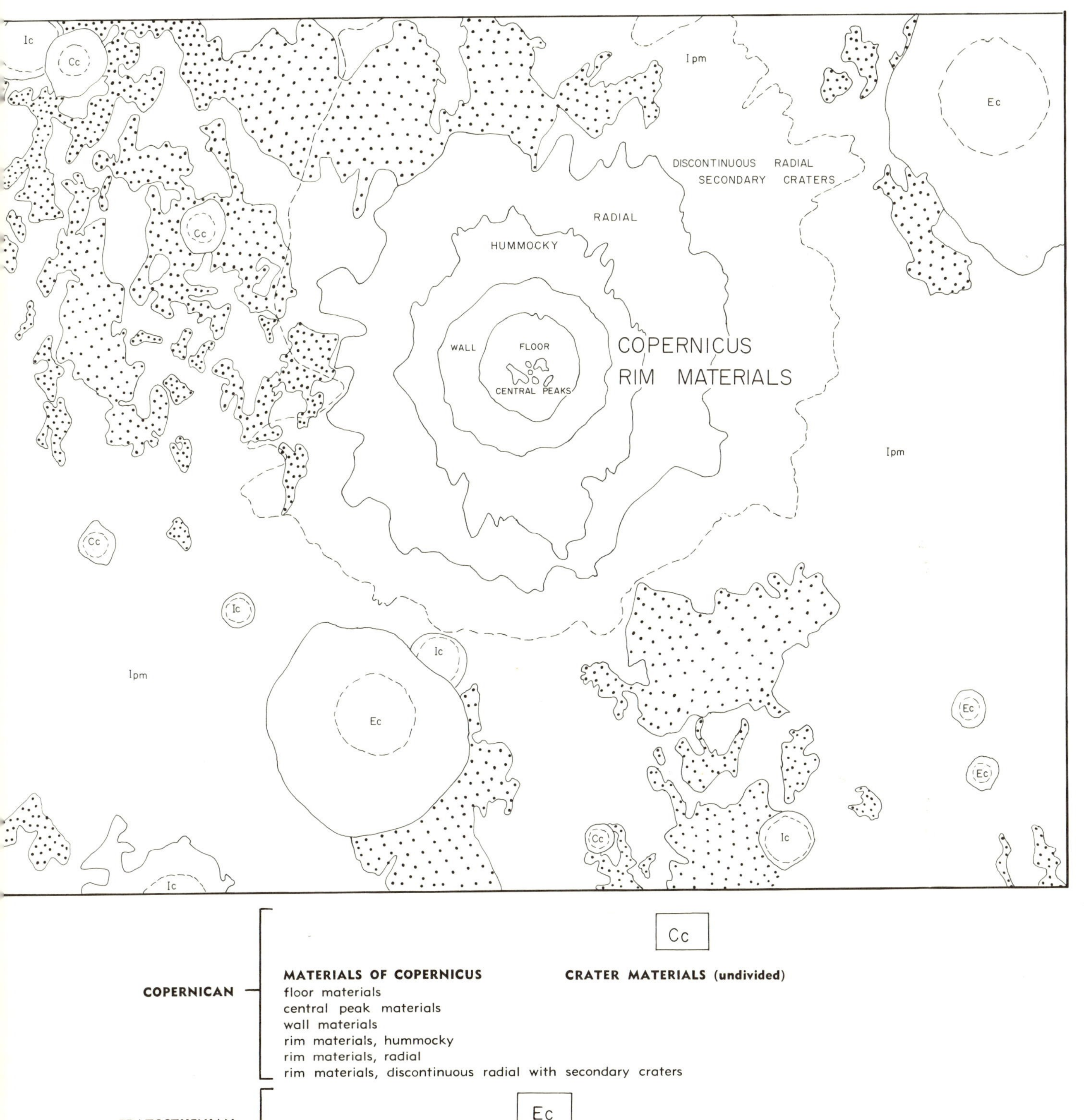
Ic
Cc
Ipm
Ec
DISCONTINUOUS RADIAL
SECONDARY CRATERS
RADIAL
HUMMOCKY
WALL
FLOOR
CENTRAL PEAKS
COPERNICUS
RIM MATERIALS
Ipm
Cc
Ic
Ic
Ipm
Ec
Ec
Ec
Cc
Ic
Ic

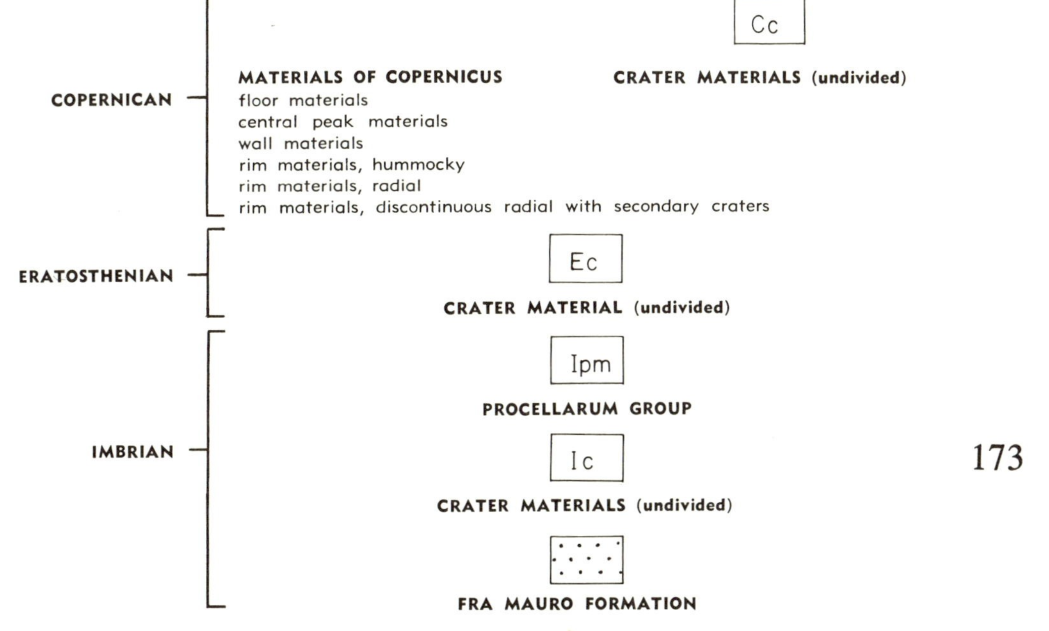
COPERNICAN
Cc
MATERIALS OF COPERNICUS
CRATER MATERIALS (undivided)
floor materials
central peak materials
wall materials
rim materials, hummocky
rim materials, radial
rim materials, discontinuous radial with secondary craters
ERATOSTHENIAN
Ec
CRATER MATERIAL (undivided)
IMBRIAN
Ipm
PROCELLARUM GROUP
Ic
CRATER MATERIALS (undivided)
FRA MAURO FORMATION

Gold (1966) has described an unusual model for lunar erosion based on the speculation that in sunlit areas of the Moon's surface, there is an electrical potential gradient extending upwards for several centimeters. He further speculates that dust grains which are dislodged by micrometeorite impact will have a significant charge, due perhaps to frictional electricity. Alternately, a charge may be induced by interaction with a hot gas associated either with meteorite explosion or with internal volcanic degassing. These charged particles will not fall back to the surface immediately but will float at that level where the combined electrical and gravitational potential is at a minimum. Floating particles with the appropriate sign will be able to glide downhill on an electron layer, finally settling to the ground when they became discharged. An interesting consequence of this mechanism is that the depositional surface would tend to be flat and featureless throughout an erosion cycle. In a large crater, for example, particles would slide down the interior walls to form a central pool of sediment. This pool would be continually smoothed out by micrometeorite bombardment, or perhaps by seismic shaking, which would cause charged particles to dance around on the surface, filling in any local depressions and planing off constructional sediment fans or cones. The level plains materials contained in the floors of many craters have an appearance consistent with this hypothetical depositional model.

Regardless of their origin many of the level surfaces are so little cratered that they must be of relatively recent origin. When they occur in relatively old craters, the plains units are assigned an age younger than the primary crater age to indicate a temporal—and probably genetic—difference.

The crater mapped in the greatest detail at the 1:1,000,000 scale is Copernicus (Figs. VII-12 and 13). Here rim materials are further subdivided into three facies: hummocky, radial, and discontinuously radial. The latter deposit is interpreted as a discontinuous blanket comprising both clots of sediment thrown out of the central crater and material evacuated from secondary craters. Still another hierarchy of subdivision involves discrimination of three albedo types within individual rim subfacies. The only explanation given for albedo differences is that they "probably reflect differences in the lithology of the rim material" (Schmitt, Trask, and Shoemaker, 1967). The equivocal wording suggests that stratigraphic subdivision has reached the limit of the interpretive base.

IN SITU SEDIMENTATION

Inclusion of crater deposits as stratigraphic units raises an interesting problem concerning the definition of sedimentation and its role in the formation of a rock-stratigraphic unit. In most terrestrial sedimentary rocks there is a significant lateral separation between source area and depositional area. Identification and correlation of sedimentary beds generally are accomplished without knowledge of the position or character of the source area. Indeed, solution of the "source problem" is a separate exercise, involving paleogeographic reconstructions and sediment dispersal patterns.

Crater deposits differ in that their identification usually goes hand-in-hand with identification of a source. Particles which make up the rim deposits have been transported significant distances from the source, but the same cannot be said for wall and hummocky floor deposits. These represent in-place modification of surficial rocks. Partial analogs are terrestrial soils which also form in place. Interestingly, the Code of Stratigraphic Nomenclature (1961, p. 654) recognizes soil-stratigraphic units but distinguishes them from rock-stratigraphic units.

> A soil-stratigraphic unit differs from a rock-stratigraphic unit in that it is formed for the most part *in situ* from underlying rock-stratigraphic units, which may be of diverse composition and geologic age . . . . Further, the characteristic features of soil-stratigraphic units are the products of surficial weathering and of the action of organisms at a later time and under ecologic conditions independent of those that prevailed while the parent rocks were formed.

In acknowledging the *in situ* generation of many crater deposits we have restated a point made earlier in this chapter. The keystone of stratigraphy is the superposition relationships of strata. It is impossible to achieve extensive superposition without extensive sediment transport. Therefore, *in situ* formation of stratigraphic units necessarily precludes a complex stratigraphy based on superposition.

Volcanic Stratigraphy

CHAPTER

TERRESTRIAL VOLCANOES

THERE are several regions on the Moon which, because of the presence of conical hills, domes, rilles, chain craters, narrow-rimmed craters, lobate scarps, and dark, smooth surficial materials, appear to be underlain by extensive floods of lava. These areas raise special problems in geologic mapping and in stratigraphic subdivision. As in the case of craters, interpretation of lunar volcanic terrains depends heavily on analogies with features observable on Earth. For that reason the discussion should be introduced with a brief review: first, of terrestrial volcanic forms, and secondly, of terrestrial volcanic stratigraphy. The stratigraphy of flood basalts deserves special attention because similar rocks may underlie large portions of the lunar surface, particularly in the maria.

Volcanoes can be morphologically classified according to the type of activity. Interrelated controls are lava composition, temperature, viscosity, and amount of contained volatiles. Broad, gently sloping shield volcanoes are formed by the relatively quiet outpouring of successive floods of liquid lava (Fig. VIII-1). Strato-volcanoes are formed as the result of alternating episodes of quiet flooding and explosive ejection of particulate material. They comprise stratified sequences of pyroclastic sediments and lava flows (Fig. VIII-2). The slopes for such volcanoes are steeper than those on shield volcanoes, both because the lavas tend to be more viscous and because the pyroclastic deposits thin rapidly away from the central source. Continuous or episodic ejection of particulate material in the absence of liquid lava leads to the formation of cinder cones (Fig. VIII-3). Finally, emplacement of highly viscous magma may result either in an intrusive spine, sheathed by previously deposited sediments, or in an extrusive dome (Fig. VIII-4).

Rittman (1962) has pointed out that volcano types are roughly related to the composition of the magma. Basic magmas tend to be relatively hot and fluid, leading to the formation of shield volcanoes, or to strato-volcanoes and cinder cones in those situations where there is rapid, explosive loss of volatiles. As magmas become more acid they also tend to be cooler and more viscous. Resulting volcanic forms include steep-sided intrusive or extrusive domes. If the viscous magma is rich in volatiles, violent explosions may take place. Accompanying volcanic ash may be blown high into the air and distributed over large areas. In some situations a highly heated gas charged with incandescent ash particles flows downhill as a *nuée ardente*, or "glowing cloud," to accumulate after cooling as an ash-flow tuff or ignimbrite.

FIGURE VIII-1. Topographic and geologic maps of Hawaii. Contour interval on topographic map is 2,000 feet. The profile is shown with the same horizontal and vertical scale (adapted from Stearns and MacDonald, 1946).

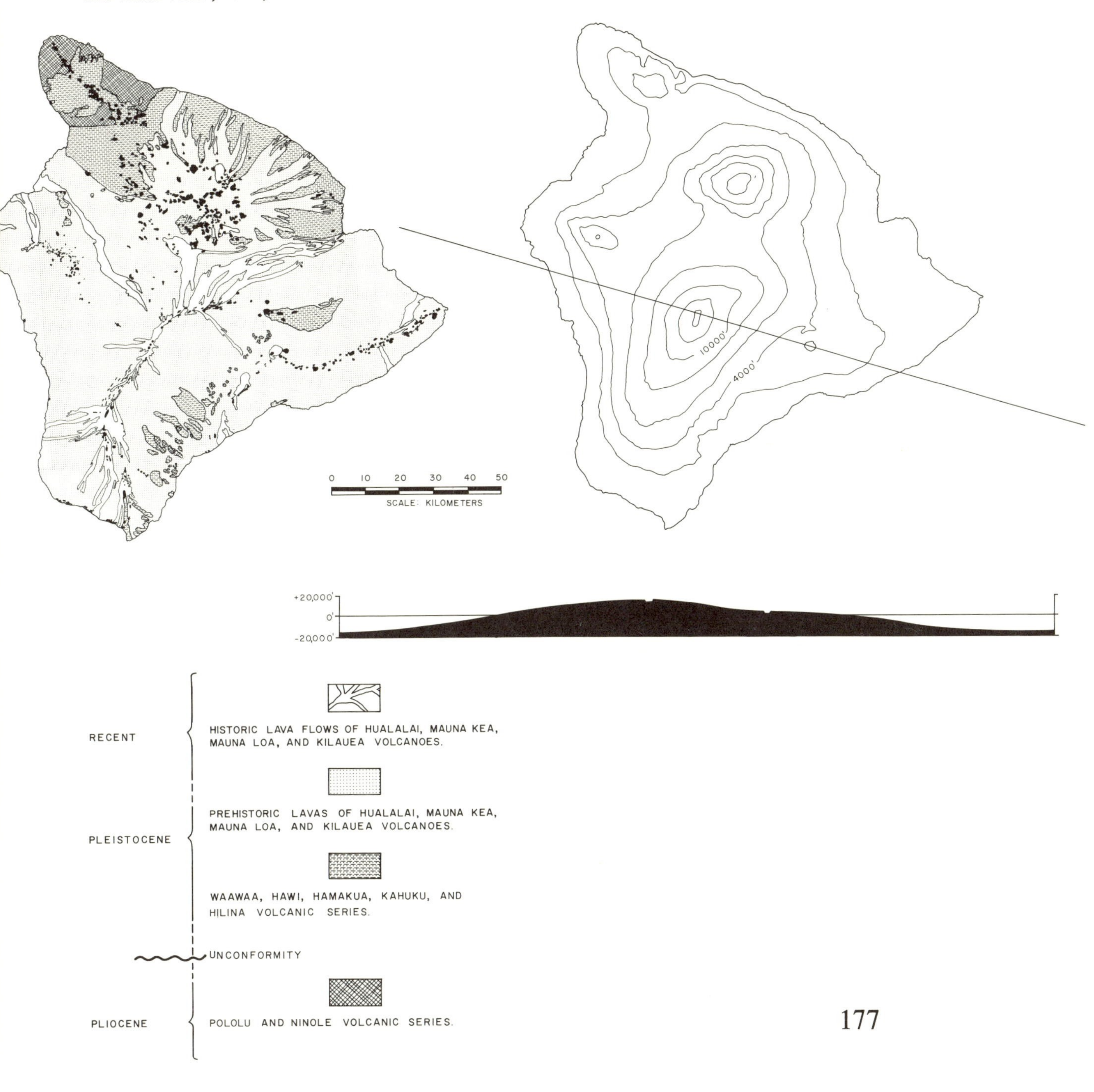

Most volcanic landforms are more complicated than the simple end members just described. Shield volcanoes display summit calderas, flank craters and rifts, and secondary cones. Alternation of magmatic episodes in strato-volcanoes may result in steep cinder cones superimposed on gently sloping shield-shaped volcanoes. The symmetry of many cinder cones is destroyed by fluid eruptions which breach retaining walls of loosely consolidated pyroclastic materials. Domes assume a variety of bulbous and platy shapes, a consequence of the pasty condition of the lava, the shape of the viaduct, the nature of the country rock, and the tendency of the magma to carry a shattered "skin" of overlying country rocks.

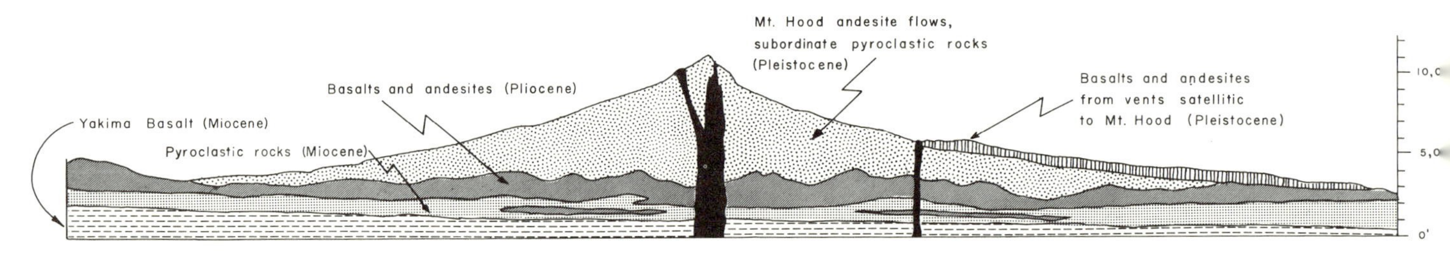

FIGURE VIII-2. Geologic cross section through the Mt. Hood area, Oregon. Horizontal and vertical scales are the same (adapted from Wise, 1969).

Fissure volcanoes are sometimes differentiated from central volcanoes (e.g., Rittman, 1962). Expectedly the principal difference is that shifting activity along a linear zone tends to produce coalescing flows, cones, and domes. On a large scale the accumulations are ridge-like, although individual central volcanoes can be differentiated at more detailed levels.

FLOOD BASALTS

A common contention is that the lunar maria are regions of vast basalt flooding. Several areas of comparable size on Earth are covered by flood (plateau) basalts, and for that reason, have attracted attention as possible terrestrial analogs for lunar maria. There is, however, a difference in topography. Maria are topographically low regions, but many terrestrial plateau basalts—as their name suggests—form topographically high surfaces on continental blocks. Regions of relatively recent activity are the Columbia River plateau and Snake River plateau of northwestern United States as well as a North Atlantic (Thulean) province comprising Greenland, Iceland, and the Hebrides. Older occurrences include the Deccan basalts of India (Cretaceous-Eocene); basalt sequences in India, Africa, South America, and Antarctica probably related in space and time (mid-Mesozoic); and several Precambrian lava sequences exposed on continental shields (Tyrrell, 1937).

One of the most extensive and best preserved areas of flood basalts is situated on the western margin of peninsular India (Fig. VIII-5). The geology of these rocks, called the Deccan lavas or traps, is summarized by Wadia (1953). The lavas attain a maximum aggregate thickness of 3,000 meters, but the usual thickness is about 600 meters.

FIGURE VIII-3. (a) Paricutin Volcano, Mexico. At this stage in its development the cone is about 500 meters above its original base. A smaller breached cone is in the left foreground. The low rounded slopes to the right of the breached cone and at the base of the main cone are the first lava flows from Paricutin, covered by ash (taken from Bullard, 1962). (b) Cinder cones in the San Francisco volcanic field, Arizona (courtesy of J. F. McCauley).

FIGURE VIII-4. (a) Andesitic lava dome in Guatemala. This dome rose from a fissure. The stiff, pasty mass reached a height of 350 meters in three years. At one point in its development a *nuée ardente* issued from a crack in the dome and rushed down an adjacent mountain side (courtesy of F. M. Bullard). (b) Rhyolite dome and surrounding pumice cone, Alaska (courtesy U.S. Geological Survey).

Individual flows average 5 meters in thickness but some reach 30 meters. The petrography of the rocks is notably uniform over the extensive area of their exposure.

Successive flow sheets are commonly separated by partings of ash or by thin detrital fossiliferous beds. The ashes attest to some explosive activity, but the bulk of lavas is thought to have flowed quietly from local vents situated along major fissures. This contention is based partly on positive evidence—the monotonous horizontality of the lavas over wide areas and the occurrence of dikes around the margin of basalt exposures where erosion has removed the extrusive rocks; and partly on negative evidence—the almost complete absence of volcanic cones and craters.

Stratigraphic division of the Deccan basalts into three vertically disposed sequences is based on the occurrence of fossiliferous interbeds present in both the upper and lower sequences but lacking in the middle sequence. Were this fossil information lacking, subdivision of the basalts on petrographic differences would be very difficult.

The most extensive flood basalts within the United States are situated on the Columbia River plateau. The rocks are Miocene in age, the chief component of a still larger volcanic terrain which covers major parts of Washington, Oregon, and Idaho and ranges in age from Eocene to Holocene. Preserved sections generally range between 300 and 600 meters in thickness (Waters, 1961).

FIGURE VIII-5. Distribution of plateau basalts in India and in northwestern United States.

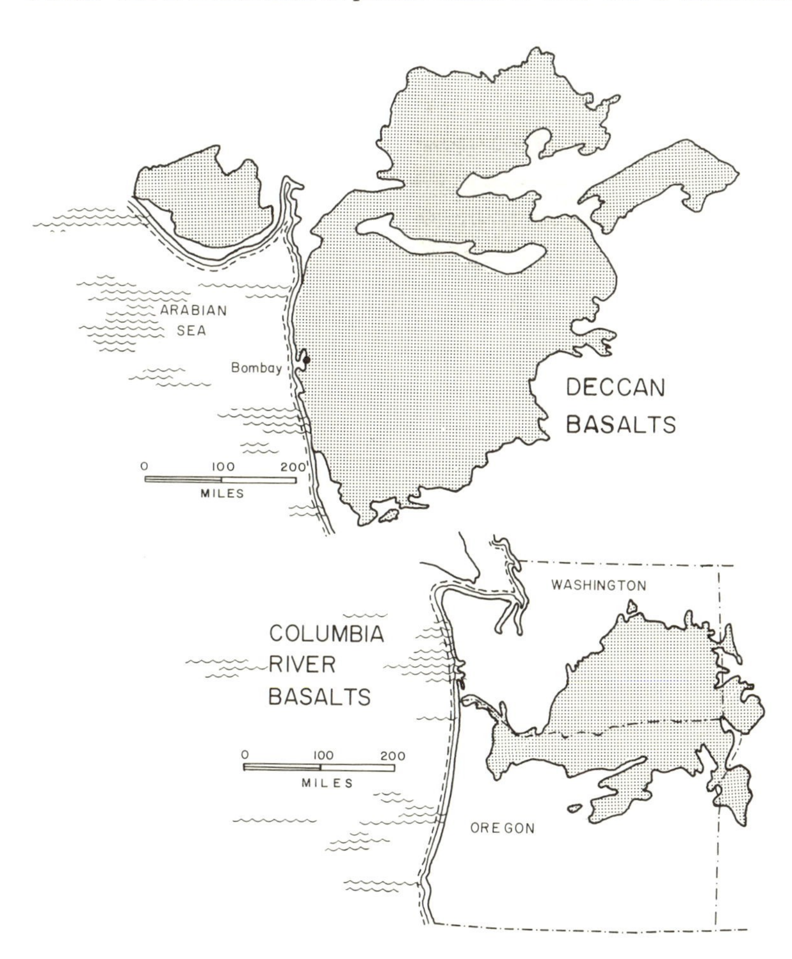

As in the case of the Deccan lavas the Columbia River basalts apparently erupted through fissures. Prominent dike swarms appear in restricted areas along the southern and western margins of the plateau. Dikes within a swarm commonly trend in a single direction (Waters, 1961). A definite connection between a dike and a flow has been described by Fuller (1927). The areal restriction of dikes to the south and west suggests that flows in the northeastern corner of the plateau may have travelled more than 150 km from their source vents. The hypothesis is supported by directional features in the flows which indicate movement towards the northeast (Waters, 1961). No shield volcanoes or cinder cones have been observed, although both structures are commonly associated with the Pliocene-Holocene flows which overlie the Columbia basalt.

Internal stratigraphic subdivision of the Columbia basalt has long posed a problem. It is a problem worth reviewing in considerable detail because it illustrates some of the possibilities—and stumbling blocks—which will probably attend on-the-ground classification and correlation of lunar volcanic rocks.

Detailed examination of twenty-eight sections and additional mapping of selected areas by Waters (1961) demonstrated no more than a rudimentary two-fold division for the Columbia basalt. The older lavas (Picture Gorge Basalt) are characterized by about 5 percent olivine and silica content between 47 and 50 percent. The younger lavas (Yakima Basalt) contain almost no olivine and have silica content of 53–54 percent.

Waters' basic subdivisions have been refined by Mackin (1961) and Schmincke (1967). Mackin presented a detailed stratigraphic section for the Yakima Basalt and the interfingering Ellensburg Formation in south central Washington (Fig. VIII-6), although he did not demonstrate the map distribution of the various units. Schmincke made a particularly thorough study of the uppermost four flows: the Umatilla, Pomona, Elephant Mountain, and Ward Gap Basalt members. All four can be traced throughout a map area approximately 200 km by 150 km in size. Each was apparently extruded as a single flow of vast extent. The Umatilla Basalt has an average thickness of 45 meters and extends over 5180 km^2. The Pomona Basalt is slightly thinner, 30 meters, but extends over a much larger area of 18,000 km^2. During flowage the lavas were more than 95 percent liquid. Variable cooling histories resulted in major variations in textures, mineral modes, and mineral compositions within a single flow. Despite these internal differences the four flows can be distinguished one from the other by primary differences in chemical composition.

The areal extents of the basalt sheets just described are by no means exceptional. Wheeler and Coombs (1967) have described a late Cenozoic olivine basalt exposed throughout Oregon, northeastern California, northwestern Nevada, and southwestern Idaho. The rock can be identified and correlated over great distances because of its unusual coarse and porous texture and uniform composition. The basalt originally covered an area of at least 3×10^6 km^2. Its thickness ranges systematically from 20 to 3 meters as it is traced from northwest to southeast. The regularly trending isopachs and the nearly constant thickness within small areas suggest that it is a single flow.

Of the several major areas of flood basalts present around the globe, Iceland is the only region where large volumes of lava have been accumulating in historic time. Several documented fissure eruptions are of special interest for the clues they contain concerning the mechanism of massive volcanic flooding.

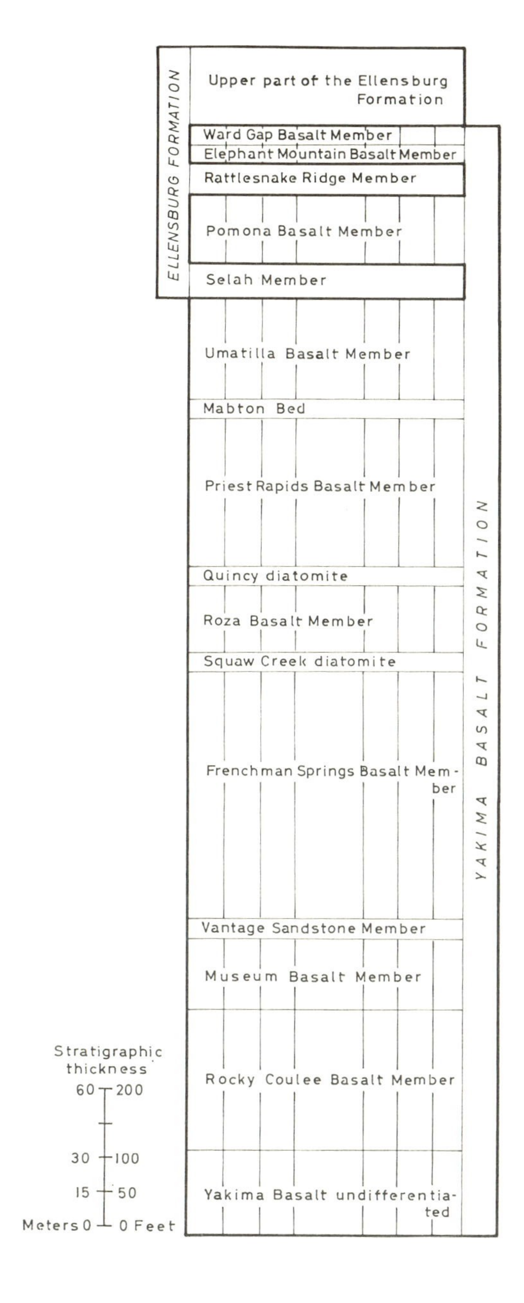

FIGURE VIII-6. Stratigraphic section of the Yakima Basalt Formation and Ellensburg Formation in south-central Washington. Basalt units are vertically ruled. Other units are clastic sedimentary rocks (from Schmincke, 1967).

FIGURE VIII-7. The Laki fissure. This linear zone of volcanic activity extends more than 40 km, trending in the same direction as northeast striking faults which are prominently displayed throughout southern Iceland. Following an initial eruption in 1783 the rift became filled with lava and a row of craters developed (from Barth, 1950).

The greatest recorded event is the Laki eruption of 1783. On June 11, following a week of severe earthquakes, floods of lava issued from twenty-two vents along a fissure 16 km long. Lava from this eruption filled the Skafta River Valley for a distance of 80 km and to an average depth of 30 meters (Bullard, 1962). Repeated outpourings continued throughout July and August. In all, a volume of 12×10^9 m^3 of lava devastated an area of 565 km^2, and 2×10^9 m^3 of ash spread over a still larger area (Barth, 1950). Eventually the rift became choked with frothy lava and a row of craters developed (Fig. VIII-7).

From studies elsewhere in Iceland, Rittman (1939) has recognized six phases in a fissure eruption. Initially a fracture develops, either in response to tectonic forces or to forces created by magma at shallow depth. Expulsion of excess gas is followed by lava flooding. The outpouring of lava leads to further decrease of pressure and to explosive gas emanation. Boiling finally abates and relatively placid lava lakes form in pre-existing craters. Some lava breaks through retaining crater walls but, being gas-poor, is too viscous to form extensive flows. In the terminal stages lava drains back into the fissures, leaving behind large cavities susceptible to collapse. Barth (1950, p. 137) has pointed out that faulting often accompanies eruption. Along one Icelandic fissure, foundering contemporaneous with the outpouring of lava is recorded in a precipitous fault escarpment more than 50 meters high.

This review of flood basalts provides a framework for speculation about the origin and mappability of lunar mare materials. First, we have seen that tremendous volumes of lava can be expelled over short periods of time and that they can travel a great distance before consolidating. Eruptions may take place along linear fractures without formation of large volcanoes. Secondly, we noted that regional stratigraphic subdivision of flood basalts is difficult because the petrography of individual flows is variable and differences between successive flows are seldom striking. Important aids in subdivision are interfingering fossiliferous sediments.

TERRESTRIAL ASH DEPOSITION

The importance of ash deposition in volcanic eruptions and the widespread occurrence of various types of ash deposits in volcanic sequences has come to light only recently (e.g., Smith, 1960*a*). Because the controlling factors are imperfectly understood, one might claim negative license to hypothesize any number of depositional situations. Geologists mapping the Moon, for example, have interpreted ash deposits as travelling great distances and as mantling large areas possessing considerable relief.

In the simplest case of ash formation particulate matter is thrown into the air in a dispersed state. The transportational history of large fragments is controlled by their initial velocities. In any event they accumulate close to the volcanic center. Smaller particles descend slowly through the atmosphere and during their descent are swept along by prevailing winds. Commonly the ejected particulate material forms a deposit with a plume-like shape—narrowest, thickest, and coarsest-grained at the source, and broadest, thinnest, and finest-grained at the downwind terminus (Eaton, 1964).

Ash falls, just described, are distinct from ash flows. The latter are composed of an intimate mixture of solid particles and hot gases, the entire mixture behaving like a fluid. The sedimentary distinction between ash falls and flows is somewhat analogous to that between suspension and turbidity current transport in water. In the first instance sedimentary particles move individually through the water (air); in the second case a turbid mixture of sediment and water (hot gas) flows beneath clearer, less dense water (air).

If ash flows are emplaced above a certain critical temperature, welding and recrystallization take place at the depositional site. The products, termed welded tuffs, superficially resemble conventional lava flows and commonly have been misidentified as such. Figure VIII-8 shows that there is a vertical succession of welding zones within a single ash flow. If the flow is sufficiently thick, a central hot zone becomes a site for dense welding. This is contained by a cooler envelope of partial welding, which is in turn bordered by marginal zones of rapid heat loss where there is no welding. A single cooling unit emplaced over irregular topography will compact during cooling, the amount of compaction being a function of the thickness of the deposit and the amount of welding. The resulting surface will be a subdued reflection of sharper irregularities present in the pre-flow rocks. Many cratered surfaces on the Moon display a "blurred" or "softened" topography which has suggested to some workers (e.g., O'Keefe, 1966) a terrain mantled with ash flows.

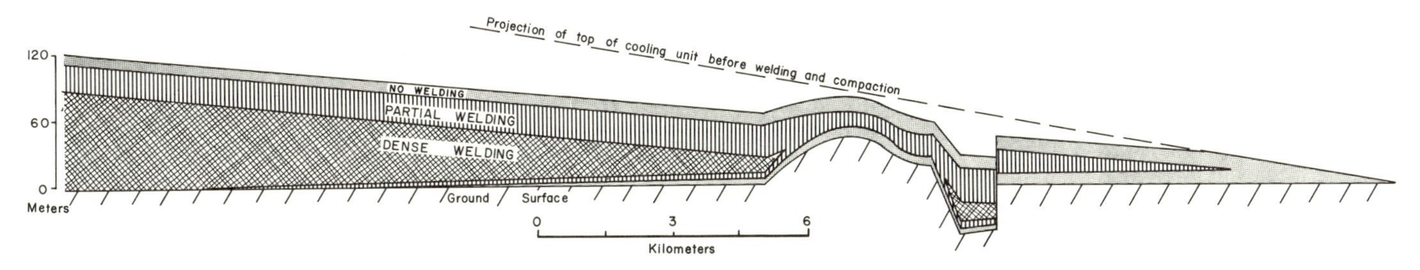

FIGURE VIII-8. Diagram showing distribution of depositional welding zones in an ash flow that has been emplaced on irregular topography. The zonal arrangements are further complicated by post-depositional crystallization (after Smith, 1960*b*).

Ash-flow deposits are generally poorly sorted, containing fragments ranging in size from large blocks to fine dust. Eight cumulative size distributions for particles at several different ash-flow localities have median values ranging from 1 mm to 1/16 mm (Smith, 1960*a*). The poor particle-sorting within most units is presumably a consequence of turbulent flow within the basal part of the *nuée ardente.* Fisher (1966) hypothesizes that drag resistance developed between a flow and the ground results in a transitional zone of low velocity at the interface. Fragments of all sizes follow irregular paths through the main turbulent portion of the flow. Consequently they enter the reduced velocity zone at random and are deposited together irrespective of size.

Although sorting within ash flows is very poor on a local level, there is some evidence suggesting that, on a regional scale, the median size decreases upward within a single flow (Tsuya and others, 1958) and laterally away from the source (Kuno and others, 1964; Fisher, 1966). The lateral variation is probably due to sorting of original sediment, but attrition of particles during transport may be a contributing factor.

There are two possible origins for the hot gases entrapped in ash flows. They may be primary volcanic gases expelled with particulate matter from the volcanic reservoir. Alternatively, they may be nothing more than atmospheric "air" trapped and heated by volcanic sediment, perhaps as it falls back to Earth close to the explosion site. In either case the gas-sediment mixture apparently maintains its integrity for long times without significant dissipation of gas in the surrounding cooler atmosphere. The behavior of similar mixtures injected into a vacuum like that on the Moon is a subject which has not yet been adequately investigated.

The analogy mentioned previously between ash flows and turbidity currents is especially appropriate when one considers the long controversy concerning the speed at which turbidity currents flow, the topographic gradients necessary for their formation and perpetuation, and the distances which they travel. Similar differences of opinion attend estimates for behavior of ash flows. One of the most extensive and well-documented depositional blankets is that around Aso caldera, Japan (Fig. IV-40). Even here there is no evidence that the flows have mantled the entire countryside. Rather they are preferentially displayed in valleys which apparently served as channels for the downward course of the flows towards sea level. The volume of ash associated with the Aso caldera is 18×10^7 m^3. Still larger volumes, up to 10^{10} m^3, are estimated to be present in certain volcanic provinces such as the Yellowstone region (Boyd, 1957) and the San Juan Mountains of Colorado (Larsen and Cross, 1956). These unusually large ash-flow fields are thought to be associated with multiple calderas (Smith, 1960*a*).

As previously mentioned, some interpreters of the lunar scene picture ash flows as having less of a digital pattern than the Aso caldera deposits and, instead, forming a continuous mantle over both highland and lowland. Perhaps this is possible in an environment where negligible atmospheric drag and relatively small gravitational forces tend to make initial velocities less susceptible to change than in the Earth environment. However it might also be argued that the gas-particle mixture would be deprived of its gaseous content immediately upon reaching the vacuum of the lunar surface and that, lacking fluidity, the particulate matter would not travel far.

MAPPING VOLCANIC TERRAINS WITH AERIAL PHOTOGRAPHS

A reasonable test of the thesis that it is possible to map lunar terrains of probable volcanic origin with Orbiter photographs is to see how much can be determined from aerial photographs of terrestrial volcanic provinces.

In some cases lava flows have individually diagnostic surface textures and superposition of successive flows is apparent (Fig. VIII-9). Typically, more complicated situations prevail, such as those shown in Figure VIII-10. In the region shown in this figure, modification of original surfaces by erosion and by covering with vegetation and snow is not great; even so, it blurs distinctions between map units. Those volcanic features which can be most easily identified are constructional forms. One gets a clear picture of their lateral arrangement but a less certain impression of the vertical—or temporal—succession of volcanic beds. In this sense the map is more a rendering of geomorphology than stratigraphy.

FIGURE VIII-9. Stereographic pair of photographs showing three overlapping lava flows, each with distinctive texture. L1 is the oldest, L3 the youngest. Arrows indicate direction of flow (from von Bandat, 1962).

Vertical sequences of lavas are clearly revealed where rivers downcut through plateau basalts (Fig. VIII-11). It is noteworthy—and somewhat disturbing—that similar layer cake successions are not commonly visible on the Moon, even though there are numerous steep escarpments both within craters and along linear scarps. It may be that outcrops on the lunar surface are rapidly modified by physical weathering: micrometeorite bombardment, thermal cracking, or seismic shaking. Consequently the original stratified

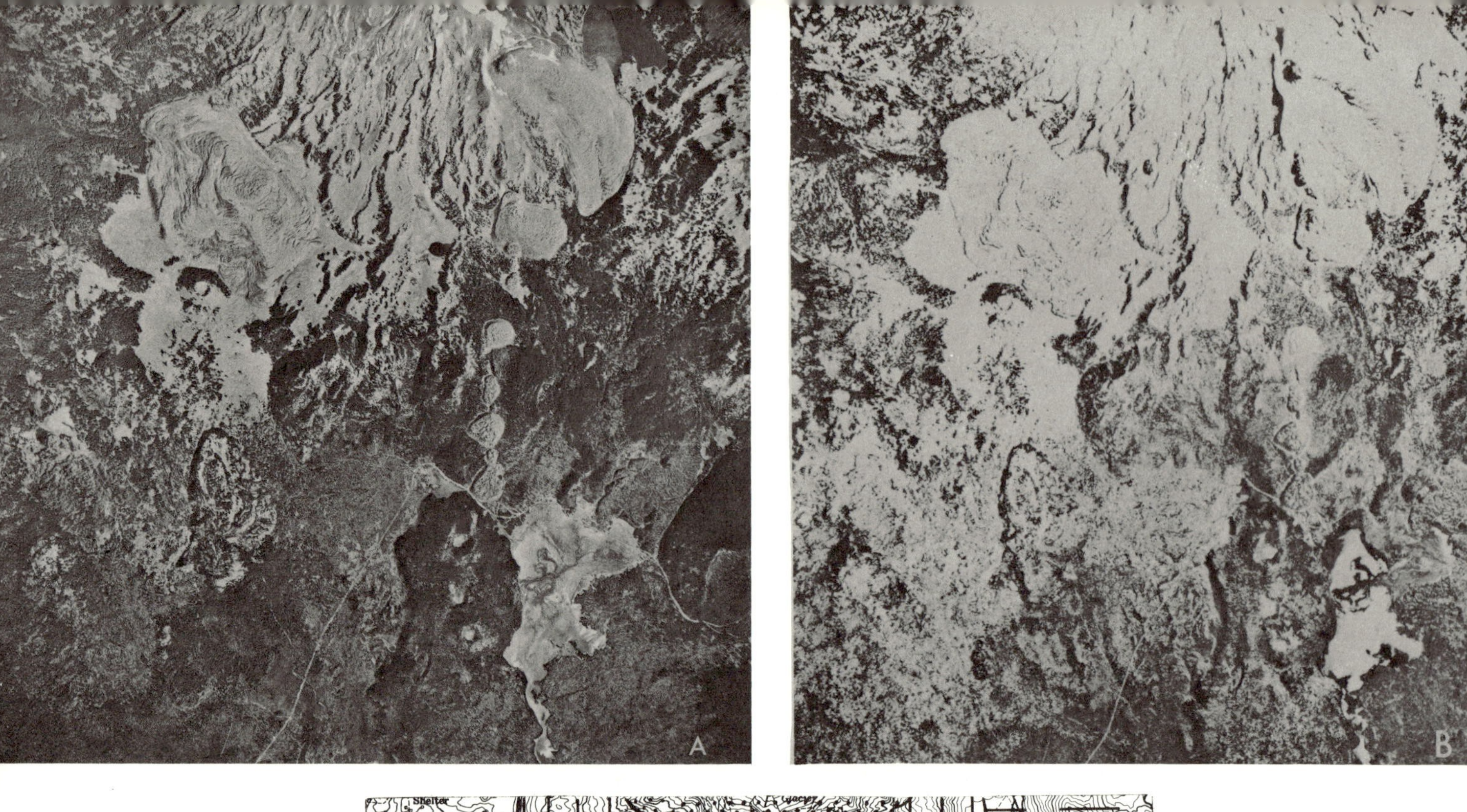

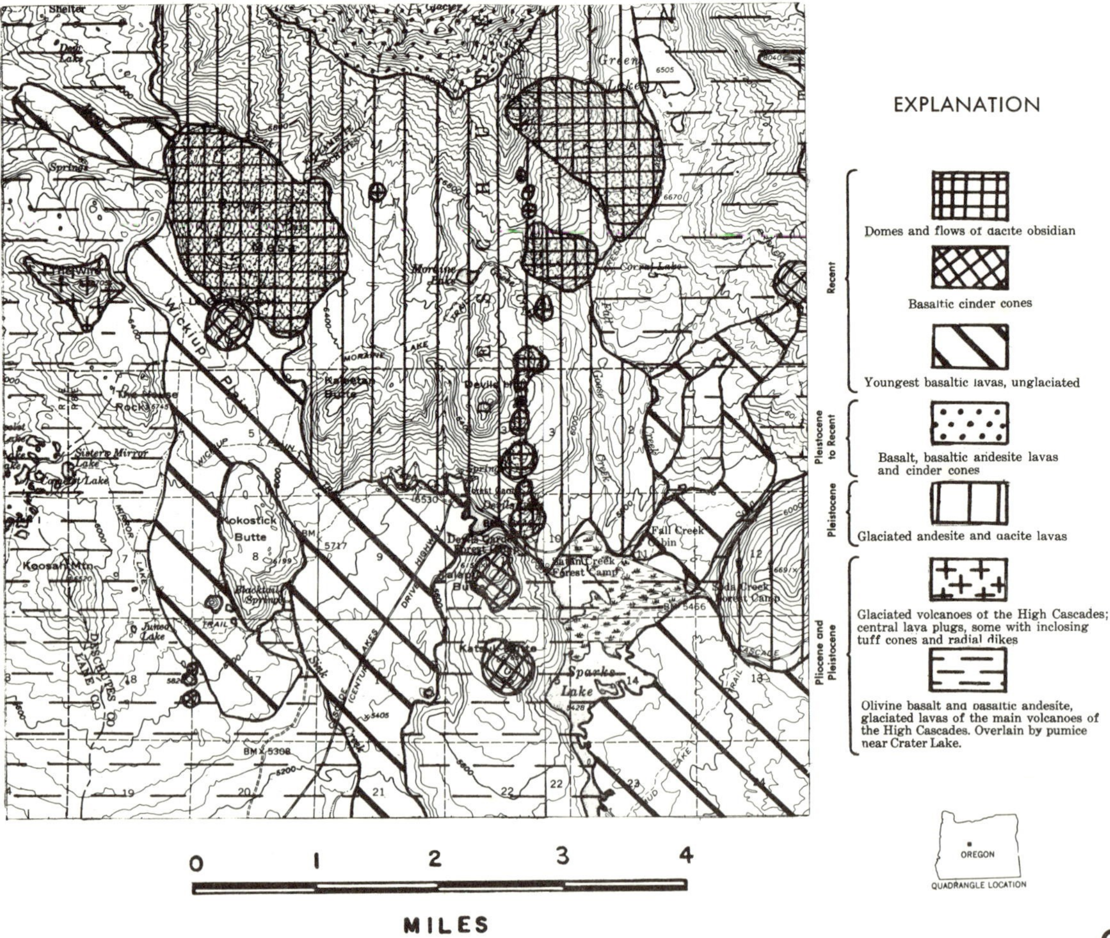

C

FIGURE VIII-10. (a) Aerial photograph of the Devils Hill–Broken Top region of the High Cascades, Oregon, showing a variety of volcanic constructional features (courtesy U.S. Geological Survey). (b) Aerial photograph of same region shown in (a) demonstrating the loss of information when the region is covered with snow (from Peterson and Groh, 1965). (c) Geologic map of volcanic terrain depicted in both (a) and (b) (from Peterson and Groh, 1965).

FIGURE VIII-11. Flows of Columbia River Basalt, Washington. Individual flows range from 10 to 50 meters in thickness. The cliff top rises 500 meters above the road in the foreground. Light-colored beds near the top of the cliff are water-laid sediments and ash falls, deposited on the basalt plain and then covered by new basalt flows (from Waters, 1967).

character of the rocks may be obscured by a thick mantle of surficial regolith. Evaluation of the evidence in support of horizontal, planar stratification is considered further in Chapter X.

LUNAR RILLES, RIDGES, AND FLOW FEATURES

The maria display a number of features which early attracted the interest of telescopic observers. More recently, in their role as the prime Apollo landing sites, selected parts of the maria have been intensively studied, not only with orbital photography (with an ingenious gravity survey thrown in as an unexpected bonus) but also with on-the-ground physical and chemical analysis by Surveyor spacecraft. This collection of many observations is something of a mixed bag, but much of the information contributes to the thesis that the maria are regions of repeated volcanic activity.

A great variety of sharp, narrow trenches occur on mare surfaces. Following Quaide (1965) and El-Baz (1968) these trenches, commonly called rilles, can be divided into four geometric types: straight, arcuate, irregularly branching, and sinuous (Figs. VIII-12 through 17). The last type has received the most attention, possibly because it has been assumed that any structure with so many unusual features should have a decipherable origin.

As the name suggests, sinuous rilles have a snake-like plan view. Some of the meanders are smooth arcs, but roughly linear segments are also common. Most sinuous rilles occur along the edges of maria, some following the boundary between highland and level mare plain. A commonly held view is that the upper, higher end of a "typical" sinuous rille issues from a crater and that, as the rille is traced downslope, the cross-

FIGURE VIII-12. The Hyginus Rille, about 180 km long, 3 to 6 km wide, and 400 meters deep. The rille is a probable graben: that is, a structurally down-faulted block. Volcanic craters are prominently arranged along the left arm of the rille, indicating a close relationship between volcanism and formation of the rille (Lunar Orbiter Photograph III 73M).

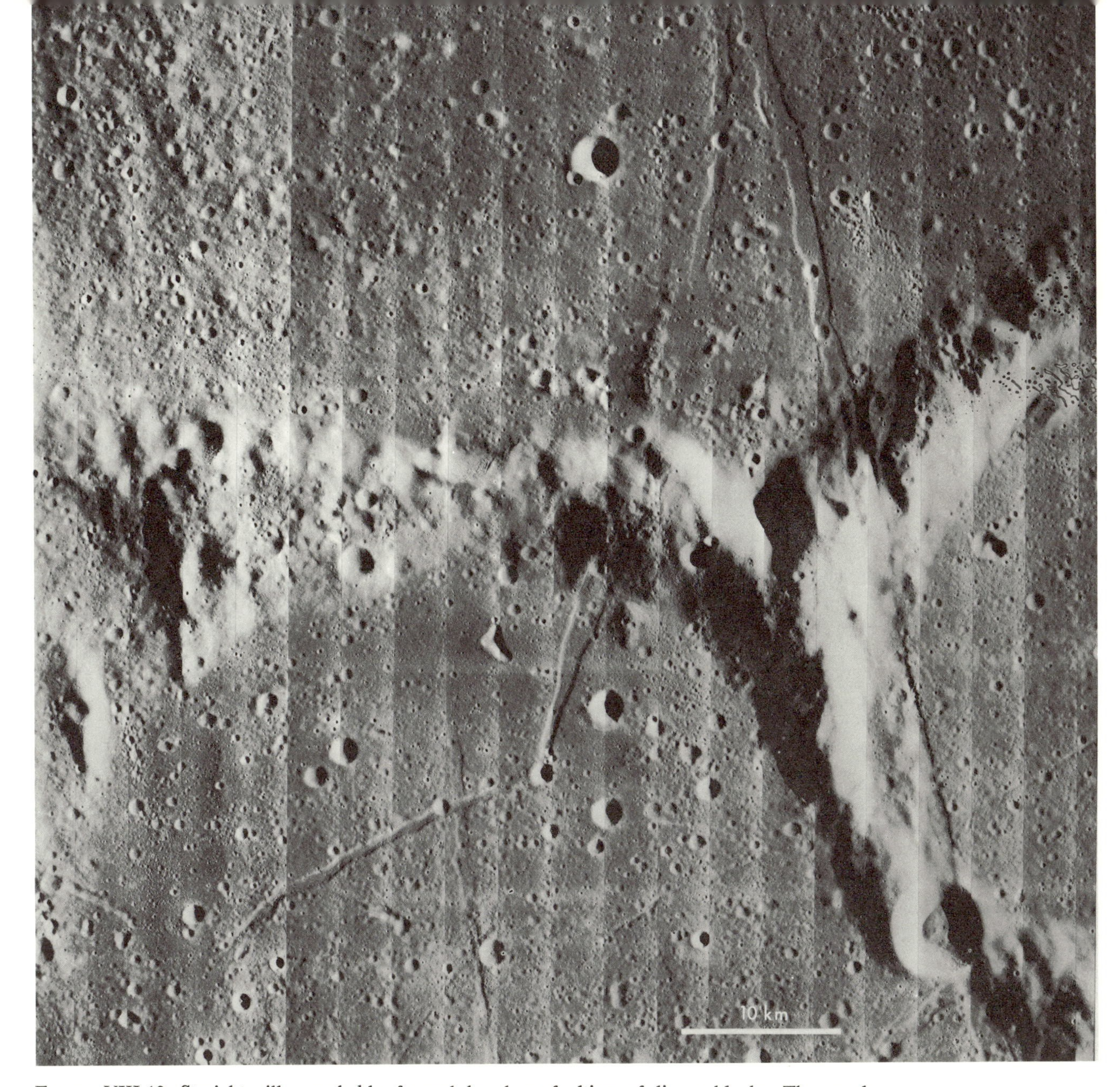

FIGURE VIII-13. Straight rilles probably formed by down-faulting of linear blocks. The regular pattern of the rilles is interrupted where they cross rim crests of two contiguous craters which are partly flooded with mare materials. However, some displacement appears to have occurred in the crater rim on the right where a straight trough follows the trend of the rilles entrenched on both sides of the rim (Lunar Orbiter Photograph V 138M).

sectional area becomes progressively less until the trench is no longer visible. This qualitative impression of flow direction is influenced by the fact that many rilles have one end close to highlands and the other end far out on a mare surface. In fact, however, the direction of slope along most rilles is difficult to determine. The possibility that many of them are essentially horizontal or sloping downward in the direction of increasing channel size cannot be discounted.

The features just described can be discerned from telescopic observations. There was

FIGURE VIII-14. Arcuate rilles entrenched in the highlands east of the Humorum basin and concentrically arranged around that basin (Lunar Orbiter Photograph IV 132H).

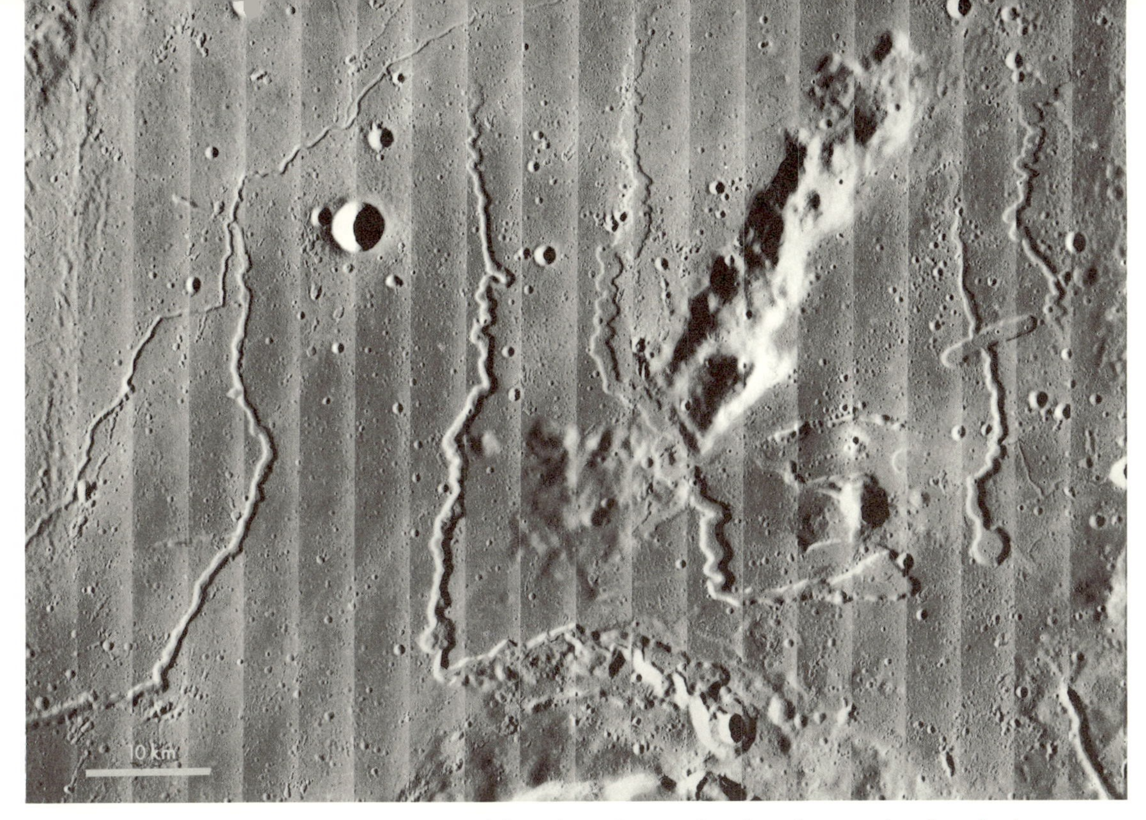

FIGURE VIII-15. Rilles exposed east of the Aristarchus plateau. Arc-shaped segments of coalescing craters impart a sinuous appearance to the rilles. The lower part of the rille in the center is made up of linear segments meeting at sharp angles (Lunar Orbiter Photograph V 191M).

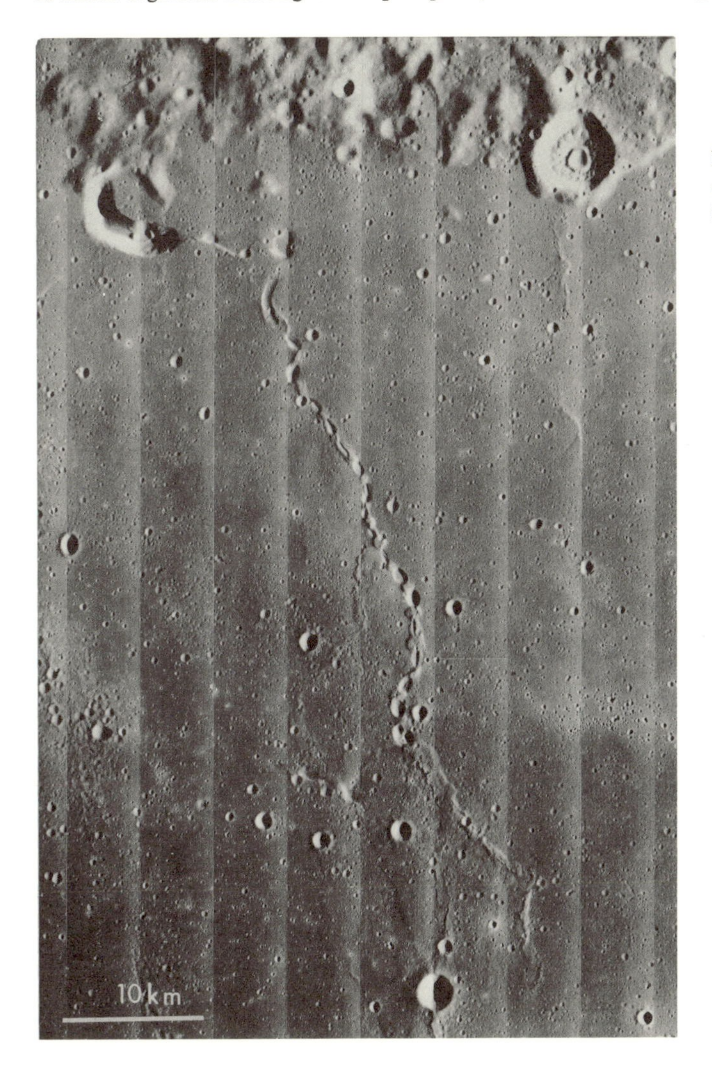

FIGURE VIII-16. Contiguous elliptical craters form a discontinuous rille (Lunar Orbiter Photograph V 182M).

FIGURE VIII-17. A sinuous rille whose pattern is difficult to explain. Note that the rille gradually disappears at both its north and south end. Part of its course appears influenced by highlands adjacent to the mare plain. An isolated volcanic cone is located on the mare in the lower right (Lunar Orbiter Photograph IV 158H).

every reason to anticipate that the Orbiter series of photographs would reveal crucial details beyond the reach of telescopic observations. Additional details are, in fact, revealed, but their significance is by no means clear. Some sinuous rilles have smooth V-shaped profiles, but more typically the floors are flat or complicated by inner channels, craters, and irregular hillocks. The presumed downslope ends of the rilles can be traced farther than with telescopic observations, but, once more, the trenches disappear imperceptibly.

Cameron (1964) put it best when she prefaced her own explanation for Schröter's Valley, one of the larger rilles, with a remark that "several more or less unsatisfactory theories have been proposed for the origin of lunar sinuous rilles." Some authors (e.g., Firsoff, 1961; Gilvarry, 1969) have argued that they are valleys scoured by actual streams associated with a former hydrosphere. From a geomorphologic point of view this explanation is difficult to believe. Admittedly, terrestrial river channels display a great diversity of patterns so that it would be unfair to test a particular rille against a particular river. However, the total suite of features which characterize terrestrial rivers—tributary systems, braids, smoothly curved meanders, channel size increasing downslope, delta fans, alluvial flood plains, levees, etc.—is absent, or at best imperfectly displayed, in lunar rilles.

As a variation on the theme of lunar rivers, Lingenfelter, Peale, and Schubert (1968) have suggested that water which is outgassed from the lunar interior and trapped beneath a permafrost layer could be released by a meteoritic impact large enough to disrupt the permafrost layer. Exposed to the near-vacuum of the lunar atmosphere, the water would boil, the latent heat being supplied from the water—principally through the formation of surface ice. As the pressure of the thickening ice overburden increased, a point would be reached where a liquid phase could be maintained beneath the ice. The authors pictured this ice skin as forming a protective blanket over water which breached crater walls and started to flow downhill. One problem in this model is that the flow characteristics are unlikely to resemble those for terrestrial rivers. The water would not be able to discharge freely at the "river" mouth, since it would at all points have to be sheathed in ice. Instead it would break through to the surface where the greatest pressure gradient existed, perhaps along the margins of the central reservoir or through the soil adjacent to the reservoir. It is difficult to see how this process could lead to development of meandering channels formed in the same way as those in terrestrial rivers.

A further possibility is that sinuous rilles are indeed channels, but channels for lava, not for water. Such channels on Hawaiian volcanoes have been noted by Wentworth and MacDonald (1953). Kuiper, Strom, and LePoole (1966) have described similarities between the curved course of terrestrial lava channels and the shape of sinuous rilles. Channels may be formed either at the time that lava is being extruded and flowing downhill to form a surficial pool, or at the time that lava from a surficial pool is draining back below the surface following the eruptive stage. In the latter situation craters which are situated at one end of a rille may be drain-back features.

Other workers have proposed that rilles were formed by some combination of faulting and subsidence (e.g., Quaide, 1965). This is a likely origin for the straight and arcuate rilles, but it is more difficult to explain the delicately contorted pattern of sinuous rilles through the same mechanism.

Final possibilities are that the sinuous rilles were formed as a result of fluidization of surficial materials by gases vented from fractures or, alternatively, that the rilles were scoured by a gas-charged medium flowing downhill. The latter point of view is developed by Cameron (1964) who compared the proposed gas-charged medium with the basal ash-flow part of the *nuée ardente*. This erosive mechanism might explain the puzzle of channels which appear to decrease in size downstream and which have no visible accumulation of sediment at their distal ends. If the gaseous component of the erosive medium escaped during its downhill course, then the erosional competency of the medium would actually decrease downstream.

Fluidization of surficial materials by venting of gases under experimental conditions has been described by Schumm (1969) and has been proposed as the origin for some sinuous rilles on the Moon (Schumm and Simons, 1969). This interpretation is supported by the frequent occurrence of small craters strung out along rilles. Indeed, the curved segments of coalescing craters result in the sinuous appearance of some rilles. A puzzling piece of counter-evidence is that many crater-like arc segments on opposite sides of rilles have parallel trends instead of opposed convexities.

Irregular ridges, discontinuous and splayed on a small scale, occur on mare surfaces. These mare ridges, sometimes called wrinkle ridges, extend brokenly for great distances, commonly paralleling the edges of the major maria. Some ridges parallel trends of structures present in the highlands. As in the case of rilles, numerous explanations for the origin of ridges have been put forth. Baldwin (1963) suggested that they were buckled by compression attending the collapse of great lava domes which formerly occupied the maria. Another possibility is that they were formed by differential compaction of sediment overlying basement ridges. Two genetically related origins which appear most reasonable are doming up of the crust over shallow intrusions or accumulation of extrusive flows along a linear fissure system. An intrusive origin for several ridges in the Ranger VII site is supported in considerable detail by Whitaker (1966). The extrusive origin is more difficult to substantiate. Orbiter photographs fail to present many compelling examples of distinctively textured and overlapping flows issuing from linear vents.

Ridges and rilles tend to be located preferentially along mare boundaries (Baldwin, 1963). The concentric arrangement of ridges around Mare Serenitatis and arcuate rilles around Mare Humorum are particularly good examples. In addition there are numerous reports of recent lunar activity—chiefly evidenced by temporary color change—for regions near mare borders (Middlehurst, 1967). These several patterns of distribution provide support for the contention that the feeder vents for mare plains are located not at the center of large basins but around the edges along concentric zones of fracture (Fig. VIII-18).

FIGURE VIII-18. Two models for basin filling showing stratigraphic relationships following (a) filling from a central vent and (b) filling from marginal vents.

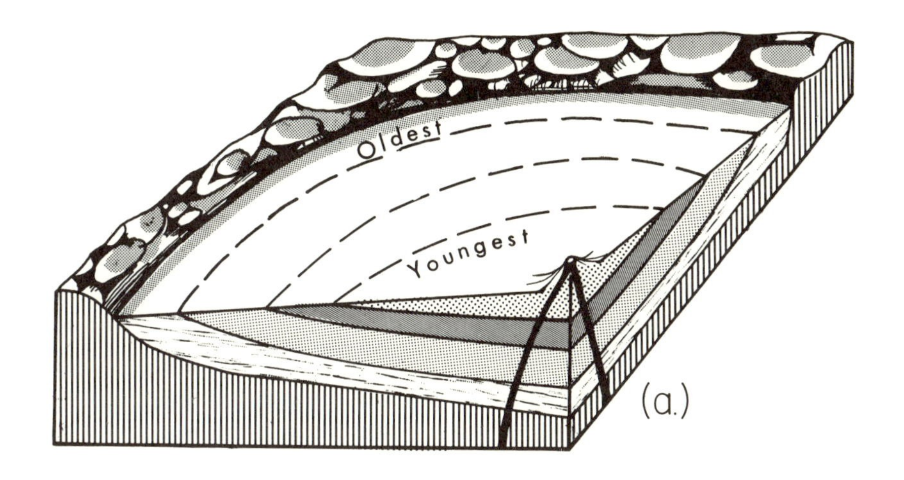

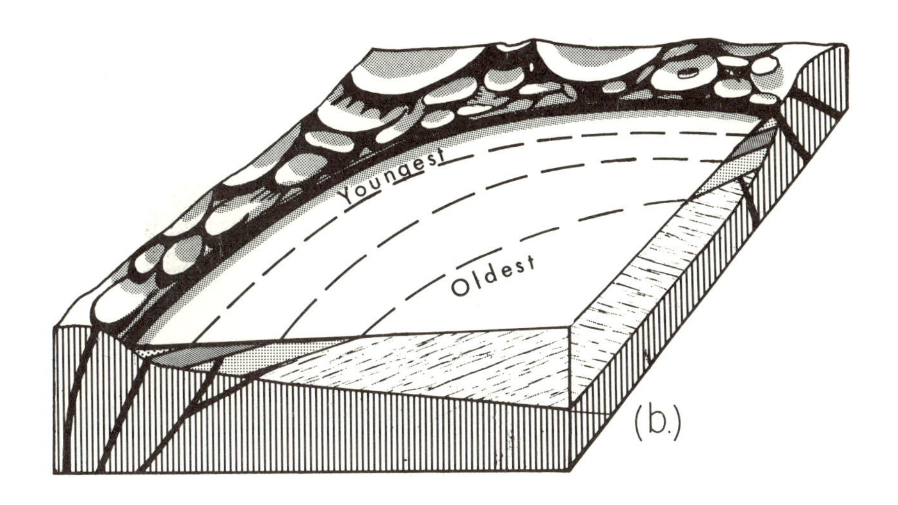

The flat mare surfaces can themselves be differentiated and analyzed according to their physical and optical characteristics. Even on telescopic photographs it is readily apparent that the albedos of various regions within the maria are not identical. Differences in brightness for surfaces within Mare Serenitatis are particularly obvious. Carr (1966) recognized these differences by dividing the Procellarum Group into four numbered facies. Figures VIII-19 and 20 illustrate that there is a general progression from lighter to darker facies as rocks are traced from the basin center to the margin. Within the Procellarum Group there is a rough correlation between increasing brightness and increasing crater density. Assuming that more cratering signifies a greater age, the darker facies is younger than the brighter. Hence the floor of Mare Serenitatis seems to be progressively younger as the margin is approached.

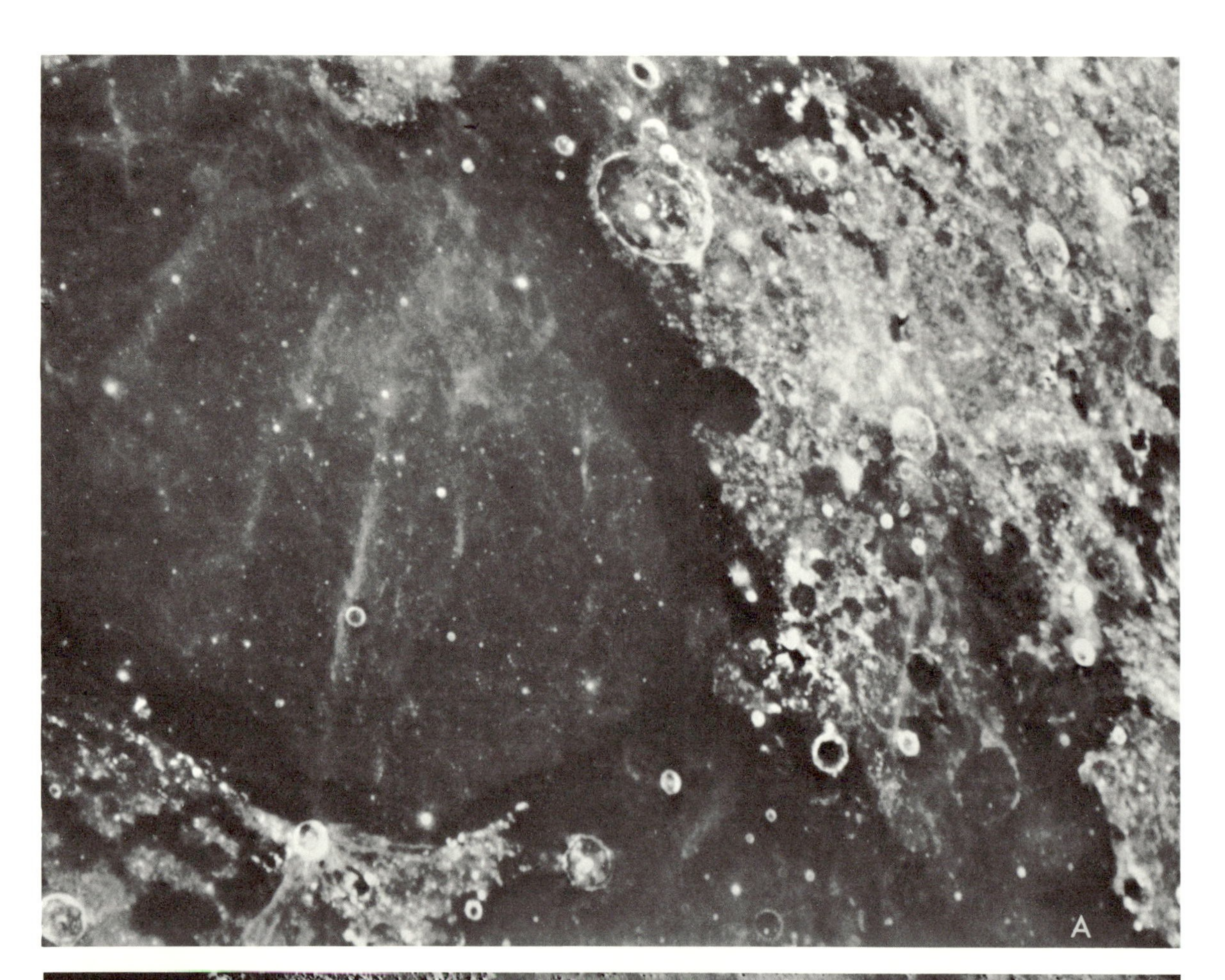
A

B

FIGURE VIII-19. (a) Full Moon telescopic photograph of Mare Serenitatis showing albedo variations within mare materials (U.S. Naval Observatory). (b) Telescopic photograph of Mare Serenitatis at low Sun angle. Concentrically arranged mare ridges are well displayed (Catalina Observatory). (c) Geologic map of the Mare Serenitatis quadrangle showing division of mare units according to albedo (after Carr, 1966).

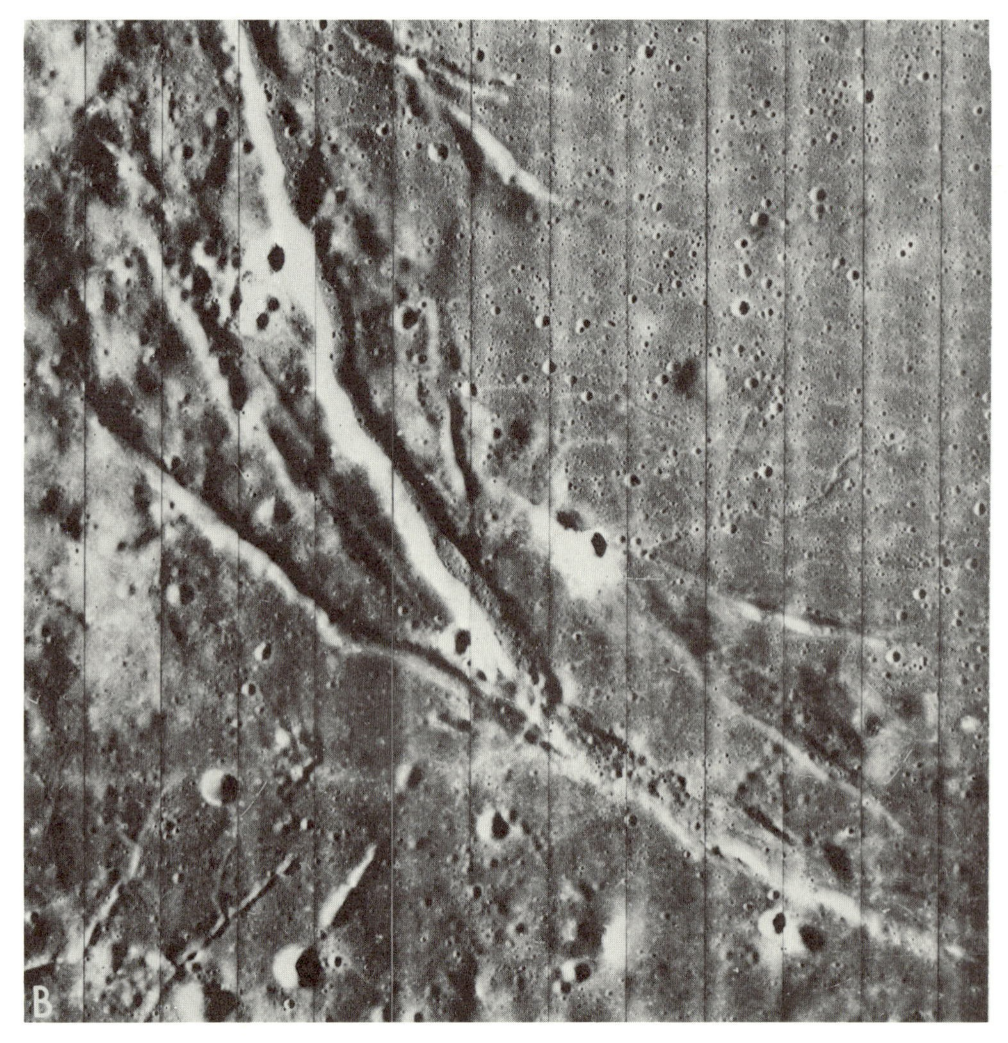

FIGURE VIII-20. (a) Telescopic photograph of the southwestern margin of Mare Serenitatis showing material with unusually low albedo situated around rilles in the near center of the photograph, just northwest of a prominent crater. (b) Orbiter view of the rilles (Rimae Sulpicius Gallus) shown in (a). The dark materials adjacent to the rilles are less cratered than mare material in the upper right. Carr (1966) interprets these dark materials as a thin veneer of pyroclastic deposits and subordinate lava issuing from the rilles (Lunar Orbiter Photograph V 93M).

Whitaker (1966) has pointed out that variations in the color of mare regions, emphasized in composite UV-IR photographs, correspond to topographic variations. In Mare Imbrium, for example, several sharp color boundaries are coincident with scarps bounding low plateaus. This suggests emplacement of fluid lavas which have successively spread across the mare plains. Orbiter photographs do not contain as much detail in support of this contention as might be expected, but one distinctively lobate flow front in Mare Imbrium has been photographed (Fig. VIII-21). The similarity of crater distributions on both sides of the flow front indicates that little time elapsed between emplacement of the two flows; the absolute crater density indicates long exposure of the surface following volcanic activity.

FIGURE VIII-21. Lobate flow front exposed on plains of Mare Imbrium. Crater densities on both sides of the flow front are similar, but the elevated surface in the lower part of the photograph shows subtle, discontinuous lineations which are absent on the topographically lower surface in the upper part of the photograph (Lunar Orbiter Photograph V 161M).

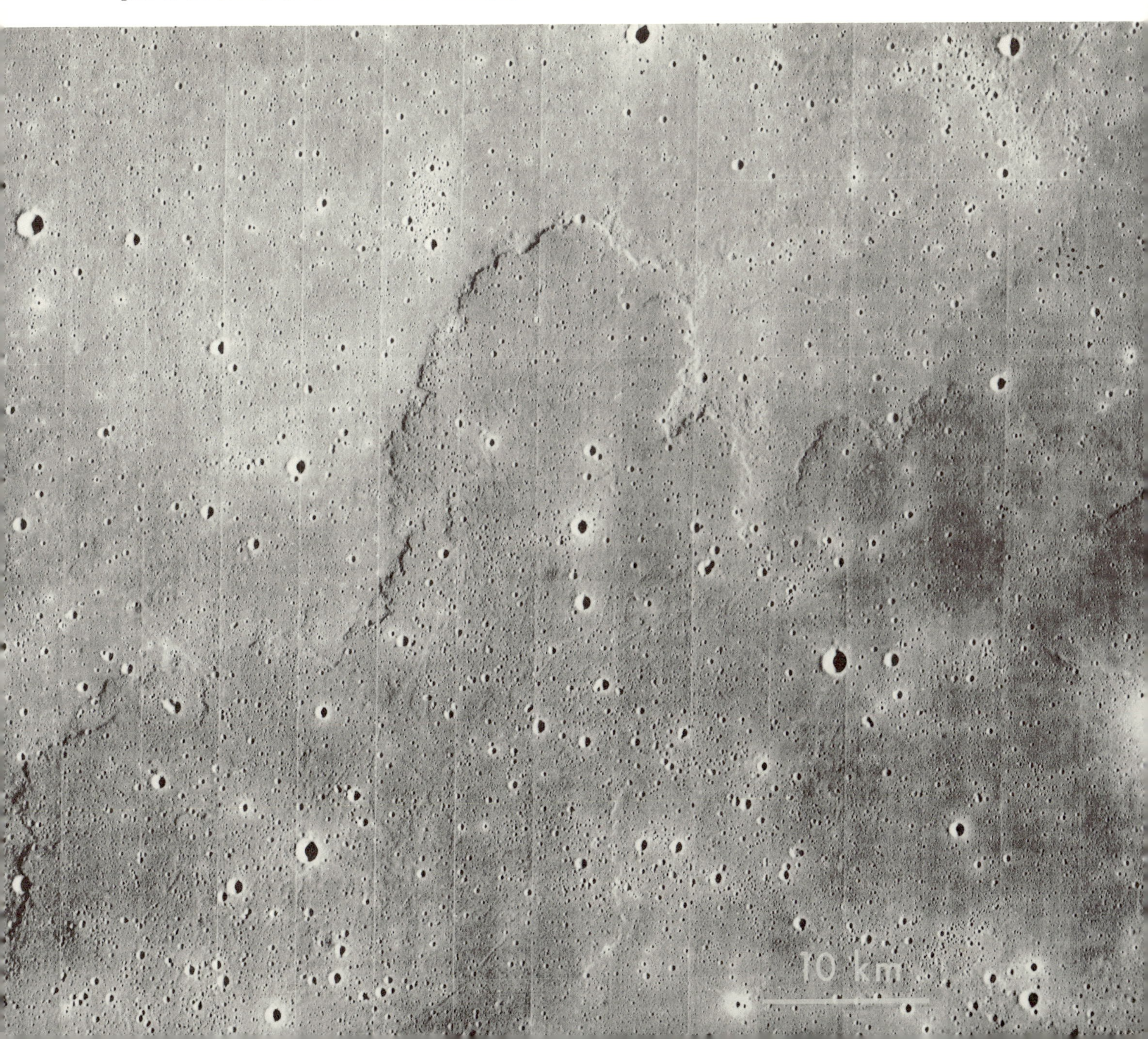

SURVEYOR RESULTS

Valuable information about the lunar maria were provided by Surveyor spacecraft (Surveyor Program, 1966, 1967*a*, 1967*b*, 1968*a*, 1968*b*; Surveyor Project Final Report, 1968*c*). In addition to returning pictures with resolutions on the order of millimeters, the spacecraft performed experiments which determined mechanical properties and chemical composition of the lunar soil. The entire spectrum of Surveyor scientific results does not appear to have been widely appreciated by either laymen or scientists. Ironically, widely publicized observations by the first men to stand on the Moon will probably in large part duplicate information gained by the unmanned Surveyors.

Even though the first four of the five successful Surveyor landings were in mare regions, the geomorphology of the four sites proved to be subtly varied. Surveyor I landed on the flat plain of Oceanus Procellarum, Surveyor III in a subdued crater south of Copernicus, Surveyor V in a rimless crater—perhaps formed by drainage of surficial debris into a linear fissure—in Mare Tranquillitatis, and Surveyor VI close to a mare ridge in Sinus Medii.[1]

Surveyor I pictures confirmed impressions gained from previous Ranger photographs. The Moon's surface was shown to be pock-marked with craters ranging downward in size to several centimeters. Particles ranging in size from 1 mm to more than 1 meter could be individually counted on photographs. Cohesion properties of the soil indicated a high percentage of clay-sized particles. Mason (1969) has suggested that the mean particle size is in the vicinity of 10 microns. Disturbance of surface materials by the spacecraft footpads indicated that the top few centimeters behaved similarly to terrestrial damp fine-grained soil, maintaining slight cohesion but also susceptible to particulate ejection away from the foot-pad depressions. Ejected material was noticeably darker than adjacent undisturbed soil, indicating a stratification within the soil—an upper light layer with thickness on the order of millimeters underlain by a darker horizon (Fig. VIII-22).

Rocks adjacent to the spacecraft were photographed by the high-resolution television camera. Definitive textures were not observed, but darkly shadowed cavities in some rocks resemble vesicles which are commonly present in terrestrial volcanic rocks (Fig. VIII-23). In this connection it is interesting to note that Dobar (1966) has experimentally solidified basaltic and granitic magma in vacuum, in both experiments producing highly porous material containing large numbers of voids.

Surveyor III confirmed most of the observations made by Surveyor I. A shovel-like scoop gouged a trench through the lunar soil, directly establishing the fact that granular material extends to depths of at least 15 cm. The irregular shape of the shovel etched delicate patterns along the walls of the trench (Fig. VIII-24). This phenomenon, in addition to demonstrating the cohesiveness of the soil, also suggested that the rubbly appearance of the undisturbed surface is caused in part by easily-broken clots of fine-grained material.

[1] Surveyors II and IV were unsuccessful missions. Both spacecrafts crashed on the lunar surface.

FIGURE VIII-22. Photograph of Surveyor VI foot pad showing dark material thrown out of a shallow depression formed by the footpad. Similar ejection of dark material was observed at all Surveyor sites.

Several boulders close to the spacecraft were partially buried with finer-grained sediment which tended to concentrate as overlapping fillets on the up-slope side of the rocks. This was interpreted as evidence for down-hill creep of the surface layer, perhaps due to shaking associated with large impact events or internal tectonic activity. Several boulders photographed by Surveyor III were distinctly tabular (Fig. VIII-25). This is not necessarily evidence of stratification in the sense of sequentially superposed layers of sediment. Platy fabric may also be formed by flow banding in volcanic rocks or by closely spaced jointing in shock-deformed rocks.

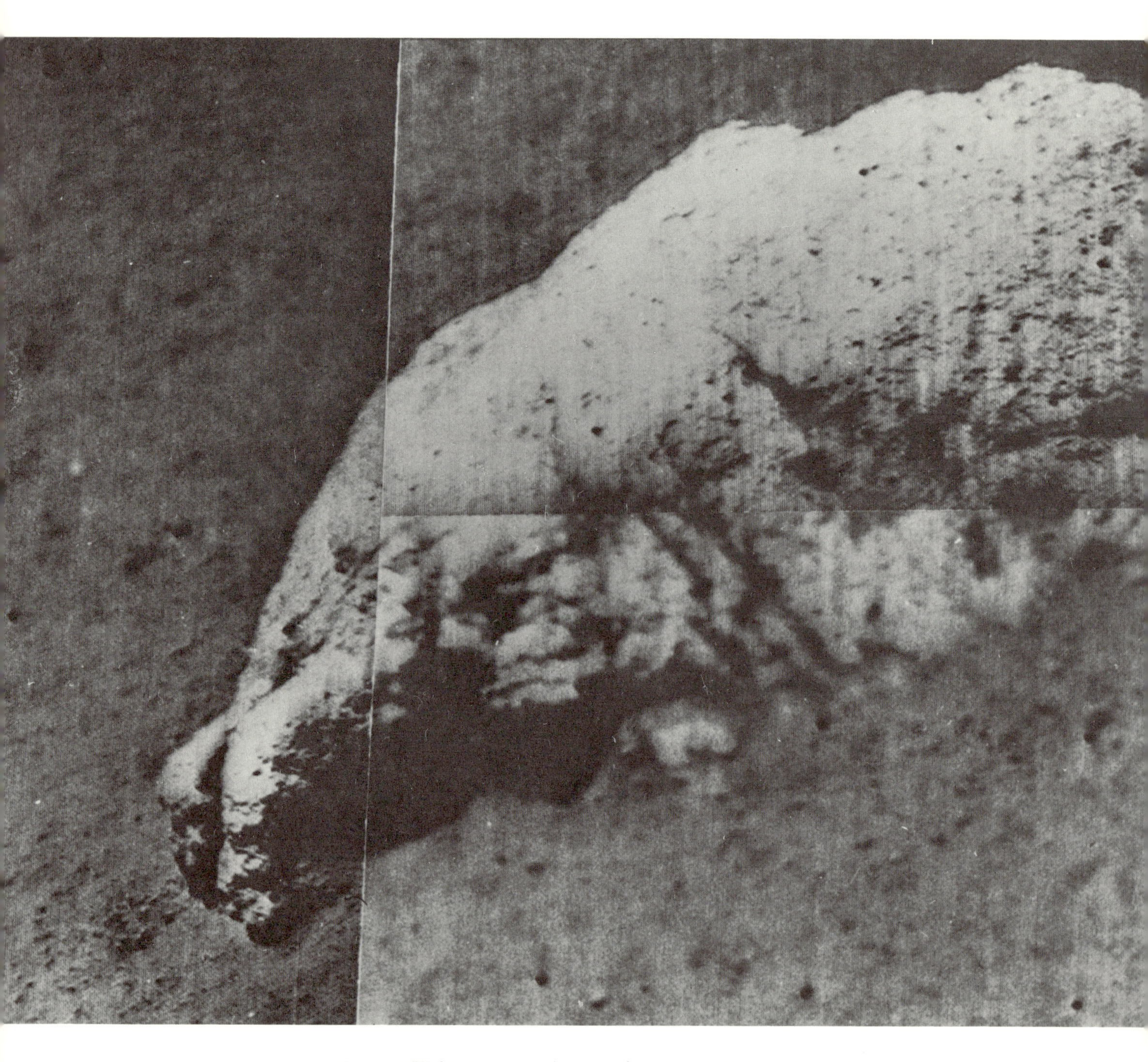

FIGURE VIII-23. Porous rock, approximately a half meter long, photographed by Surveyor I.

FIGURE VIII-24. Smooth surface along the wall of a trench dug by the Surveyor III shovel. Linear grooves on that surface faithfully record irregularities in the edge of the scoop.

Surveyor V was the first lunar spacecraft to return a rudimentary chemical analysis of the lunar surface material. The analytical instrument contained a radioactive source emitting alpha particles. When these particles struck the lunar surface, some were reflected back on a detector. The number and energy of the particles were governed by the elemental composition of the surface. Additional chemical information was gained by monitoring protons split off nuclei by the alpha particle bombardment. Two samples of the uppermost lunar soil were measured by Surveyor V, and another analysis was made by Surveyor VI (Turkevich and others, 1968*a*, 1968*b*). In all three cases the analyses were very similar (Table VIII-1). The elemental abundances closely resemble

TABLE VIII-1. Average chemical analysis for Surveyors V and VI experiments.

Oxide	*Percent, by Weight*
SiO_2	50
Al_2O_3	14
Fe_2O_3, FeO	16
CaO	15
M_gO	5
Na_2O	not definitely established, could be present up to 3%

those found in terrestrial basalts. Equally important is the fact that they are significantly different from several other rock types which, before the mission, had been championed as likely candidates for lunar surficial material. These discredited rock types include chondrites, granites, and tektites.

FIGURE VIII-25. Blocky and platy rocks situated near Surveyor III. The large rock on the right appears to be laminated.

Prior to the Surveyor V mission several workers had entertained the possibility that some small rimless lunar craters arranged in parallel rows may be an indication of localized downward drainage of fragmental debris along steeply dipping fissures. Surveyor V landed in just such a rimless crater, 9 x 12 meters, the largest in a linear array of craters trending northwest. This particular crater appears to have a compound shape with its long axis trending northwest. Within its boundaries there is a row of smaller rimless craters ranging from 20 to 40 cm in diameter which, again, trend northwest (Fig. VIII-26). All of these relationships point to the existence of a set of linear fissures trending northwest. The size of the drainage craters probably varies with the depth and width of particular fissures within the set.

FIGURE VIII-26. The view northwest from the Surveyor V landing site. Chain of small craters 20 to 40 cm in diameter extends from center to lower right.

Surveyor VI landed in Sinus Medii about 500 meters north of an east-west trending mare ridge (Figs. VIII-27). The general appearance of the mare surface resembles that of earlier Surveyor landing sites. The most important scientific experiment on the Surveyor VI mission was a repetition of the chemical analysis first made by Surveyor V. The remarkable resemblance of these two analyses from widely separated localities in Sinus Medii and Mare Tranquillitatis gives one some confidence in assuming a common chemical composition for all mare surfaces.

Midway through its on-the-ground mission the vernier engines of Surveyor VI were fired, causing it to rise vertically about 4 meters before settling down on the surface at a point a little more than 2 meters from its original position. During this hop forcibly expelled exhaust gases caused considerable transportation of surficial materials. Pre-hop and post-hop pictures permit one to identify a variety of erosional and depositional features. The probable framework for this sedimentological experiment is shown in Figure VIII-28. Two different types of deposition are represented. Some coarse fragments directly below the vernier engine were thrown into ballistic trajectories. Other finer soil particles mixed with exhaust gases to form a turbid mixture moving along the surface. This model closely resembles that developed in previous chapters to explain the distribution of ejected sediment around large terrestrial and lunar craters.

FIGURE VIII-27. Mosaic of Surveyor VI photographs showing a mare ridge about 500 meters southwest of the spacecraft.

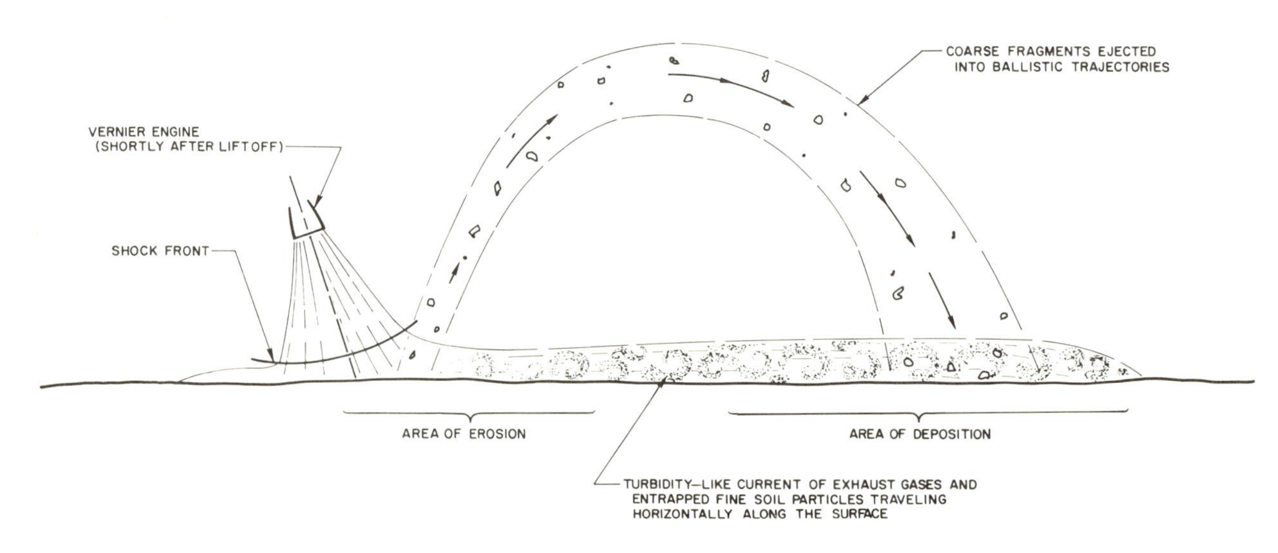

FIGURE VIII-28. Probable pattern of erosional, transportational and depositional events during firing of Surveyor VI vernier engine (taken from Surveyor Program, 1968*a*).

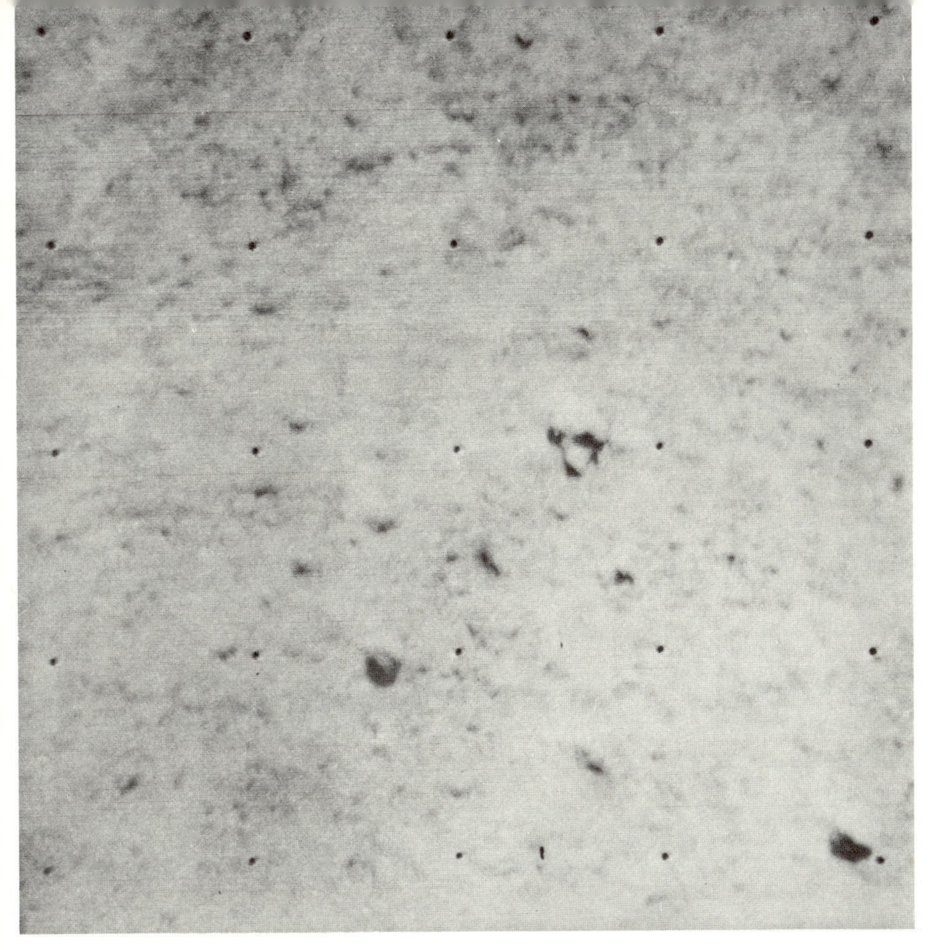

FIGURE VIII-29. Pre-hop picture of area near Surveyor VI spacecraft.

FIGURE VIII-30. Post-hop picture of area near Surveyor VI spacecraft. Comparison of this picture with same area shown in FIGURE VIII-29 illustrates that the larger rocks have been partially buried with fine detritus.

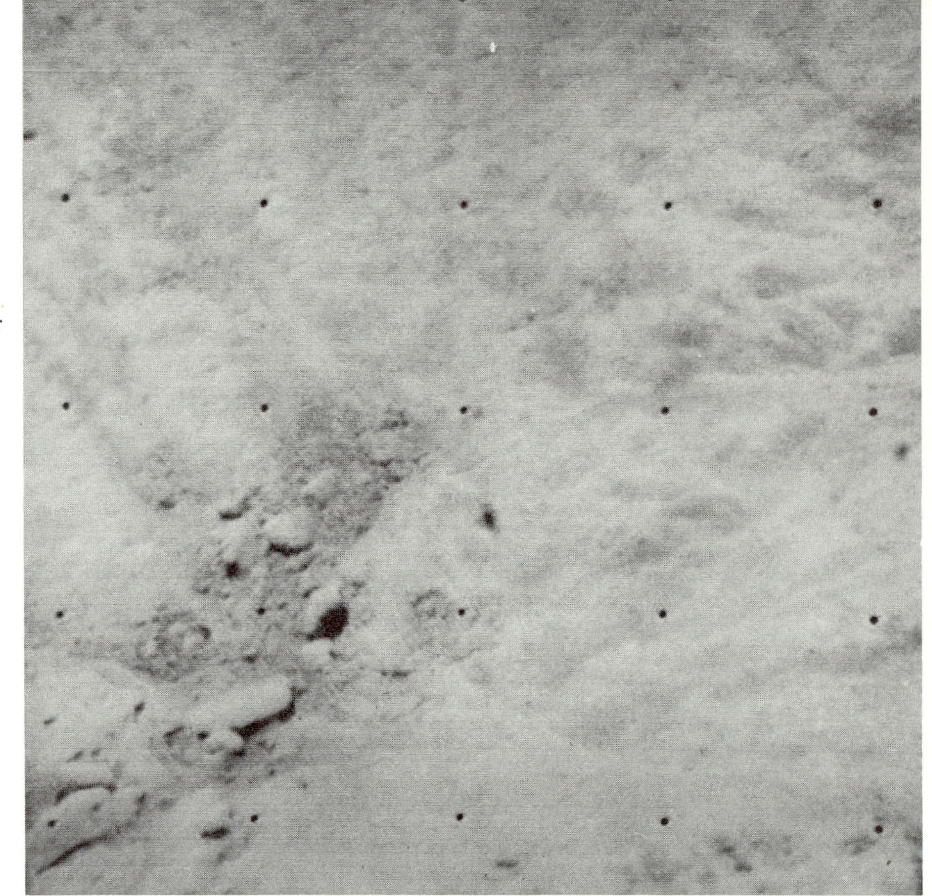

FIGURE VIII-31. Rippled surface near Surveyor VI. Ripples were formed by transportation of sediment during firing of the vernier engine.

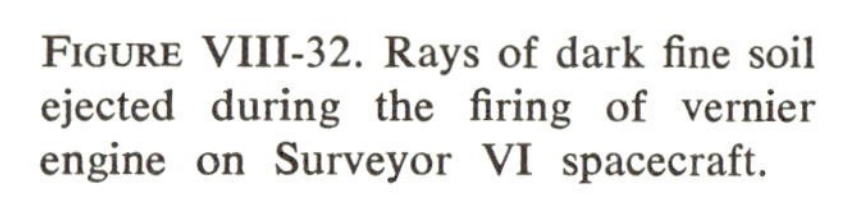

FIGURE VIII-32. Rays of dark fine soil ejected during the firing of vernier engine on Surveyor VI spacecraft.

After the test, areas of fine soil close to the vernier engines were littered with coarser fragments either due to addition of coarse materials or perhaps due to selective winnowing of fine sediment. Elsewhere shallow depressions were partially filled with fine soil which mantled and buried larger stationary rocks (Figs. VIII-29 and 30). Some surfaces were covered with irregular ripples (Fig. VIII-31). In other areas dark soil from several millimeters beneath the surface had been sprayed across lighter topsoil to form a plume of diverging rays (Fig. VIII-32).

MAPPING THE MARIA AT DETAILED SCALES

The knowledge that men will be landing on the Moon and that they will be obliged, initially at least, to gain their geologic insights by roaming no more than a few meters, or at best a few kilometers, from the lander has engendered the production of detailed geologic maps designed to direct the astronauts to a variety of closely spaced materials differing in composition and age. As part of this program the U.S. Geological Survey Branch of Astrogeologic Studies is preparing maps of six possible early Apollo sites at scales of 1:100,000, 1:25,000, and 1:5,000.

In critiquing these maps one could fairly attribute a certain bias of optimism to each mapper. If he determined that the entire surface is uniform—or uninterpretably disordered—this would preclude definition of any mappable units. A more satisfying result would be to identify several map units, each with a provocative geologic interpretation. When one considers that this exercise is carried out on parts of the mare surface selected by engineers on the basis of unusual smoothness, he cannot help but entertain some uneasiness about the reality of any geologic differentiation.

Certainly it would be a mistake to think that units mapped at a scale of 1:5,000 have the same significance as those mapped at 1:1,000,000. At the latter scale most map units have unequivocal differences in topography and albedo. Superposition relationships are commonly observable. The inference is that the several mapped units differ in age, in mode of origin, or in both. At the 1:5,000 scale these major physical distinctions are lost. Just as a randomly chosen small area on Earth is unlikely to contain any major stratigraphic boundaries, so is the same true for the Moon. At best one can expect to map subtly different facies of the same major unit. For example, the underlying rocks at all of the early Apollo sites are probably basaltic lavas. There is little reason to think that these lavas differ widely in age or chemical composition within the small area of any given site. However, there may be subtle differences in crater density, lineament density and orientation, and regional elevation which suggest a slightly different sequence of events for adjacent areas. The temporal interpretation of these events is usually hazardous (Fig. VIII-33).

Apollo site maps do have one important point in common with 1:1,000,000 maps. In both instances mapping of crater deposits proceeds more or less independently of definition of regional background units. In the case of Apollo site mapping Trask (1970) has developed a numerical classification for craters ranging in size from 10 meters to 10 km. In this classification youngest craters of the Copernican System are numbered 8

FIGURE VIII-33. (a) Mosaic of Lunar Orbiter III photographs showing a level mare landing site between Mare Cognitum and Oceanus Procellarum. The Surveyor III site is shown by a cross in the upper center of the figure. The Apollo XII site is shown by a circle close to the Surveyor site.

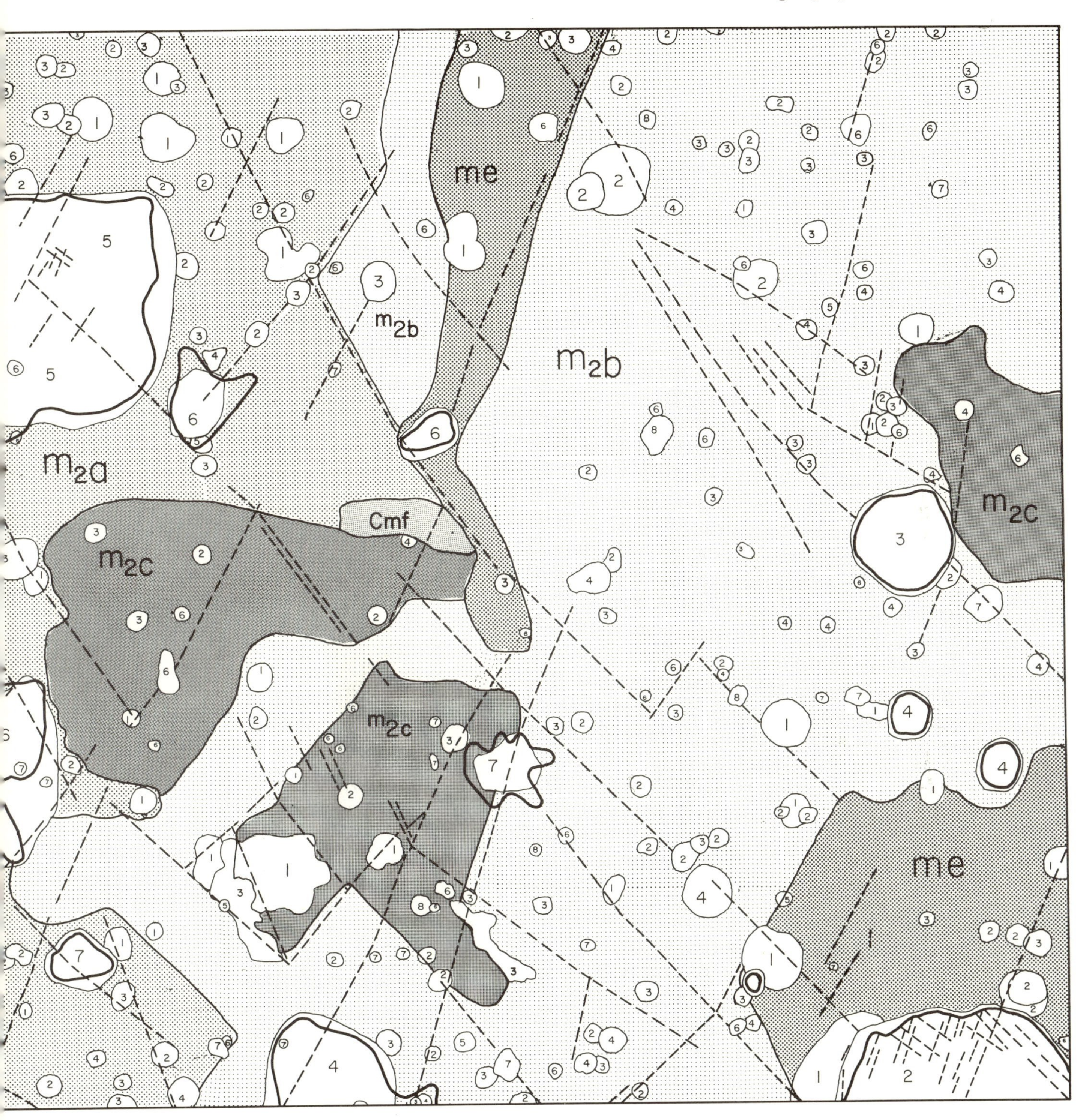

FIGURE VIII-33. (b) Geologic map of the same region shown in (a) on facing page (after Saunders and Mutch, 1967). Legend details for this map appear on the page which follows.

Legend for map (Fig. VIII-33, b) on preceding page.

Cmf

FLOW MATERIAL

CHARACTERISTICS:
Level surface broken by sharp craters less than 25 meters in diameter. Few craters larger than 50 meters in diameter.

INTERPRETATION:
Volcanic flow probably most recent within the map area.

ELEVATED MARE MATERIAL

CHARACTERISTICS:
Occurs in elongate, polygonal areas bounded by lineation segments and standing higher than adjacent areas. Ridge tops have strong to gentle undulations with wavelengths of approximately 100 meters. Subdued craters smaller than 50 meters are less common than in unit m_{2b}.

INTERPRETATION:
Material upwarped by pressure from shallow intrusions. More prominent mare ridges in the Riphaeus Mountains quadrangle which have the same northerly trend as the ridge in the center of this area, exhibit features suggestive of doming over shallow intrusions.

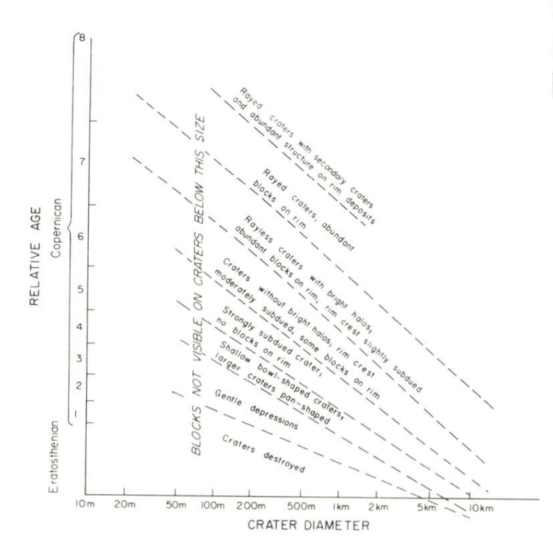

CRATER MATERIALS (undivided)

(for description of numbered craters, see insert graph)

8

7

6

5

4

3

2

1

m_{2c}

LEVEL MARE MATERIAL

CHARACTERISTICS:
Surface level to slightly undulating; abundant fresh craters smaller than 50 meters; strongly subdued craters 50-200 meters more numerous than on unit Cmf, less numerous than on unit m_{2b}.

INTERPRETATION:
Volcanic flows with mantle of fragmental debris thicker than Cmf.

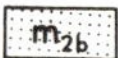

MARE MATERIAL

CHARACTERISTICS:
Gently undulating to level material, covering approximately half the map area. Sharp-rimmed craters, 10-50 meters, cover less than 5 percent of the terrain and are superposed on a background of larger subdued craters which cover no more than 20 percent of the surface. Lineaments are present but are relatively less common than in m_{2a}. Low sinuous scarps occur throughout. Visible blocks absent except for local fields around large craters.

INTERPRETATION:
Volcanic flows covered, at least in part, with ejected debris from Copernicus and ejecta from superposed craters. Small indistinct sinuous scarps may mark flow fronts.

PITTED AND LINEATED MARE MATERIAL

CHARACTERISTICS:
Appearance variable, but all surfaces show more small-scale relief than m_{2b}. Clusters of subdued overlapping craters smaller than 50 meters are common. Several areas contain overlapping subdued craters with diameters approximately 200 meters. Patterned ground is conspicuous, with lineations trending northwest and northeast, similar to lineaments throughout the map area.

INTERPRETATION:
Similar to m_{2b} except that surface material is older and has been subjected to more cratering. Patterned ground is produced by repeated periods of stress concentrated along regional joints. The stress is induced by local impact events (seismic shaking).

LINEAMENT – – – – – – – – –

CHARACTERISTICS:
Shallow linear depressions. Some lineaments display relief of several meters. Craters smaller than 50 meters may be alined along some lineaments; commonly craters are slightly extended or polygonized in the direction of the lineament. Closely spaced lineaments are locally present, especially on walls and rims of craters greater than 400 meters in diameter. Patterned ground occurs where lineament sets are most closely spaced.

INTERPRETATION:
Joints determined by a regional stress field forming the lunar grid. Vertical displacement of several meters may have occurred along some fractures, particularly those with the greatest lateral continuity. Periodic seismic shaking, caused by internal movement or impact, produces and accentuates these joints in young materials as well as older basement rock. Conspicuous close-spaced joints on the large crater rim in the lower right corner are evidence of the correlation between cratering events and surficial propagation of the joint system.

Abundant blocks present greater than 1 meter; 10-meter blocks not uncommon; 40 meters maximum size.

and oldest are numbered 1. Expectably, young craters are those with sharp rims, bright halos, and clearly recognizable rim deposits. Old craters are expressed only as shallow pan-shaped depressions.

An important consideration in detailed mapping of mare sites is the knowledge that the actual surface materials are not bare rock, but rather soil which probably ranges in thickness from 3 to 16 meters (Oberbeck and Quaide, 1968). Granted that the thickness of the soil will vary with the age of the surface and that large craters will penetrate the soil and throw out blocks of the underlying solid rocks, the fact remains that an astronaut picking up grab samples on several of the previously mapped units is apt to bring back to Earth approximately similar mixtures of lava fragments, meteoritic debris, and shock-lithified rocks.

MASCONS

Following completion of the Lunar Orbiter missions the tracking data were examined in detail. Unexpected local accelerations and decelerations in the spacecraft courses were revealed. These perturbations are related to local variations in the Moon's gravitational field. Integration of the tracking data, particularly for the Orbiter IV and V tracks which covered the entire Earthside hemisphere along successive longitudinal paths, led to production of a gravimetric map (Muller and Sjogren, 1968). This two-dimensional display of local gravity anomalies indicated strong positive features (mascons, for mass concentrations) located under circular maria (Fig. VIII-34). The strongest anomaly is associated with Mare Imbrium, but distinctively localized highs also occur under Maria Serenitatis, Crisium, Humorum, Nectaris, and Orientale. A possible mascon or combination of mascons occurs under Aestuum-Medii.

This documentation of relatively dense materials at shallow depth associated with mare basins has been variously interpreted in support of previously held views of lunar history. Urey (1968) for example, has maintained that the mascons are iron-nickel masses emplaced by meteoritic collision. He argues that the presence of high-density objects at shallow depth indicates that the Moon was so cold at the time of collision that the heavy masses did not sink towards the deep interior. According to his theory the Moon has been sufficiently rigid to maintain a non-isostatic condition over long periods of time.

O'Keefe (1968) comes to contradictory conclusions. Taking into account the regional topographic effects he reasons that the anomalies can be adequately explained by an isostatic condition. There may be either regional differences in crustal thickness of constant density material or lateral variations in crustal density. In his opinion the gravity variations can be explained by some combination of greater thickness of low-density material in the highlands and a lesser thickness of high-density rock in the maria. The mare materials are considered to be lava generated at distant points, either downward or sidewise, and poured out on the mare surfaces. A similar interpretation involving isostatic adjustment of crater floors coupled with infilling by high-density lava has been proposed by Baldwin (1968).

FIGURE VIII-34. Gravimetric and acceleration contour map of the lunar nearside. Numbers are in tens of milligals. Note areas of large positive acceleration over Maria Imbrium, Serenitatis, Crisium, Nectaris, and Humorum (from Muller and Sjogren, copyright 1968 by the American Association for the Advancement of Science).

Still another explanation has been put forth by Gilvarry (1969). Elaborating on his formerly held views that the mare basins were formed by underwater meteoritic impacts at a time early in lunar history when the Moon was inundated with water to a depth of 1.2 miles, he proposed that the mascons are deposits of sedimentary rocks carried into the basins of lunar seas by rivers and that the deposits were exposed when the seas dried up. This explanation finds little support in the evidence assembled during Orbiter and Surveyor missions. Both in texture and chemical composition the rocks of the maria resemble terrestrial volcanic rocks. The character and location of presently visible rilles does not suggest that they were the channels through which the proposed flood of detritus reached the maria basins.

Using Orbiter tracking data it is not possible to construct a detailed gravimetric map for the farside of the Moon, since the spacecraft were hidden from line-of-sight Earth observation. However, utilization of the spherical harmonic expansion of the Moon's gravitational potential calculated from nearside tracking data permits one to plot gross spatial variations in mass density for the entire lunar surface. Campbell and others (1969) tentatively define a large mascon about 1,000 km in diameter near the center of the farside which may be related to a multi-ring basin revealed in Orbiter V photographs. Unlike the mascon basins on the nearside, this one contains virtually no mare material.

Even though the precise genetic significance of mascons is uncertain, there are a few facts of general agreement. First, the several interpretations all must start from the fact that the circular mare regions are basins of considerable depth filled with relatively dense rock. Although this may be either lava or a meteoritic mass, both situations are compatible with the assumption that the original basinal cavity was formed by a huge impact.

It is significant that mascons do not underlie all regions covered with mare plains. For example, the large expanse of Oceanus Procellarum which lies between Mare Humorum and Mare Imbrium contains no distinctive gravity highs. This confirms the independent observation that Oceanus Procellarum is a mare region with irregular boundaries which do not outline either a single large basin or several fused basins. Instead, both the topography and the gravimetrics suggest that Oceanus Procellarum was flooded with a shallow veneer of lava, perhaps emplaced during a late overflow stage in adjacent circular basins.

MARIUS HILLS

Perhaps the most instructive area on the Moon for studying problems of volcanic stratigraphy is the Marius Hills. This region is situated on the western side of the Earthside disc, just north of the equator. An oblique photograph from Orbiter II (Fig. VIII-35) reveals an array of domes and cones which bring to mind many similar constructional features present in terrestrial volcanic terrains.

FIGURE VIII-35. An oblique aerial photograph of the Marius Hills region showing a field of domes believed to be of volcanic origin. The large crater in the background is Marius. The same region photographed from a near-vertical perspective is shown in FIGURE VIII-36. For orientation refer to the dome and crater indicated by arrows on both figures (Lunar Orbiter Photograph II 213M).

On his 1:1,000,000 map of the Hevelius quadrangle of the Moon, McCauley (1967*b*) recognized that the Marius Hills are situated on a southward-sloping plateau several hundred meters higher than the adjacent mare (Fig. VIII-36). The boundary is particularly clear in the Seleucus quadrangle where it is marked by a sinuous scarp. McCauley also noted that crater density on the plateau surface in the diameter range 1-2 km is about one-tenth that of the mare surface situated to the east in the Kepler quadrangle.

All rocks exposed on the plateau are included within a rock stratigraphic unit called the Marius Group. The group is divided into four mapping units. The first is smooth plateau materials, the second low domes, the third steep domes, and the fourth clusters of domes. Obviously, this is a geomorphic differentiation of the terrain reminiscent of the terrestrial map shown in Figure VIII-10.

FIGURE VIII-36. The Marius Hills region, showing a variety of domical features, rilles, and ridges. The dashed line indicates the boundary between the plateau underlying the Marius Hills and lower, more densely cratered, and presumably older mare material situated to the left (west). The area in the lower right is also shown in FIGURE VIII-35. For orientation refer to the dome and crater indicated by arrows on both figures (Lunar Orbiter Photograph IV 157H).

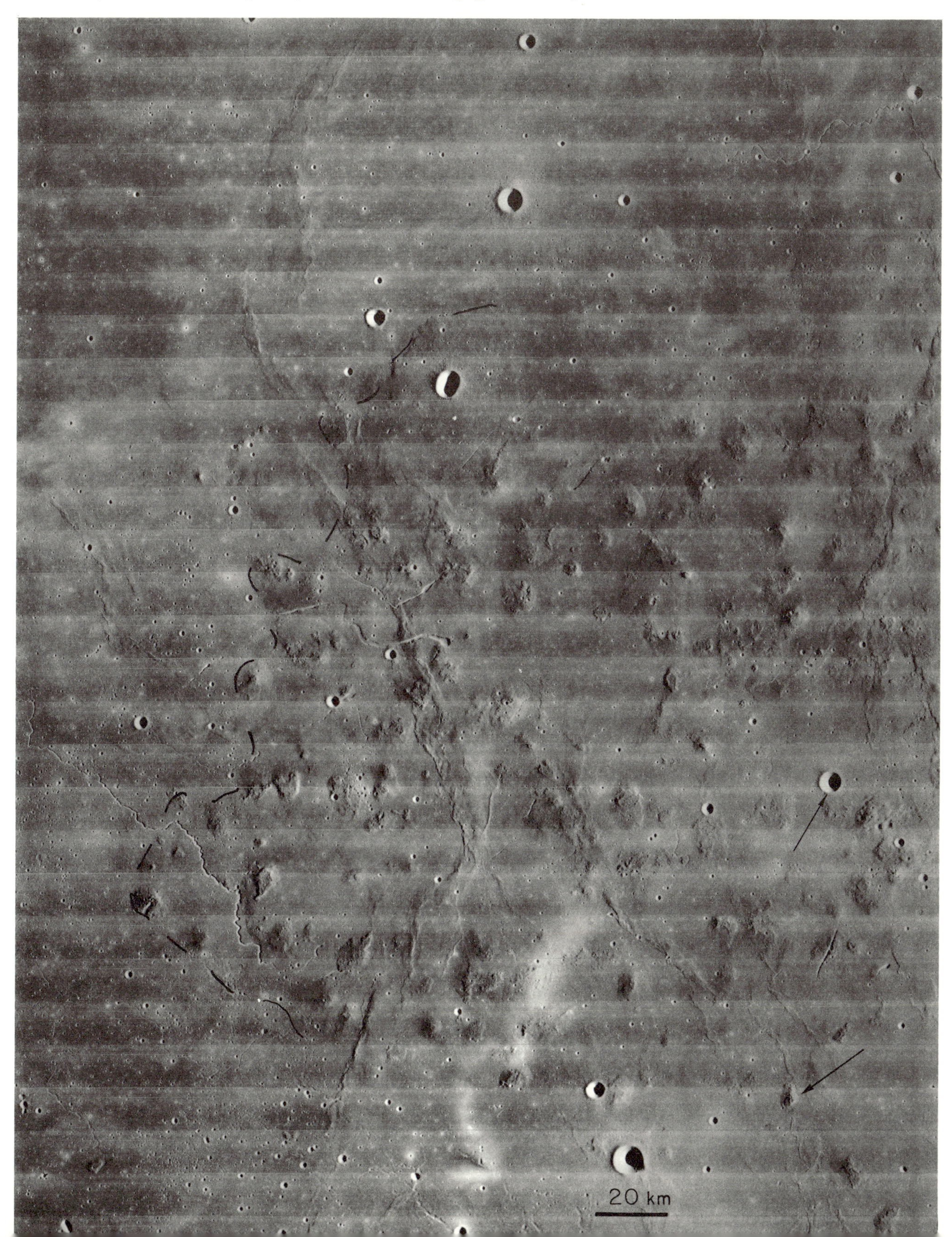

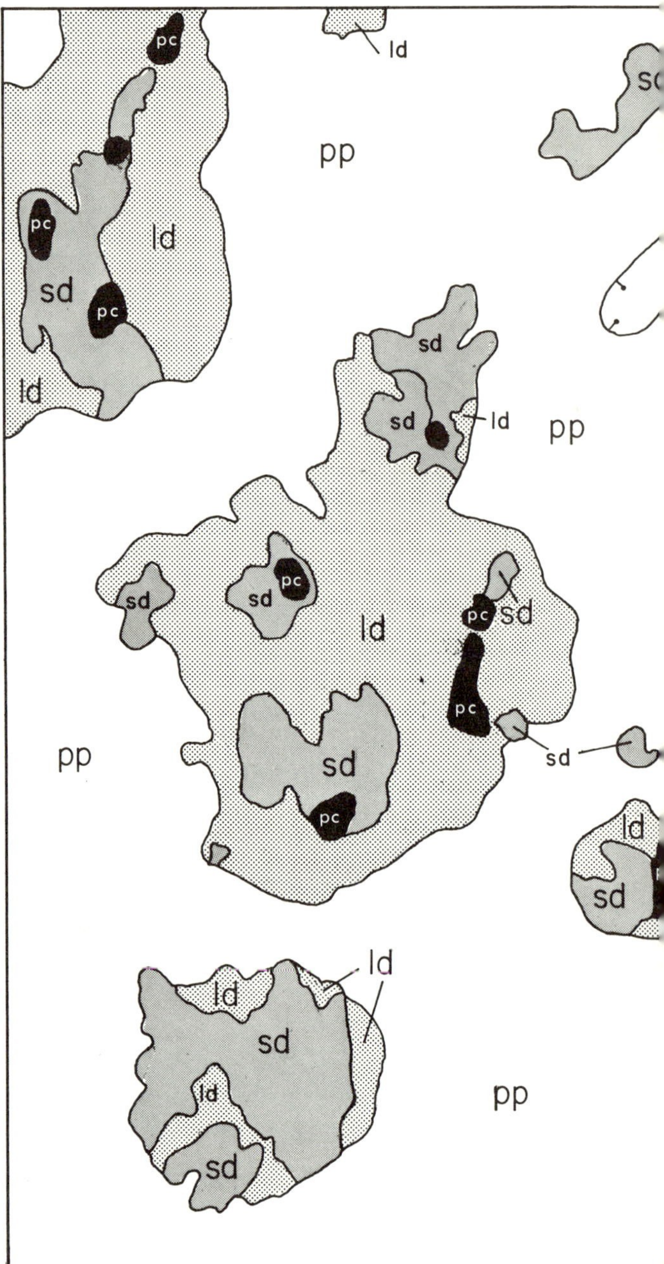

(A)

pc **PUNCTURED CONE MATERIAL**

CHARACTERISTICS:
Material of steep-flanked relatively smooth, conical structures having single or multiple linear depressions near the summits. Heights up to 300 meters above surrounding plains. Generally superposed on steep-sided domes.

INTERPRETATION:
Pyroclastic deposits surrounding structurally controlled vents.

sd **STEEP-SIDED DOME MATERIAL**

CHARACTERISTICS:
Material of steep-sided rough-textured domes generally perched upon low domes. Heights from 200 to 300 meters above the surrounding plains.

INTERPRETATION:
May be extrusive structures resulting from differentiation of the more primitive magmas that produced mare material and low domes.

ld **LOW-DOME MATERIAL**

CHARACTERISTICS:
Material of smooth-textured gently tumescent structures as much as 100 meters higher than surrounding plain.

INTERPRETATION:
Lack of flow front morphology suggests they are laccoliths (i.e. intrusive bodies that have domed up overlying rocks).

pp **PLATEAU-PLAIN MATERIAL**

CHARACTERISTICS:
Material of smooth to gently undulating cratered plains lying between domes and cones of Marius Plateau. Average elevation on the order of several hundred meters above adjacent maria.

INTERPRETATION:
Probably consists of volcanic flows, interbedded pyroclastic deposits, and a moderately thick impact regolith.

A temporal sequence of events can be reconstructed, based partially on direct observation and partially on circumstantial evidence. Because the Marius Group is less cratered than the surrounding Procellarum Group and because it appears to be topographically superposed on this unit, it is younger than Imbrian—assuming the Procellarum to be uppermost Imbrian. Because it is overlain by rays from Kepler and Aristarchus, it is older than these Copernican craters. According to these arguments, then, the age of the Marius Group must be Eratosthenian. Rudimentary temporal arrangement of map units within the group depends on observation of superposition relations and on interpretations of magmatic sequences. For example, McCauley suggested that the low domes might be underlain by basaltic lavas, relatively fluid at the time of their emplacement, and that the steep domes which commonly occur on the low domes represent more felsic rocks, viscous and pasty when they were extruded. This is reminiscent of the classic magmatic evolutionary sequence in which emplacement of felsic rocks follows that of mafic rocks (Bowen, 1928). Consistent with this sequence, McCauley's map explanation schematically shows the steep domes younger than the low ones.

McCauley's map explanation also indicates that the domes are in part contemporaneous with, in part younger than, the background plateau materials. This conclusion, entirely reasonable in the general sense, does not enable one to make superposition determinations for particular domes where there are no objective boundaries beyond the topographic ones at the base of the hill. It is possible that flows from some domes covered large areas. In this case the uppermost layers of plateau materials are, in fact, the same age as dome materials. However, rocks a few meters below the surface may be much older. The essential point is that complex vertical interfingering of wedge-shaped volcanic flows can take place across a plateau which is morphologically undistinguished (Fig. VIII-38).

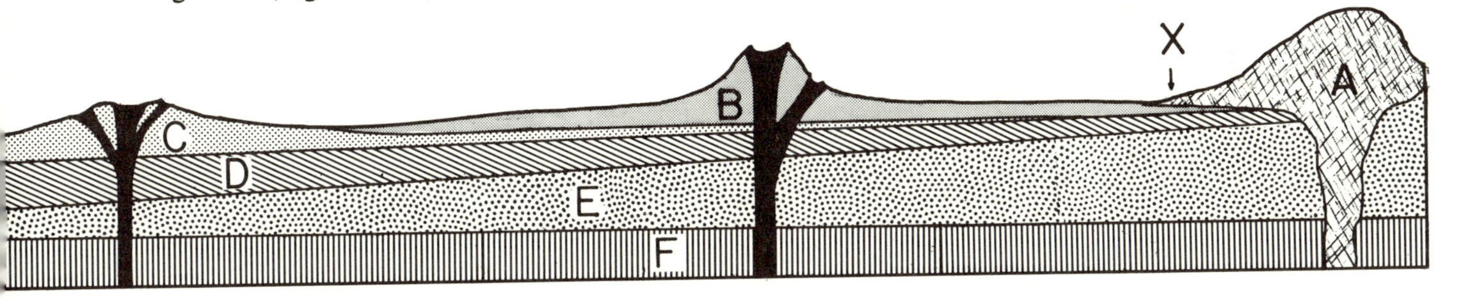

FIGURE VIII-38. Schematic cross section showing the potentially complicated stratigraphic situations which may not be reflected by distribution of volcanic constructional forms. F is the oldest volcanic stratum, A the youngest. Three volcanic domes are identifiable, but the areal extents of lava flows associated with each dome do not have clear topographic expression. At point X surficial rocks are young, but flows associated with an older dome B are encountered at shallow depth. Flows associated with dome C wedge out to the left of X so that the next unit encountered in a vertical section is formation D. The thinning trend of D suggests that these flows issued from a vent to the left of the section. Flows of formation E occur next, closer to the surface than at any other point in the cross section.

Opposite page:
FIGURE VIII-37. (a) A photograph showing, in detail, part of the Marius Hills region (Lunar Orbiter Photograph V 215M). (b) A geologic map of the area shown in (a). Volcanic structures are classified according to their morphology, and a temporal sequence of events is inferred (after McCauley, 1968*c*).

Rilles and related chain craters are particularly difficult features to classify stratigraphically; they more closely resemble structural elements. For that reason McCauley, in common with most other lunar mappers, did not assign them a position in the stratigraphic column but, instead, described them in that portion of the legend devoted to structural and other symbols which are not arranged according to stratigraphic position. However, the rilles must be younger than the rocks into which they are channelled. As in the case of the domes they may be only slightly younger. Indeed, material within a rille may be part of a larger sheet which flowed out across the adjacent countryside.

In reconstructing the history of the Moon it would be of great interest to determine whether or not igneous differentiation has taken place. Of all the regions so far studied in any detail the Marius Hills appear to offer the best chance for a positive answer to this question of differentiation. For that reason they are frequently mentioned as one of the more important sites for early on-the-ground exploration.

There are a number of hilly and elevated regions on the Moon with an origin probably similar to that of the Marius Hills. In these other regions the topography is more irregular and complicated; accordingly, deciphering the presumably complex volcanic history is more difficult. For example, in the eastern part of the Rümker quadrangle north of the crater Gruithuisen, large bulbous domes stud the highlands close to the mare boundary (Fig. VIII-39). In describing these protuberances R. E. Eggleton (personal communication) makes a distinction between certain low smooth domes which he interprets as shield volcanoes, and hackly domes and plateaus which he believes might be underlain by extrusive and intrusive plugs of intermediate or felsic rocks intercalated with subordinate pyroclastic rocks.

Another probable volcanic province with complex topography is the Aristarchus plateau situated to the north and west of the crater Aristarchus (Fig. VIII-40). Moore (1965) has named the plateau-forming materials the Vallis Schröteri Formation after the prominent rille which cuts through the western part of the plateau. In a geologic map of the Seleucus quadrangle (1967) he has divided the formation into five units: a dark smooth member, hummocky member, dome material, cone-crater material, and low-rimmed crater material. The dark smooth member which characteristically forms level surfaces within local basins is interpreted as a combination of lava flows, ash flows, and ash falls. The hummocky member may have a similar origin, with the undulating appearance due to underlying topography. Dome materials are either extrusive flows or intrusive laccoliths. The cratered-cone material forms rims and flanks of irregular high-rimmed craters, presumably volcanic vents. The low-rimmed crater materials, characterized by unusually low albedo, are thought to be maar deposits.

These five members doubtless have complex temporal and areal relationships, but criteria of superposition and topographic simplicity suggest that the dark smooth member and low-rimmed crater material may have been emplaced late in the sequence of volcanic events. The entire Vallis Schröteri Formation is lower in albedo than the adjacent mare materials. Because of this and also because it is topographically higher than

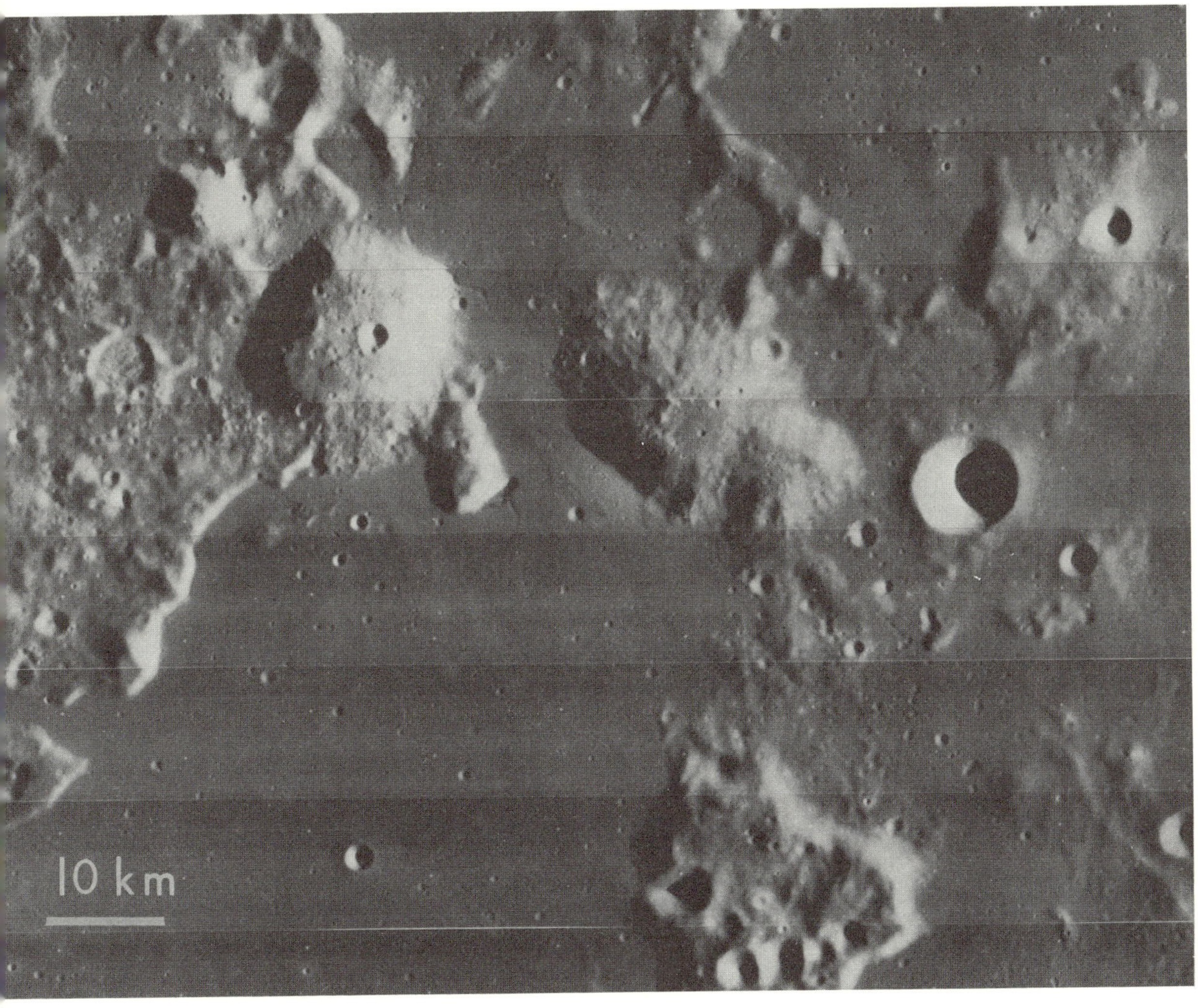

FIGURE VIII-39. Bulbous domes of probable volcanic origin north of the crater Gruithuisen (Lunar Orbiter Photograph IV 145H).

mare materials, a fact suggestive of superposition, it is assigned a younger Copernican or Eratosthenian age.

Extensive plateaus of probable volcanic origin are not confined to the maria but also occur in the highlands (Fig. VIII-41). Milton (1968) has mapped such a terrain, the Kant plateau, just west of the crater Theophilus. The plateau displays a regionally smooth to gently undulating surface which distinguishes it from adjacent, presumably older, terra with irregular texture. At lower elevations Milton mapped level plains with intermediate albedo as the Cayley Formation. Although interpreting both the Cayley Formation and the Kant plateau materials as volcanic, he made a distinction between viscous plateau-forming materials which tend to support elevated surfaces and less viscous lava and ash flows which spread across low plains.

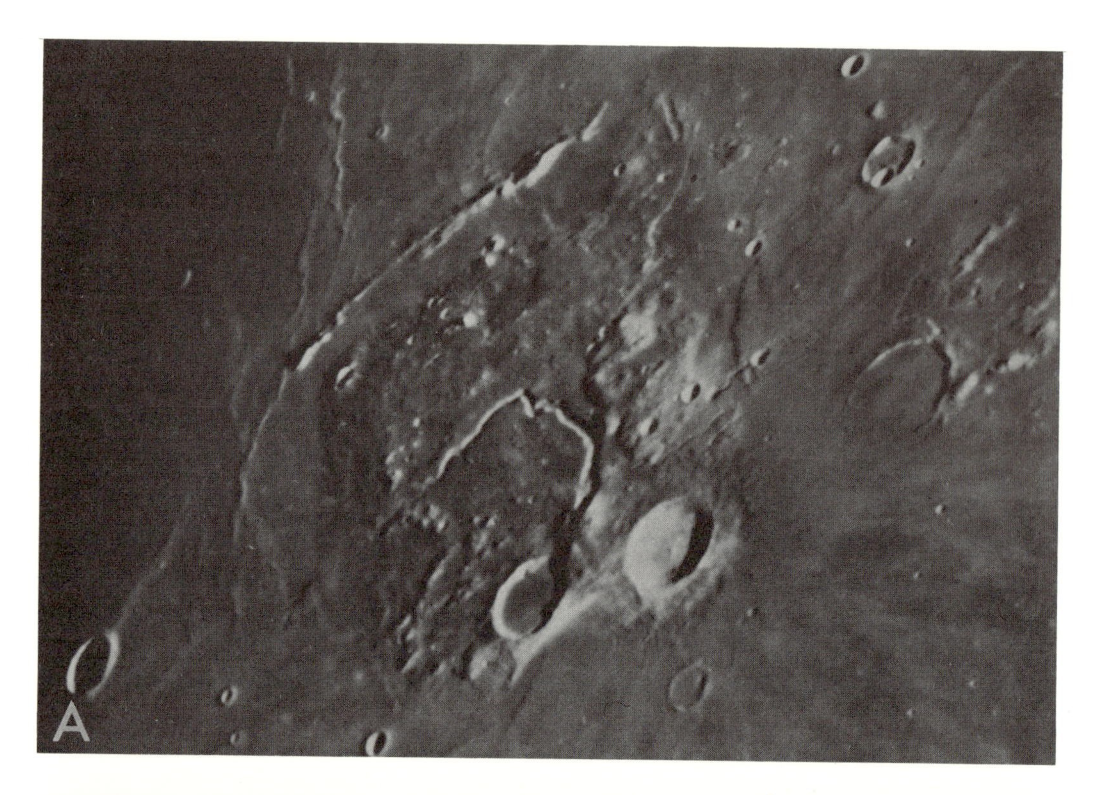

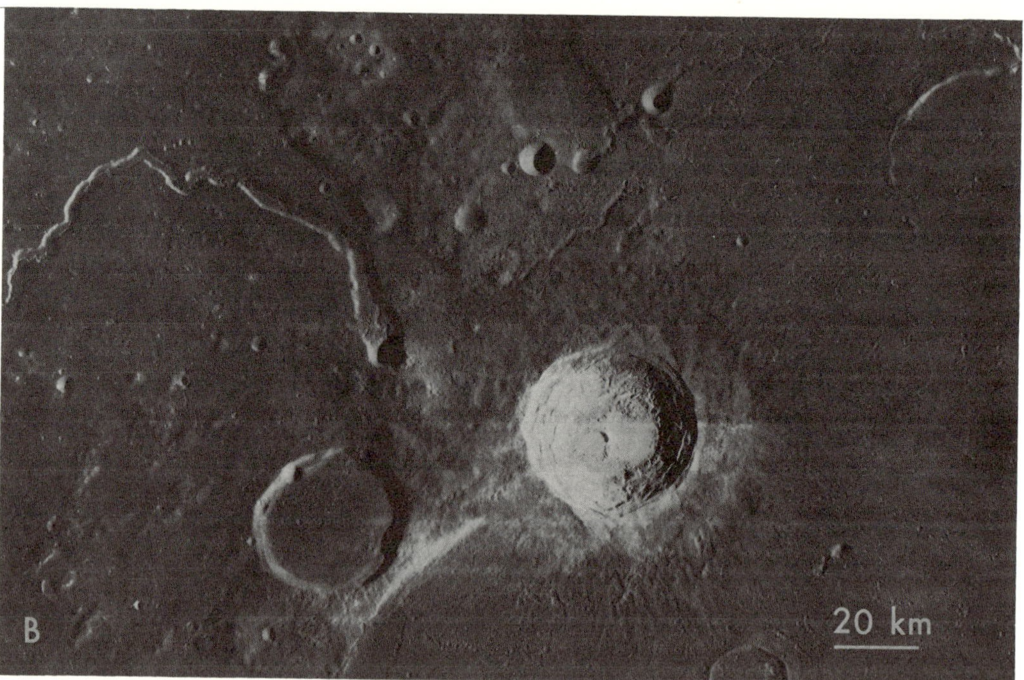

FIGURE VIII-40. (a) Telescopic photograph of the diamond-shaped Aristarchus plateau rising above Oceanus Procellarum (Catalina Observatory Photograph). (b) Orbiter photograph of eastern margin of the Aristarchus plateau. The prominent high-albedo crater is Aristarchus. Large rille at the left is Vallis Schröteri (Lunar Orbiter Photograph IV 150H).

FIGURE VIII-41. Displayed in the central part of this photograph is an elevated region with irregularly wormy texture. Materials in this region appear to overlie and fill in the adjacent terrain. The crater Descartes, indicated by arrows, displays this sort of post-crater modification along its northern margin. These in-filling materials are thought to represent viscous volcanic extrusions (Lunar Orbiter Photograph IV 89H).

FIGURE VIII-42. (a) Telescopic photograph showing, from north to south, the craters Ptolemaeus, Alphonsus, and Arzachel (Catalina Observatory Photograph).

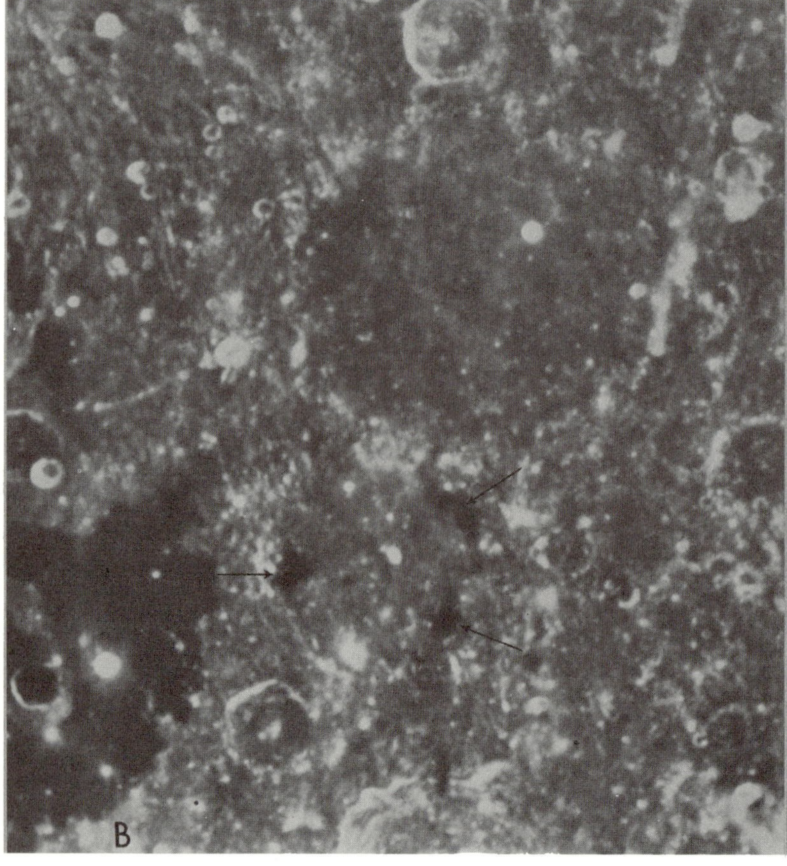

(b) Full-Moon telescopic photograph showing Ptolemaeus, Alphonsus, and northern part of Arzachel. Arrows point to three areas of dark albedo around the edge of the floor of Alphonsus. These are also visible in (a). Note that the floor materials of Alphonsus, which resemble mare material as photographed in (a), have markedly lighter and more mottled albedo than do the dark mare materials situated in the lower left (U.S. Naval Observatory Photograph).

Volcanic Stratigraphy

This review of possible constructional volcanic forms, starting with the Marius Hills and proceeding with increasing complexity through the Gruithuisen domes, the Aristarchus plateau, and the Kant plateau, indicates, on the one hand, the folly of equating volcanism with the formation of symmetrical cones topped by central craters, and on the other hand, the difficulty in positively identifying volcanic forms once one moves away from ideally conical forms. It should be of some comfort to note that the situation on Earth is very similar. In areas like the San Francisco volcanic field of northern Arizona, the individual volcanic cones are distinguished by their form alone. However, other volcanic provinces have a complex morphology reflecting a long and complex history. A good example is Iceland, lying astride the northern part of the mid-Atlantic ridge. Were the water drained from the Atlantic ocean, the polygonal block of volcanic rocks comprising Iceland would have an appearance not unlike that of the Aristarchus plateau rising above adjacent mare surfaces.

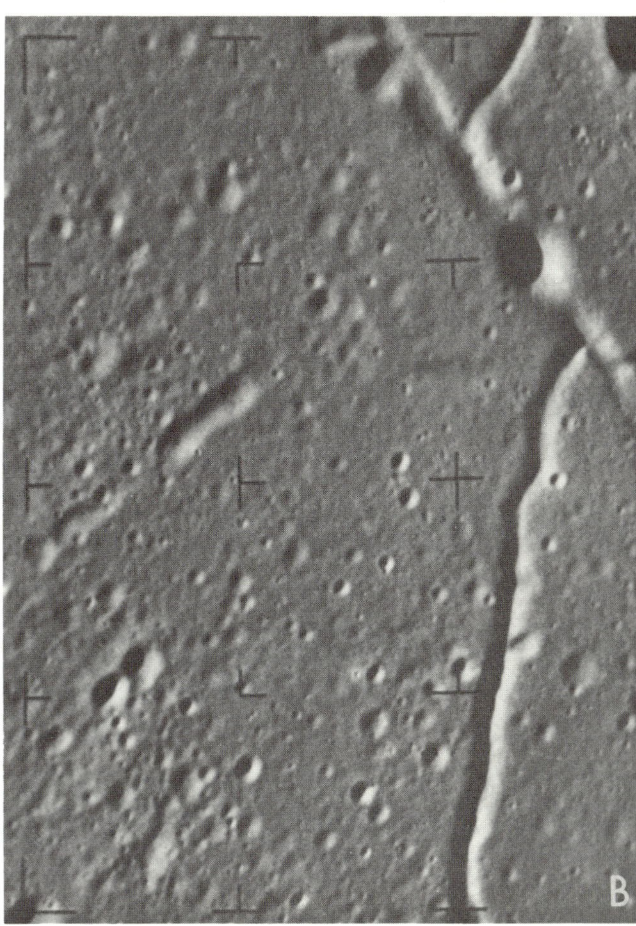

FIGURE VIII-43. Ranger IX photographs showing parts of the area geologically mapped in FIGURE VIII-44.

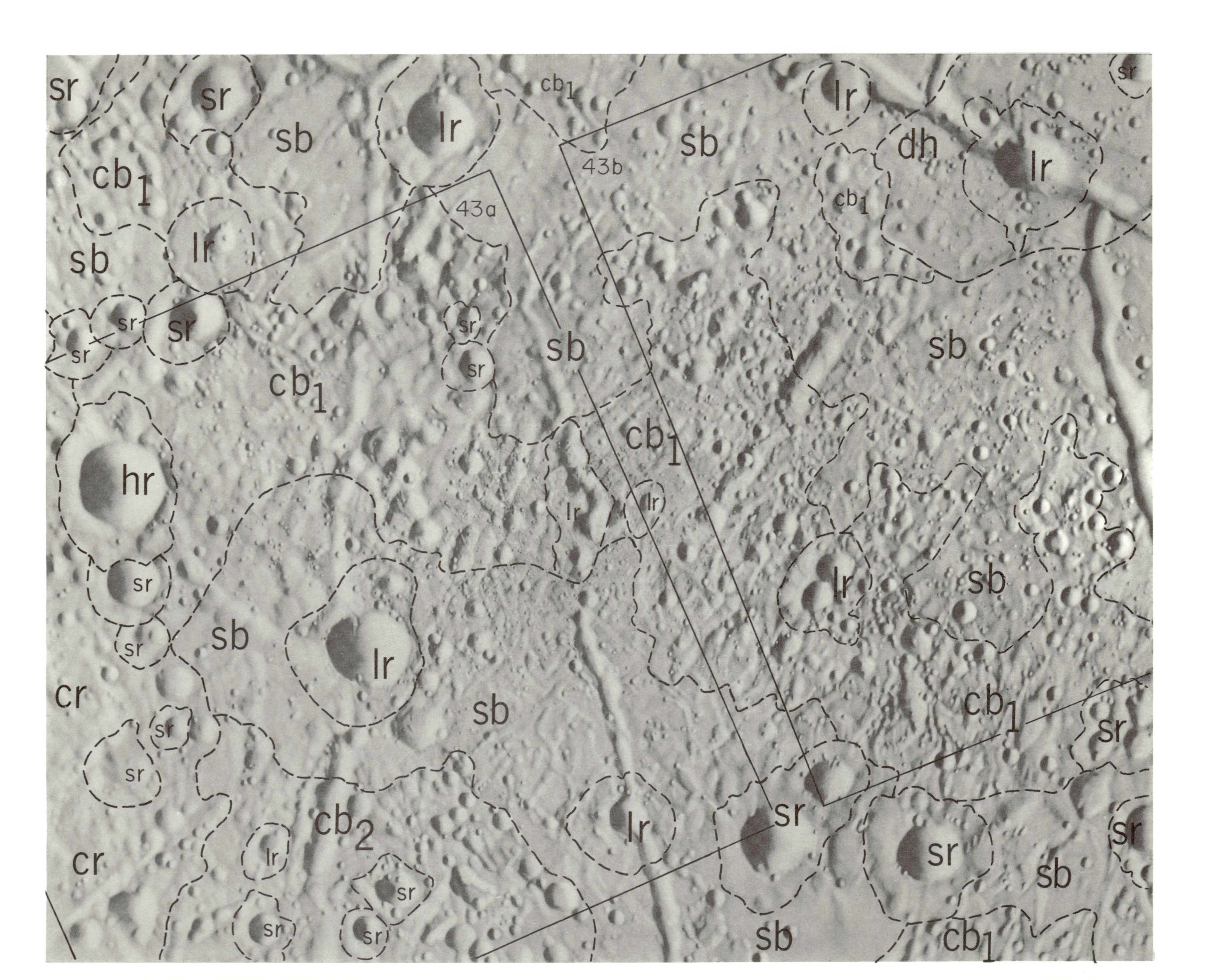

sr
lr
dh
sb
cb_1
43b
43a
hr
cr
cb_2

CRATER UNITS

hr

MATERIALS OF HIGH-RIMMED CRATERS

CHARACTERISTICS:
Occur in and around craters with steep interior slopes. Rim crest of associated crater is sharp and high.

INTERPRETATION:
Impact crater ejecta, fallback, and talus.

lr

MATERIALS OF LOW-RIMMED CRATERS

CHARACTERISTICS:
Rim of associated crater smooth. Profile is gently convex upward. (Numerous small low-rimmed craters mapped by McCauley are not included in this simplified version).

INTERPRETATION:
Mostly ash-fall deposits surrounding volcanic vents.

sr

MATERIALS OF SUBDUED RIMMED CRATERS

CHARACTERISTICS:
Occur in and around craters most of which have moderate to steep interior slopes. Rim crest of crater is rounded. (Numerous small subdued-rimmed craters mapped by McCauley are not included in this simplified version).

INTERPRETATION:
Mostly impact crater ejecta similar to that described for unit hr. Mass wasting has reduced original relief.

BASIN FILL UNITS

dh

DARK HALO MATERIAL

CHARACTERISTICS:
Albedo low; topography relatively smooth; crater density low.

INTERPRETATION:
Volcanic ash derived from structurally controlled vent.

sb

SMOOTH BASIN FILL

CHARACTERISTICS:
Topography smooth to gently rolling. Less cratered than units cr, cb1, and cb2; more cratered than dh.

INTERPRETATION:
Mostly volcanic ash-fall and possibly ash-flow deposits.

cb1 | cb2

CRATERED BASIN FILL

CHARACTERISTICS:
Topography moderately rough. Numerous superposed craters and subtle northeast and northwest-trending ridges and troughs are present. Unit cb1 shows a pronounced northeast lineation trend; unit cb2 shows both a northeast and a northwest lineation.

INTERPRETATION:
Older volcanic material extensively reworked by impact.

cr

CENTRAL RIDGE MATERIAL

CHARACTERISTICS:
Topography rolling to gently hummocky. Individual highs and lows have a general north-south orientation.

INTERPRETATION:
Structurally deformed crater floor material preferentially uplifted along north-south fractures by the Arzachel impact.

FIGURE VIII-44. Geologic map of a part of the floor of Alphonsus (simplified from McCauley, 1969). Areas covered by the two Ranger IX photographs of FIGURE VIII-43 are outlined.

ALPHONSUS

Slightly different problems in volcanic mapping are illustrated by relationships within the crater Alphonsus. This is a large crater, 115 km in diameter, a member of a prominent triplet of craters including Ptolemaeus and Arzachel (Fig. VIII-42). For a variety of reasons these craters have often been interpreted as volcanic structures (e.g., Fielder, 1961; 1965). They are alined in a north-south direction and display polygonal outlines, both features suggesting tectonic control. Ptolemaeus and Alphonsus have flat, shallow interiors, suggesting filling from internal sources. Finally, Alphonsus has been the site for a number of reported transient events (Middlehurst, 1967). This evidence is balanced by that for impact origin. Alphonsus has a raised rim with rough topography similar to that of other impact craters but quite unlike that of terrestrial calderas (Smith, 1966). Furthermore, it is possible that the shallow interior is in part a consequence of rebound of the crater floor following impact (Masursky, 1964). If the floor is restored to its assumed original position, the crater shape closely resembles that of impact craters such as Copernicus. Whatever the origin of the crater itself, any number of structures within the interior of the crater attest to a certain amount of volcanic activity following initial formation.

Unlike the Marius Hills, the floor of Alphonsus is a terrain dominated by negative features. Evidence for volcanic activity is found in numerous rilles, linear depressions, and crater chains which trend north and northeast, apparently controlled by internal fracture zones. A large proportion of the craters shows features different from those of typical impact craters. Some have rims that are wide, smooth, and elongate or irregular in outline, and resemble rims around some terrestrial maars. Moreover, some rims in Alphonsus are convex upward, whereas all known impact crater rims are concave. Some craters are surrounded by smooth dark halos not seen around known impact craters.

A stratigraphic sequence worked out by McCauley (1969) is shown in Fig. VIII-44. Four widespread units are recognized on the basis of topographic relief, crater density, surface texture, and in one case, albedo. The unit interpreted as oldest includes hummocky, lineated material underlying a high ridge. Proceeding up the stratigraphic section the remaining units are progressively smoother, less cratered, and are exposed at progressively lower elevations. The stratigraphic analysis is based in part on the structural argument that successive horst and graben-like uplifts elevated rocks just deposited and created new depressions for succeeding lava lakes. Those rocks raised the highest are also the oldest. At first glance this may appear to be an inverted stratigraphy, but the model yields a normal vertical section at any point (Fig. VIII-45). The "inversions" are actually systematic post-depositional structural displacements.

McCauley's map of Alphonsus is noteworthy for including one of the first attempts to use crater statistics as an aid in stratigraphic mapping at a detailed scale, in this case 1:50,000. For each of the regional units, crater densities in the diameter range 0.2-2 km were measured. Craters were further subdivided according to the percentage of the

FIGURE VIII-45. Schematic model showing the type of structure and stratigraphy represented by volcanic rocks in the floor of Alphonsus. For explanation of stratigraphic units see FIGURE VIII-44. The favored interpretation is that faulting and volcanic activity occurred concurrently. The block on the far left was upraised prior to the deposition of cb_1 and cb_2, so that the edge of the block formed a retaining wall for this pool of lava. Similarly, the middle block was uplifted following deposition of cb_1 and cb_2 but prior to deposition of sb. An alternate interpretation, which cannot be discounted considering only the relations of this schematic section, is that all faulting occurred following deposition of sb and that erosion has subsequently stripped the younger units from the elevated blocks.

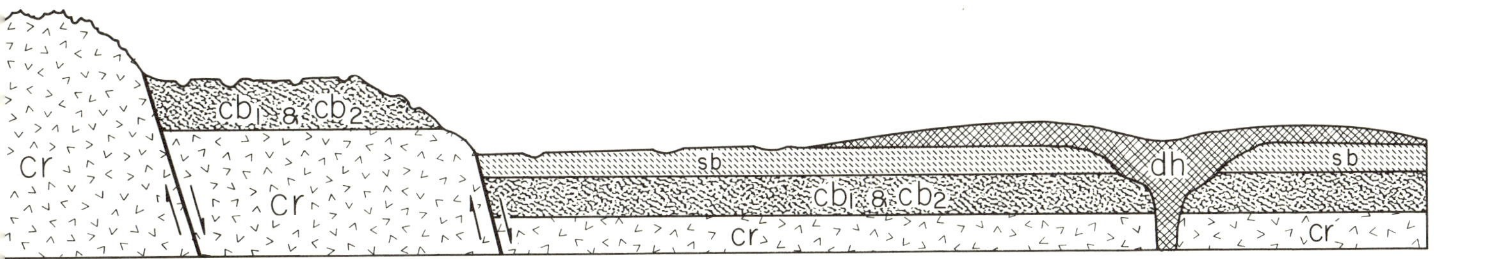

interior in shadow, this being a measure of crater depth and shape. The results of this analysis are helpful but not unequivocal. Crater populations are so small that the statistical errors are certainly large. Because the craters could have been formed by processes of volcanic eruption or by drainage as well as by impact, the genetic significance of relative crater densities is difficult to evaluate. Nonetheless, cumulative curves are noticeably different for the four regional units and do corroborate the thesis of decreasing crater density with decreasing age.

VOLCANISM AND CRATER FORMATION

The possibility that certain craters may have a volcanic origin has been mentioned both in this chapter and in preceding chapters. Distinctions have been drawn between craters which resemble calderas and those which display impact features. In addition, evidence for internal flooding of certain craters—such as Alphonsus—has been presented.

Recent studies suggest that the roles of volcanism and impact-induced evacuation may be more closely entwined than the preceding discussion would suggest. The arguments favoring a close relationship are well illustrated by the geology around Tycho as revealed by Orbiter V and by Surveyor VII. The review which follows is based on data and interpretations contained in Surveyor VII Mission Report (1968*b*).

One of the most remarkable features of the landing site shown in Surveyor VII pictures is the abundance of large, angular rock fragments which litter the surface (Fig. IV-15). The rocks show a diversity of textures and structures, many of which suggest volcanic processes (Fig. VIII-46). Some rocks are covered with irregularly shaped, indistinctly bounded white spots. These spots may represent mechanically broken fragments cemented together by darker matrix material. Another possibility is that they reflect variable crystallinity within a partially crystallized volcanic rock.

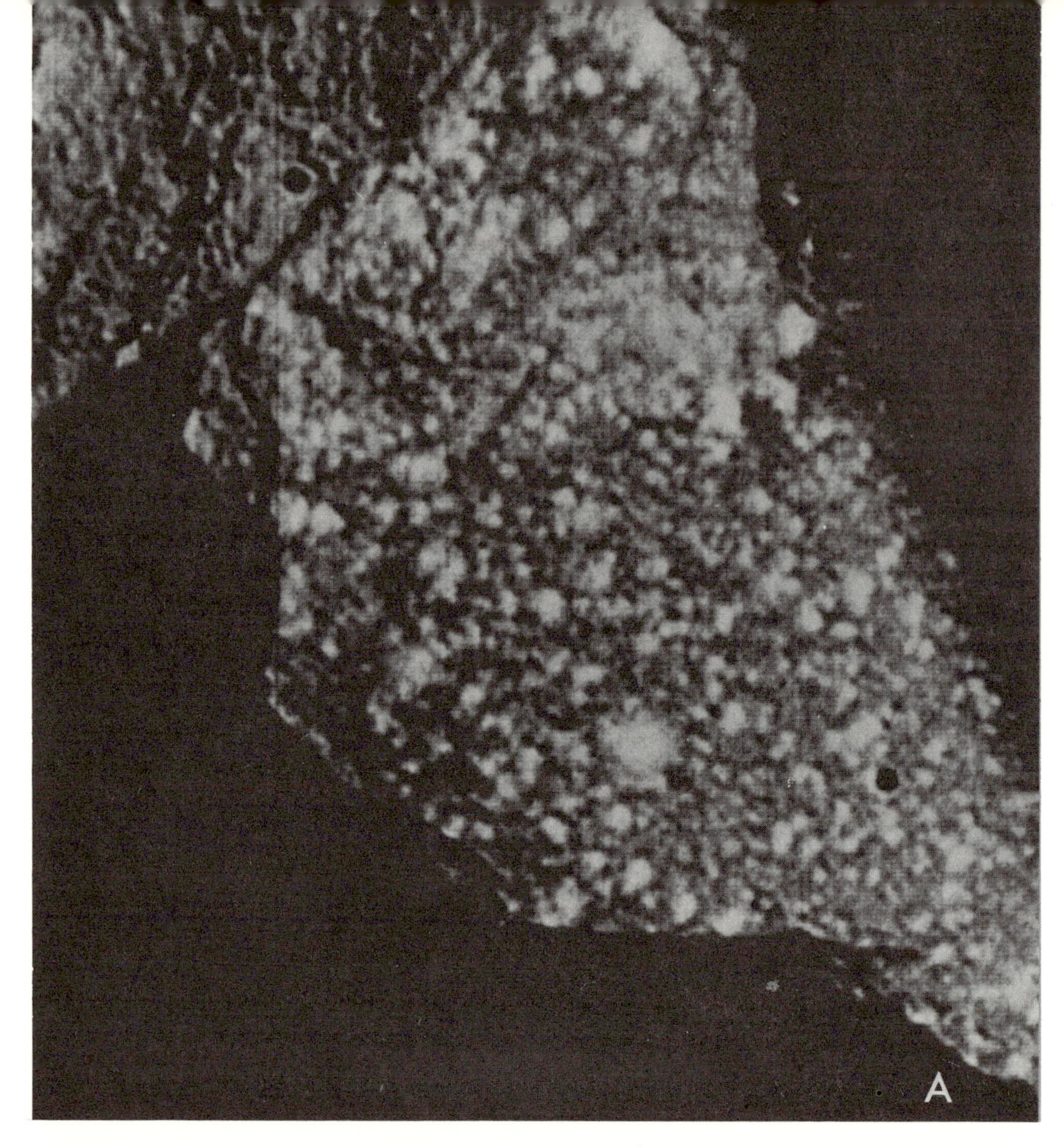
A

D

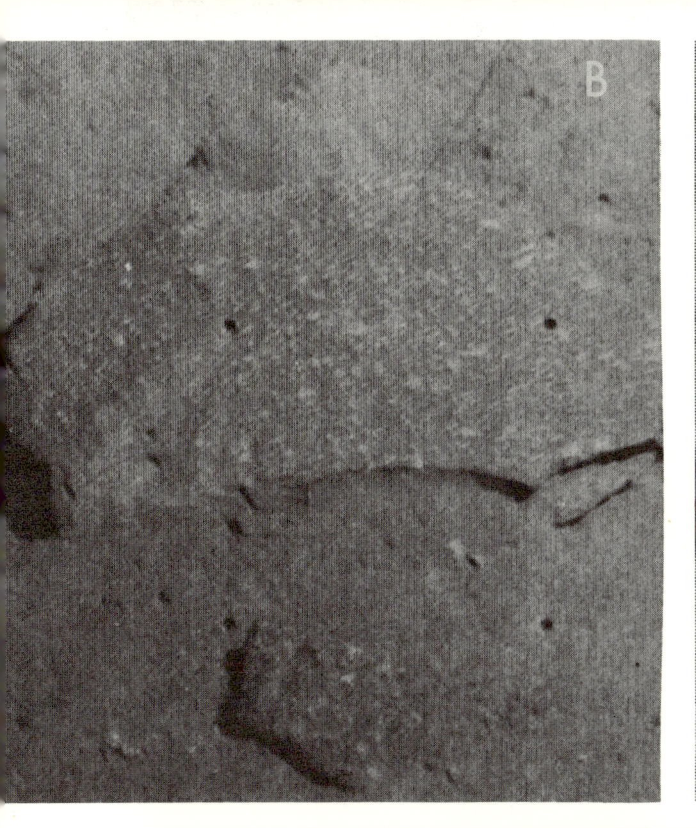

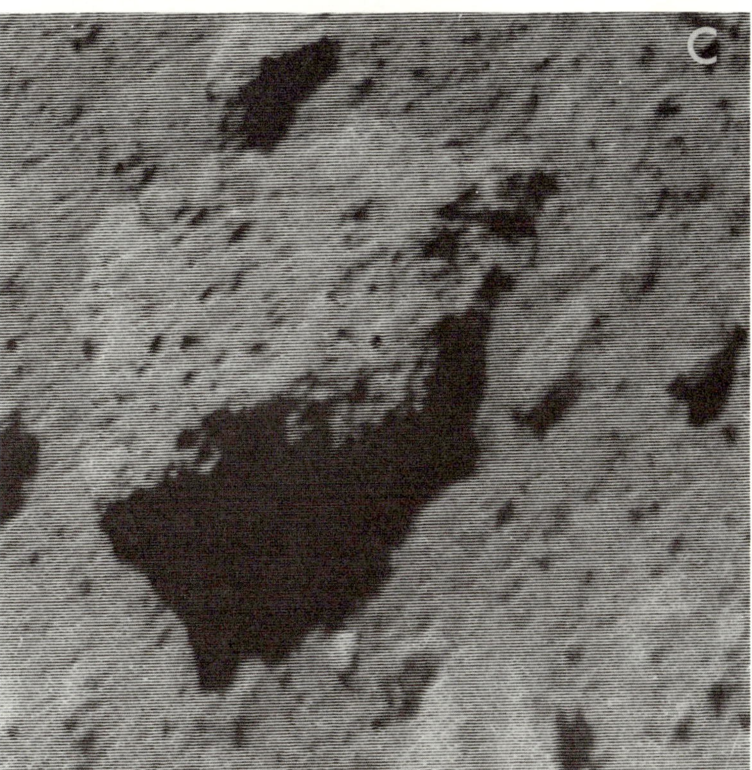

FIGURE VIII-46. Rocks photographed by Surveyor VII. (a) Spotted rock about 25 cm across. The spots range in size from less than 1 mm to 3 cm. Their origin is not clear, but their indistinct boundaries and irregular shapes suggest they are not phenocrysts. They may be determined by a fragmental texture within the rock. (b) Block about 18 cm across. Bright elongate spots about 1 mm wide and up to 10 mm long are roughly parallel, a fabric similar to that found in terrestrial volcanic rocks as a result of flow lineation. (c) Vesicular rock about 10 cm across. (d) Vesicular fragment about 35 cm across. The rock shows subdued banding from lower left to upper right. The vesicles are elongate and define a second lineation set which trends from upper left to lower right, forming an angle of about 70° with the banding. (e) A broken rock about 30 cm across. The cracks may be due to thermal expansion and contraction.

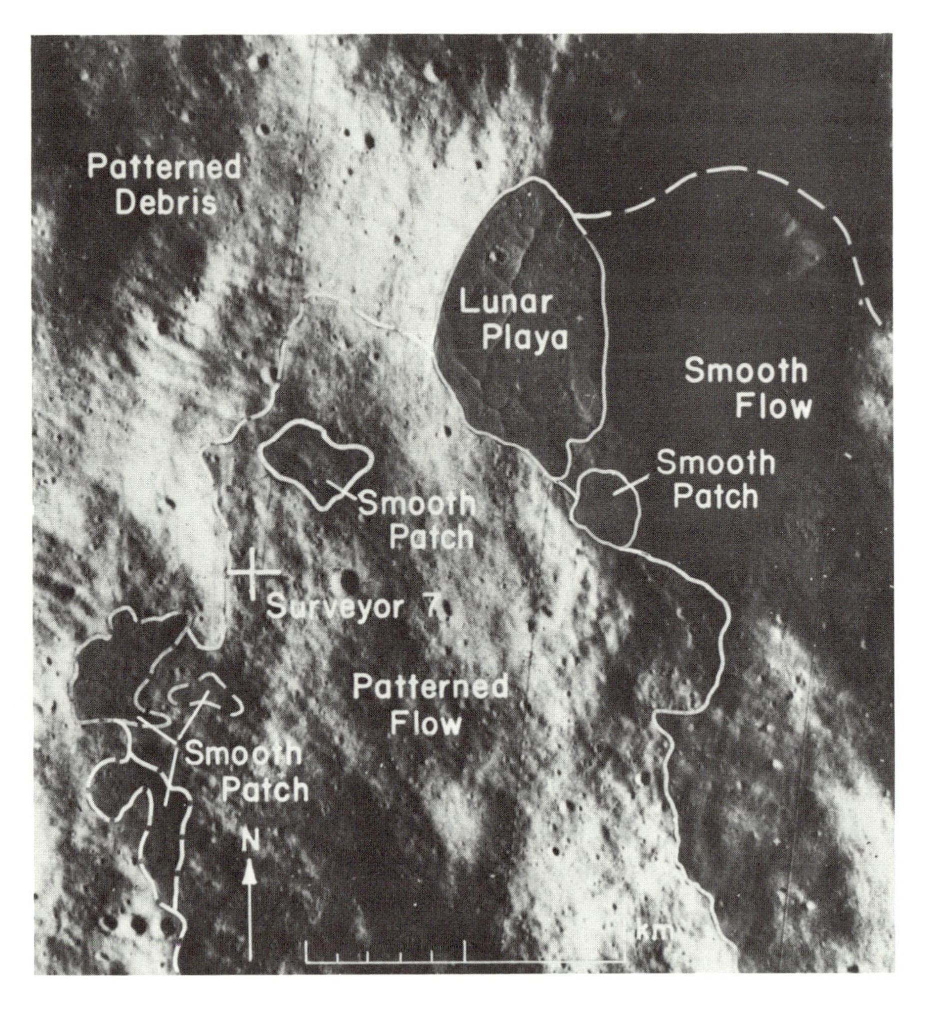

PATTERNED FLOW

CHARACTERISTICS:
Surface composed of irregular low hills and depressions, ranging from 100 meters to several hundred meters across, studded with smaller blocks and craters. Abundant fissures.

INTERPRETATION:
Emplaced as extremely viscous fluid flow of hot partially melted debris ejected from Tycho. Fissures formed by spreading of flow.

SMOOTH FLOW

CHARACTERISTICS:
Smooth undulating surface with rimless flow channels and flow lineations.

INTERPRETATION:
Emplaced as fluid flow of melted ejecta from Tycho.

LUNAR PLAYA MATERIAL

CHARACTERISTICS:
Smooth, level surface on floor of closed depression Branching fissures.

INTERPRETATION:
Emplaced as very low viscosity fluid flow of charged ejecta from Tycho. Fissures are cooling contraction cracks.

SMOOTH PATCH MATERIAL

CHARACTERISTICS:
Relatively dark, smooth, level surface. Occurs in small depressions and on benches.

INTERPRETATION:
Emplaced as low viscosity flow of melted ejecta from Tycho. Similar to lunar playa material but deposits are thinner.

PATTERNED DEBRIS

CHARACTERISTICS:
Smooth surface with pattern of gentle ridges and shallow grooves superposed on larger hillocks and swales.

INTERPRETATION:
Fragmental debris ejected from Tycho.

FIGURE VIII-47. Annotated Orbiter photograph of the Surveyor VII landing site showing different geologic units (simplified from Shoemaker and others, *in* Surveyor Program, 1968*b*).

Other rocks display small elongate spots whose long axes tend to be oriented parallel with one another. Similar textures are extremely common in terrestrial volcanic rocks in which early-formed phenocrysts become alined during later stages of magma flow. The sub-prismatic shape and relative brightness of the elongate spots suggest that they may be feldspar crystals. Many rocks in the vicinity of the Surveyor VII spacecraft are pitted. The cavities are probably formed by exsolution of a volatile phase at the time the material was molten. A few rocks display two sets of linear structures. One set may correspond to primary flow banding formed during igneous crystallization; the second set may include structural cleavage planes formed during subsequent shock metamorphism.

Interpretations of the Surveyor VII observations are enhanced by coupling them with analysis of Orbiter V photographs. The latter permit a regional geologic survey within which the detailed relationships revealed by Surveyor VII can better be appreciated.

The interior of Tycho contains extensive flows present both on the terraced walls and also on the floor where they form a fissured, cracked surface (Fig. IV-12). Beyond the rim crest three concentric rings can be defined on the basis of topography and albedo (Figs. IV-10 and 11). The inner ring is marked by relatively high albedo, hummocky topography, and concentrically lineated texture (Fig. IV-13). The second ring has a relatively low albedo and is marked by radial ridges and valleys. The third ring has intermediate albedo and is characterized by numerous secondary craters.

Superposed on the first two rings are small flat areas confined within local depressions. The surfaces of the flat areas are locally marked by systems of branching grooves. Also scattered across the two rings are large bifurcating lobes and small regions of positive relief with alternating sinuous ridges and complexly textured depressions (Figs. IV-13 and 14). These features may have been formed by any one of several flow mechanisms (Shoemaker and others, *in* Surveyor Program, 1968*b*). First, they may be truly volcanic, formed by extrusion of lava from depth. Second, they may represent flow of cold debris fluidized by addition of liquid water or gas. Third, they may be hot debris flows mobilized by shock melting and by concurrent generation of gas.

The last possibility clearly associates the flows with the impact event which formed Tycho. The first possibility could be interpreted as extrusion of lava in response to forces which formed the crater, but it is equally reasonable to argue that the volcanism occurred some time after crater formation.

The consequences of the choice one makes between the three possibilities just mentioned are considerable. Figure VIII-47 shows that the Surveyor spacecraft landed on a patterned flow unit believed emplaced as fluid flow of melted ejecta from Tycho (Shoemaker and others, *in* Surveyor Program, 1968*b*). At this site the alpha-scattering instrument analyzed three samples: undisturbed lunar surface materials, a small rock, and a disturbed area exposing subsurface material. The chemical analyses resemble those obtained at mare sites except for a marked decrease in iron. The significance of that decrease depends on the model accepted for evolution of this part of the highland surface. The Surveyor VII site was originally selected for a landing because the fresh appearance of Tycho suggested that its ejecta blanket was recently emplaced and should therefore contain fragments of subsurface highland rock little modified by addition of meteoritic material or by surface alteration. This strategy appears reasonable but leaves open the additional question of how heterogeneous in chemical composition are rocks within the highlands. In particular, if one accepts the thesis that Surveyor VII landed on a blanket of mixed and melted surface or near-surface debris, then there is reason to assume that the chemical analyses are representative for that part of the crust in which the crater formed. If, however, one maintains that the Surveyor VII flow formed by melting at depth and extrusion of lava, then there is no reason to expect that the chemical analyses represent anything more than a special case of chemical differentiation during partial melting of the crust.

Accepting the former point of view that the analyses are representative of the highland rock, one can convert the lower iron values into possible variations in silicate mineralogy, and from that point, into differences in rock density. Making no allowance for porosity, the density for highland rocks is 3.0 and for maria rocks is 3.2 (Shoemaker and others, *in* Surveyor Program, 1968*b*). Since the mean density of the Moon is 3.34, this result suggests not only lateral differentiation between low-density materials in the highlands and high-density materials in the maria but also some compositional differentiation within the deeper interior of the Moon.

Highland Stratigraphy

CHAPTER

EARLY WORK

THE dark maria dominate that half of the Moon which faces the Earth, and consideration of their origin has been strongly emphasized in most geologic studies. Intensive study of the cratered highlands which occupy the southeastern quadrant of the nearside was initiated only following receipt of Orbiter IV and Surveyor VII results. The logic in this tacit decision to concentrate attention first on the maria is undeniable—the simple should precede the complex—but the recent revelation that nearly the entire farside of the Moon is highland (Figs. IX-1 and 2) dramatizes the need to understand the geology of this complex terrain which extends over more than three-quarters of the Moon's surface.

The units which have proved so useful in deciphering a stratigraphy for the mare basins are presently unrecognized or are, in fact, lacking in the highlands. Pools of dark mare-like material are so restricted that it is impossible to make extensive use of the preflooding-postflooding distinction. Notwithstanding the abundance of large craters it is difficult to identify blanketing ejecta deposits, let alone extend them over great distances. The reason for this is that the rugged background tends to distort and mask any superimposed deposits possessing distinctive but subtle patterns. The problems are something like separating a particular electronic signal from a penetrating background of random noise. A few large highland craters, notably Tycho, do have clearly resolvable "signals," but only because the cratering events are very recent. Their value as stratigraphic discriminators is minimal since their ejecta blankets and secondary craters are superimposed on the entire spectrum of adjacent stratigraphic units.

First efforts to unravel the stratigraphy of the southeastern highlands included attempts to establish a series of overlapping stepping stones leading from mare basin margins into the heart of the highlands. A series of six regional ejecta blankets, several associated with specific basins and others identified only according to source direction, was described (Elston, 1965). Although conceptually attractive, the postulated stratigraphic units lacked objective criteria necessary for their unique identification. Consequently it proved impossible to use the model as a guide for further mapping.

FIGURE IX-1. The farside of the Moon (Lunar Orbiter Photograph V 5M).

URE IX-2. The farside of the Moon. The crater flooded with dark mare material is Tsiolkovsky (Lunar Orbiter Photograph III 121M).

The stratigraphic arrangements for most maps of the highlands have been based on a judgment regarding the relatively ancient (pre-Imbrian) or recent (Imbrian-Copernican) character of the terrain. For example, the surface of the Rupes Altai quadrangle was interpreted to be largely pre-Imbrian (Rowan, 1965). However, in the Theophilus quadrangle, situated directly north of Rupes Altai, Milton (1968) interpreted the absence of surficial features attributable to the Imbrian impact event as an indication that the earliest Imbrian surface has been covered by younger materials, many of which are themselves assigned an Imbrian age. Milton subdivided units according to topographic position and form, with emphasis more on implied differences in origin than on probable temporal successions. Many of the units are interpreted as igneous materials: viscous lava flows underlying plateaus, pyroclastic materials and subordinate lava flows forming deep fill in basins, and inferred intrusive plutons at shallow depths.

Fortunately there is one type of regional superposition widely displayed throughout the highlands. This involves flooding of craters and intercrater depressions with light materials which form markedly level plains. Surfaces of the plains are variably pitted by craters but never in sufficient density to obscure their level, homogeneous character. First interpretations of such deposits where they occurred adjacent to large basins were similar to the interpretation of the Apennine Bench Formation in the Imbrium basin—younger than the basin ejecta deposits but older than the dark mare materials. However, many occurrences of light plains materials in the highlands are restricted to crater interiors and spatially unrelated to larger basins. Some of these isolated pools are so smooth and uncratered that a Copernican or Eratosthenian age is suggested.

As Wilhelms (1970) has pointed out, each new map demonstrates a little more of the real complexity contained in units first grouped together as Cayley (Apennine Bench)-like plains materials. Additional local subdivisions are necessary, and temporal correlation over great distances remains hazardous.

In summary, the stratigraphic column which emerged from early work in the highlands is approximately as follows: undivided pre-Imbrian materials and degraded crater materials, ejecta deposits from basin-forming events, light plains-forming materials, darker mare materials, and relatively fresh crater materials. Individual maps show many more units but they all are hung on this basic framework.

A CLOSER LOOK

The problems inherent in highland mapping are best illustrated by looking in detail at a single area, in this case the Hommel quadrangle. The eastern and southern parts of the region are rugged, with sets of linear hills trending in several directions (Fig. IX-3). Some of these boldly dissected surfaces stand adjacent to large craters, notably Hommel, Pitiscus, and Mutus. A first conclusion might be that the hills are formed by hummocky ejecta deposits, now somewhat subdued and degraded just as their inferred parent craters are. Closer examination provides little support for this contention. The hills are not concentrically arranged around one or several large craters. Instead, lineations which form the valleys between adjacent ranges march across crater rims, walls,

FIGURE IX-3. The region around the crater Pitiscus in the eastern part of the Hommel quadrangle. This rugged highland surface is transected by several sets of linear hills. A prominent set trends from lower right to upper left. No rim deposits are identifiable around Pitiscus. Presumably they have been modified beyond recognition by more recent structural events (Lunar Orbiter Photograph IV 88H).

and floors without deflection. Assuming the lineations mark the position of faults it seems likely that the crosshatched fabric of the highlands is principally structural, not depositional, in origin.

The western part of the quadrangle contains a variety of flat plains, some relatively uncratered, others densely pockmarked (Fig. IX-4). The presence of particularly smooth, uncratered plains in the floors of the majority of large craters is striking. Beyond that, subdivision is unexpectedly difficult. What at first examination appears to be a real boundary within plains materials is no sooner drawn than one is tempted to move it or even to remove it. Distinctions on the basis of crater densities, so clear in the abstract or for selected areas, are difficult to apply systematically throughout the quadrangle.

FIGURE IX-4. Variably cratered plains in the western part of the Hommel quadrangle. Note the presence of relatively uncratered plains inside the craters, particularly inside the crater in the upper far left (Lunar Orbiter Photograph IV 94H).

The relation between materials of the rugged lineated highlands and materials of the plains is perhaps the central stratigraphic problem in the Hommel quadrangle. At first glance it seems intuitively clear that the smooth plains are younger than, and distinctly different from, the lineated highlands. However, there are several inconsistencies in this simple model. First, the crater densities on many of the plains surfaces are greater than for the highlands. Even the most generous counting of ill-defined craters in the highlands and the most restrictive numeration of sharp-rimmed craters on the plains fail to erase the difference. Secondly, the junction between plains and highlands is seldom a knife-edged contact. The vagueness of the boundary is well illustrated in the central part of the quadrangle. Most of the plains display faint lineations and, traced from west to east, merge imperceptibly with surfaces which take on more and more of the attributes of the highlands.

There is one model which well explains the observed relationships. The basic idea is that repeated vertical movement has taken place along two sets of fractures trending northeast and northwest (Figs. IX-5 and 6). Some lineations, presumably the surface reflections of joints or faults, can be traced continuously across the quadrangle. They divide the region into several square or diamond-shaped blocks. Differential subsidence of blocks during the geologic development of the area has had important stratigraphic consequences. First, the depressed blocks have become sedimentary basins. Secondly,

FIGURE IX-5. Tectonic map of the Hommel quadrangle. High structural blocks are underlain by structured terra material of probable pre-Imbrian age. Intermediate blocks are underlain by topographically subdued terra material partially mantled by volcanic flows. Low blocks are completely flooded with cratered plains material of volcanic origin.

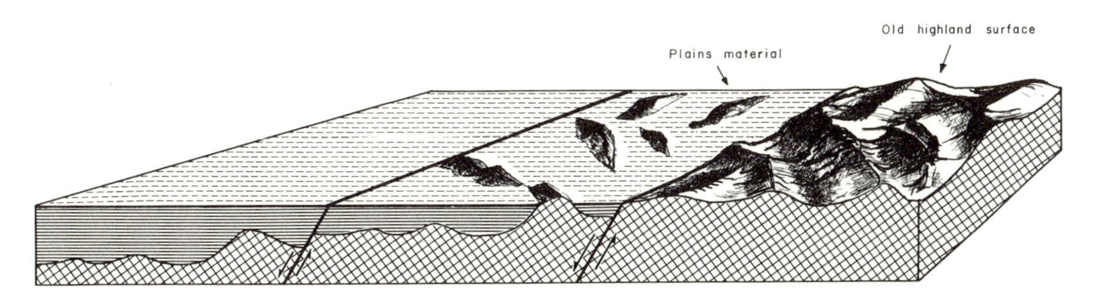

FIGURE IX-6. Schematic block diagram illustrating movement of structural blocks and superposition of materials resulting in the plan view shown in FIGURE IX-5.

boundary faults have formed natural conduits for the upward surge of lava. Indeed, evacuation of an underground magma chamber, foundering of the overlying block, and emplacement of lavas on top of the depressed block may be causally related events. If there is no lateral subsurface movement of magma, then the total thickness of the lava beds must be approximately equivalent to the amount of block subsidence.

Lavas would have to accumulate to a depth of a thousand meters or more before the underlying topography was completely covered. Sequences several hundred meters thick would flood local depressions eliminating the minor relief elements but not submerging the major promontories. But in this latter case associated ash falls and flows might mantle the entire region, forming smooth, gentle slopes on formerly steep escarpments. This sort of a history explains the observed continuum of surface textures. Where block subsidence has occurred and attendant volcanic flooding has been great, all that is seen is a flat, cratered plain. Where flooding of intermediate magnitude has taken place, the underlying topography is partly preserved either as steep hills rising above the lava plains or as gentle undulations in a cover of ash.

The model just described is at variance with the traditional view that the highlands are underlain by ancient crust modified principally by swarms of overlapping impact craters. Instead, volcanic processes assume central importance. Admittedly the argument is circumstantial. There are perhaps fewer indisputable examples of volcanic features than there are in the Marius Hills. But there are a number of smoothly convex-upward hills, some of which are situated on lineaments. In the center of the quadrangle where several major lineaments intersect there is a unique waffle-like network of intersecting ridges with smooth plains between. These appear to be linear extrusion ridges building along intersecting fractures (Fig. IX-7).

The failure to find distinct, unmodified volcanic features is at least consistent with the absence of displaced marker elements along the presumed block-bounding faults. A reasonable explanation is that both faulting and volcanic activity occurred relatively early in lunar history and that the features which might uniquely distinguish these processes have been degraded beyond recognition.

The emphasis on volcanism supplies a possible explanation for the anomalously different crater densities between lineated highlands and level plains of the highlands. If many of the smaller craters are maars, calderas, or drainage features, then one might expect to find an unusual concentration of craters within the flooded regions. Unfortunately, there is little positive evidence that can be brought to bear on this hypothesis. None of the small craters are in any way extraordinary. They fail to reveal deep interiors, dark halos, chain-like arrangements, or an association with rilles.

The geologic history envisioned for the Hommel quadrangle can be set in a larger context. Moving westward from the Hommel quadrangle through the Clavius and Schiller quadrangles there is a progressive increase in area covered by plains units (Cummings, 1968; Offield, 1968). This may be analogous to a transgressive sedimentary situation where the seas (in this case "lava seas") encroach on a continent either be-

FIGURE IX-7. Linear sets of intersecting ridges near the middle of the Hommel quadrangle. One set trends towards the upper right, a second towards the upper left. They are believed to be volcanic extrusive ridges built up along crustal fractures (Lunar Orbiter Photograph IV 94H).

cause of the rise of sea level or because of subsidence of the continent. The hypothesized transgression is, of course, from west to east.

A final word might be said about the credibility of block tectonics in terms of what we know about the evolution of the Earth's crust. Volcanic activity related to block faulting and largely confined to grabens is by no means exceptional. Triassic sedimentary rocks of eastern United States were deposited in a system of grabens extending from North Carolina to Nova Scotia. Sedimentary prisms include nonmarine clastic rocks interbedded with basaltic flows and intruded by diabasic sills and dikes. In Pennsylvania and New Jersey a concentration of intrusions close to the normal fault bounding the graben on its northwest side leaves little doubt that this fracture zone served as a conduit for the lava.

A great thickness of basaltic lava, as much as 3,000 meters, is interlayered with Devonian clastic rocks of the Midland Valley, England (Wells and Kirkaldy, 1966). Volcanic rocks of the same age are found throughout England and Europe, but large-scale volcanic activity was confined to a narrow graben. An inescapable conclusion is that the outpouring of lava contributed to the foundering of the graben floor.

A more complicated example of block tectonics is provided by Tertiary basins of Wyoming and Colorado. These basins are located some distance to the east of the mobile geosynclinal belt which was the site of major deposition, igneous activity, and tectonic compression periodically throughout the Paleozoic, Mesozoic, and Cenozoic.

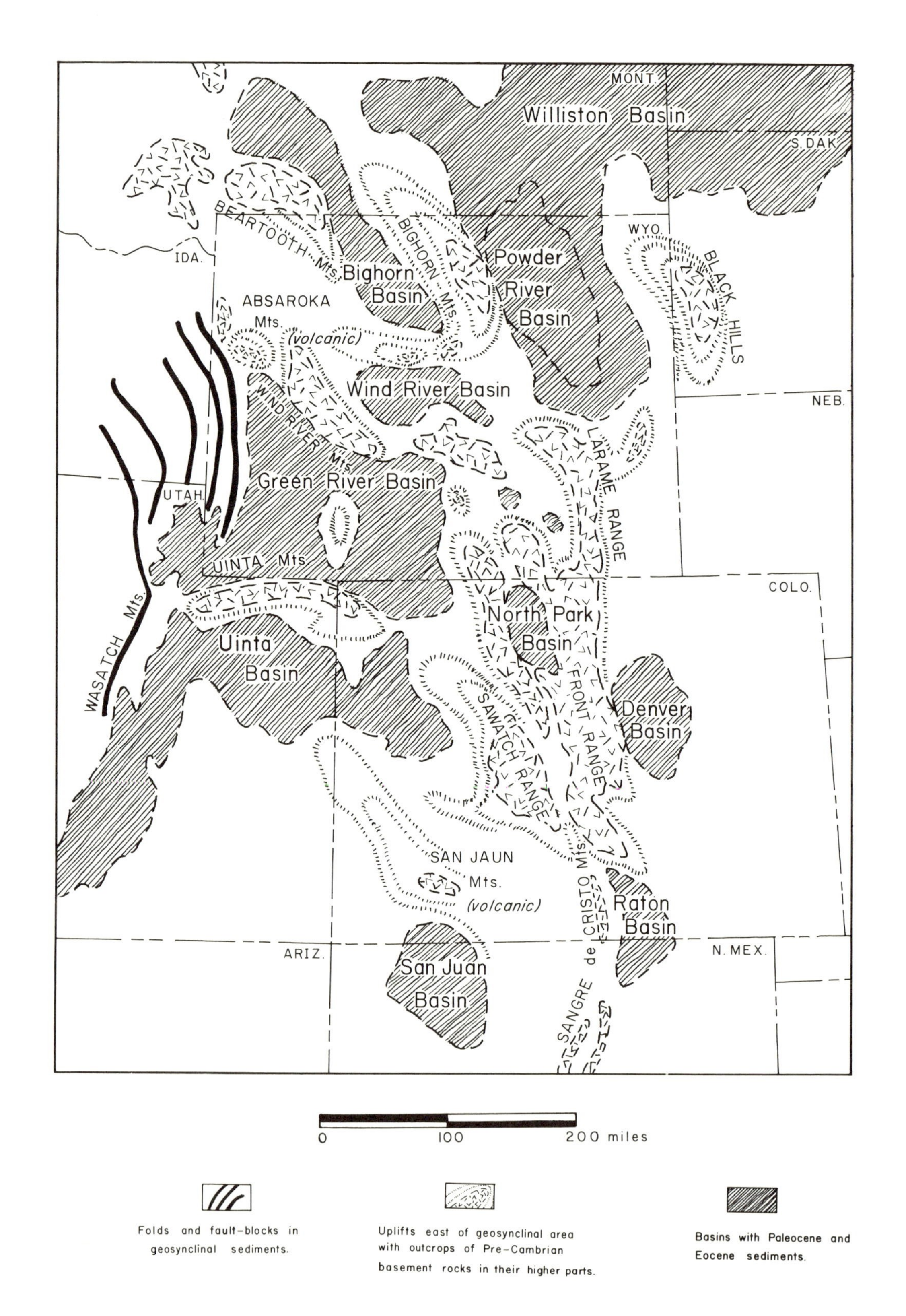

FIGURE IX-8. Map of Central and Southern Rocky Mountains showing uplifts and basins of Tertiary age (modified from King, 1959).

Nevertheless vertical displacements of "stable" crust beneath the Wyoming and Colorado basins are impressive. In the Bighorn Basin, for example, late Cretaceous and early Cenozoic deformation produced structural relief of at least 17,000 feet (Van Houten, 1952). Each basin has a unique history of subsidence and sedimentary infilling. The general distribution of landforms and sediments is shown in Figure IX-8. The jigsaw pattern of Precambrian uplifts and Tertiary basin fill indicates the presence of vertically shifting basement blocks, an idea first developed in detail by Thom (1923).

COMPOSITE STRATIGRAPHIC UNITS

The preceding discussion of the Hommel quadrangle emphasized the difficulty of defining conventional stratigraphic units within the highlands. The mapped units are actually provinces characterized by a complex sequence of sedimentary, igneous, and structural events. The nature of these units is such that it may not be appropriate to assign them a single stratigraphic age. At best, one can talk about the probability that certain surficial materials will be encountered. For example, structural blocks that have been greatly depressed will be largely covered by relatively recent volcanic materials. Blocks that have been only slightly depressed will have a partial veneer of younger lava flows contained in local depressions and a mantle of ash on upland slopes. The remaining surface will be underlain by a complex suite of crater and volcanic materials which make up the basement block. Raised structural blocks will have almost no post-basement volcanic cover. Such blocks can be identified by their pattern of unmodified structural lineations, but determination of internal stratigraphic divisions is ambiguous, if not impossible.

This sort of mapping is reminiscent of the delineation of structural provinces on Earth. Within the United States the Basin and Range province, briefly described in the preceding section, displays a distinctive topography which, in turn, is associated with a distinctive suite of sedimentary rocks, igneous intrusions, and structural elements. Other representative geologic provinces include the central stable region, the Gulf coastal plain, the folded Appalachians, and the western coastal ranges. Each of these regions can be easily defined and delimited on aerial photographs. Further, the style of structural deformation can be determined and the character of the rocks inferred. Detailed stratigraphic subdivision based on the information contained in aerial photographs is in some places possible, but it is almost always a secondary exercise most easily accomplished following interpretation of the geologic provinces and their composite origins.

THE LUNAR GRID

The northeast- and northwest-trending sets of linear features in the Hommel quadrangle are part of a larger array seen throughout the nearside of the Moon and commonly referred to as the lunar grid (Spurr, 1944). Parallel but independent studies by Fielder (1963) and Strom (1964) resulted in the conclusion that if one deletes radial lineament systems associated with mare basins, three prominent systems remain, trending NW-SE, NE-SW, and N-S. Included in approximately 10,000 lineaments measured by Strom are linear ridges on both terrae and maria, valleys, polygonal crater rims, linear scarps, crater chains, and linear portions of central peaks. The azimuthal distributions for two regions in the northern and southern hemispheres are shown in Figure IX-9. The areas were selected so as to be fairly free from radial systems; the only significant contribution appears to be from the Imbrium radial system in the southern hemisphere. Both Strom

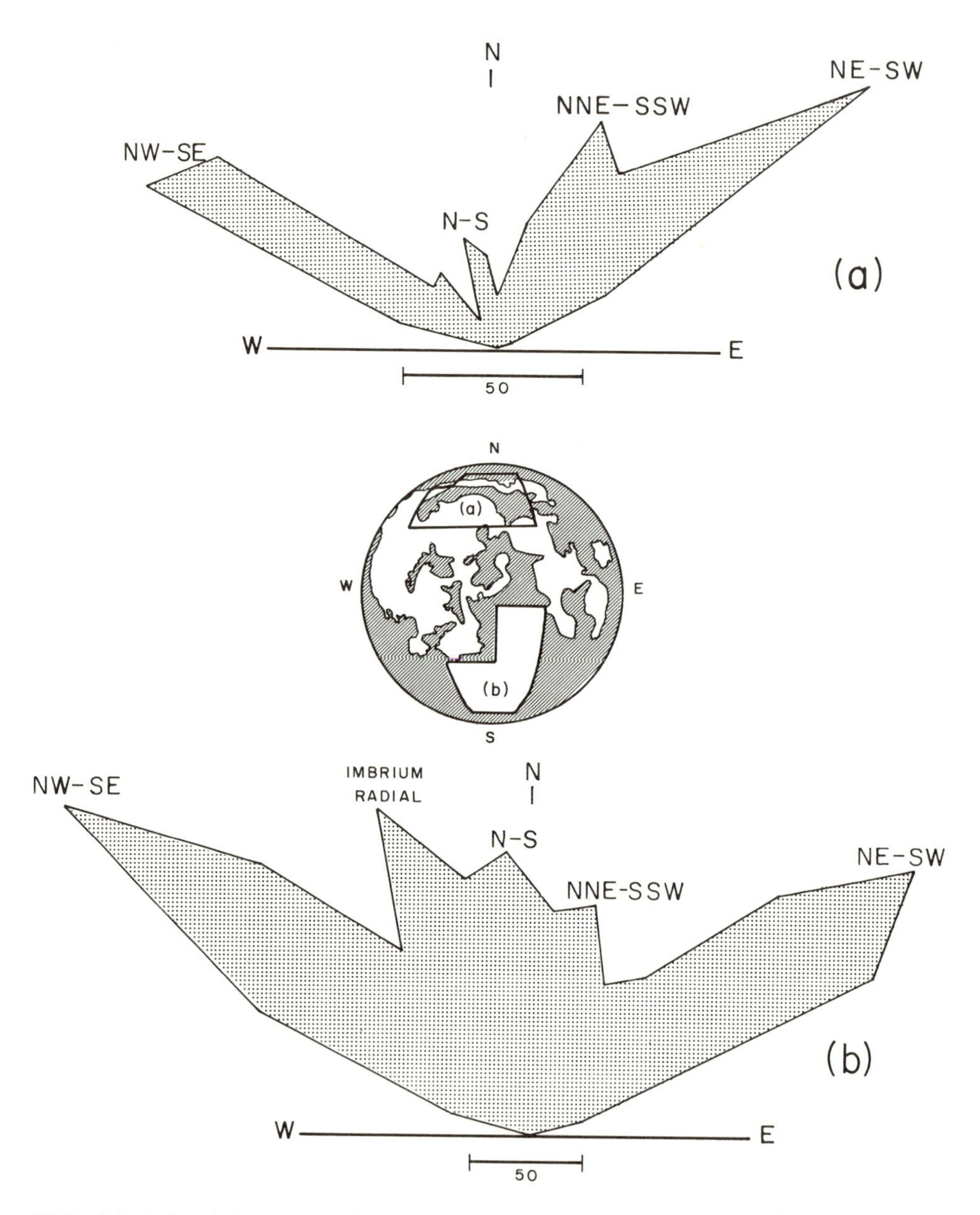

FIGURE IX-9. (a) Azimuth-frequency diagram of lineaments in a part of the Moon's northern hemisphere shown on the index map. (b) Azimuth-frequency diagram of lineaments in a part of the Moon's southern hemisphere shown in the index map. Both diagrams represent the apparent azimuths as viewed in orthographic projection (from Strom, 1964).

and Fielder interpreted their results as evidence for a global orientation of meridional compressional forces, with strike-slip faults forming in directions inclined to the principal stress axis and tension faults developed at right angles to the least principal stress axis. (In experimental rock deformation shear sets are oriented approximately 30° to the principal stress axis, and extension fractures are parallel to that axis. However, consideration of the entire Moon as an analogously homogeneous "laboratory sample" represents an oversimplified view.) Strom lists five possibilities for generation of stress within the lunar crust: "convection currents within the Moon . . . 'tidal bulge' as the Moon recedes from the Earth . . . [thermal] expansion and contraction of the Moon . . . a change in the possible axis of flattening due to a shift of the Moon's axis of rotation . . . and lunar body tides due to librations in longitude and latitude, and the changing distance between the Earth and the Moon" (p. 216). Vening Meinesz (1947) has made quantitative determinations of the stress fields which might exist in the Earth's crust as a result of a change in the axis of flattening. These calculations indicate that northeast and northwest trending shear sets could predominate locally. However, the global pattern is considerably more complicated, with several loci of radiating shears and a general absence of shears in the polar regions.

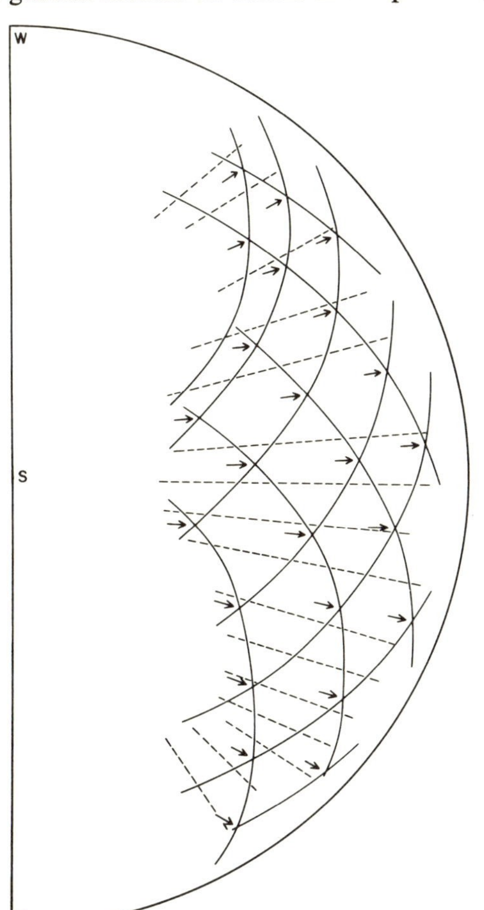

FIGURE IX-10. South polar projection of the lunar lineament system showing generalized directions of the NW-SE, NE-SW, and N-S (dashed lines) systems. The arrows bisect angles formed by NW-SE and NE-SW systems, and are interpreted by Strom (1964) as axes of maximum compression.

A more complicated structural situation is documented by Offield (1966) for a region close to the center of the visible disc. He measured more than eight hundred linear structures including linear valleys, escarpments, rilles, chain craters, mare ridges, and polygonal crater rims. The distribution of all lineaments with lengths of more than 25 km is shown in Figure IX-11. There is one concentration of N0°-15°E. Otherwise, none of the peaks correspond closely with those shown by Strom and Fielder unless one contends that the angle between northeast and northwest elements of the grid is more acute in the equatorial belt than in the polar regions.

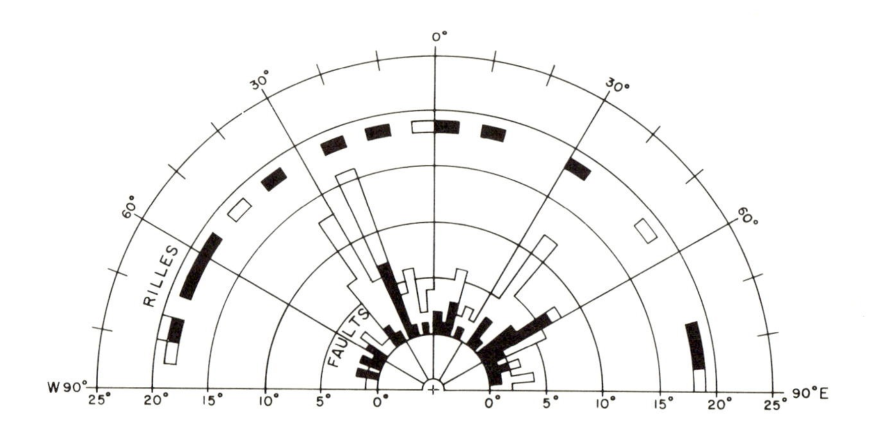

FIGURE IX-11. Azimuth frequency of lineaments measured in the Triesnecker and Hipparchus regions near the center of the earthside hemisphere of the Moon. Lineaments shown in black are more than 50 km long. Others are between 25 and 50 km in length (from Offield, 1966).

Offield raises a cogent objection to the compression model of Strom and Fielder. According to their strain analysis, both northeast and northwest systems should be shear faults. However, only four or five of the linear structures in Offield's study display possible lateral offset. Similar ambiguity attends the great majority of structures measured in the large samples of Fielder and Strom.

FAULTS

The Moon would seem to be an ideal place for identifying faults and determining differential movement. An important aid in determining fault movement is the presence of marker horizons or zones. In the terrestrial situation separation of sedimentary strata permits accurate calculation of vertical movement, but the magnitude of strike-slip movement commonly is difficult to determine in the absence of distinctive horizontally displaced features. The craters on the Moon provide a closely-spaced net of ideal marker

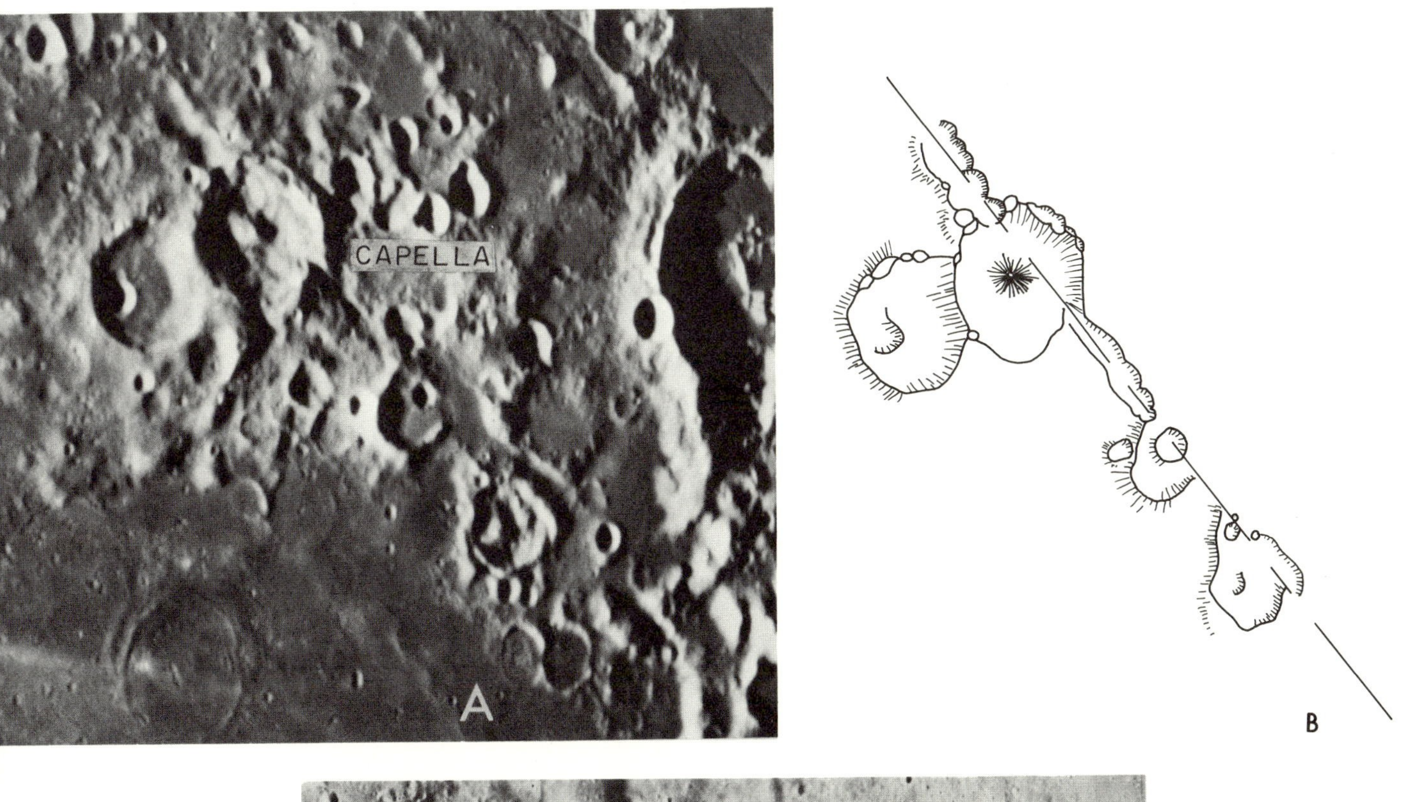

FIGURE IX-12. (a) Telescopic photograph of region near the crater Capella showing a prominent lineament trending from lower right to upper left and cutting across Capella (Catalina Observatory Photograph). (b) Sketch map according to Fielder (1965) showing possible strike-slip movement along the lineament shown in (a). (c) The region near Capella (Lunar Orbiter Photograph IV 72H).

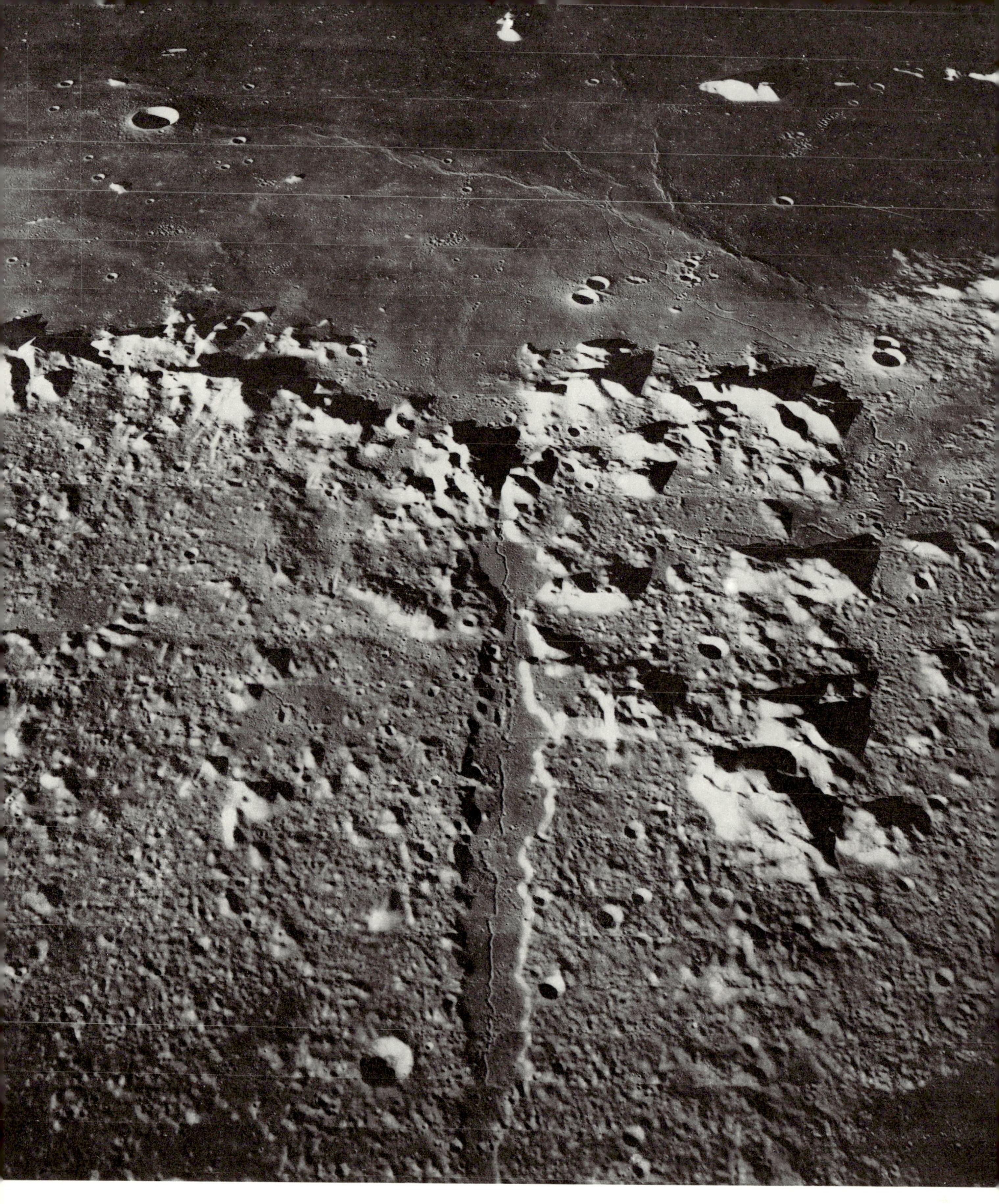

FIGURE IX-13. View looking west along the Alpine Valley and across the northeastern edge of the Imbrium basin. The peak located on the mare plain along the trend of the Alpine Valley is Agassiz Promontory. The peak located to the right is Pico (Lunar Orbiter Photograph V 102M).

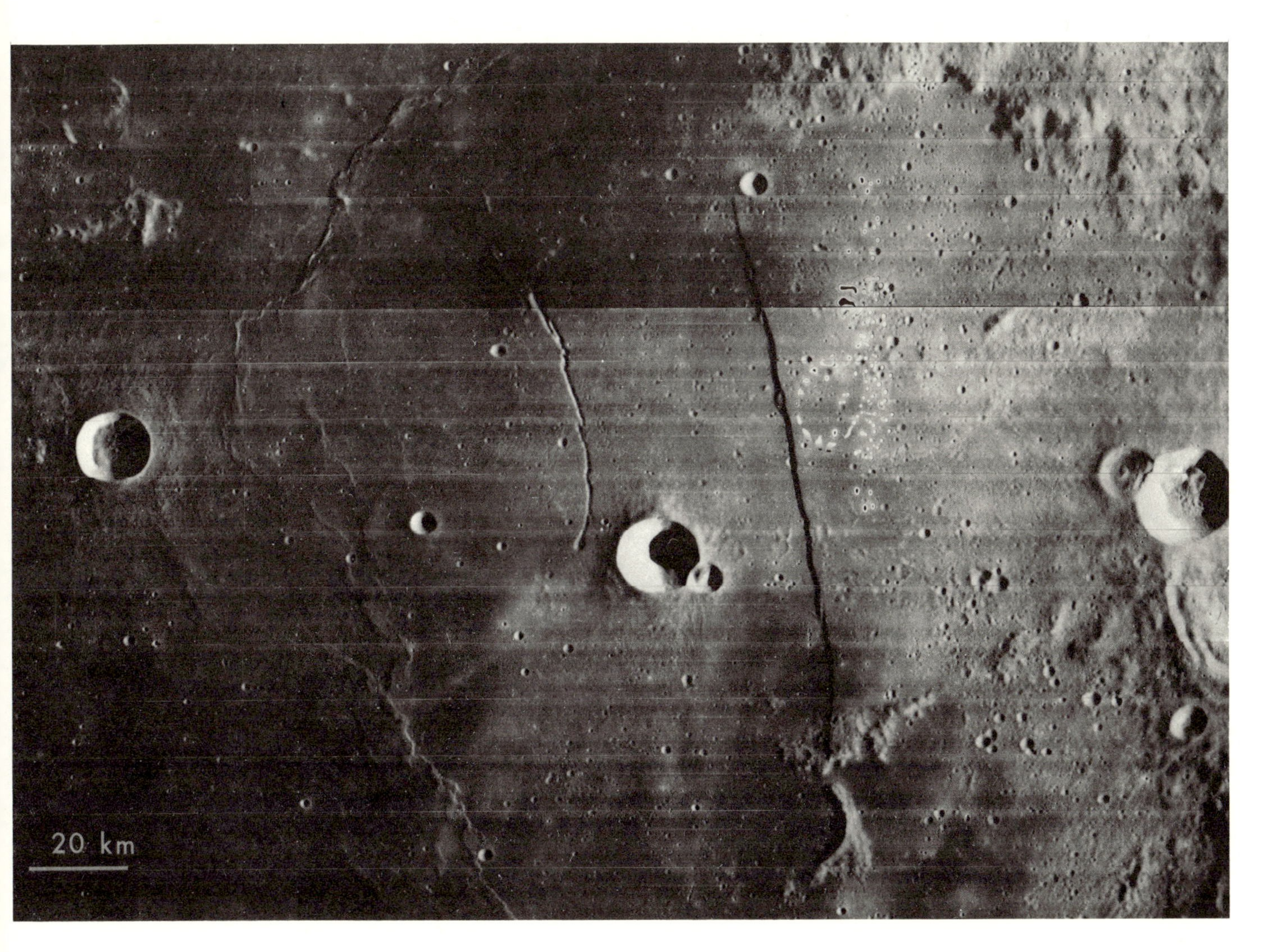

FIGURE IX-14. The Straight Wall, located in the eastern part of Mare Nubium (Lunar Orbiter Photograph IV 113H).

elements, especially for strike-slip faults. Considering this fact, it is surprising that there is so little evidence for lunar faults. The lack of demonstrable displacement is doubly puzzling in light of the previously noted profusion of linear features, especially in the highlands. These are defined by shallow depressions, nicks in crater walls, straight crater wall sections, linear swells, and albedo changes. All of the lineaments are topographically expressed, but none of them show indisputable evidence for lateral movement.

Fielder (1965) summarizes the evidence favoring strike-slip movement along certain lineaments. One of his most convincing examples is shown in Figure IX-12. Even this might be explained as volcanic craters located along a linear vent, with apparent offset being the consequence of nonsymmetrical volcanic subsidence and erosion along the vent.

FIGURE IX-15. Fault-line scarp in Precambrian rocks of the Canadian Shield. Although no movement has taken place along this fault for many millions of years, recent differential erosion has produced a prominent cliff 300 meters high. Only a short segment of the 500 km long fracture is shown in this photograph (courtesy of the Canadian Air Force).

There are several cliffs and steep-walled valleys of remarkably large dimensions which are interpreted as normal faults and grabens. Examples are the Straight Wall which forms a cliff 250 meters high and 100 km long, also the Alpine Valley which extends 130 km and is flanked by peaks towering over 3,000 meters (Wilkins and Moore, 1955) (Figs. IX-13 and 14). However, the possibility remains that many linear features mark joints, which are defined as fractures along which there has been no relative movement. Prominent relief is then a reflection of differential erosion along lines of weakness in the lunar crust. Still another possibility is that the lineaments mark faults which were active during the first stages of lunar history. The cliffs which we now see are the product of later erosion. The absence of internal displacement in craters which straddle the faults simply indicates the temporal sequence of early faulting and later cratering. Prominent fault-line scarps on terrestrial Precambrian shields provide persuasive analogs for this point of view. Even though movement on many of these faults occurred a billion or more years ago, the fault traces are presently marked by prominent topographic lineations (Fig. IX-15).

Relative and Absolute Ages of Lunar Materials

CHAPTER

STRATIFICATION

THE very word "stratigraphy" suggests a fact central to the study of sedimentary rocks: namely, that these rocks occur in stratified or layered sequences. It is the successive superposition of sedimentary layers which permits one to define relative ages for a sequence of sedimentary events. Stratification, in addition to being present in sedimentary sequences, may also exist deep within a planetary body or in the topmost few centimeters of surficial soil. Before considering the more conventional sedimentary stratigraphy, it may be worthwhile briefly to examine evidence for the presence of these two other types of stratification on the Moon.

We have already noted that chemical determinations by three Surveyor spacecraft suggest that the maria are underlain by basalt and that the highlands may be underlain, in part at least, by slightly less dense materials. The density of normal terrestrial basalt is about 3.0. The mean density of the Moon is known to be 3.34. If the surficial rocks are, in fact, less than this mean value, it follows that some of the rocks in the interior must have densities exceeding 3.34. Probable candidates would be mixtures of olivine-rich materials (dunite) and basalt. Within the Moon, then, there may be a gross density stratification of material comparable to the core-mantle-crust stratification which is present in the Earth. Layering of this type would demonstrate that the Moon has gone through a period of large-scale differentiation following initial accretion.

A differentiation model has been developed in considerable detail by McConnell and others (1967). Making assumptions about the initial temperature of the Moon, its uranium content, and the ratios Th/U and K/U they estimated that melting would have begun about 1.5×10^9 years after the moon's formation (assumed to be 4.5×10^9 years ago), that it would have reached its greatest intensity in less than 10^8 years, and that it then would have declined exponentially with a half life on the order of 3×10^8 years.

There will be no opportunity for directly observing petrographic boundaries within the Moon's interior, but there should be an abundance of indirect information easily available. Petrographic determinations for rocks at the Apollo landing sites will either confirm or disprove the model based on Surveyor data. If there was a major pulse of volcanic activity which had a peak intensity about three billion years ago—1.5 billion years after the Moon's formation—then radiometric age determinations should point to this fact. Certain volcanic provinces such as the Marius Hills may contain a variety of igneous rocks—from granite to basalt—which will demonstrate the existence of local melting and differentiation.

Some large impact craters may intersect several crustal layers and contain in their ejecta blankets representative specimens from each of these layers. Some craters of internal origin may mark the presence of diatremes extending deep into the interior. Rock fragments which have moved upward from great depths along these conduits, transported within an explosive mixture of gas and magma, may be collected at the Moon's surface. The geometry of internal stratification may be indicated by wave velocity measurements made by an array of seismometers which will be deployed at the several Apollo landing sites. Finally, the possibility that unfoundered parts of the primeval crust are still exposed at the lunar surface should not be discounted.

Another type of stratification mentioned at the outset of this section involves surficial soil overlying solid rock. This layer, often termed a regolith, is fragmental debris of relatively low cohesion which overlies a more coherent substance. It is inferred to have been produced primarily by repetitive bombardment and impact comminution of coherent rock (Surveyor Program, 1967*a*).

The upper part of the lunar regolith has been scoured by Surveyor shovels, and the total thickness has been determined by the use of several models. According to Shoemaker and others (Surveyor Program, 1968*b*) the observed frequency distribution of small craters corresponds to an equilibrium or steady-state distribution. (The character of steady-state distributions is discussed in a later section of this chapter.) The upper crater diameter limit of a steady-state population varies from site to site and can be used to deduce regolith thickness. Experimental studies show that the upper crater size limit is about thirty times larger than the size limit for craters whose aggregate area would just cover the surface. The depth of these smaller craters represents the approximate average depth affected by impact comminution. At the Surveyor I site the upper limit of the steady-state crater distribution is 200 meters; accordingly the depth of the regolith should correspond to the depth of a crater 7 meters in diameter—about 2 meters. This theoretically determined depth is consistent with the observation at this site that no crater less than 8 meters in diameter contains blocky rock fragments along its rim.

The apparent thickness of the regolith at the four Surveyor mare sites ranges from 1 to 10 meters. According to Shoemaker and others (Surveyor Program 1968*b*) it is very fine-grained. By extrapolated particle count 90 percent of its fragments were determined to be less than 1 mm.

The upper crater-diameter limit of the steady-state population on the patterned flow on which Surveyor VII landed is about 10 meters (Surveyor Program, 1968*b*). From this it can be inferred that the surface materials have been completely turned over once by craters with diameters of 30 cm or greater. The associated prediction is that the depth of the regolith is about 10 cm. In contrast with this prediction the smallest crater which displays a blocky rim is 3 meters in diameter, implying a regolith depth of about 1 meter. The contradiction is best explained by assuming that the patterned flow unit on which the regolith is formed is itself a coarse-grained fragmental unit, so that craters which penetrate its surface form blocky rims only when they happen to encounter widely spaced detached blocks. The slight thickness of regolith relative to that for mare sites is consistent with the assumption that the regolith layer thickens with time and that the surfaces around Tycho are younger than those on the maria.

In addition to being thinner than mare regolith, the Tycho fragmental debris is also coarser. Again, this is a reasonable difference if one assumes that the regolith is formed by repetitive meteoritic bombardment. With passage of time large blocks will be repeatedly broken and abraded to form homogeneous fine-grained debris which characterizes a thick, mature regolith.

Another survey of regolith thickness based on variations in crater geometry has been carried out by Oberbeck and Quaide (1968). They utilized the crater-shape model described in Figure IV-37. With increasing crater size the internal geometry changes from concentric to flat-bottomed. The transition size includes those craters just deep enough to intersect the regolith-rock interface. The median soil thickness for three Orbiter sites in Oceanus Procellarum, three sites in Mare Tranquillitatis, and one site in Sinus Medii was determined to be about 4.6 meters. Two sites in the southern part of Oceanus Procellarum have a regolith thickness of about 3.3 meters. Greater thicknesses, about 7.5 meters, are recorded for regolith in northern Sinus Medii and Mare Imbrium. The greatest value, 16 meters, was calculated for the inner wall of the crater Hipparchus.

An ingenious remote analysis of the lunar surface layer was performed by Tyler (1968). He used the communication system of an Explorer satellite as a lunar-orbiting radar beacon. Quasi-specular bistatic-radar measurements were made of the oblique scattering properties of the lunar surface along the track of the orbiting satellite. Echo intensities for maria were found to be approximately 30 percent greater than those for highlands. An unusually high reflectivity was documented in the vicinity of the crater Flamsteed. Tyler proposed that the reflectivity variations are related to depth of lunar regolith which, according to his model, ranged from 20 meters in the highlands to as little as 1 meter on the maria. His maria values are generally comparable to those obtained by Shoemaker and by Oberbeck and Quaide. Comparative determinations for the highlands regolith have not, as yet, been made.

There is an internal stratigraphy within the regolith which has been recorded by all Surveyor spacecraft. The topmost few millimeters are always underlain by darker material. Shoemaker and others (1968) have speculated that particles in the shallow subsurface become coated with a dark material, a hypothetical substance called "lunar varnish" which is "scrubbed" off on the surface by cosmic ray-micrometeorite scouring.

FIGURE X-1. View looking north over Copernicus. The south crater rim is in the foreground, central peaks in the middleground, and north crater rim in the background. Note a prominent dark stratum with nearly vertical attitude exposed on one of the central peaks. This may be either an igneous dike or a sedimentary layer in a structurally deformed central uplift. Downslope mass movement of material is shown at the far left side of the north wall. The material displays flow lineations and the approximate shape of a fan spreading out on the crater floor (Lunar Orbiter Photograph II 162H).

The authors admit that the origin of lunar varnish is unknown. Going further, one could ask whether there is sufficient understanding even to use the genetically suggestive terms "varnish" and "scrubbing."

Perhaps the chief contribution of Shoemaker's discussion is that it accentuates how little is presently known about the chemical and physical processes which modify particles within the regolith. Our ignorance on this point is further emphasized by the fact that several darkening and lightening trends for lunar surface materials seem to require contradictory—or at least separate—explanations. For example, bright halos and rays are confined to young craters. Shoemaker and others (*in* Surveyor Program, 1968*b*) attempt to explain this by hypothesizing that the scrubbing process is more than compensated by a gross build-up of lunar varnish within the regolith. As the soil fragments age, they become increasingly dark both at the depth and at the surface, even though a relative contrast may remain between the topmost particles and those which are buried. The bright ray craters have a high albedo because the ejected materials are excavated from depths that have been protected from lunar-surface alteration. As time passes, the particles become coated and the ray materials darken.

Another type of albedo change, this one an apparent trend in lightening of volcanic materials with increasing age, has been mentioned in previous chapters. Some fresh-appearing volcanic craters are surrounded by materials with unusually low albedo. Perhaps they are covered with a substance which is deposited from associated volcanic gases and is gradually removed by radiation erosion (i.e., scrubbing).

Setting aside consideration of gross density stratification and surficial regoliths, what is the evidence for "normal" sedimentary stratigraphy on the Moon? This type of stratification is easy to recognize on the Earth, be it in the many-colored horizontal beds of the Grand Canyon, the tilted snow-covered ledges of the Canadian Rockies, or the sinuous ridges of the folded Appalachians. Although it is clearly displayed on Earth, sedimentary layering is difficult to discern on the Moon. Some possible examples are shown in Figures X-1 through 3. Even these, selected for their unusually suggestive detail, are not beyond dispute.

It should be noted that many Orbiter photographs do show closely spaced linear ridges and depressions which, at first glance, suggest horizontal layering, especially where the lineations happen to parallel topographic contour lines. However, unlike the bounding surfaces of horizontal strata, these linear features can be observed both on level surfaces and on steep slopes. The fact that the lineations maintain a constant trend throughout large areas with variable topography suggests that they are formed by the intersection of vertical planes with the lunar surface. Most probably they mark the position of closely spaced joints in the lunar crust. This texture has been termed "patterned ground" or "tree bark structure" by Shoemaker (*in* Ranger VIII and IX experimenters' analyses and interpretations, 1966).

One can think of any number of reasons why stratification should be lacking in lunar rocks. First, there is no indication of regional compression and folding in the lunar crust. Consequently all layered rocks should be approximately horizontal except where

FIGURE X-2. View of crater wall inside Aristarchus. Slope is down toward lower right. Prominent layered appearance of wall materials may be the result either of depositional stratification or of parallel structural displacements. Note the pool of dark fissured materials, probably of volcanic origin, in upper left (Lunar Orbiter Photograph V 199H).

they are locally tilted around volcanic vents or impact craters. Large areas of the Moon are approximately level. In such areas only one or—at the most—several horizontal layers will be exposed at the surface. Opportunities for seeing a more elaborate "layer cake" exist only on steep, high cliffs, and many of these appear covered by mass-wasted debris.

Although the end product of stratification is easy to describe on Earth, the process whereby the regularly layered pattern is achieved is surprisingly difficult to understand. More often than not, deposition is gradual and slow with numerous hiati. Conversely, where deposition is rapid and unsystematic the resulting assemblage of sediments has a chaotic appearance. The Apennines of northern Italy afford an excellent example. Many of the sediments were completely churned up and reshuffled by huge submarine slides following their initial deposition (Maxwell, 1959). So confusing are the relationships that the name "chaos" is actually applied to one of the mappable units. The general lack of stratification on the Moon argues for a rapid, disorganized, "Apennine" type of deposition as opposed to a regional "grain-by-grain" accumulation.

FIGURE X-3. The Hadley Rille. The shadowed wall displays faint lineations and bouldery ledges which may mark stratification in the rocks cut by the rille (Lunar Orbiter Photograph V 105H).

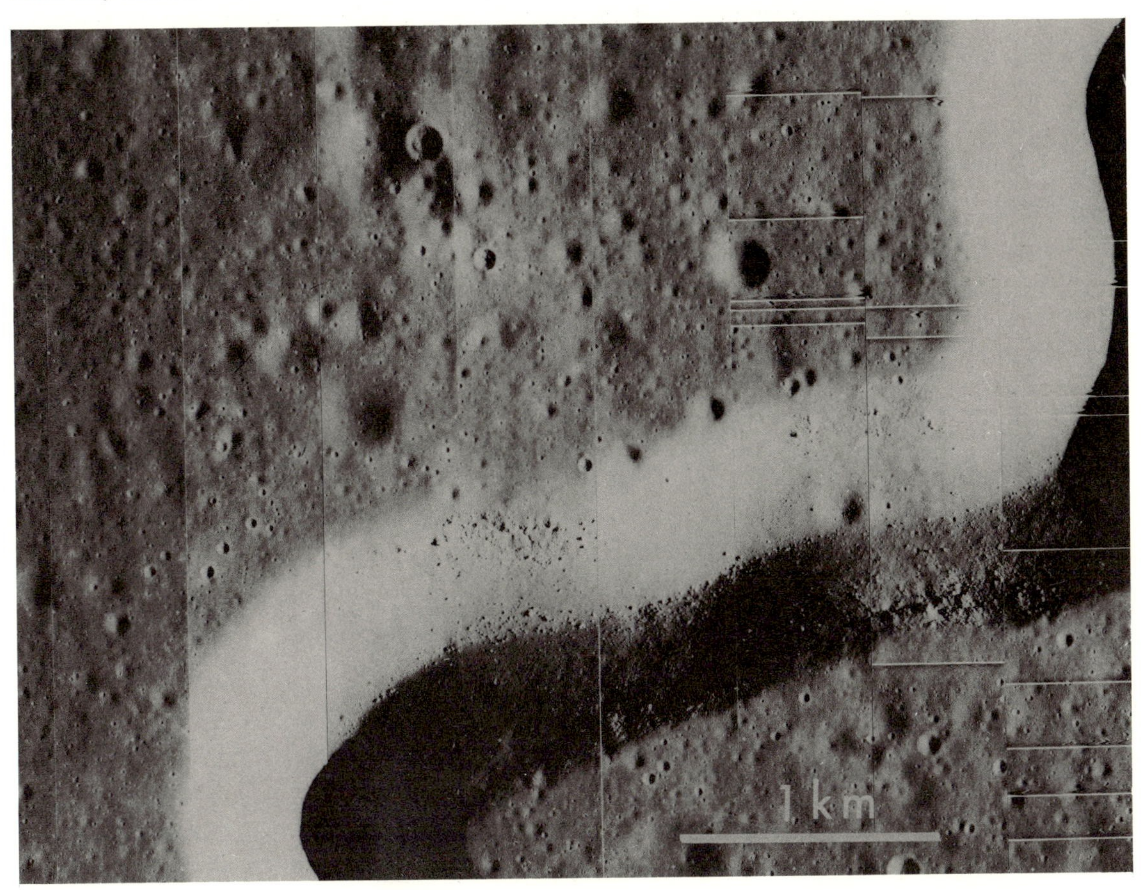

Differential weathering and erosion play important roles in emphasizing sedimentary strata on Earth. In the Grand Canyon resistant limestones and sandstones form bold cliffs; the more easily weathered shales underlie gently sloping benches. In the folded Appalachians many of the linear ridges owe their relief to the presence of massively bedded quartzite. Different soils which formed over various rocks lead to a distinctive separation of vegetation. A particular bed may be indirectly but precisely delimited by a line of closely spaced conifers. Clearly, erosional processes on the Moon are much different from those on Earth. It may be that the processes reworking the lunar surface are relatively ineffective in etching distinctive relief on the different superposed layers.

A consideration of the nature of stratification is central to an understanding of age relationships. It is often assumed that a traverse across a steep-sided lunar crater will reveal a vertical stratigraphy analogous to that in a terrestrial canyon. Unlike the situation on Earth, however, one cannot expect to differentiate rocks by fossil content. Further, if the crater is located on a mare surface it may intersect a sequence of volcanic flows that are lithologically similar.

It is difficult to anticipate in detail what a typical lunar section will look like or how it can best be differentiated and interpreted. One possible arrangement is shown in Figure X-4. The assumption here is that the same concepts used in the mapping of terrestrial volcanic terrains can be applied on the Moon. The Moon apparently lacks a magnetic field. Consequently, it is nearly certain that paleomagnetic reversal techniques similar to those used on Earth (Irving, 1964; Cox, Dalrymple, and Doell, 1967) cannot be used in stratigraphic analysis of lunar rocks. However, periodic variation of some other physical effect might well take place, inducing a characteristic pattern in rocks

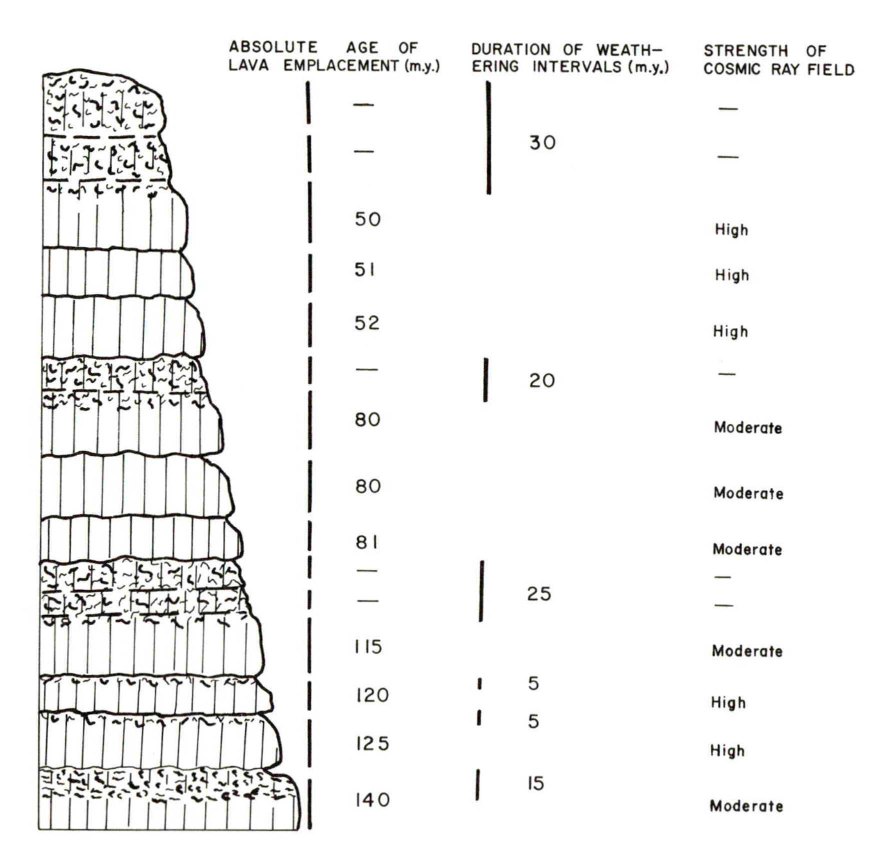

FIGURE X-4. Schematic vertical section of lavas exposed on the Moon, showing several techniques for determining temporal relationships. Ruled areas indicate lava flows; a total of fifteen are revealed. Stippled areas indicate weathering zones; a total of six are revealed. Contacts between lavas that are included within a weathering zone are indicated by dashed lines. Absolute age of lava emplacement could be measured with several conventional radiometric clocks, notably K^{40}-A^{40}. Meaningful ages will not be obtained from those lavas which have been extensively weathered. Duration of weathering intervals is shown related to depth of weathering profile. It is also related to intensity of weathering at a given point in the profile. Arbitrarily, the cosmic ray field is considered to fluctuate from moderate to high on a 25 m.y. cycle. Note that there is insufficient evidence in the section to establish this fact. Also note that several of the high strength phases are not represented by a depositional event.

as a function of time. For example, either cosmic or solar radiation may systematically rise and wane. An unusually high flux of solar protons might have a detectable effect on oxidation states of rocks crystallizing from a cooling lava. The Fe^{3+} : Fe^{2+} ratio might be lowered; metallic iron, magnesium, and silicon might be produced (Roberts, 1967). Although the proton flux would tend to alter solid rocks in the same way, the effect should be greatly intensified during the crystallization stage.

Turning to reactions of solid surficial materials with cosmic rays and meteoritic particles it is clear that the changes in the "weathering zone" increase with passage of time. If one assumes that cosmic radiation and meteoritic flux are more or less constant, then the additive effects of impact and irradiation are directly proportional to time. Of course, the technique does not necessarily date the time of rock formation. It only records the amount of time the rock has been exposed to surficial weathering.

Parenthetically it should be noted that the entire problem of dating the lunar surface rocks on the basis of the first returned samples is one that is too often discussed in a manner expansive but vague. Little detailed attention has been given to the matter of obtaining a vertical sequence of samples through more than the upper few inches of the surficial fragmental layer. Yet we are told, rather simplistically, that radiometric dating of samples collected near craters will date them and will, indeed, allow us to order the individual cratering events temporally. Just how this will be accomplished is not clear. Unfortunately, what is clear is that we have only the sketchiest idea of the geochemistry of the lunar surface, the isotopic fractionation in lunar volcanic rocks, and the redistribution of isotopes during impact or radiation events.

CRATERS AND CORRELATION OF GEOLOGIC TIME

The phenomenon which most obviously differentiates parts of the Moon's surface is crater density. The rugged highlands with their swarms of overlapping craters are easily distinguished from the more level maria. Intuitively it seems clear that the complexly cratered terrain of the highlands is older than the relatively unblemished mare surfaces. A further intuition is that quantitative measurements of crater density might provide a sensitive indicator of age for the entire spectrum of lunar stratigraphic units.

Unfortunately, elevation of these intuitions to objective certainties is an elusive goal. In treating the problem quantitatively one needs a great deal of supplementary information. Let us assume that we wish to determine the absolute age of a surface on the basis of impact crater distribution. Solution of the problem demands

1. Ability to differentiate primary impact craters from secondary impact craters, volcanic craters, collapse craters—and any other type of crater that might be present.
2. Knowledge of the processes of crater degradation, especially for craters located on slopes.
3. Understanding the character of saturation crater distributions.

4. Knowledge of gravitational interactions between meteoroids and the Earth-Moon system.
5. Knowledge of distribution of meteoroids in space.
6. Knowledge of distribution of meteoroids throughout geologic time.
7. Correlation of meteoroid mass and impact velocity with crater size.

Before we become bogged down with the rather frustrating task of evaluating these seven unknowns, let us review a solution of the central problem, necessarily accomplished by means of shortcuts and simplifications. This particular solution was published by Shoemaker, Hackman, and Eggleton in 1962.

Assume a constant supply of meteoroids in space. Further, assume that meteoroids are produced by break-up of asteroids at approximately the same rate they are destroyed by collisions with planets. Unusually dense clusters of meteoroids will form around planetary bodies as a result of gravitational interactions. From considerations of conservation of angular momentum and energy it can be shown that the ratio of frequency (F) of impacts per unit area for two planetary bodies—in this case the Earth and Moon—can be expressed as

$$\frac{F_{\text{Earth}}}{F_{\text{Moon}}} = \frac{2g_e{}^r e + V^2}{2g_m{}^r m + V^2} = 2.15$$

where r is the radius of the planetary body, g is gravitational acceleration at r, and V is the velocity of the meteoroid at an infinite distance from the planet. Astronomical observations suggest a velocity of 15 km/sec for a meteorite entering the Earth's atmosphere; the equivalent velocity at infinity will be about 10 km/sec. Values for g and r are expressions of the mass and size of the Earth and Moon, parameters that are well known. The ratio F indicates that impacts will be slightly more than twice as great per unit area on the Earth as on the Moon.

Determination of impact frequency on Earth establishes a standard with which lunar data can be compared. There are two sets of evidence which can be used in determining this standard. Brown (1960) has estimated the influx or meteoritic bodies by evaluating frequency of observed falls and mass distribution for these falls. To express his data in terms of crater distributions it is necessary to calculate crater diameters as a function of energy released. But to estimate energy released it is necessary to know not only meteorite mass but also impact velocity. Assuming a spectrum of velocities allows construction of the curves shown in Figure X-5. These curves can be checked by a second body of information, the distribution of large meteorite craters on Earth. Information here is very uncertain, but examination of an area within the United States reveals ten possible impact structures, each greater than 3 km, contained within limits of space and time such that an impact frequency of $60/(10^6\text{km}^2)(10^9\text{yr})$ is suggested. This compares favorably with a predicted frequency of 30 to $100/(10^6\text{km}^2)(10^9\text{yr})$ from the data of Figure X-5.

FIGURE X-5. Cumulative frequency of craters as a function of size (from Shoemaker, Hackman, and Eggleton, 1962).

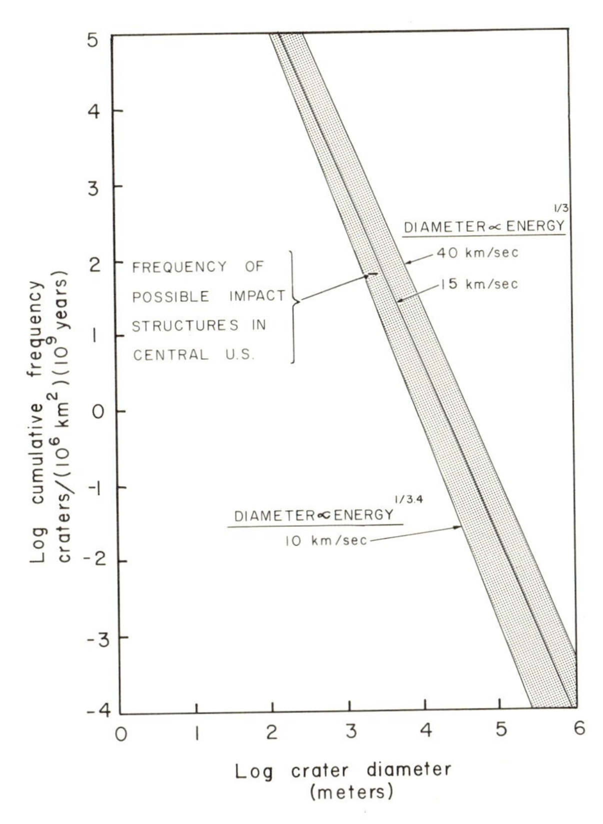

The next step is to measure crater distributions on lunar surfaces, in this case maria. In order to compare these distributions to the terrestrial standard, two adjustments must be made. First, it is necessary to use a terminal lunar velocity of 10.2 km/sec, the equivalent for a terminal velocity of 15 km/sec on Earth. As a consequence of this difference, impact of a particular mass will cause a smaller crater on the Moon than on the Earth. As a second adjustment, the ratio of impact frequencies, previously determined to be 2.15, must be applied. No account is taken of the fact that release of the same energy will form a slightly larger crater on the Moon than on Earth due to lower gravitational acceleration on the Moon.

Incorporating the two adjustments of the preceding paragraph and assuming the age of the maria to be 4.5 billion years, the crater frequency is as shown in Figure X-6. If our assumptions to this point had been correct and sufficient, the curve would be coincident with the terrestrial standard. The similarity of the two curves can be improved by assuming a focusing factor of 2 on the side of the Moon facing the Earth, but even then the maria show a surplus of large craters. There are any number of highly speculative reasons for this residual difference. Most reasons invoke significantly different populations of meteoroids as a function of space, time, or both.

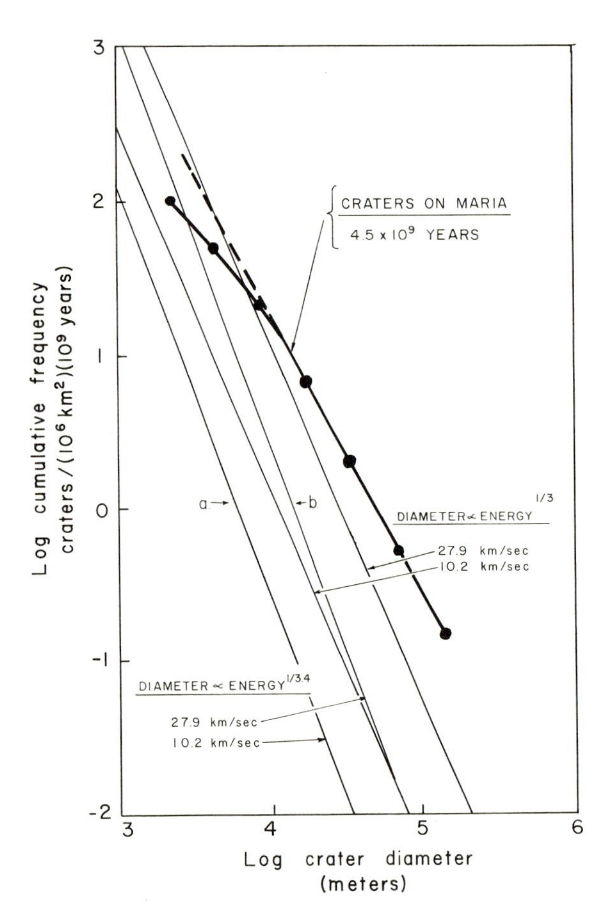

FIGURE X-6. Predicted frequency of impact craters on the Moon compared with observed crater frequency distribution on the maria (from Shoemaker, Hackman, and Eggleton, 1962).

The age of 4.5 billion years for the maria is not randomly chosen. It represents a view that these surfaces were formed very early in the history of the Moon, approximately at the same time as the formation of all the planets within our solar system. Radiometrically determined meteorite ages which cluster in the vicinity of 4.5 billion years suggest that this was the time of planetary formation.

In 1960, the same year that Shoemaker, Hackman, and Eggleton orally presented their calculations for the age of the maria, four other independent analyses appeared. Opik (1960) determined the age of the western part of the Mare Imbrium to be 4.5 billion years. By contrast Kreiter (1960) estimated an age between 1 and 500 million years for Mare Imbrium. Even younger ages, between 2 and 46 million years for Mare Tranquillitatis and between 4 and 94 million years for Mare Humorum, were calculated by Lyubarski (1960). The differences in the various analyses lie not in the empirical determination of crater distributions but rather in theoretical assumptions about meteoroid distribution. Within the widespread spectrum of possible solutions two diametrically opposed points of view are seen to emerge: first, that the maria formed shortly after the formation of the Moon itself, and that the Moon has been internally inactive since that time; secondly, that the maria represent recent internal activity—no older than Paleozoic by Earth standards and possibly as recent as Tertiary. In this connection it should be noted that recent studies by Shoemaker of comet and meteoroid distribution in the vicinity of Earth lead to determination of flux rates greater than those contained in his 1962 paper. According to this new information his reported maria ages should be reduced, perhaps by a major amount.

NUMBERS DO NOT LIE—OR DO THEY?

Crater distributions are usually presented as a cumulative frequency curve plotted on log-log paper. The log of N, the number of craters per unit area larger than diameter D, is plotted against the log of D. The function has the form $N = AD^B$ or, in logarithmic form, $\log N = \log A + B \log D$. B defines the slope of the curve, always a negative number. A defines the position of the curve relative to the N axis. Most crater distributions approximate a straight line when plotted on log-log paper. Accordingly, a single value of B defines the slope and permits extrapolation beyond the region of observation.

Log-log plots are commonly used in situations where the variables (in this case crater diameter and crater density) have a wide range of values. At the smaller crater sizes it is important to distinguish between 1 meter and 10 meters. This requirement dictates scale divisions no greater than 10 meters. But, since many distributions include craters with diameters of 10 km or more, 1,000 ten-meter divisions would be necessary on an arithmetic scale. On a logarithmic scale the situation can be compactly handled.

The usual procedure for counting craters is to divide the population into a small number of classes, generally less than ten. Increments are larger for large-diameter classes than for small-diameter intervals so that each class contains a significant number of individuals. The number of craters in successive classes are cumulatively figured and plotted as points on log-log paper. Finally, a line is fitted to the points.

Scientists will tell you that log-log plots are graphical presentations. The unwary reader might better consider them optical illusions. Each day we are confronted with a flood of graphs, both in scientific publications and in popular magazines and newspapers. Interpretation of these graphs becomes almost instinctive. We assume that sets of points, bars, or lines which are close together are similar. We tend to equate "similar" with "essentially identical" and do not make a further quantitative analysis. But two lines which appear "similar" or "close together" on a log-log plot are far from identical. In Figure X-6 the two lines labelled *a* and *b* define crater distributions whose cumulative frequencies differ by a factor of 10. To dramatize the meaning of such a difference we have only to state it in terms of mare ages. A reduction in age by a factor of 10 would mean, for example, a change from 4.5 billion to 450 million years—a radical revision.

A second optical illusion of log-log plots involves slope determinations. Based on visual measurements a single curve may be assigned a constant slope or two curves may be assigned identical slopes when, in fact, the slopes are significantly different. Mathematically significant differences commonly do not make much impression on the eye. Quantitative determinations of slope should never be determined visually but should be figured by statistical treatment of the component points.

Chapman and Haefner (1967) have pointed out that the cumulative frequency presentation contains inherent inconsistencies which can be avoided if the data are analyzed by an incremental frequency technique. A number of workers have chosen to adopt this approach, but the cumulative frequency display remains the most familiar and, for that reason, the most commonly used.

STEADY-STATE DISTRIBUTION

An important feature in crater distributions is the saturation or steady-state distribution. If we consider an impact counter-surface which is very large relative to the combined area of all craters, it is reasonable to suppose that all craters will be preserved, regardless of their size and age. If we discount any weathering processes, then surfaces of greater age will have more craters in all size classes. Close-up pictures of the Moon make it clear that the surface does *not* resemble this idealized counter. Instead, a myriad of overlapping craters are preserved in all states of degradation. Indeed, it is easy to demonstrate that, for any reasonable rate of crater production, the total surface will be covered with craters in a few thousand years. Of course, this calculation is not sufficient to describe the more complicated problem of crater destruction. This involves a consideration of the interaction between craters of all sizes as well as degradation and filling of individual craters over a period of time.

If crater production and destruction both proceed systematically, then a combination of the two relationships should provide a steady-state relationship. Demonstration of this steady-state condition can be attempted in a number of different ways. The problem can be attacked theoretically, setting up a model and testing it against known distributions (Marcus, 1964, 1966*a*, 1966*b*, 1966*c*, 1966*d*, 1967; Chapman, 1968). Another approach is experimental, involving actual production of a surface with constant crater distribution. The most direct approach is observational, measuring natural crater distributions and observing the limits toward which they tend. Employing this latter technique, Trask (1966) derives a steady-state distribution of $N = 10^{10.4}\ D^{-2}$ for the diameter range of 1 meter to 100 km where N is the cumulative number of craters per 10^6 km^2 with diameter greater than D (in meters).

If steady-state distributions are achieved, it is reasonable to assume that the saturation condition will exist first for small craters with short life-spans. Accordingly it is not sufficient to say that a surface is saturated with craters. Instead, it is necessary to define the upper limiting crater size. Pursuing this argument, one might expect to encounter many crater distributions which evidence a break in slope at the upper crater size limit for the saturation condition. As we shall see presently, distributions with discontinuous slopes are fairly common, but other possible explanations preclude a rigorous test of the steady-state hypothesis. Trask (1966) has noted a general tendency for crater distributions measured from all three Ranger missions to converge at small crater diameters and he interprets this as support for the steady-state hypothesis.

CRATER DISTRIBUTIONS AS A STRATIGRAPHIC TOOL

Crater distributions can serve a variety of stratigraphic roles. The ultimate use would be to date units absolutely, a technique analogous to radiometric dating. If absolute calibration is not possible, relative densities might indicate a hierarchy of age in the same way that fossils indicate relative age. Finally, a particular crater distribution might serve to characterize the surface material of a particular area even though the significance of the distribution was poorly understood.

Any dating tool has increased value if it can be used to define the age of a small volume or area. Fossils fulfill this requirement admirably. If the fauna is well preserved, examination of a few cubic feet of rock is sufficient to define the relative age. One or two good specimens of an index fossil are adequate. There is no need to collect hundreds.

Unlike fossils, craters have to be present in fairly large numbers before significant statistics emerge. Many small regions of stratigraphic interest do not have enough craters of different sizes to allow construction of a frequency curve. Even a count of several hundred craters does not insure against the possibility of a spurious anomaly. A single swarm of secondary craters might double the crater count of a small area. For these reasons the most reliable crater statistics are generally prepared for large areas—maria and highlands.

A summary of data for the highlands and maria based on telescopic observations has been prepared by Chapman and Haefner (1967). Expectably, mare surfaces are less highly cratered than highland surfaces. Slopes for mare crater distributions tend to be steeper than average for both the small-diameter and large-diameter classes in the observed range. Highland distributions show a notable steepening of slope for distributions of larger craters. A similar steepening is noted by Dodd and others (1963) for a crater distribution measured in the central part of the Southern Highlands. They interpret this break to be the result of a regional blanketing which buried all pre-existing craters with diameters less than 12 km. In another portion of the Southern Highlands the distribution curve shows a general change in slope but no sharp discontinuity. This they interpret as the result of random aging of craters, a trend towards the steady-state distribution.

There have been several attempts to use Orbiter photographs in determining absolute age by crater density techniques. Hartmann (1968) determined ages of the craters Copernicus, Aristarchus, and Tycho to be 2×10^9, 1×10^9, and 2×10^8 years respectively. By contrast, Gault and others (1968) conclude that Copernicus is younger than 10^9 years and that Tycho was formed 10^6 to 10^7 years ago. The discrepancies can be traced to different values for the meteoritic flux chosen in the two studies.

PATTERNS IN CRATER DISTRIBUTION

Spacing of craters on lunar surfaces is seldom regular. Proponents of internal origin have pointed to apparent clustering and to crater alinements as features inconsistent with a random impact mechanism. However, Arthur (1954), Lenham (1964), and Marcus (1966*a*, *b*) have pointed out that even if crater centers were originally distributed randomly, obliteration of older small craters by younger large ones would create an apparent clustering in the distribution of small crater centers.

Fielder and Marcus (1967) circumvented this problem of obliteration clustering by studying craters with diameters of 1 to 4 km in the floor of Ptolemaeus. Craters this size and larger are widely spaced, so that apparent clustering by obliteration is negligible. Utilizing a distance test for randomness first developed by ecologists to study the way in which plants spread, they determined that the craters are definitely clustered. A further test for linear arrays showed preferred alinement in two directions. The authors contend that clusters could be produced either by secondary impacts or by volcanism. However, they add, the absence of a radial pattern around a larger primary crater suggests that the observed craters are of internal origin.

Statistical tests for randomness are time consuming. In many cases straightforward observation of a photograph reveals a line of craters, a radial pattern, or a cluster. Linear patterns can be enhanced by viewing the photograph either from a low oblique perspective or through a linear moiré net (H. A. Pohn, personal communication).

A great deal remains to be done in the area of pattern analysis. Certainly, more studies like that of Fielder and Marcus will reveal significant information concerning the balance of external and internal processes modifying the lunar surface. A further possibility is that certain stratigraphic units have distinctive signatures spelled out in the arrangement of craters. However, it takes imagination to envision a situation in which adjacent units with different patterns would not also have noticeably different crater densities. If simple standard area counts fail to reveal any differences, perhaps one is following a will-of-the-wisp in using sophisticated statistical techniques to search for subtle differences in patterns.

CRATER DEGRADATION

The view that craters are subdued and degraded with age is qualitatively accepted by most geologists studying the Moon. General correlations of morphology with age have been proposed by Dietz (1949), Hackman (1961), Baldwin (1963), and Trask and Rowan (1967). Most of the 1:1,000,000 geologic maps have been compiled with this assumption in mind. Recently, however, there have been several attempts to recast the idea in a more quantitative and objective mold.

One of the first of these was by Trask (1970) who proposed that craters smaller than 3 km within Apollo landing sites be classified according to size, morphology, and age—the last indicated by a sequence of dimensionless numbers. Central to Trask's classification is the idea that the morphological lifetime of a crater is a function of its size. Accordingly, morphological trend lines run obliquely to isochronous lines (i.e., lines separating numbered fields). A crater 100 meters in diameter with a bright halo and a rim crest only slightly subdued would be assigned a more recent age than a 1-km crater with similar features.

The slopes of the morphological trend lines are determined empirically by examination of crater populations thought to have the same age. The morphological boundaries of numbered age fields are chosen chiefly for their operational convenience. The numbered age categories divide the total crater population into groups of approximately equal size.

For certain purposes a crater classification of this type is an end in itself, but, for the stratigrapher, it is a means to another end, the identification of regional surfaces on which characteristic crater populations occur. Assume that craters are assigned numbers from 10 (recent) to 1 (ancient). Old surfaces should have a complete spectrum of craters numbered 1 to 10. Younger surfaces might have no craters older than 5, very young surfaces no craters older than 10 or 9. Even if the assumptions about crater morphology being chiefly a function of age are incorrect, the objectively determinable spectrum of crater types might continue to have stratigraphic significance. For example, impact craters formed in a rubbly surface layers may tend toward a shallow pan shape, while craters formed in consolidated rock may retain sharp outlines for long periods of time. Although both surfaces may have the same age of formation, the consolidated rock will have a concentration of craters higher numbered than those on the rubbly surface.

Pohn and Offield (1969) extended and modified Trask's classification to include craters larger than 8 km. They described three age sequences: one for craters with diameters from 8 to 20 km, a second for craters from 20 to 45 km, and a third for craters larger than 45 km. Craters are assigned dimensionless age numbers from 0.0 (oldest) to 7.0 (youngest). In an effort to standardize assignment of numerical ages they designated type craters which showed the variety of features believed to be diagnostic (Fig. X-7). Very young craters display rays and a hummocky ejecta facies. Terraces are present in craters with ages both young and intermediate. Interior radial channels are confined to older craters. Expectably, rim crests of craters of all sizes are progressively more subdued with increasing age.

The size dependency of obliteration by degradation is not as apparent for large craters as for those less than 1 km in diameter. Further, the wealth of complex detail in any large crater makes it difficult to establish precise correlations either with specifically designated craters or with some abstractly defined form. For this reason subdivisions such as those by Pohn and Offield can never be treated mechanistically. Evaluation of

FIGURE X-7. Type craters chosen by Pohn and Offield (1969) to illustrate crater modification with increasing age. Craters are divided into three classes according to size. Possible age index numbers range from 7.0 (recent) to 1.0 (ancient).

crater ages must draw heavily on experience and intuition. Differences in analysis of the same scene by two observers, both of them experienced, may be great.

Offield and Pohn (1970) employed their classification to reach some extremely provocative stratigraphic conclusions. Examining many of the surfaces previously mapped as part of the 1:1,000,000 program, they assigned them ages based on the oldest superposed crater. For convenience the surfaces can be divided into three categories: highland plains, large basin blankets, and mare fillings (Fig. X-8). The oldest basin is determined to be one on which the crater Schiller is situated. Others, in order of decreasing

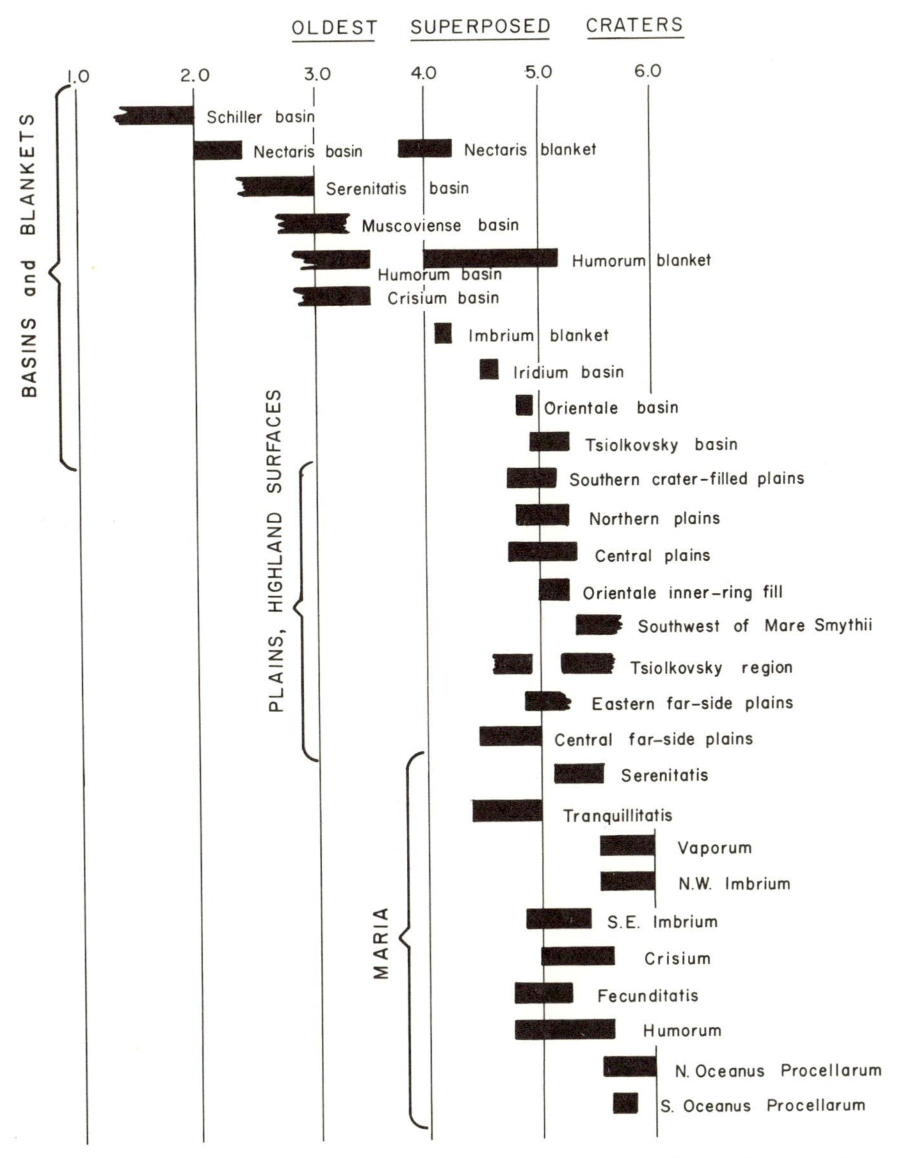

FIGURE X-8. Relative ages of selected lunar regions determined according to the age of the oldest superposed crater (from Offield and Pohn, 1969).

age, are Nectaris, Serenitatis, Muscoviense, Humorum, Crisium, Imbrium, and Orientale. The surface units in the maria are younger than the basins they occupy. The oldest mare surface unit is slightly older than the Orientale basin. There is a suggestion that the western mare units are younger than those of the eastern equatorial belt. Surface units of the light terra plains all appear to have formed in the interval 4.5–5.6. This supports a conclusion reached in the previous chapter on highland stratigraphy that much of the primeval highland crust has been subsequently flooded with volcanic materials. The age of this flooding is seen to be similar to that for some mare regions.

One anomaly deserves special note. Some of the major basins are surrounded by inner "structural" rims and outer units once thought to be depositional ejecta blankets. Following the line of reasoning developed in previous chapters the two should have identical ages. However, the oldest superposed crater on the Nectaris rim is 2, the oldest on the Nectaris "blanket" is 3.5. The age of the Humorum rim is 3–3.5, the age of the "blanket" 4–5. Offield and Pohn suggest that the hummocky outer-rim deposits around these two basins include relatively young volcanic blanketing units which are not related to the formation of the basins themselves.

CHRONOLOGY FOR LUNAR EVENTS

Several models for a chronology of major lunar events have been proposed, each with different genetic implications. At one end of the spectrum Urey has repeatedly argued that the Moon is a primitive "dead" body; at the other end Shoemaker's definition of stratigraphic systems implies a continuing crustal evolution analogous to that for the Earth's crust.

Urey's (1969) proposal is that the Moon went through a molten stage very early in its history. Following that, after it became cold and rigid, it was subjected to intense meteoritic bombardment. Based on the qualitative observation that there is no evidence that sedimentary rocks on Earth as old as 3.5 billion years have been extensively fragmented by meteoritic bombardment, Urey argues that formation of most large lunar craters and basins antedated 3.5 billion years ago. According to him large meteoritic masses lying close to the surface beneath basins produce the positive gravity anomalies noted by Muller and Sjogren (1968). The mare plains, formed after the basins, are interpreted as huge lake beds. The accretion of cold particulate matter is consistent with Urey's requirement that the Moon be a cold, rigid body throughout the major part of its lifetime. Although he does not specifically present an explanation for the basaltic chemical composition of mare materials, the implication is that previously extruded basalt has been eroded from highlands and deposited in the mare basins to form sediments made up of basaltic detritus.

At the other end of the spectrum the argument for a continuum of melting, meteoritic impact, and gradual stratigraphic evolution was implied by Shoemaker and Hackman (1962) and has been repeatedly utilized in the construction of U.S. Geological Survey maps of the lunar surface. The results of this program have been described and critiqued throughout the text.

Between the two polar points of view just noted, there is a middleground. Hartmann (1966), among several others, has proposed a model involving intense early cratering. According to his calculations, during the first seventh of lunar history the cratering rate was approximately two hundred times the average post-mare rate. Included in this pre-mare cratering interval was the formation of large basins. Hartmann estimates that this period of intense bombardment was followed, approximately four billion years ago, by a period of extensive crustal melting and volcanic activity.

Although Hartmann's opinions run contrary to the idea that the lunar surface has developed uniformly with respect to time, an interesting point is that the *relative* stratigraphic relations shown on most geologic maps of the Moon are consistent with either a uniform or a non-uniform sequence of events. It is the *absolute* age significance that is in question. According to one view a temporally condensed stratigraphic section provides a record for an early brief period of lunar history. According to the other point of view temporally expansive stratigraphic sections provide information for events ranging from the time of the Moon's formation to the present.

Hartmann's model has some interesting consequences for models involving interaction of the Earth-Moon system. He favors the view that early intense bombardment of the Moon was by circumterrestrial debris swept up by the Moon as tidal friction forced it outward from the Earth. Wells (1963) observed different groupings of daily growth lines on corals from the recent, Pennsylvanian, and Devonian. The variations suggest a progressive increase in the length of the day as the present is approached. One explanation is that the increase is related to changing tidal interaction between the Earth and Moon, this in turn related to progressive change in the Earth-Moon distance. In interpreting these results Lamar and Merifield (1968) conclude that the Moon became an Earth satellite between 0.5 and 2.0 billion years ago. The possible geologic effects on the Earth arising from such proximity were explored in some detail, but surficial effects on the Moon were ignored. Whatever origin one assumes for the Moon it seems probable, as Hartmann suggests, that the time of its close approach to the Earth was also a time of intense bombardment. But if that event occurred less than 2.0 billion years ago, as the paleontologic evidence suggests, then the previously cited absolute ages of about 4.0 billion years for the mare surfaces must be in error by at least a factor of 2. Shoemaker's previously noted revised ages for mare surfaces lends additional weight to the possibility that there has been a major period of melting more recent than four billion years ago.

A recent review of the problems outlined in the preceding paragraphs is provided by Alfvén and Arrhenius (1969). They present two models for lunar history, one termed catastrophic and the other noncatastrophic. According to the noncatastrophic model the Moon was formed within the solar system at the same time as the other planets, about 4.5 billion years ago. It subsequently was captured by the Earth but never approached the Roche limit, the distance at which gravitational forces between the Earth and Moon would cause breakup of the Moon.

According to the catastrophic model the Moon approached the Earth within the Roche limit some 0.5 to 0.7 billion years ago. As a result of this close passage the Moon broke up and many of the pieces were subsequently recovered during an impact accretion stage. In support of the proposed time for this catastrophic event Alfvén and Arrhenius point to cosmic ray exposure ages and degassing ages for certain types of meteorites which cluster in the range 0.5 to 0.7 billion years. They contend that these meteorites were formed during the Moon's breakup and were ultimately captured by the Earth. The authors suggest that dating of lunar igneous rocks will allow one to choose between the catastrophic and noncatastrophic models. In the former case, ages should be in the vicinity of 4 billion years. In the latter case, they should be 0.7 billion years or less, corresponding to a partial melting which accompanied the Moon's deformation during its close approach to Earth.

Lunar Stratigraphy Reconsidered

CHAPTER

WE began this book with the proposition that geological investigation of the Earth has provided us with a backlog of experience which should influence and expedite our understanding of the Moon. Throughout the book we have sought out similarities between terrestrial stratigraphy and the geology of the Moon. But now that much of the evidence has been presented, how persuasive does this argument by analogy appear?

Central to any evaluation of the evidence is consideration of the concepts of uniformitarianism and catastrophism. These two terms have figured prominently in the development of terrestrial geology. Most students are told rather simplistically that early geologists, notably Werner, held the incorrect view that the surface of the Earth was shaped by one or several catastrophic events. Hutton, they are told, corrected this error by pointing out that the past can be explained in terms of the same processes that are observable today and that these have shaped a gradually evolving world. However, as Simpson (1963) notes, the idea of uniformitarianism does not rule out the existence of catastrophies. Neither is it necessary to assume that the same processes have always acted at the same scales and the same rates. The only valid uniformity principle is that which rests on the distinction between immanence and configuration.

> The unchanging properties of matter and energy and the likewise unchanging processes and principles arising therefrom are *immanent* in the material universe. They are non-historical, even though they occur and act in the course of history. The actual state of the universe or of any part of it at a given time, its configuration, is not immanent and is constantly changing. It is *contingent* . . . or *configurational.* . . . History may be defined as configurational change through time, i.e., a sequence of real, individual, but interrelated events. These distinctions between the immanent and the configurational and between the non-historical and the historical are essential to clear analysis and comprehension of history and of science (Simpson, 1963, pp. 24–25).

Viewed at this level of sophistication the geology of the Moon can be interpreted in the same uniformitarian sense as the geology of the Earth. But there remains a difference in the pattern of changing configurations which we might describe in terms of *gradualism.*

Geology of the Moon

The marvel of the terrestrial wave-cut cliff, the polished boulder, the deep canyon is that the configurations remain essentially constant from day to day, from year to year. Even so, gradual change is occurring—if only in knife-edge increments. The same holds true for the deposition of most sedimentary rocks. Admittedly many sediments reach their final place of deposition in a catastrophic surge, carried by earth slides, floods, and turbidity currents. But many more are deposited or precipitated grain by grain. It is an infinitely gradual process—sand swept back and forth by ceaseless currents; minerals precipitated in delicately bedded layers, each bearing witness to a single year.

This pattern of gradualism and systematically repeated configurations results in a terrestrial rock record which gives significant emphasis to each interval of geologic time. An exception, of course, involves the period from the time of the Earth's assumed formation 4.5 billion years ago to the time of the oldest radiometrically dated rocks 3.4 billion years ago. But for major time intervals more recent than 3.4 billion years, sequences of sedimentary rocks record the passage of time. Some gaps in the sedimentary rock record are almost certainly present in the Precambrian. During these gaps there was either no deposition or deposition followed by erosion. Durations may be on the order of millions of years. In the Paleozoic rock record it is probable that no time interval greater than 1,000 years is unaccompanied by a significant sedimentation event. For the Mesozoic unrecorded intervals are probably no more than 100 years, and for the Cenozoic the number is perhaps an order of magnitude less. In summary, then, the Earth displays a gradually accumulating sedimentary rock record which permits one to decipher a detailed sequence of events starting 3.4 billion years before the present. As the present is approached the rock record is progressively more complete and the matrix of recorded events becomes increasingly elaborate.

For the Moon it is difficult to state dogmatically that gradually applied forces have shaped the surface as they have on Earth. The lunar regolith does appear to be churned systematically and gradually by micrometeoritic bombardment, and particles may be gradually altered by radiation. However, impacts of large meteorites represent sharp pulses of geologic activity. As noted in the preceding chapter there is good reason for thinking that meteoritic bombardment was concentrated during an early period in the Moon's development and that the formation of large basins was also restricted to that period.

Much of the lunar surface probably is underlain by volcanic rock. It is possible that molten materials have gradually and continually flowed out on the Moon's surface from shortly after the time of the Moon's accretion to the present. However, there is little evidence presently available to prove this contention. Both observational dating of mare surfaces by crater densities and development of models for lunar differentiation suggest, to the contrary, that regional volcanic activity was concentrated during one or several limited temporal periods. In contrast with the Earth, then, there is little evidence for gradualism in development of the lunar rock record, or for the presence of a rock record which becomes progressively more comprehensive with passage of time from ancient to recent.

Some of the more exciting imaginable results of manned lunar exploration would be identification of rocks of different ages, discovery of distinctive stratigraphic units traceable over great lateral distances, and compilation of a composite stratigraphic section covering a significant part of geologic time. The first result would at least verify the idea of an evolving Moon; the second result would hold out the hope of temporal correlation using key stratigraphic horizons as datum planes; and the last result would enable one actually to decipher a reasonably complete history.

The central place of paleontology in the development of terrestrial stratigraphy imparts to this discipline a peculiar flavor lacking in lunar stratigraphy. At one operational level it is obvious that fossils provide an important means for correlation of rocks over great distances and for identification of depositional environments. Perhaps it is even more important to remember that sedimentary rocks comprise a time-space continuum within which a record of the evolution of life is abundantly, though not completely, displayed. The deciphering of this record poses a problem of special relevance to men trying to understand their place in a changing universe. It is possible (Sagan, 1961), but not likely, that remains of living forms are preserved in lunar rocks and that an understanding of their distribution and evolution will be aided by knowledge of the sedimentary sequence in which they are contained.

Given all the uncertainties mentioned throughout this book a reasonable question is: Why don't we wait until we get to the Moon to unravel its geology? Two quite different answers are possible. The first is pragmatic and easily formulated. Engineering and financial constraints militate against a comprehensive field study of the Moon analogous to that conducted on Earth. The several manned landings will result in collection of invaluable geologic information at a few selected points. Study of large numbers of stratigraphic sections and the recovery of drill cores to great depth may not be feasible for many years. Admitting this fact there is no great value in deferring an attempt at a stratigraphic synthesis. The majority of the relevant information—high quality photographs—is already in hand. Astronauts sampling the Moon's surface will be able to make critical tests of pre-existing stratigraphic ideas. However, it is unlikely that information gained on manned excursions will be sufficient to provide substance for a new stratigraphy based on conventional "outcrop" and "hand sample" analysis.

The second argument against deferring stratigraphic study of the Moon is more elusive. It involves timeliness of scientific questions, permanence of ideas, and man's natural curiosity. In that context three categories of scientific research can be defined: first, that in which the question is important but untestable; secondly, that in which the question is testable but is not of general importance; and thirdly, that in which the question is both important and testable. The second category with its inherent limitations makes the fewest intellectual demands and, in turn, promises the least in terms of significant results. The choice between the first and third categories is more difficult. One is naturally inclined to give greatest credit to those scientists whose work has a direct and lasting applicability. So it is that Newton's laws of motion are well known to any student of physics. But what of those scientists who wrestle with questions of continuing importance, making influential contributions but still falling short of successful solutions? Here we have to judge incalculables. How much credit do they deserve for recognizing the critical problems, if not their solutions? More importantly, to what extent did their efforts stimulate—or provoke—others who followed them? Since answers to these questions are generally indeterminate, the contributions of such scientists must always remain uncertain. At best, however, their work has value only slightly less than that of those who break through to definite solutions.

Where should one place studies of lunar geology and stratigraphy in this threefold classification of scientific inquiry? Clearly they do not fall in the second category where the problems are not of general importance. The questions which are being asked relate to the origin of the Moon. This, in turn, is closely related to the question of total solar system origin. It is reasonable to contend that this latter problem is one of unique importance for twentieth-century man, especially because the exploration of space gives promise of critical experimental data which was previously lacking.

But does the information in hand permit formulation of intellectually satisfying answers to the geologic questions being asked? The answer is not easy. It must be admitted that there is a great deal we cannot know without actually travelling across the face of the Moon. Furthermore, it is unthinkable that such traverses will leave our present stratigraphic ideas essentially unchanged. In that sense, weighed against the possibility of intensive lunar exploration in the distant future, our present ideas are impermanent and inconclusive.

The question posed in the preceding paragraph can be rephrased. Does analysis of the information in hand make possible a stratigraphic synthesis precise enough to provide new understanding of the Moon's origin and evolution? To this question the answer is cautiously affirmative. Already, geologic studies suggest the Moon's great age and its continuing crustal development. The modifying roles of impact and volcanism are rudimentarily understood. Depositional and structural mechanisms are tentatively defined. Lunar processes and products are dimly perceived. Perhaps a small amount of additional on-the-ground information is all that is necessary for a response to the central question: the origin of the Moon and its relationship to the rest of the solar system.

It may be that the ultimate justification for man's interest in the Moon is also the simplest. The pages of history abound with accounts of exploration: new continents, new frontiers, mountains, and ocean depths. The forces which drive man to an exploration of the world around him are so elemental that they defy rational analysis. In turning our attention to the Moon, is it not likely that we are really inexorably driven to take part, however vicariously, in yet another journey of exploration?

Apollo Results

CHAPTER

SO much has been written about the rocks collected on the several Apollo missions that anyone who follows the literature conscientiously sometimes feels more overwhelmed than informed. Preliminary reviews of results from each mission have been published by the Lunar Sample Preliminary Examination Team (LSPET) in *Science.* The most comprehensive report on Apollo XI is a three-volume supplement of *Geochimica et Cosmochimica Acta.* Included there are expanded versions of papers presented at the Lunar Science Conference in Houston, January 1970. Briefer summaries of most of these papers appear in a special issue of *Science* (January 1970). A second Lunar Conference was held in January, 1971 and the proceedings, emphasizing Apollo XII results, again have been published as a supplement to *Geochimica et Cosmochimica Acta.*

Other articles appear in many journals, but especially in *Science* and the *Journal of Geophysical Research.* An accessible and clearly written book, *The Lunar Rocks* by Mason and Melson, integrates much of the literature through March 1970. Two books with excellent selections of annotated orbital photographs should also be noted. These are *Lunar Panorama* by Lowman (1969) and *The Moon as Viewed by Lunar Orbiter* by Kosofsky and El-Baz (1970). Complete citations for the several publications just mentioned and all of the other articles mentioned in this chapter are contained in a supplementary listing of references immediately preceding the index.

To summarize all the Apollo results knowledgeably would require not another chapter, but another book. Instead, it will be more convenient to emphasize those results most closely related to the data and interpretations of previous chapters. Specific questions we will address are:

1) Do the field observations of the astronauts confirm or contradict previous geologic interpretations?

2) Do the lunar rocks resemble particular terrestrial rocks or meteorites?

3) Are volcanic and impact-ejected materials the two most important surface materials, as previously speculated?

4) Are the rocks underlying the highlands different from those underlying the maria?

5) Is there evidence for differentiation and stratification of the planet—whether on a centimeter scale wtihin the regolith, on a meter scale within the near-surface rocks, or on a tens-of-kilometers scale to great depths within the Moon?

6) Has widespread melting of the Moon ever taken place?

7) Can an absolute time scale be established? If so, does that time scale indicate over what period of its history the Moon has been geologically active? Can a time span for emplacement of mare materials be determined?

FIELD GEOLOGY

The selection of Apollo sites is determined by a combination of engineering and scientific constraints. The strategy has been to direct Apollo XI and XII to near-equatorial sites which present the fewest topographic hazards and navigational difficulties. Pre-Apollo mapping of these sites from Lunar Orbiter pictures suggested a simple geologic situation—lavas overlain by regolith materials. Successive Apollo missions have been scheduled for sites more rugged topographically and more complicated geologically. Concurrently, the requirement to land in the middle part of the Earthside hemisphere has been relaxed.

Apollo XI

Apollo XI landed in the southwestern part of Mare Tranquillitatis (Fig. XII-1). The site is approximately 25 km south of the Surveyor V landing site and 70 km southwest of the Ranger VIII impact site. Light north-trending rays, perhaps associated with the crater Alfraganus situated 160 km to the southwest or with Theophilus 320 km to the southeast, cross this part of Mare Tranquillitatis. The landing site lies between major rays.

The Apollo XI site is pockmarked with craters and is more abundantly littered with boulders than any of the four Surveyor mare sites. A sharp-rimmed ray crater (West crater) approximately 180 meters in diameter and 30 meters deep is located 400 meters east of the landing point (Fig. XII-2). Blocks up to 5 meters in size occur on the rim crest of this crater, and rays of blocky ejecta extend to the west of the landing site. The more prominent strewn-fields of coarse blocks pass to the north and south of the landing site.

Approximately 60 meters east of the landing site is a shallow but fairly steep-walled crater 33 meters in diameter and 4 meters deep. A pavement of blocky rocks occurs in the floor (Fig. XII-3). Several other craters of the same size have shallow, flat floors at about the same depth. This suggests that the regolith thickness is in the range of 3 to 6 meters, and that the underlying rocks have been excavated by numerous craters in the vicinity of the Lunar Module (LM). In the immediate vicinity of the LM the mare surface is pockmarked with smaller craters ranging in size from several centimeters upward. Directly southwest of the LM is a strongly subdued double crater 12 meters long, 6 meters wide, and 1 meter deep.

FIGURE XII-1. Oblique picture taken from the Apollo XI Lunar Module (LM) during descent to the lunar surface. View is toward the west. Large crater in the lower right is Maskelyne. Numerous mare ridges are present, including one in echelon system trending northwest from Maskelyne. In the middleground a sinuous rille, approximately 100 km long, can be seen. One of the attitude engines on the LM appears in the lower left of the picture (NASA photograph AS 11-37-5437).

FIGURE XII-2. Mosaic of Lunar Orbiter V photographs showing the location of the Apollo XI LM, West crater, and 33 m crater.

It is important to note that the description of the Apollo XI site just given is based largely on post-mission analysis. During the actual landing maneuvers the astronauts were forced to direct the LM to a position several hundred meters west of the projected landing site in order to eliminate the risk of landing on a dangerously block-littered surface. As a result, they had considerable difficulty orienting themselves and identifying lunar landmarks previously studied in orbital photographs. During a single

FIGURE XII-3. A crater 33 meters in diameter located approximately 60 meters east of the Apollo XI LM. Note the pavement of blocky fragments in the crater floor. The same crater can be seen in Figure XII-2 (NASA photograph AS 11-40-5956).

period of extravehicular activity (EVA) lasting 2 hours and 40 minutes, Armstrong and Aldrin restricted their field excursion to a distance of about 50 meters from the LM. Although 22 kg of samples were collected, most of the rocks were picked up hurriedly during a 4-minute period late in the EVA. This first brief geologic field investigation of the Moon stands in contrast to the progressively more ambitious and sophisticated field programs of successive Apollo missions.

Without in any way downgrading the drama of man's first description of the landscape from the surface of another planet, it should be noted that many of the astronauts' descriptions closely resembled those postulated after studying Surveyor photographs. Aldrin, for example, said: "It looks like a collection of just about every variety of shape and irregularity, every variety of rock you could find. There doesn't appear to be too much color except that it looks as though some of the boulders are going to have some interesting color." Armstrong continued: "It's a relatively flat plain with a lot of craters of the 5- to 50-foot variety. . . . Some small ridges 20 to 30 feet high. . . . Thousands of little 1- and 2-foot craters. . . . Some of the surface rocks in close look like they might have a coating on them. Where they're broken they display a very dark grey interior. It looks like it could be country basalt."

The observations made by the astronauts as they stepped outside the LM and explored the surface again confirmed—and extended—observations previously made from Surveyor experiments. One of Armstrong's first remarks was that "a lot of the hard rocks appear to have vesicles on the surface"—a strong indication of volcanism. A few minutes later he said that what he first thought to be vesicles might be instead pockmarks resulting from micrometeorite erosion. Then, during his last few minutes of rock collecting, Armstrong again commented on the vesicular character of several rocks.

Aldrin obtained two samples of the lunar soil with a hand-driven coring tube (Fig. XII-4). The tube was pushed easily through the first few centimeters of surface material but was difficult to drive beyond a depth of 15 cm. This phenomenon, and additional observations that the topmost layer of soil was "very, very fine-grained" and that the soil had cohesion resembling that of damp sand, all bring to mind determination of the size distribution, depth, and bearing strength of lunar surface materials made by Surveyor experiments.

Apollo XII

The Apollo XII LM landed at a mare site approximately 300 km south of Copernicus, a region in the eastern part of Oceanus Procellarum. A broad, high-albedo ray of Copernicus crosses the area. The landing was on the northwest rim of the same crater in which Surveyor III had landed 2.5 years previously (Fig. XII-5). The site contains a cluster of four subdued craters ranging in diameter from 50 to 400 meters. Superposed on these are numerous smaller sharp-rimmed craters. Age assignments for craters on the basis of shape and ejecta characteristics indicate a sequence from oldest to youngest as follows: (1) Middle Crescent, (2) Surveyor, (3) Head, (4) Bench, and (5) Sharp, Halo, and Block craters (Fig. XII-6).

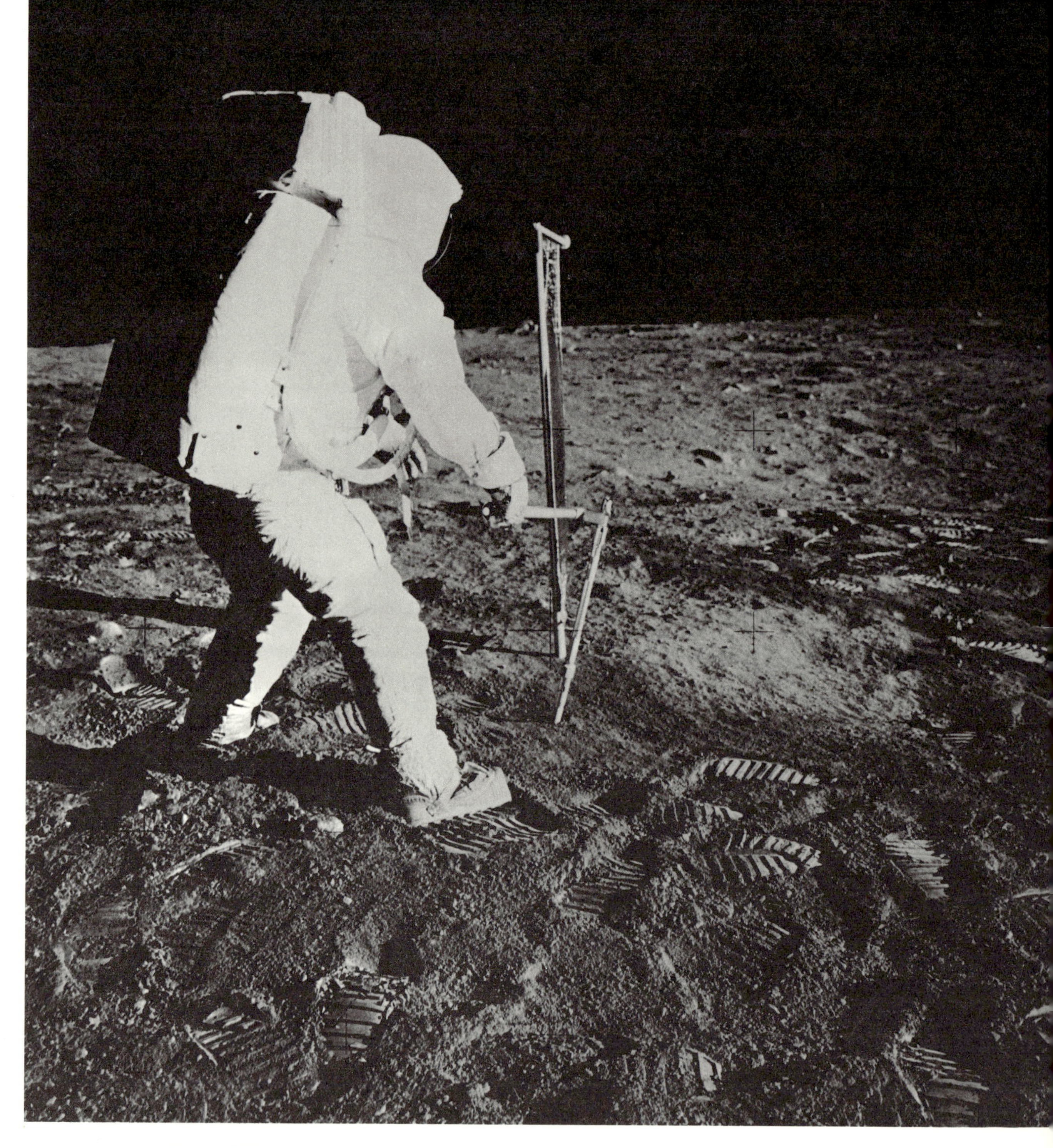

FIGURE XII-4. Apollo XI activity. Aldrin is driving a core tube into the unconsolidated surficial material (NASA photograph AS 11-40-5963).

The accomplishments of the Apollo XII field program were significantly greater than those of Apollo XI. The LM landed close to the preselected target, and less than 200 meters from Surveyor III. During two 4-hour EVA periods the astronauts completed a traverse of almost 1 km. Contrasted with the 22 kg returned by Apollo XI, more than

FIGURE XII-5. View of the Surveyor III spacecraft in the foreground and the Apollo XII LM in the background. The distance between them is about 180 meters. Surveyor III is located within Surveyor crater, and the Apollo LM is located on the northern rim of that crater. To the right of the LM the S-band antenna and solar wind experiment are visible, silhouetted against the sky. To the left of the LM the silhouette of a mound can be seen. This mound is interpreted as a clump of regolith material, slightly indurated by impact, and ejected from a nearby crater (NASA photograph AS 12-48-7091).

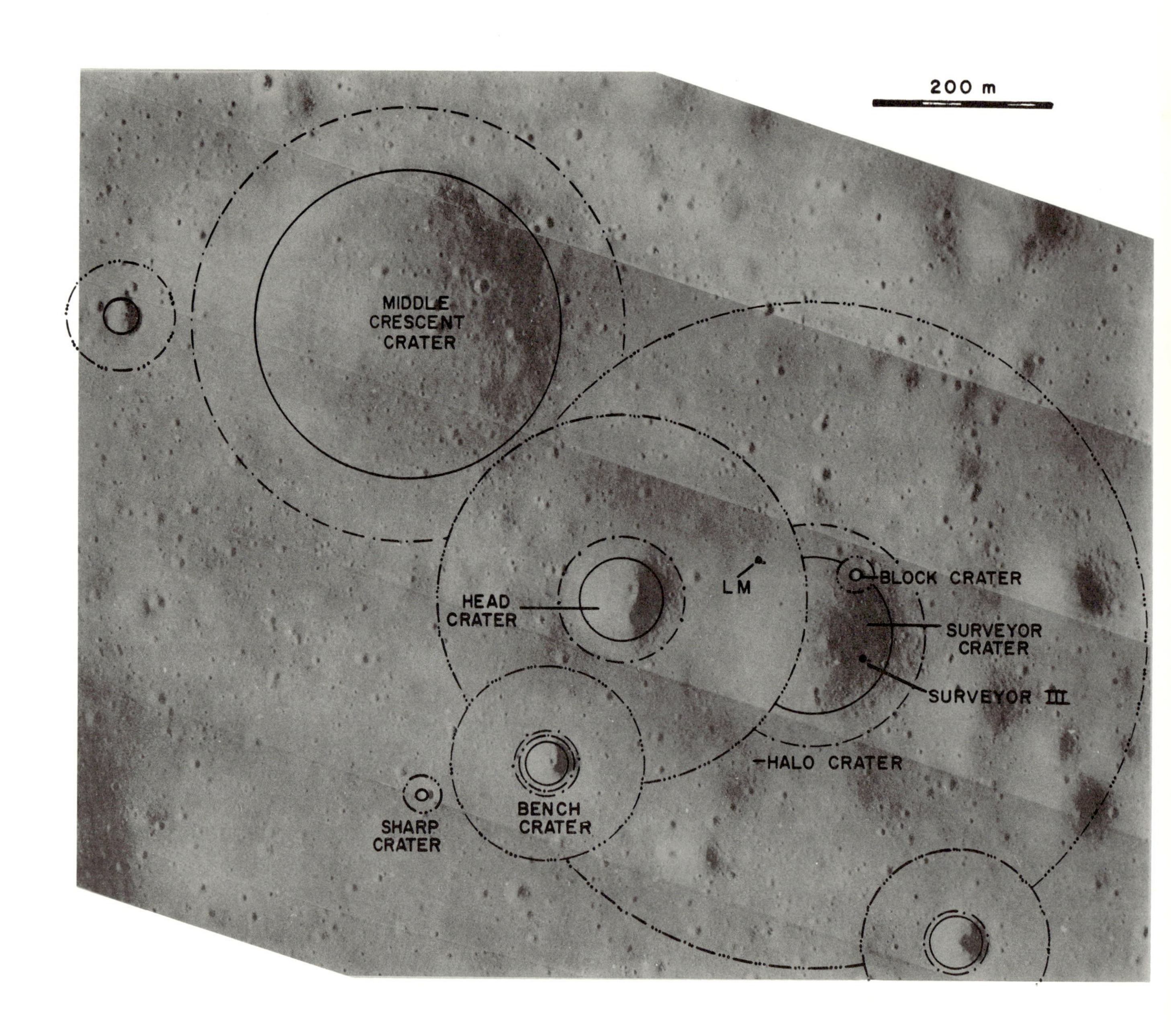

CRATER RIM UNITS:

———	Inner rim; top of the steepest crater wall. (This coincides with the rim crest in small sharp craters.)
·—·—·	Rimcrest; top of upper wall slope (shelf). (Note: Bench Crater shows an additional rim to denote the upper bench.)
—···—	Generalized limit of probable continuous ejecta, drawn 1-crater-diameter from the crater crest. (Drawn to show overlapping age relations of the different craters and their ejecta.)

FIGURE XII-6. Mosaic of Lunar Orbiter photographs showing Apollo XII field area. The inferred overlap relationships for crater deposits are taken from a map by R. L. Sutton.

34 kg of rocks and fines were collected. In response to the requests of scientists, greater emphasis was placed on the collection of large rocks from photographically documented sites.

Almost all the rocks collected by the Apollo XII astronauts were crystalline. Only half of the Apollo XI rocks were crystalline; the other half were microbreccia. The regolith at the Apollo XII site is approximately half as thick as at the Apollo XI site. These differences suggest that the Apollo XII regolith is immature and that many of the rocks, collected on or near the rims of craters, have been excavated from a near-surface layer of crystalline rock. At the Apollo XI site the regolith is older and more mature. Most of the rock fragments have been recycled by multiple impact events to form microbreccia. Many of the rocks now on the surface have been ejected from craters too small to penetrate bedrock.

Postmission classification of Apollo XII samples and consideration of their original position on the lunar surface has revealed a vertical stratigraphic sequence (Schmitt and Sutton, 1971). The fundamental assumption is that the ejecta blanket of an impact crater contains an inverted sequence of the units in which the crater was emplaced. The ejecta are presumed to be subhorizontally truncated so that progressively deeper pre-impact units are encountered as one moves toward the crater. In other words, the distance of a fragment from the rim crest is inversely related to its original stratigraphic depth. Utilizing this assumption, Schmitt and Sutton tentatively identified: (1) a regolith layer, (2) an upper differentiated flow, (3) a sill or dike within the flow unit, and (4) a lower differentiated flow (Fig. XII-7).

Apollo XIV

With the flight of Apollo XIV the term "Fra Mauro Formation" suddenly emerged from stratigraphic obscurity. The Apollo XIV site is in a rough terrain north of the

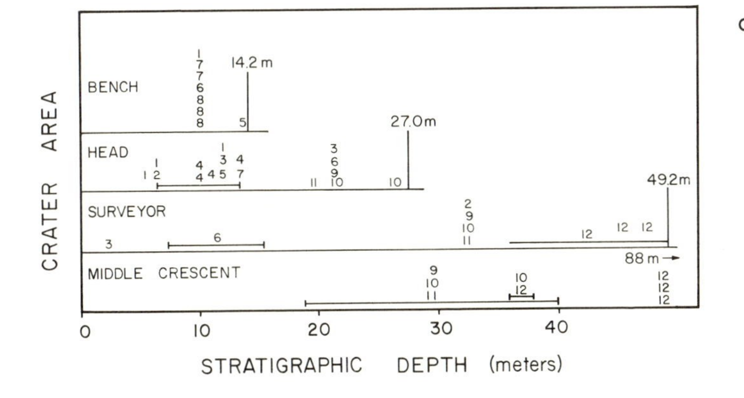

CODE	ROCK TYPE
1	Indurated regolithic breccia
2	Welded felsic breccia
	——— Flow Top ———
3	Gridded olivine/clinopyroxene basalt porphyry
4	Granular olivine/clinopyroxene basalt porphyry
5	Olivine-rich clinopyroxene/olivine basalt porphyry
6	Tridymite/pyroxferroite gabbro
7	Cristobalite basalt
8	Olivine gabbro
9	Skeletal olivine basalt porphyry—sill or dike
	——— Flow Top ———
10	Plumose clinopyroxene/olivine basalt porphyry
11	Cristobalite pyroxenite
12	Porphyritic cristobalite basalt

FIGURE XII-7. Reconstruction of stratigraphic depth of rock fragments at the Apollo XII site. The number code for rock types is shown in the table, which is a probable stratigraphic sequence, oldest at the bottom, youngest at the top (adapted from Schmitt and Sutton, 1971).

crater Fra Mauro, a region that Eggleton (1965), six years previously, had mapped as a remnant of the vast ejecta blanket thrown out from the Imbrium basin at the time of its formation (Fig. XII-8). The relative antiquity of the surface is indicated by the fact that craters between 400 meters and 1 km in diameter are approximately four times more abundant than at the Apollo XI site and approximately six times more numerous than at the Apollo XII site (Swann and others, 1971). The exact landing target was a smooth slope between two sets of overlapping craters, Doublet craters and Triplet craters. Approximately 1 km to the northeast stands Cone crater: youthful, sharp-rimmed, surrounded by a field of large blocks, and rising more than 100 meters above the

FIGURE XII-8. Oblique photograph of Apollo XIV landing site, looking northeast. Landing ellipse is shown by white outline. The hummocky terrain with ridges trending from lower right to upper left is underlain by the Fra Mauro Formation. The massive hills in the upper right are probably underlain by pre-Imbrian materials. Light plains-forming materials of the Cayley Formation are visible along the right edge. Darker, less cratered mare materials are visible along the left edge. The northwest corner of the crater Fra Mauo is in the lower right (NASA photograph 69-H-1863).

FIGURE XII-9. View of Apollo XIV site showing position of LM, Doublet craters, Triplet craters, and Cone crater (Lunar Orbiter Photograph IV 122H).

surrounding terrain (Fig. XII-9). A traverse to Cone crater was the principal geological objective of the field excursion. Rocks collected from close to the rim crest were expected to provide samples of units originally 50 meters or more below the surface.

The field program proceeded almost exactly as conceived by the planners of the mission. The LM landed a scant 25 meters from the preselected site. During a first EVA of close to 5 hours the astronauts deployed an array of surface instruments, including a seismometer utilizing an astronaut-activated thumper device and an Earth-commanded mortar package as energy sources. In their second EVA the astronauts hiked to Cone crater. The journey, over undulating terrain, through thick dust, and around large boulders, proved more exhausting than expected, and they turned back about 20 meters short of the rim of the crater.

Approximately 43 kg of rocks and fines were collected on the two EVA's. Thirty-three rocks larger than 50 grams were returned and, of these, all but two are breccia. The

remaining two are basaltic rocks. One of these is a large clast collected from the side of a boulder (Sutton and others, 1971). The other is considerably smaller than the largest clasts photographed on the lunar surface, and is lithologically identical to clasts in returned fragmental rocks. Therefore, it appears probable that all of the returned rocks are breccia fragments.

Many of the boulders in the vicinity of Cone crater show well-developed planar structures (Fig. XII-10). Most of these are interpreted as sedimentary discontinuities in a large ejecta deposit, although some transcurrent planar sets may be shock foliations (Swann and others, 1971). The knobby protuberances on some boulders were probably formed by small-impact erosion, which rapidly modifies the poorly indurated and friable breccia blocks.

The character of the Apollo XIV landing site completely confirms the anticipations of geologists, Eggleton in particular, who had predicted that this would be a region underlain by deposits of ejecta from the Imbrium basin. The results of the active seismic experiment indicate a regolith approximately 8.5 meters thick. Beneath this, a higher velocity zone between 38 and 76 meters thick is tentatively interpreted as the total thickness of the Fra Mauro Formation (Kovach and others, 1971). Although some breccias were formed by local cratering events, it is most unlikely that the overall character of the rocks is related to *in situ* formation of an unusually thick regolith.

FIGURE XII-10. Photographic mosaic showing the "white rocks" photographed by Apollo XIV astronauts close to the rim of Cone crater. The large rock on the far left is a little over 2 meters in diameter. The large rock second from left is almost 3 meters high. Irregular, angular fragments are common within the boulders. The rock on the right shows stratification from upper left to lower right. A second set of planar surfaces trends from lower left to upper right. The two rocks on the left display bright zones which may be injection dikes or contorted sedimentary beds. Note that most of the boulders are surrounded by fillets of fine, unconsolidated material. However, the boulder on the left lacks these overlapping fines. It appears to have been recently placed in its present position. Perhaps it split off one of the boulders to its right (NASA photographs AS 14-9948-9451).

Apollo XV

The Apollo XV landing area is located on mare materials of the Palus Putredinis embayment, along the southeastern rim of the Imbrium basin. Immediately to the south and east are the towering escarpments of the Apennine Mountains, rising more than 3,400 meters above the mare plains. Hadley Rille is located less than 1 km west of the LM (Figs. XII-11, 12).

The Apollo XV astronauts took full advantage of the spectacular geology and scenery in the vicinity of the landing site. Traveling across the surface in a small battery-powered Lunar Roving Vehicle (LRV), they covered more than 28 km on three EVA's. On two of these sorties they visited the eastern rim of Hadley Rille; on the third they reached the lower slopes of Hadley Delta (Figs. XII-13,14). They brought back to Earth 77 kg of lunar materials, including more than 350 individual samples from ten sampling areas.

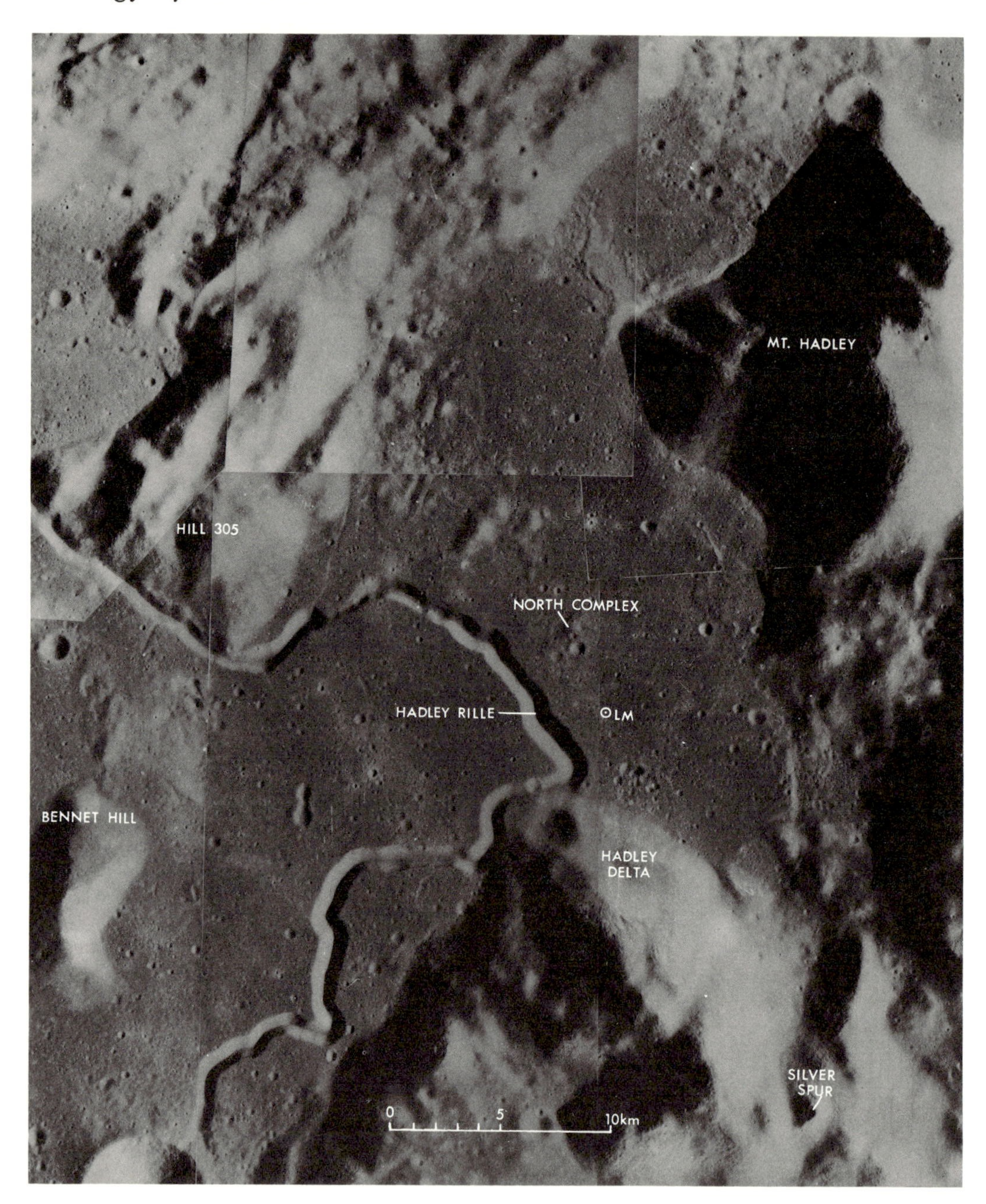

FIGURE XII-11. The Apollo XV landing site. This mosaic is made from Apollo XV panorama photographs.

(*Opposite page*)

FIGURE XII-12. Geologic map of the Apollo XV landing site, approximately the same area shown in Figure XII-11 (adapted from a map by Howard, 1971, which was prepared before the mission).

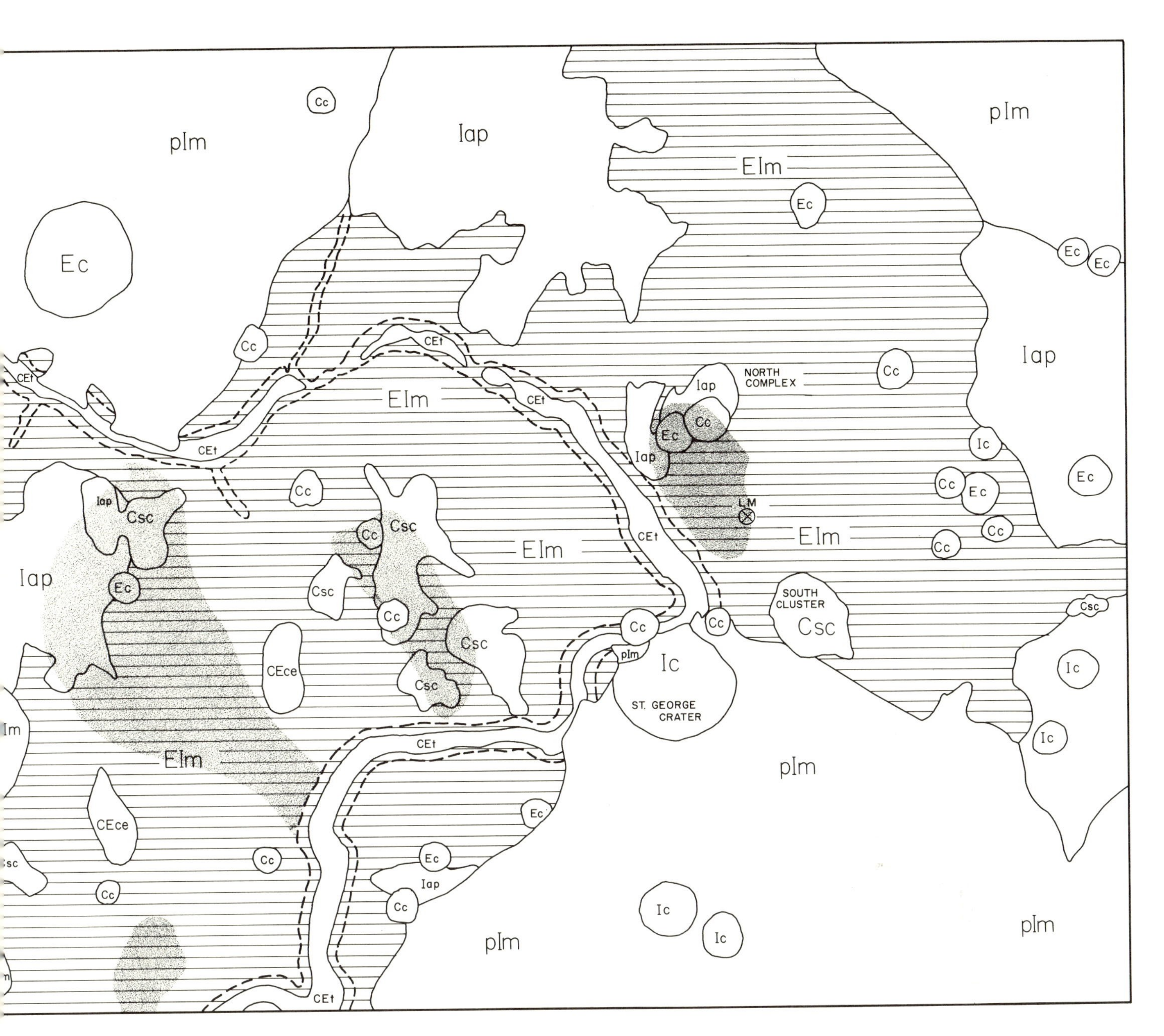
pIm
Iap
EIm
Ec
Cc
Csc
CEt
CEce
Ic
Im
NORTH COMPLEX
LM
SOUTH CLUSTER
ST. GEORGE CRATER

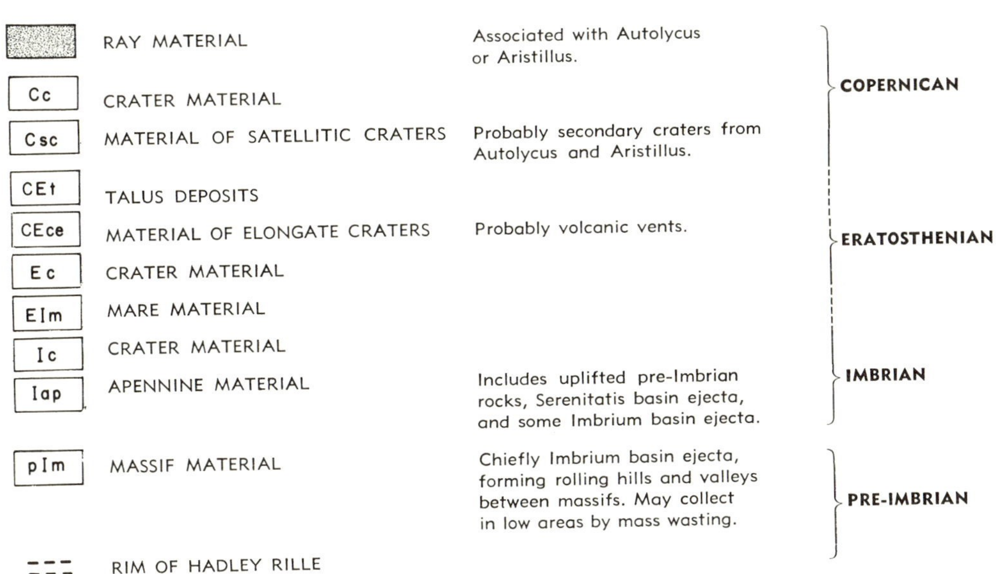
RAY MATERIAL
Associated with Autolycus or Aristillus.
Cc
CRATER MATERIAL
Csc
MATERIAL OF SATELLITIC CRATERS
Probably secondary craters from Autolycus and Aristillus.
COPERNICAN
CEt
TALUS DEPOSITS
CEce
MATERIAL OF ELONGATE CRATERS
Probably volcanic vents.
ERATOSTHENIAN
Ec
CRATER MATERIAL
EIm
MARE MATERIAL
Ic
CRATER MATERIAL
Iap
APENNINE MATERIAL
Includes uplifted pre-Imbrian rocks, Serenitatis basin ejecta, and some Imbrium basin ejecta.
IMBRIAN
pIm
MASSIF MATERIAL
Chiefly Imbrium basin ejecta, forming rolling hills and valleys between massifs. May collect in low areas by mass wasting.
PRE-IMBRIAN
RIM OF HADLEY RILLE

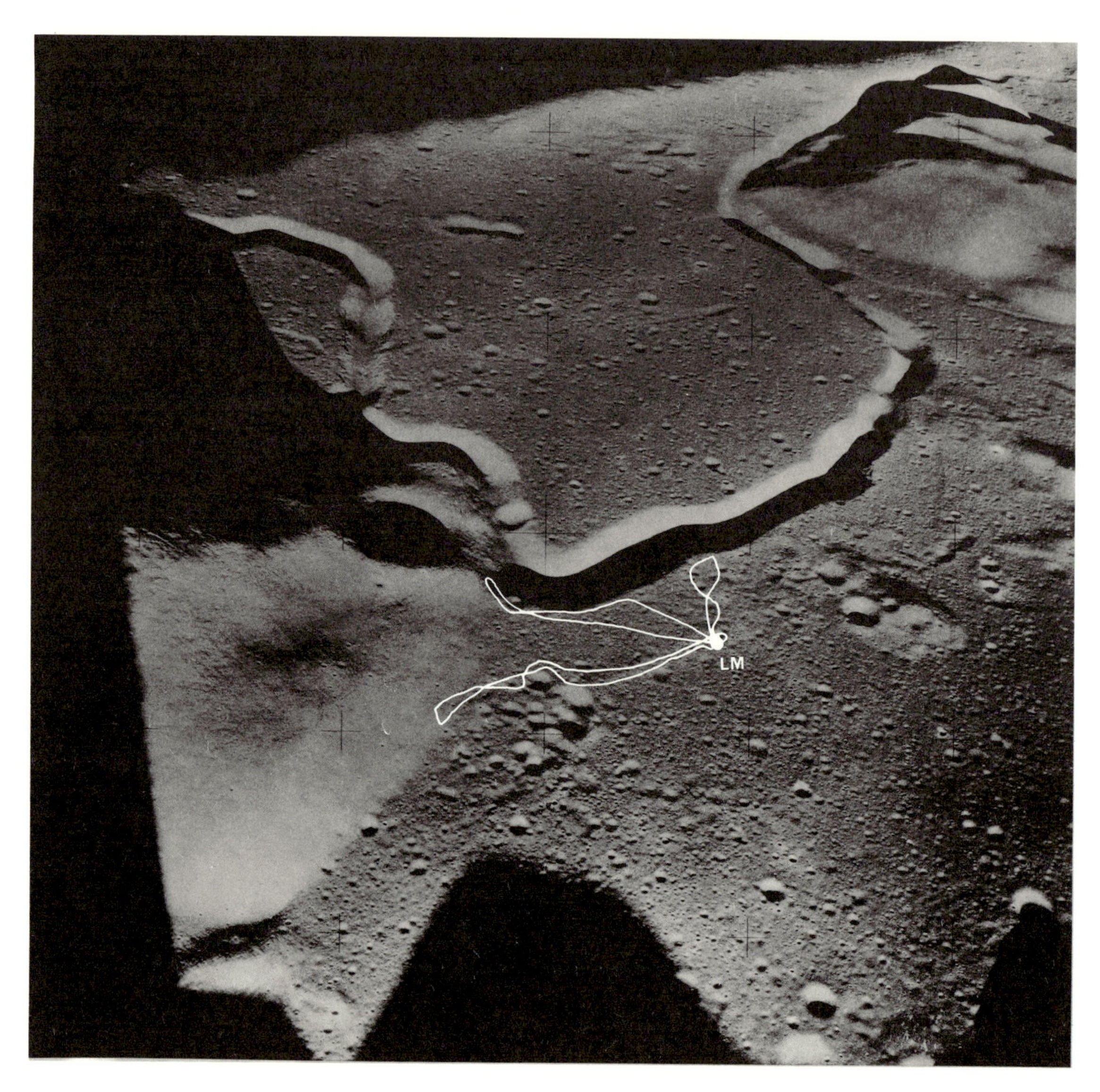

FIGURE XII-13. Telephoto picture (NASA photograph AS 15-84-11718) of the Apollo XV site taken from the Command Module before landing. Approximate routes for the three EVA's are indicated.

Approximately one-third of the rock specimens are mare basalts displaying a wide range of texture and mineralogy. The remaining two-thirds comprise a bewildering assortment of breccias, many of which are glass-coated or glass-cemented. Breccias from the plains contain numerous fragments of mare basalt and particles of abundant opaque minerals. Breccias collected along the Apennine front are characterized by fragments ranging in composition from gabbroic to anorthositic, and by a relatively low amount of opaque minerals (LSPET, 1972).

The astronaut crew was tremendously impressed with what they termed the "organization" of rocks displayed in the Apennine front. This organization consists of several

FIGURE XII-14. Telephoto picture (NASA photograph AS 15-84-11324) taken on second EVA of Apollo XV mission, looking north from the lower slopes of Hadley Delta. The LM is visible in the distance.

intersecting sets of linear features. On the southwestern slopes of Hadley Mountain one set of linear markings dips steeply to the left. A second set is nearly horizontal and forms regolith-mantled benches (Fig. XII-15). Farther to the south, on Silver Spur, the lineations define planes that apparently are close to horizontal (Fig. XII-16). One is strongly tempted to ascribe a sedimentary origin to this "stratification," especially since pre-mission analysis indicated that ejecta deposits of the Imbrium and Serenitatis basins might be present along the Apennine front (Fig. XII-17). However, a more detailed analysis raises some doubts. The linear features are not confined to areas where bedrock

(a)

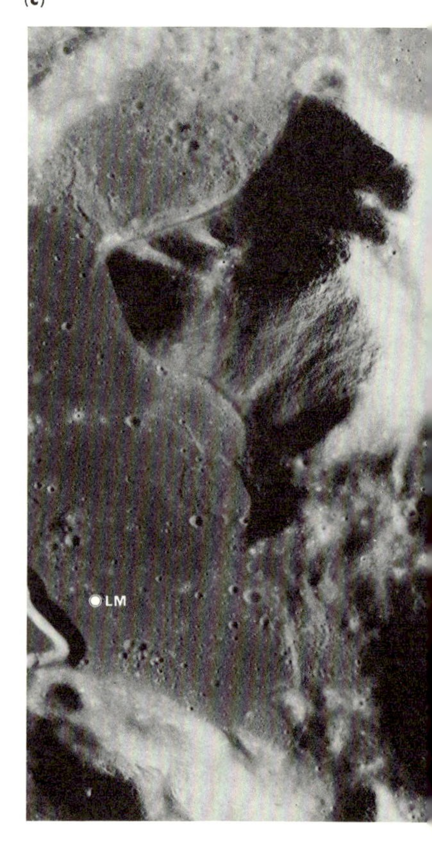

(c)

FIGURE XII-15. (a) View looking northeast from Apollo XV LM toward Mount Hadley. Prominent lineations dip steeply to the left. The dark band at the base of Mount Hadley may delineate the former "high water mark" of mare basalt flows (NASA photograph AS 15-90-12208). (b) Telephoto picture of the upper slopes of Mount Hadley showing regular lineaments dipping steeply to the left and a second set of approximately horizontal sediment-veneered benches (NASA photograph AS 15-84-11321). (c) Vertical view of Mount Hadley showing northeasterly trending lineaments which can be correlated with steeply dipping lineaments visible in (a) and (b). The dark band is visible at the base of Mount Hadley's western slopes (panorama photograph 9798).

(b)

(a)

FIGURE XII-16. (a) View looking southeast from Apollo XV LM toward Silver Spur. Prominent lineations appear to be caused by ledges which are nearly horizontal (NASA photograph AS 15-84-11250). (b) Vertical view of Silver Spur (shown by arrow). "Stratification" visible in (a) can be correlated with lineaments striking slightly east of north (panorama photograph 9430).

(b)

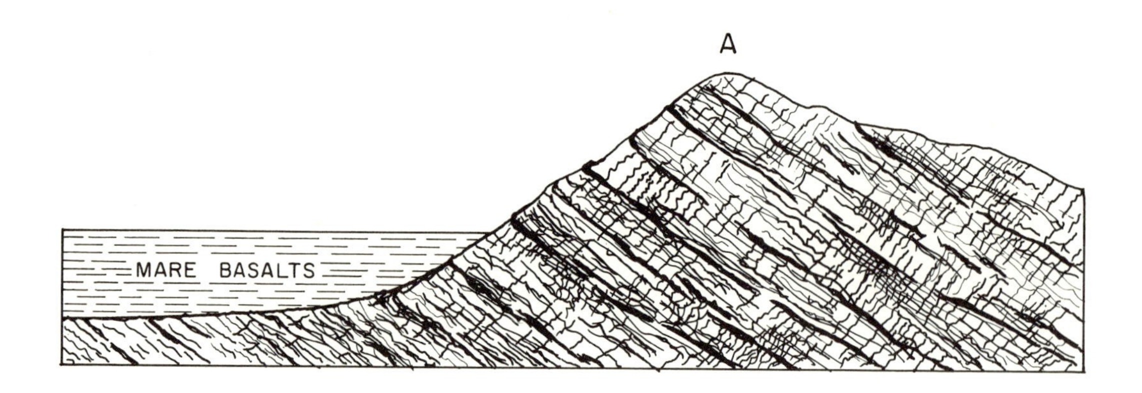

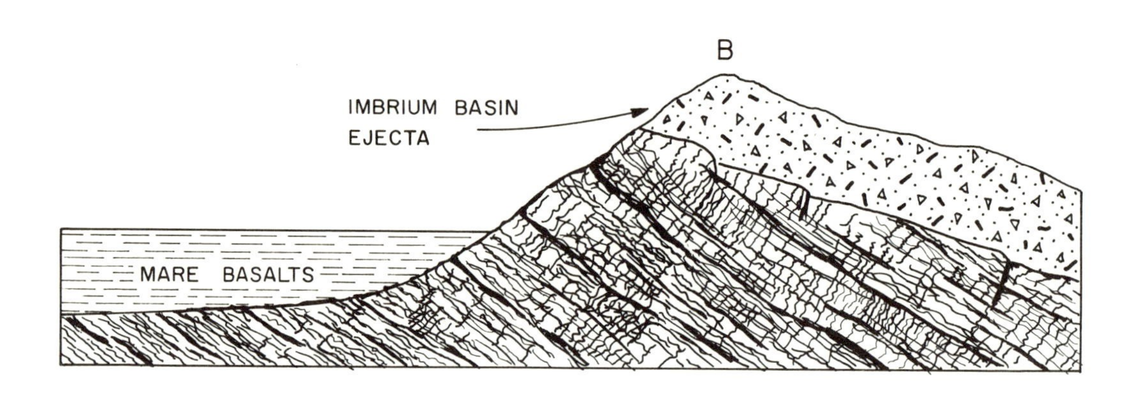

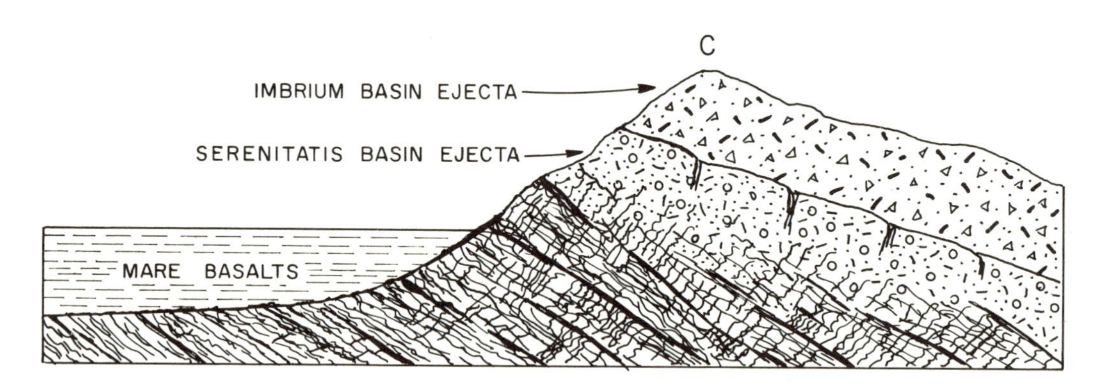

FIGURE XII-17. Three schematic cross sections showing possible geologic relationships along the Apennine front. (a) Pre-Imbrian rocks are exposed, uplifted by the Imbrium basin event. (b) Imbrium basin ejecta deposits (Fra Mauro Formation) are exposed above pre-Imbrian rocks. (c) Serenitatis basin ejecta deposits are exposed as upper unit of pre-Imbrian rocks.

is close to the surface. For example, they continue across Hadley Delta, a feature interpreted as a talus apron. Orbital photographs show that the steeply dipping lineations on Hadley Mountain are part of a northeasterly trending set of lineations. The markings that delineate apparently horizontal layers on Silver Spur are clearly revealed as a set of lineations striking slightly east of north. Foreshortening in the photograph taken on the lunar surface has produced a misleading impression of vertical superposition. Certainly one cannot discount the possibility that at least some of the lineaments visible in the Apennine front record base-surge stratification in basin ejecta deposits. But orbital pictures suggest that we are once more dealing with the pervasive and enigmatic lunar grid.

There was one particularly dramatic moment in the mission when Scott, picking up a light-colored rock, exclaimed: "I think we found what we came for." His remark was prompted by the fact that scientists had briefed him on the importance of finding anorthositic rocks which might represent the pre-mare or highland crust. When Scott saw the brightly reflecting plagioclase crystals, he correctly surmised that this was the anorthosite for which he had been encouraged to search. This rock, subsequently labelled the "genesis rock" in the popular press, contains over 98 percent plagioclase, and yields a K–Ar age of 4.1 billion years (Husain and others, 1972). The rock apparently has had a complex history, for its original cumulate texture is obscured by recrystallization textures and shock structures. For this reason, its crystallization age may be significantly greater than 4.1 billion years.

The most provocative feature in the Apollo XV landing area is Hadley Rille. In an article published just before the mission, Greeley (1971) presented a compelling comparison of the rille, terrestrial lava channels, and partly collapsed lava tubes (Fig. XII-18). None of the observations made by the astronauts were definitive, but it remains likely

FIGURE XII-18. (a) Hadley Rille, issuing from an irregular crater in the lower part of the picture and trending downslope toward the north. The rille is emplaced in mare materials. The rugged terrain in the right part of the picture is the Apennine Mountains. Note that the V-shape of the rille is interrupted at points "1" and "2." Greeley (1971) suggests that these may be roofed segments of a lava channel (Lunar Orbiter Photograph IV 102H). (b) Terrestrial lava channel, partly roofed to form a tube at "A," originating from the southwest Rift Zone ("B") of Mauna Loa, Hawaii: (from Greeley, 1971).

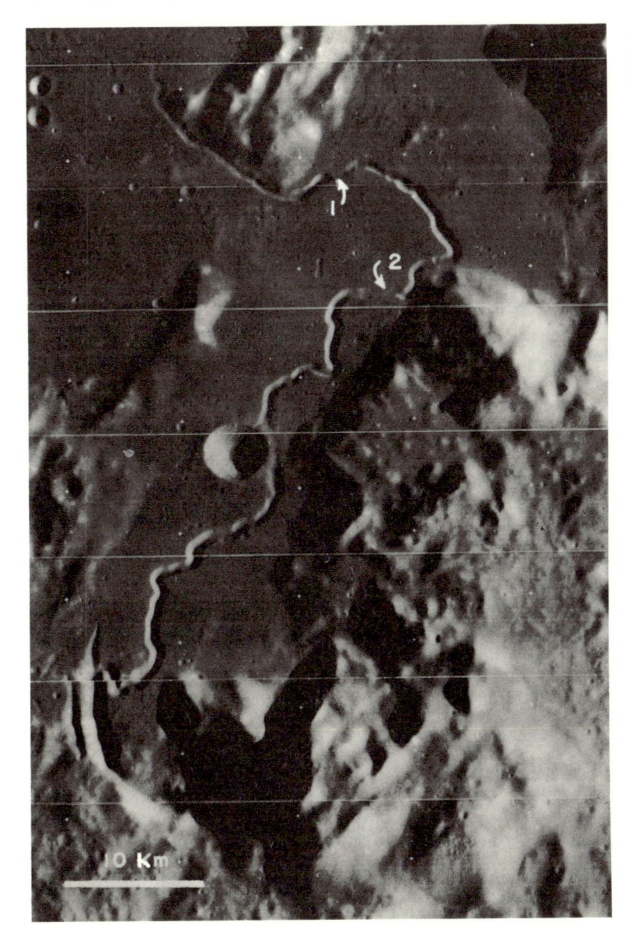

FIGURE XII-19. Telephoto picture taken from the eastern edge of Hadley Rille showing outcropping strata on the western side. A unit with irregular fractures dipping steeply to the right (1) overlies a horizontally layered unit (2) the total thickness of the two units is about 35 meters (NASA photograph AS 15-89-12157).

that the rille is an ancient lava channel. Telephoto pictures of the western wall reveal unmistakable outcrops of layered rocks, almost certainly a succession of mare basalts (Fig. XII-19). No data were collected to support the view that the rille had been cut by flowing water.

As on previous missions, the Apollo XV astronauts deployed several scientific instruments on the lunar surface. Of particular importance was a probe designed to measure heat flow from the interior of the Moon to the surface. Preliminary results from this experiment indicate an anomalously large heat flow which, if extrapolated to great depths, would imply complete melting. This is clearly not the actual situation. Instead, it is probable that the outer layers of the Moon are relatively enriched in radioactive elements, and that the heat from their decay accounts for the observed heat flow.

As was immediately apparent to the millions who followed the course of the Apollo XV mission on television, it was a tremendously successful scientific venture. The single most important element in its success was the performance of the astronaut crew. They made sophisticated scientific observations and collected samples with unprecedented intelligence and care. So many puzzling features are displayed in the numerous rocks collected, that it will take many years to unravel completely their geological significance and history.

Luna 16

The Russians have been successful in using an unmanned spacecraft to obtain a sample of lunar material. Luna 16 was launched in September 1970, landed in Mare Fecunditatis (0° 41′ S, 56° 18′ E), and returned to Earth with a 100-gram sample of regolith material. The sample was collected with the help of a drill which penetrated to a depth of 35 cm. The character of the sample, as reported by Vinogradov (1971), closely resembles the regolith material collected on the Apollo missions. Particles include gabbro, basalt, anorthosite, breccia, cinders, and glass. Main minerals are plagioclase, pyroxene, ilmenite, and olivine. The major-element chemical composition is shown later in this chapter, in Table XII-2. Rb–Sr analyses for the fine fractions yield isochrons of 4.45 and 4.65×10^9 years.

Future Apollo Missions

The first few Apollo missions focus attention on the formation of the maria and the attendent problem of basin formation. But most of the lunar surface comprises highlands, not maria. Therefore, it is obvious that there will be increasing scientific pressure to select one or several sites in the highlands as the Apollo program proceeds. Making this decision in principle is a good deal easier than actually selecting a site. What are "typical" highland materials and where does one go to sample them? The rocks which one would like to collect are those which formed the primeval highland crust, presumably in a period of extensive differentiation immediately following the Moon's formation 4.6 billion years ago. However, these rocks have been extensively mantled with basin ejecta and with younger volcanic deposits. A mission to an ejecta-covered region might do little more than repeat the observations of the Apollo XIV mission. Sampling the younger volcanic terrain will at least permit one to compare differentiated igneous sequences of the highlands with those of the maria. Following this latter point of view, the site for Apollo XVI is a region just north of the crater Descartes where a large dome of presumed volcanic materials is superposed on pre-Imbrian rocks (Figs. XII-20, 21). The landing site is located on flat, Cayley-like plains material. A few kilometers to the east is the edge of the volcanic dome. Two small rayed craters just north and south of the site will expose rocks ejected from considerable depth. Farther to the north, but still accessible with the LRV, linear ridges showing Imbrian sculpture may expose the primitive highland crust.

A large number of sites have been considered for Apollo XVII, the final mission planned by NASA. These include Tycho, Copernicus, Gassendi, Alphonsus, Marius Hills, Davy crater chain, and the Littrow rilles. From the scientific point of view two paramount factors dictate the choice of these sites. First, there is interest in sampling deep crustal rocks, especially in highland localities far removed from previous Apollo sites. An excursion across the rim of Tycho, for example, would provide opportunity to collect rocks excavated from great depths as well as opportunity to study impact-related volcanism. The central peaks of Copernicus and Gassendi may be underlain by rocks that have been structurally uplifted from a point many kilometers below the original lunar surface. The Davy crater chain may have been the site for maar-type volcanic activity that has brought to the surface rocks formed at depth (Fig. XII-26).

Exploration of the Marius Hills, Alphonsus, or the region around the Littrow rilles gives promise of unraveling lunar history not by probing further into its early stages, but by investigating volcanic activity, which is probably different in character and more recent in time than mare activity documented on early Apollo missions. The Marius Hills and Alphonsus have already been discussed in Chapter VIII. The Littrow region, on the eastern edge of Mare Serenitatis, is covered by extremely low albedo materials, probably pyroclastic volcanics (Carr, 1972).

At the time of this writing, the leading candidate sites appear to be Alphonsus, Littrow, and Gassendi. Although the Marius Hills or Tycho are more attractive scientifically, they are ranked lower because of operational difficulties in effecting a safe landing.

One of the important but generally unpublicized elements of all Apollo missions is the science carried out from orbit. A primary objective is to gain good stereoscopic coverage of potential landing sites for subsequent missions. Parenthetically, it is interesting to note that availability of high-quality photographs is one of the most important requirements for final selection of an Apollo site. Because so little of the lunar surface beyond the maria has been photographed at high resolution, the number of certified sites is exceedingly small and can be enlarged only by receipt of additional Apollo photographs. The Apollo XV orbiting command module carried, for the first time, two automated film camera systems. An optical bar panorama camera took overlapping pictures

FIGURE XII-20. View north of the crater Descartes. The arrow shows the location of the Apollo XVI landing. Areas visited by the astronauts include: the Cayley plains, on which the spacecraft landed; North Ray crater, about 5 km north of the LM; and Stone Mountain, an upland area south of the LM and east of South Ray crater. Premission analysis suggested that the Cayley plains and the highlands to the southeast were underlain by volcanic deposits. However, almost all the collected rocks are light-colored breccias superficially resembling Apollo XIV breccias. An interesting problem, awaiting more detailed analysis, is whether some of the rocks are volcanic tuff breccias instead of impact breccias. A more detailed view of the landing area is shown in Figure XII-21 (Lunar Orbiter Photograph IV 89H).

FIGURE XII-21. A high-resolution view of the Apollo XVI landing site, showing routes followed by the astronauts on their three EVA's (Apollo XVI metric photograph 0440).

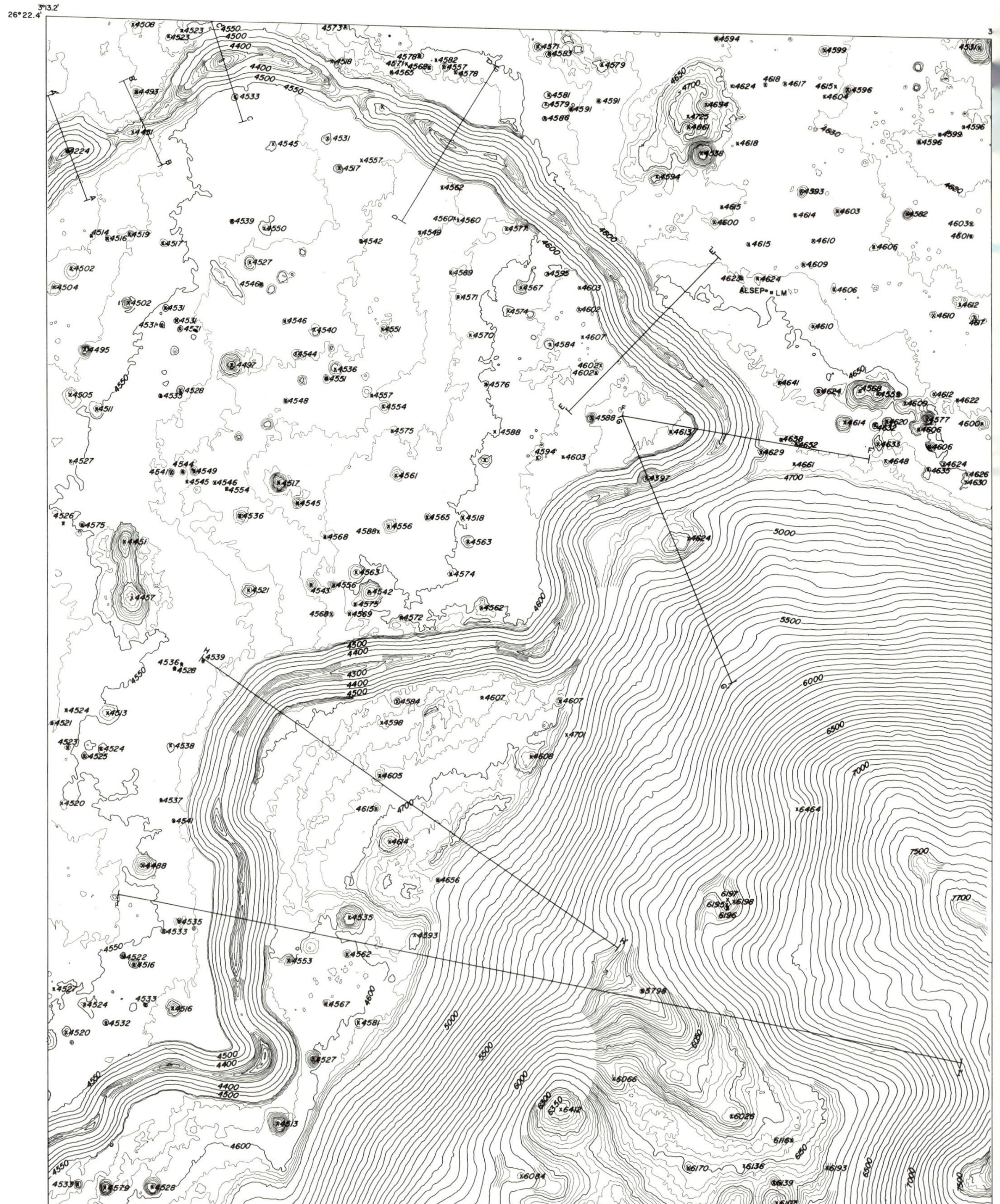

3°13.2'

26°22.4'

25°35.2 N
3° 04.5'E

1000 0 1000 2000 3000 METERS

Contour Interval: 10 Meters

N

with ground resolution of 1 to 2 meters. A metric camera, used in conjunction with a stellar camera and laser altimeter, provided information not only for topographic maps (Fig. XII-22), but also for establishment of a lunar geodetic network.

Other experiments flown on the Apollo XV command module include an X-ray fluorescence spectrometer and a gamma-ray spectrometer. The former provided a regional geochemical survey. Preliminary analysis indicates Al/Si ratios of approximately 0.67 for maria, compared with values of 1.13 for highlands (Adler and others, 1972). The Al enrichment in the highlands suggests an anorthositic composition.

Some of the more spectacular orbital pictures taken on Apollo Missions X-XV with a handheld Hasselblad camera are shown in Figures XII-23 to 32. Oblique photography, at times of high sun elevation, provides some particularly dramatic views.

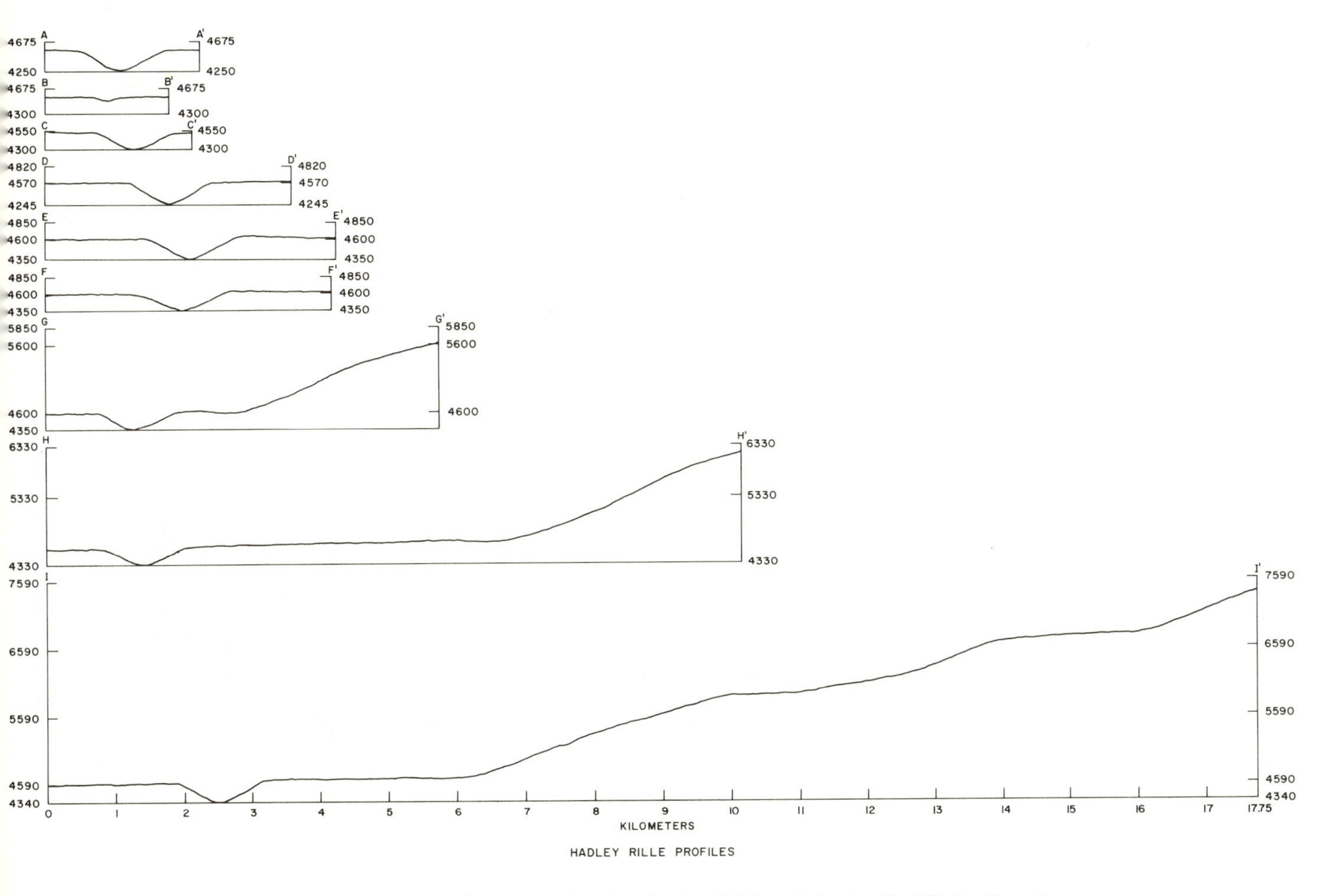

FIGURE XII-22. Contour map (opposite page) of region in the vicinity of the Apollo XV landing site. Topographic cross sections are across Hadley Rille. This map was constructed from overlapping panorama photographs. Similar maps can be constructed from metric photography (from Wu and others, 1972).

FIGURE XII-23. Region west of La Hire, in approximate center of Mare Imbrium. North is toward the top. Sinuous scarps may delineate volcanic flow fronts (NASA photograph AS 15-M3-1558).

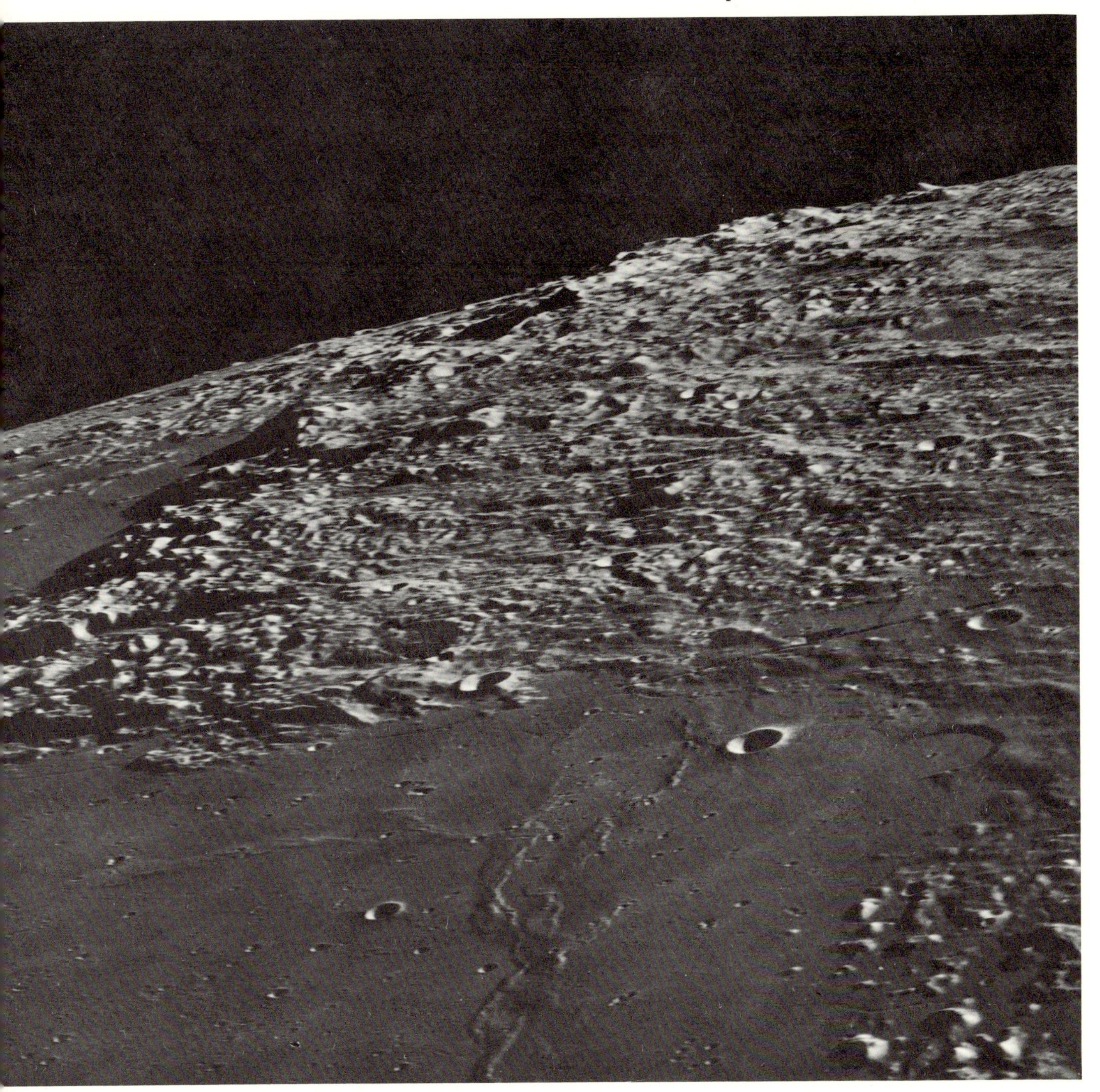

FIGURE XII-24. Looking north across Sinus Aestuum. The Apennine Mountains are silhouetted on the horizon. Note the gridded character of the mountains, caused by intersecting sets of lineaments. The prominent straight rille separating mare materials of Sinus Aestuum from the rugged terrain beyond is Rima Bode. The crater Eratosthenes is just to the left of this view (NASA photograph AS 12-50-7434).

FIGURE XII-25. View looking south across the crater Davy. Crater in the middleground, south of Davy, is Lassell. Ropy material of probable volcanic origin fills Davy. Arrows point to two scarps with Sun-facing slopes appearing bright (NASA photograph AS 12-51-7478).

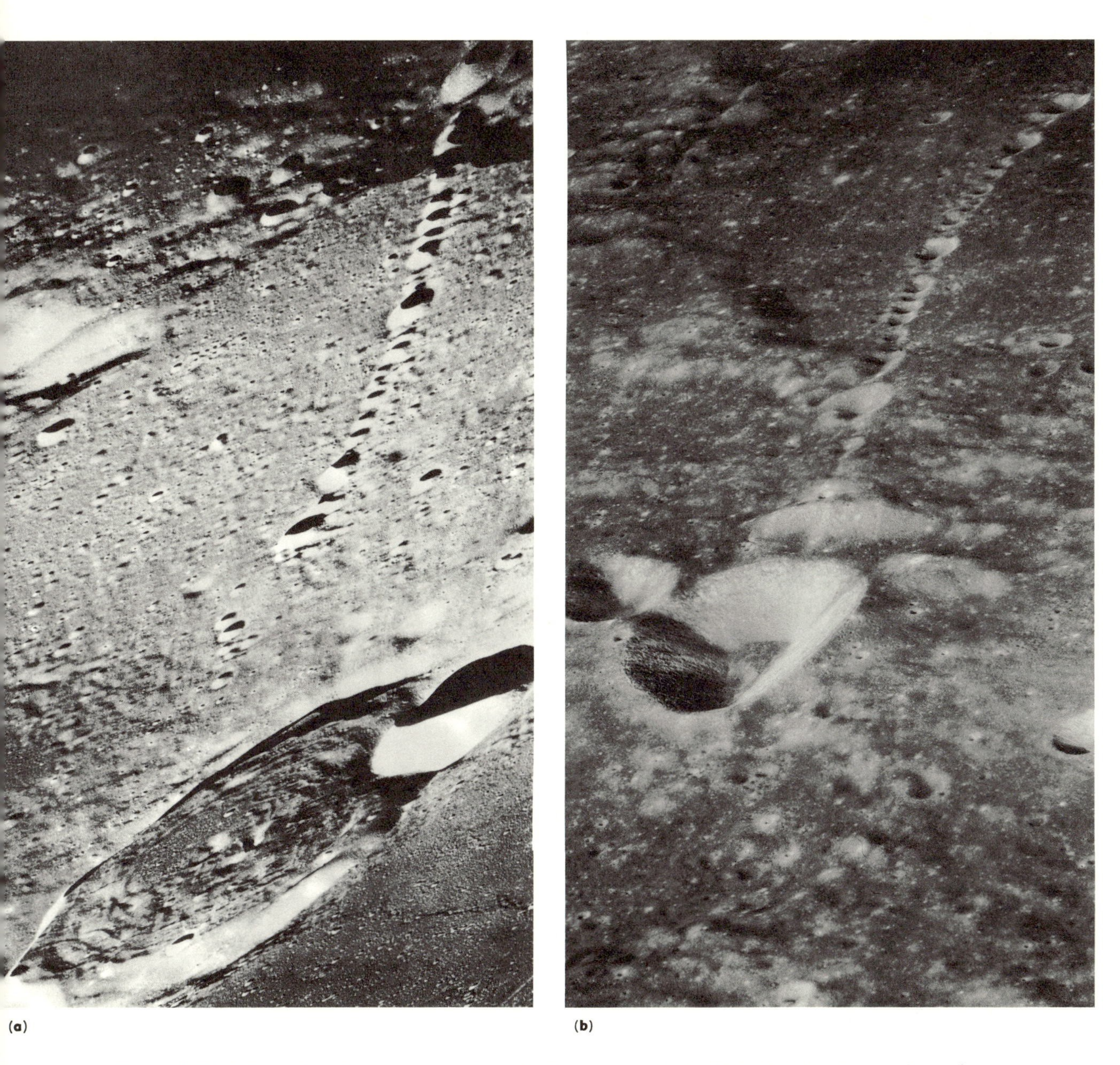

FIGURE XII-26. Two views looking west along the Davy crater chain. Note that the large polygonal crater (Davy G) at the east end of the chain and visible in part (b) is elongated in the same direction as the string of small craters. It seems unlikely that the "string-of-beads" pattern of the crater chain could have resulted from impact. Instead, the craters are interpreted as volcanic maars, aligned along a deep crustal fracture (NASA photograph AS 12-51-7485).

FIGURE XII-27. Extreme accentuation of topographic detail close to the terminator. The crater Gambart is at the top (north) of the picture. The Apollo XIV landing site is located in the shadowed highlands along the left side of the picture. All the uplands in this area are mapped as Fra Mauro Formation (Eggleton, 1965) (NASA photograph AS 12-50-7438).

FIGURE XII-28. View looking north across Mare Crisium. The high Sun elevation imparts a luminous softness to the highlands that bound the mare plains (NASA photograph AS 10-31-4418).

FIGURE XII-29. The floor of the crater Humboldt. The high Sun elevation accentuates brightness contrasts. The system of cracks in floor materials is probably related to the emplacement and later disruption of lava pools. Dark plains of Mare Australe are visible just below the horizon at the left (NASA photograph 12-50-7415).

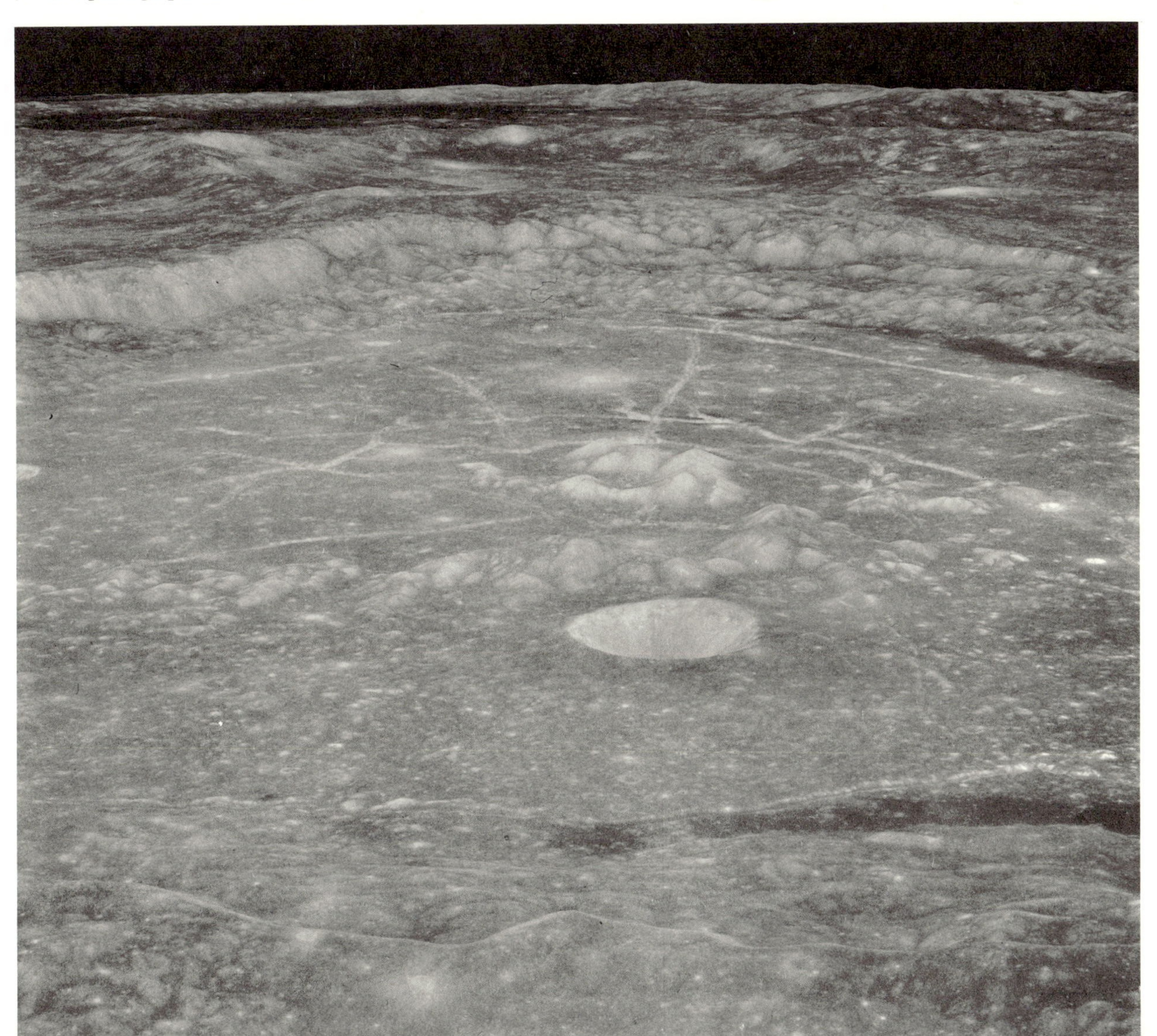

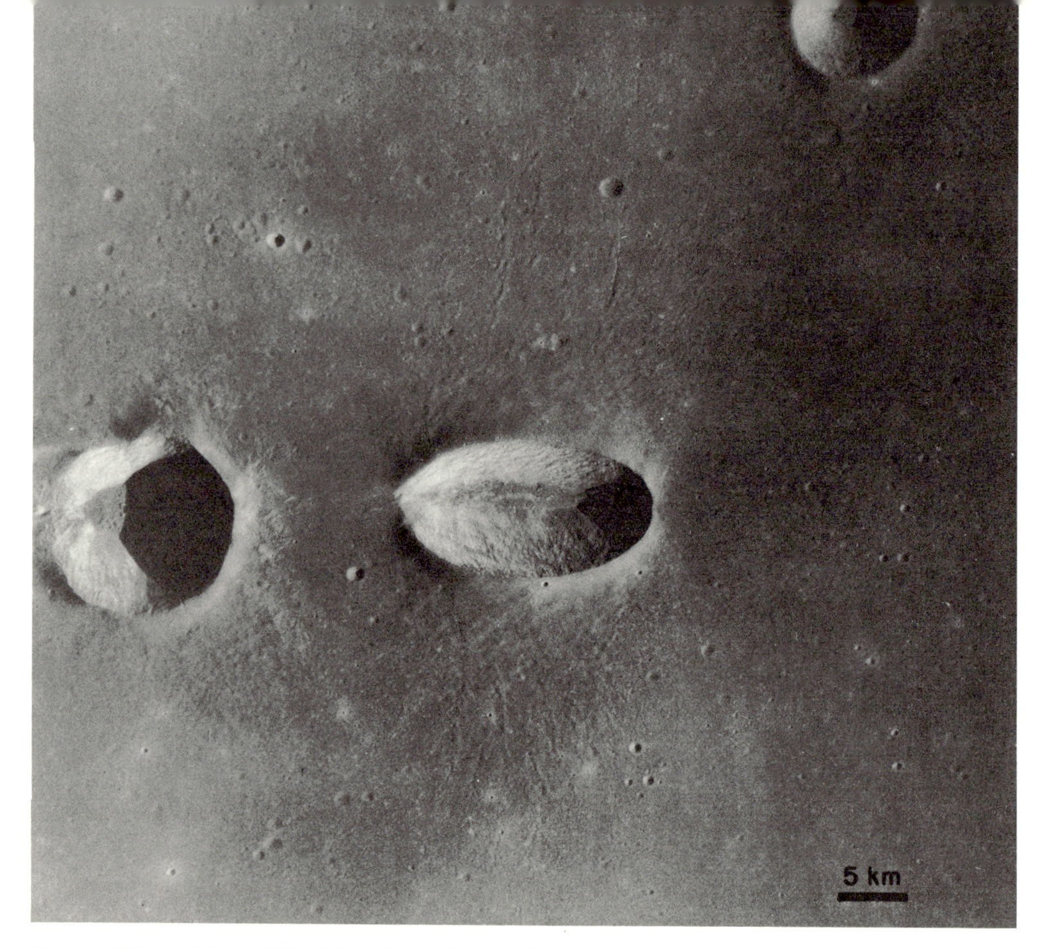

FIGURE XII-30. View within Mare Fecunditatis. The elliptical crater is Messier. Its grooved interior, bilateral lobes of ejecta, and association with a double crater, aligned along the trend of Messier's long axis, all suggest a highly oblique impact (NASA photograph AS 10-33-4906).

FIGURE XII-31. Crater in Mare Fecunditatis. High Sun elevation reveals brightness differences in surficial materials, and shows downslope creep of variable-albedo material within the crater. Flat floor may be caused by a pooling of unconsolidated sediment (NASA photograph AS 10-29-4254).

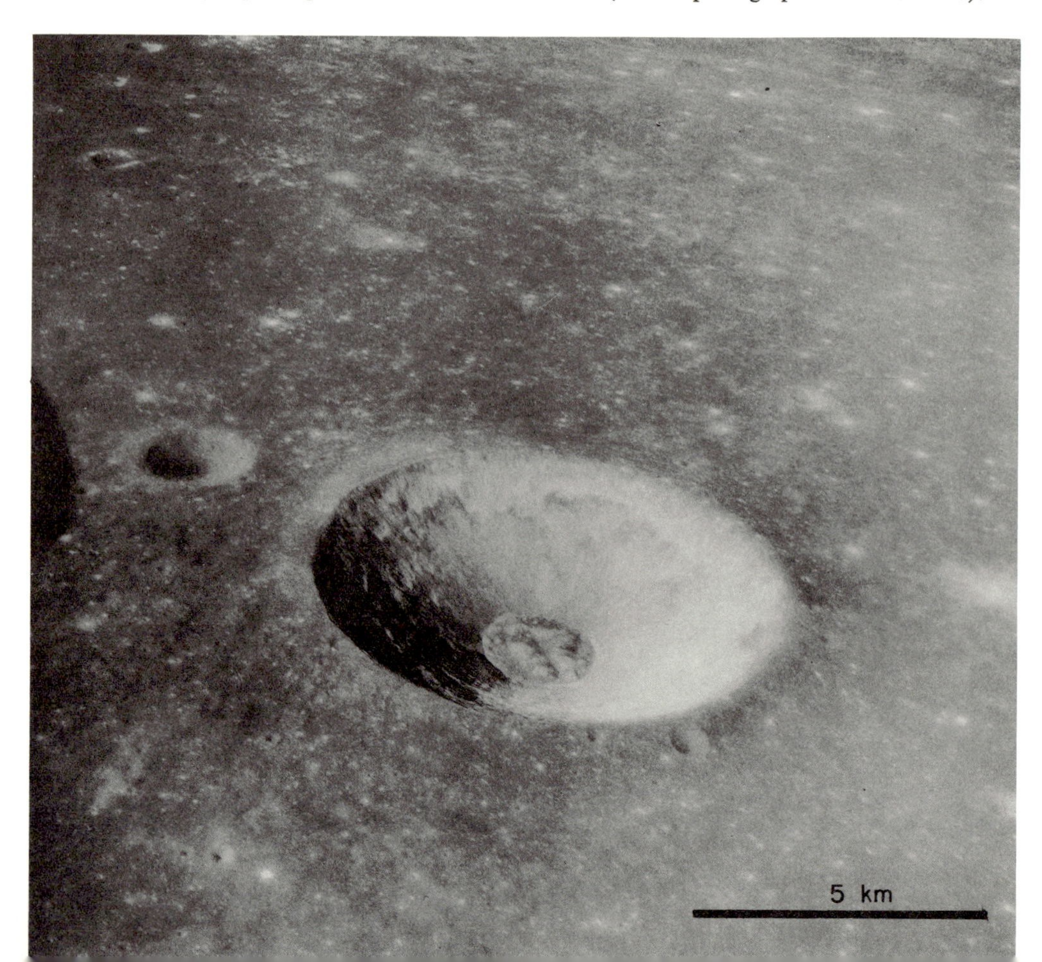

FIGURE XII-32. Cratered terrain on the farside of the Moon showing the strong influence of the lunar grid. One set of lineaments trends from lower right to upper left. A second set trends from left to right and is particularly prominent in the lower left of the picture (NASA photograph AS 11-41-6152).

PETROLOGY

Mineralogical and chemical analyses of Apollo rocks indicate two broad categories: crystalline rocks and breccia. The crystalline rocks resemble terrestrial basaltic lavas, but they have distinctive features which set them apart from "common" basalt. The breccias are formed by lithification of mixed fragments from crystalline rock, pre-existing breccia, and glass. A third type of material includes the "fines," all unconsolidated particles less than one cm in diameter. These samples from the lunar regolith contain a complex mixture of diverse mineral, rock, and glass fragments.

Igneous Rocks

Of the 36 rocks larger than 4 cm returned on the Apollo XI mission, 16 were characterized as igneous, the remaining 20 as breccia. Three groups of igneous rocks have been identified by Schmitt and others (1970): basalts, olivine basalts, and cristobalite basalts. Basalts and olivine basalts are fine-grained and vesicular. Cristobalite basalts are coarser-grained and contain irregularly shaped crystalline cavities, termed vugs. Because of their grain size and texture the basalts are thought to have crystallized near the edge of a lava flow. The olivine basalts crystallized slowly enough to permit forma-

FIGURE XII-33. (a) Apollo XI fine-grained basalt. Spherical vesicles are lined with brightly reflecting groundmass minerals. (b) A photomicrograph of the same rock taken with transmitted light and crossed nicols (NASA photograph S 69-45209).

tion of euhedral crystals of plagioclase embedded in a more finely-grained matrix of pyroxene (ophitic texture). The coarse-grained cristobalite basalts crystallized most slowly of all, either in the interior of a thick flow or in a shallow intrusive body. Inasmuch as fundamental chemical differences accompany the textural differences, it would be an oversimplification to assign all rocks to a single flow derived from one magma source.

Minerals common to all Apollo XI crystalline rocks are clinopyroxene, plagioclase, and ilmenite. These three constitute over 90 percent of the total volume, but many other minerals have been identified as accessories (Table XII-1).

Of the 45 rocks collected by the Apollo XII astronauts, all but two are of probable igneous origin. In further contrast with Apollo XI samples, the igneous rocks show a broader range in texture and in mineralogy (Figs. XII-34, 35). Grain sizes range from 0.05 to 35 mm. Many rocks are equigranular, but some display ophitic texture. Others have large phenocrysts and feathery sheaves of pyroxene-plagioclase intergrowths. Mineralogically, the rocks range from pyroxene-rich peridotite to one unusual rock (No. 12013) which consists of a "granitic" vein of alkali feldspar, quartz, and plagioclase injected into a basaltic microbreccia.

TABLE XII-1. Minerals in Apollo XI samples*

	Name	*Formula*
Major (> 10%)	Pyroxene	$(Ca,Fe,Mg)_2Si_2O_6$
	Plagioclase	$(Ca,Na)(Al,Si)_4O_8$
	Ilmenite	$FeTiO_3$
Minor (1–10%)	Olivine	$(Mg,Fe)_2SiO_4$
	Cristobalite	SiO_2
	Tridymite	SiO_2
	Pyroxferroite	$CaFe_6(SiO_3)_7$
Accessory (< 1%)	Copper	Cu
	Iron	Fe
	Nickel-iron	(Fe,Ni)
	Cohenite	Fe_3C
	Schreibersite	$(Fe,Ni)_3P$
	Troilite	FeS
	Potash feldspar	$KAlSi_3O_8$
	Quartz	SiO_2
	Armalcolite	$(Fe,Mg)Ti_2O_5$
	Ulvöspinel	Fe_2TiO_4
	Chromite	$FeCr_2O_4$
	Spinel	$MgAl_2O_4$
	Perovskite	$CaTiO_3$
	Rutile	TiO_2
	Baddeleyite	ZrO_2
	Zircon	$ZrSiO_4$
	Apatite	$Ca_5(PO_4)_3(F,Cl)$
	Whitlockite	$Ca_3(PO_4)_2$

* Taken from Mason and Melson (1970).

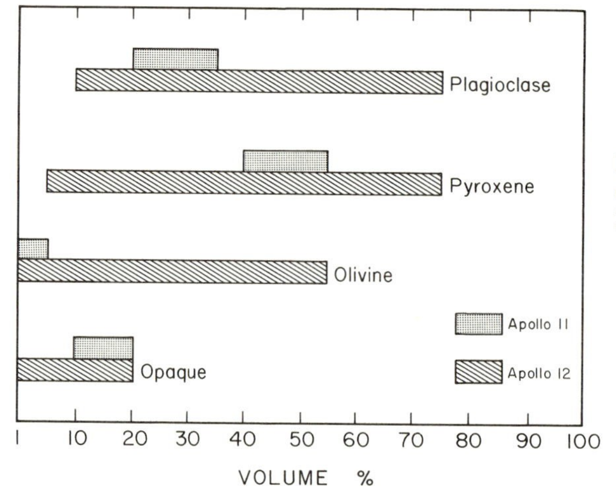

FIGURE XII-34. Comparison of range of modal mineralogy of Apollo XI and XII rocks (from LSPET, 1970).

Warner (1971) has developed a common textural classification for Apollo XI and XII crystalline rocks. Three textural types have been identified: (1) intergranular or intersertal, (2) granular and ophitic, and (3) porphyritic. The first type generally corresponds to the basalts and olivine basalts identified by Schmitt and others (1970); the second type corresponds to their cristobalite basalt. The last textural type is peculiar

(*Opposite page*)

FIGURE XII-35. Photomicrographs of Apollo rocks, all taken with plane light.

(a) Apollo XI medium-grained basalt. Plagioclase 35 to 40 percent, lath shaped, clear to white; pyroxene 4 to 50 percent, elongate crystals, dark brown; ilmenite, 10 to 15 percent, opaque (NASA photograph S-69-59291).

(b) Apollo XI coarse-grained cristobalite basalt. Plagioclase, 50 percent; pyroxene, 30 percent; ilmenite, 20 percent (NASA photograph S-69-47907).

(c) Apollo XII olivine basalt. Clear grains are plagioclase; transluscent grains with cleavage, pyroxene; transluscent grains without cleavage, olivine (NASA photograph S-69-63409).

(d) Apollo XII porphyritic gabbro. Phenocrysts are pigeonite (variety of clinopyroxene). Radiating laths are an intergrowth of pyroxene and plagioclase (NASA photograph S-70-20749).

(e) Apollo XI microbreccia (NASA photograph S-69-59843).

(f) Apollo XII microbreccia. Note spherules in matrix. A contact between two large fragments with matrices of different texture is indicated by arrows. This is evidence for multiple-impact events recorded within a single rock (NASA photograph S-69-63407).

(g) Apollo XIV breccia. A dark, annealed breccia clast (a) occupies the central part of the photograph. Within this clast are several light, annealed breccia clasts (b). White spots in (a) and (b) are plagioclase porphyroblasts.

(h) Apollo XV breccia, the "black-and-white rock" (15455) collected near Spur Crater. Lower half of the photograph is a cataclastically deformed norite. Upper half is dark, vesicular breccia. White clast in upper right (a) is recrystallized plagioclase.

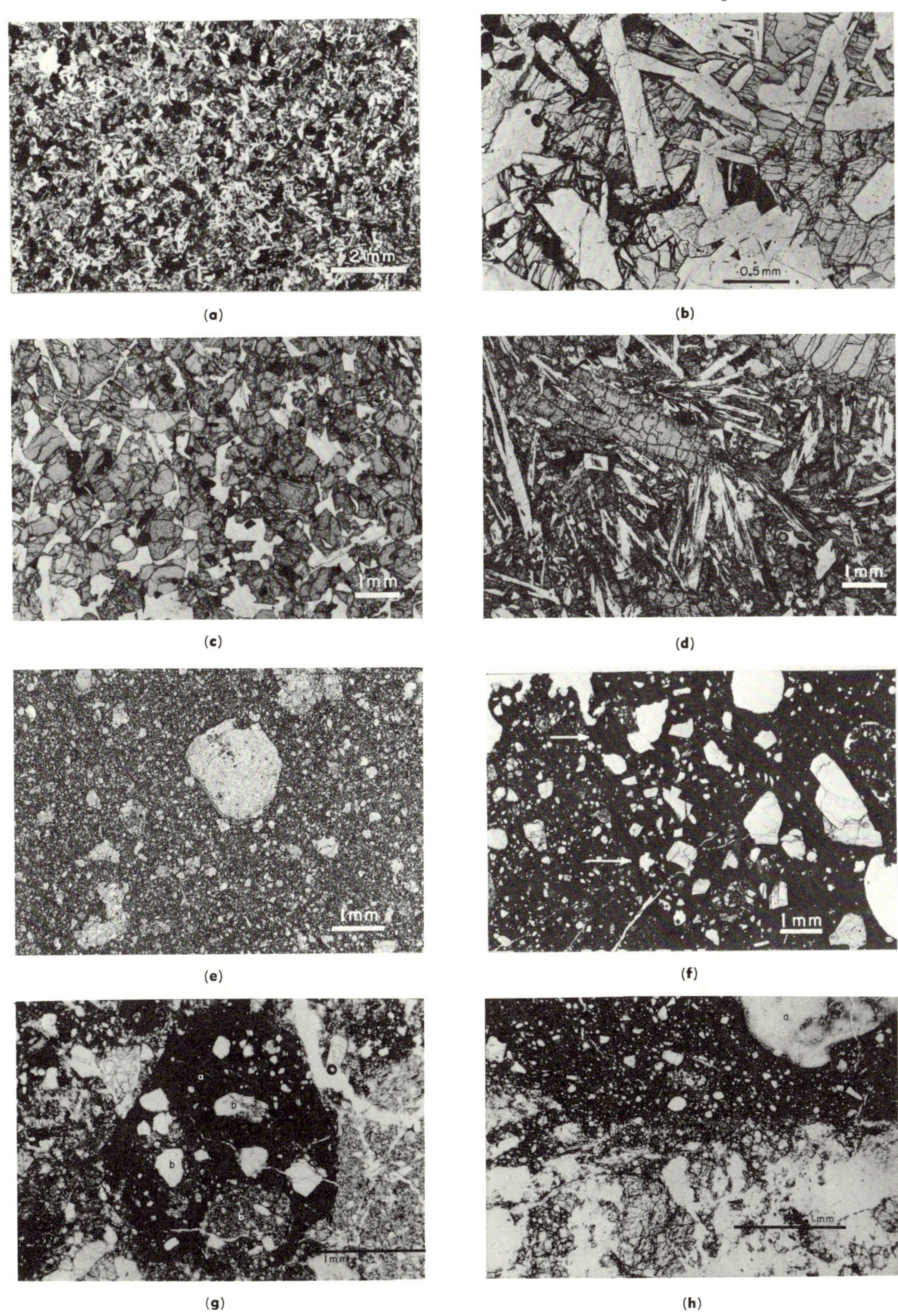
2 mm
0.5 mm
(a)
(b)
1 mm
1 mm
(c)
(d)
1 mm
1 mm
(e)
(f)
a
b
b
b
1 mm
a
1 mm
(g)
(h)

		TEXTURES		
		INTERGRANULAR	GRANULAR AND OPHITIC	PORPHYRITIC
CHEMICAL TYPES	HIGH ALKALI HIGH TITANIUM	APOLLO-11 TYPE A		
	LOW ALKALI HIGH TITANIUM		APOLLO-11 TYPE B	
	LOW ALKALI LOW TITANIUM		APOLLO-12 TYPE 2	APOLLO-12 TYPE 1

FIGURE XII-36. Relation between textures and chemical types of Apollo XI and XII rocks (taken from Warner, 1971).

to Apollo XII rocks. The correlation between textural and chemical groups is shown in Figure XII-36.

A notable feature of many Apollo crystalline rocks is the occurrence of small amounts of residual glass. Two phases have been identified: (1) a high-silica "granitic" component, and (2) a low-silica Fe-enriched fraction. These glasses apparently were formed from immiscible liquids late in the crystallization sequence of the parent magma (Roedder and Weiblen, 1970). A further possibility is that, elsewhere on the Moon, these immiscible liquids have been concentrated by filter pressing during the terminal stages of crystallization in a large magma chamber to yield a segregated granite body of substantial size. The previously mentioned granitic vein of rock 12013 may have been derived from such a magma body.

Laboratory heating of Apollo crystalline rocks indicates that, at 1 atm. total pressure, complete melting occurs in the vicinity of 1175°C. Complete solidification occurs in the vicinity of 1075°C (Anderson and others, 1970; O'Hara and others, 1970, Ringwood and Essene, 1970). The approximate order of crystallization, from first to last is: ilmenite, olivine, pyroxene, and plagioclase. The presence of tridymite and cristobalite provide additional information about the temperature-pressure environment at the time of crystallization. Both minerals are high-temperature (867° to 1713°C) low-pressure (<5 kb) polymorphs of SiO_2 (Fig. IV-27). Apparently the basalts crystallized at depths less than 100 km, the corresponding depth for a lithostatic pressure of 5 kb. Theoretically, cristobalite is unstable at temperatures below 1470°C. It may occur as a metastable phase in a rapidly cooling lava or near-surface intrusion.

A complete absence of water-containing inclusions within Apollo crystalline rocks, and a virtually complete lack of hydrated minerals indicate that the rocks crystallized from a magma essentially without water, and suggest that no surficial water (e.g., oceans, rivers, groundwater, permafrost) has ever existed on the Moon.

Based on the minerals present, and their chemical compositions, numerous inferences can be made about the composition of the parent magma. The presence of ilmenite indicates a high iron-titanium content. Inasmuch as ilmenite enrichment is much more prominent in Apollo XI rocks than in Apollo XII rocks, it may be the result of local crystal differentiation similar to the process that occurs in stratiform sheets on Earth. The occurrence of native iron and the absence of ferric iron indicate partial pressure of oxygen three or four orders of magnitude lower than in terrestrial basaltic magmas. The highly calcic character of the plagioclase indicates probable depletion of sodium in source rocks.

Even though all the Apollo XIV rocks are thought to be breccias, the crystalline rock fragments contained in the breccias can be analyzed and classified. Two groups are present: one with basaltic texture and a second with fine-grained granulitic texture. The basaltic rocks differ from Apollo XI and XII specimens in having more plagioclase, lighter colored pyroxene, and less ilmenite. One subgroup contains 40 to 50 percent plagioclase; another has 60 to 70 percent plagioclase (LSPET, 1971).

Preliminary examination of Apollo XV rocks confirms the existence of two broad categories of basalt—mare and non-mare. Mare basalts have textures ranging from porphyritic to highly vesicular and modal mineralogies closely resembling those of Apollo XI and XII basalts. Non-mare basalts, occurring as walnut-sized fragments in the regolith and clasts in breccia, resemble the "Fra Mauro" basalts of Apollo XIV. Plagioclase content is relatively high, pyroxene is light-colored and shows no zoning, and there are no vugs or vesicles. Mineralogical variations are accompanied by chemical differences. Mare basalts are much richer in FeO and lower in Al_2O_3 than non-mare basalts (LSPET, 1972).

The emplacement site for non-mare basalts has not been identified, but their occurrence as clasts in presumed Imbrium basin ejecta suggests that they composed part of the lunar crust which was excavated to form the large basins. Accordingly, they are widely separated from mare basalts not only in mineralogy and chemistry, but also in age.

Breccias

Breccias are composed of angular fragments of rocks, minerals, and glass in a groundmass of comminuted material and glass. The presence of both igneous and fragmental rock fragments in most breccias indicates a history of repeated fragmentation and consolidation. The rocks probably formed by compaction and welding within hot ejecta deposits. Microbreccias contain fragments generally less than 1 mm, and probably were formed by shock lithification of unconsolidated regolith.

A few of the Apollo XI and XII breccia mineral fragments show evidence of shock deformation, but this phenomenon is much more widespread in Apollo XIV and XV breccias. Features include shock-induced twinning, fracturing, vitrification, melting, and recrystallization (Chao and others, 1970; Sclar, 1970; Short, 1970). Most crystalline rocks do not show shock features, although several Apollo XII rocks display fracturing, vitrification, and development of lamellar structures within plagioclase (LSPET, 1970). The general rarity of shock features is not as anomalous as it appears. Similar situations exist on Earth where only a small percentage of the fragments ejected from impact craters have textures or mineral compositions which point uniquely to shock modification.

As previously mentioned, all but possibly two of the rocks observed and collected by Apollo XIV astronauts are breccias. Boulders observed in the field are up to 6 meters in diameter and contain clasts many centimeters in diameter. Preliminary hand-specimen and thin-section analysis by the LSPET (1971) indicates several types of breccia, separated according to coherence, size of clasts, color of lithic clasts, and degree of matrix crystallinity. One type, characterized by friable matrix and abundance of small clasts, is interpreted as indurated soil. Two other types, distinguished by their ratio of light-colored to dark-colored clasts, are thought to be derived from the Fra Mauro Formation. Clast types include glass, mineral fragments, recrystallized mineral grains, basalt, gabbro, anorthosite, metaclastic rocks, metagabbros, and metabasalts. The last three clast types are particularly significant because they record metamorphism, probably shock-related, which predated the metamorphic event responsible for forming the breccia in which the fragments are encased. Some breccias contain compound clasts which record as many as four brecciation events (Wilshire and Jackson, unpublished manuscript).

Many of the Apollo XIV breccias show post-consolidation metamorphic reactions between clasts and matrix. These effects are more marked in breccia collected close to Cone Crater and presumably ejected from relatively great depths. This suggests that the lower zone of the ejecta blanket (Fra Mauro Formation) was hot and well insulated when deposited. The resulting thermal metamorphism was somewhat analogous to the annealing which takes place in a terrestrial welded tuff. There can be little doubt that the breccias were well lithified before formation of Cone Crater, since glass-filled fractures, presumably related to that impact event, cut through both matrix and clasts of some rocks (Wilshire and Jackson, unpublished manuscript).

Apollo XV breccias are complicated in composition and diverse in origin, but several generalities appear valid (LSPET, 1972). Distinction can be made between mare plains breccia, rich in mare basalt fragments, and Apennine front breccias, which have a greater abundance of gabbroic rocks. Many breccias are cemented and extensively coated by glass. They differ from Apollo XIV breccias in showing less post-depositional recrystallization and in containing fewer clasts of pre-existing breccias. Some, though by no means all, breccias collected close to the front contain large fragments of a gabbroic to anorthositic suite of igneous rocks. No obvious correlation of breccia type with particular stratigraphic or structural units exposed along the Apennine front has yet been established.

Both crystalline rocks and breccias from all Apollo missions show distinctive surface features. Small semi-spherical pits, generally less than 1 mm in diameter and lined with glass, are formed by impact of micrometeorites (Figs. XII-37, 38). Glass-spatter crusts up to 1 cm in diameter occur on some soil surfaces (Fig. XII-39). They are interpreted as gobs of shock-melted material thrown out from nearby impact craters.

Fines

Lunar fines contain a mixture of mineral fragments, rock fragments, and glass (Fig. XII-40). About half the Apollo XI fines are glass which occurs in a variety of forms: as dark globular fragments, as colorless to brown angular fragments, and as spheroidal, dumbbell-shaped, and teardrop-shaped beads ranging in color from red to yellow to

FIGURE XII-37. Surface of Apollo XI breccia fragment. Numerous pits formed by micrometeoritic bombardment are observable. In the upper left the rock is coated by a blob of glassy material (NASA photograph S-69-47905).

green. The remaining half of Apollo XI fines consists chiefly of plagioclase, clinopyroxene, ilmenite, and olivine (LSPET, 1969).

Apollo XII fines are distinctly lighter in color than Apollo XI samples. The major constituents, in decreasing order of abundance, are pyroxene, plagioclase, glass, and

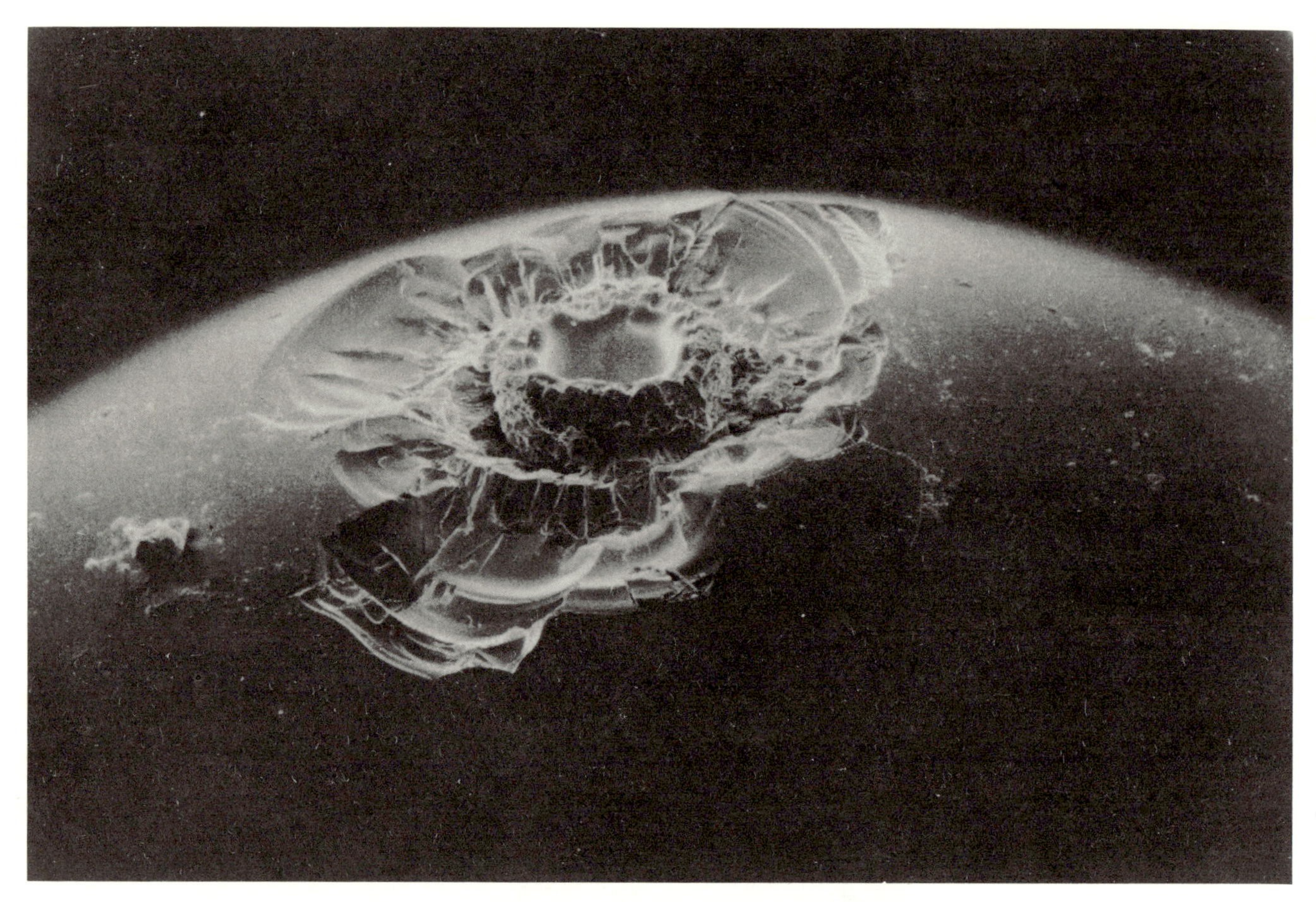

FIGURE XII-38. Scanning electron microscope photograph of a hypervelocity impact crater with raised rim and surrounding spalled zone. Diameter of inner rimmed crater is 0.06 mm. Crater appears on a dark brown glass sphere from Apollo XI fines (from Frondel and others, 1970).

olivine. Glass fragments, spheroidal to angular and colorless to brown, make up approximately 20 percent of the fines. The color difference between Apollo XI and XII samples is attributed to the marked decrease in dark and opaque glass in Apollo XII fines (LSPET, 1970). Apollo XIV soil samples resemble those of previous missions but show greater internal variation in grain size and model mineralogy (LSPET, 1971). Apollo XV soils also vary according to sampling locality and, in addition, contain a distinctive new component, green glass spheres which comprise as much as 11 percent of some samples (LSPET, 1972).

The Apollo glasses show a wide variety of chemical composition, indicating that they were produced by melting of various proportions of minerals. Almost all the glass fragments were probably formed by shock melting during impact events. A variety of processes have been proposed: droplet condensation, subdivision of a liquid jet, and vesiculation within glass (McKay and others, 1970). Although some of the glass fragments have shapes similar to microtektites, their chemical composition is generally dissimilar. Silica, for example, is much lower than in most tektites.

Two unusual fragment types within the Apollo fines have been singled out for special attention. The first includes anorthite-rich fragments, commonly referred to as anorthosite. Several investigators have speculated that the lunar highlands are underlain by similar rocks and that fragments from this terrain have been mixed into the mare regolith by impact-ejection (Wood and others, 1970*a*; Smith and others, 1970*a*; Short, 1970). This conclusion is supported by the Surveyor VII analyses obtained on the rim of the crater Tycho. Both calcium and aluminum abundances are high, in contrast with other Surveyor analyses at mare sites, but in close accord with analytical results for Apollo anorthite-rich fragments. King and others (1971) have shown that the percentages of plagioclase fines in three Apollo XII samples and one Apollo XI sample remain constant at about 5 percent, even though the percentages of other components vary widely. They interpret this as added support for the suggestion that the anorthositic fragments are ejected from the highlands and are deposited over large regions in approximately constant amounts.

The second unusual fragment type, first identified in Apollo XII soil and apparently even more abundant in Apollo XIV fines, is dark brown glass termed "KREEP" [high

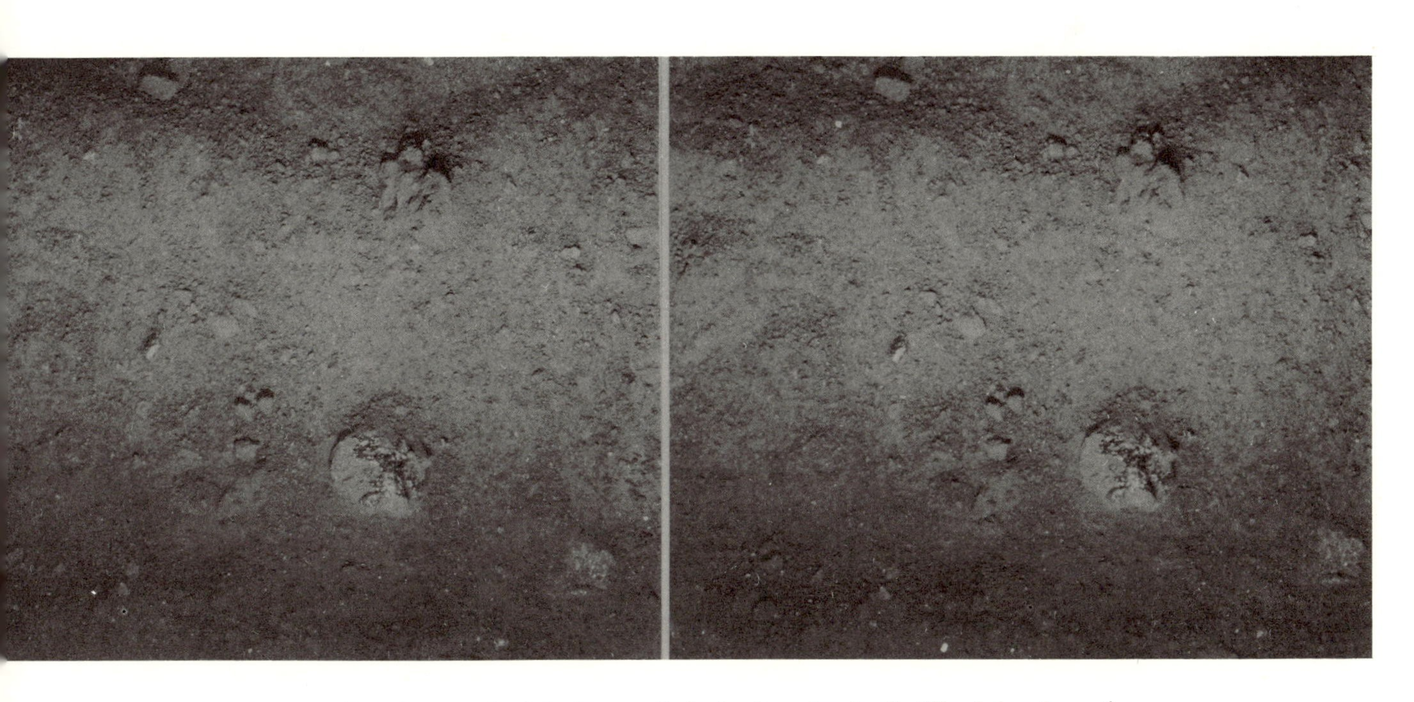

FIGURE XII-39. A detailed photograph of the lunar soil obtained, on the Apollo XI mission, by resting a camera mounted on a walking stick directly on the lunar surface. An area approximately 8 by 8 cm is photographed in stereo, with flash illumination, and at a fixed distance. Note the clump of lunar soil about one centimeter in diameter which has a crown of brightly-reflecting glassy material. This is probably splash material associated with a nearby impact.

(a)

(b)

(c)

(d)

in potassium (K), rare earth elements (REE) and phosphorus (P)] (Meyer and others, 1971). Certain noritic breccia fragments have similar composition, suggesting that the glass may have been derived from the norite by shock-melting. As will be discussed in the section on radiometric ages, the selective elemental and isotopic enrichment within KREEP particles suggests that they may be derived from the primeval lunar crust by shock-melting and ejection.

As more and more exotic rock types are identified—norite seems to be especially prominent in Apollo XIV breccias—it becomes apparent that reconstructing global rock distributions from a few samples is a hazardous business. Put a little differently, the willingness of investigators to make such reconstructions appears inversely related to the amount of available information. It is a near certainty that successive missions will lead to the construction of more complicated models for highland rock distributions, not to the confirmation of earlier and simpler models.

Contrary to what one might expect, the meteoritic addition to the lunar regolith is apparently small. Of course, most infalling meteorites will strike the surface at high velocity, and probably will be completely melted or vaporized. In particular, silicate materials from stony meteorites will be reconstituted and difficult to distinguish from basaltic materials in the soil. The breccias and fines do contain a few fragments of meteoritic nickel-iron. On the basis of a comprehensive trace element study, Ganapathy and others (1970*a*) conclude that the total meteoritic addition is no more than 1 or 2 percent, mostly in the form of carbonaceous chondrites. The average meteoritic influx for the Moon appears to be about 4×10^{-9} grams per square centimeter per year (Ganapathy and others, 1970*b*).

A number of size analyses of lunar fines have been published (Fig. XII-41). The median sizes and sorting values are approximately the same for both Apollo XI and XII samples. King and others (1971) conclude that Apollo XII fines are slightly

(*Opposite page*)

FIGURE XII-40. Some photographs of lunar fines (courtesy of John A. Wood). (a) Representative assortment of coarse fines from Apollo XI sample, including black irregular glass (g), glass spherules (s), breccia (br), basalt (b), and one fragment of anorthositic glass (a). (b) Coarse fines from Apollo XII sample. Fragments include basalt (b), breccia (br), homogeneous glass (g), ropy glass (r), and norite (n). The ropy glass particles have been referred to as KREEP because of their enrichment in potassium (K), rare earth elements (REE), and phosphorus (P). The constant composition of KREEP glass and its enrichment in some soil samples suggests that it may be Copernicus ray material added to the Apollo XII soil (Meyer and others, 1971). Norite particles are present in all Apollo XII soil samples in approximately constant amounts. They have a composition similar, but not identical, to KREEP glass. For those reasons Wood (personal communication) disassociates the two particle types and ascribes a local source to the norite. (c) Fines from Apollo XII sample. This sample contains some norite (n) but no KREEP glass. Agglutinated glass fragments (g) have compositions very similar to the bulk composition of the local lunar soil. One fragment of potash rhyolite (rh) is present in this sample. (d) Thin-section photomicrograph of anorthositic fragment in Apollo XI fines. Crossed nicols. Olivine crystals with high relief are dispersed in a matrix of coarser grained calcic plagioclase. Almost all anorthositic fragments are recrystallized breccias. If they are derived from primary igneous anorthositic bodies, then they have been subsequently modified by comminution and shock metamorphism.

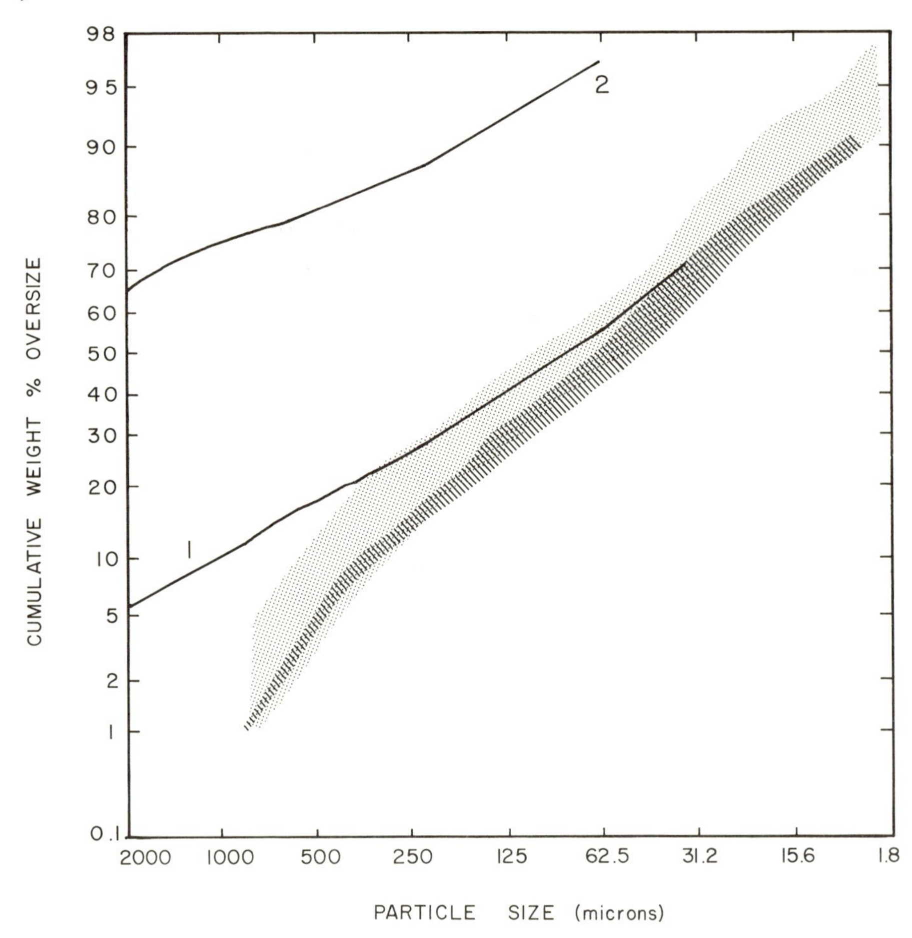

FIGURE XII-41. Size distributions in lunar fines. Sizes are plotted against a probability scale showing cumulative weight percent. The stippled zone represents the range of ten Apollo XII samples measured by King and others (1971). The lined zone shows the range of three Apollo XI samples measured by the same authors. Curve 1 is the distribution in an Apollo XII core sample (LSPET, 1970). Curve 2 is the distribution in a coarse-grained layer of olivine crystals, gabbroic rock fragments, and glass observed within an Apollo XII core (LSPET, 1970).

coarser than Apollo XI samples. They regard this as evidence that the Apollo XII regolith is relatively young and has not been so finely comminuted as Apollo XI soil. However, these results should be interpreted with caution. McKay and others (1971) point out that there is a wide variety in size distribution and modal composition for Apollo XII samples, probably related to a locally complex history of ejecta overlap, and to constructional welding of sediment clots. Apollo XIV samples are even more diverse. For example, glass content varies from 10 to 75 percent (LSPET, 1971).

In connection with a study of powdered coal, Rosin and Rammler (1933) described a crushing distribution for mechanically-ground material. Since the lunar regolith is presumably formed by breaking up of bedrock, one might expect the Rosin-Rammler function to describe accurately the size distribution for lunar fines. However, this is

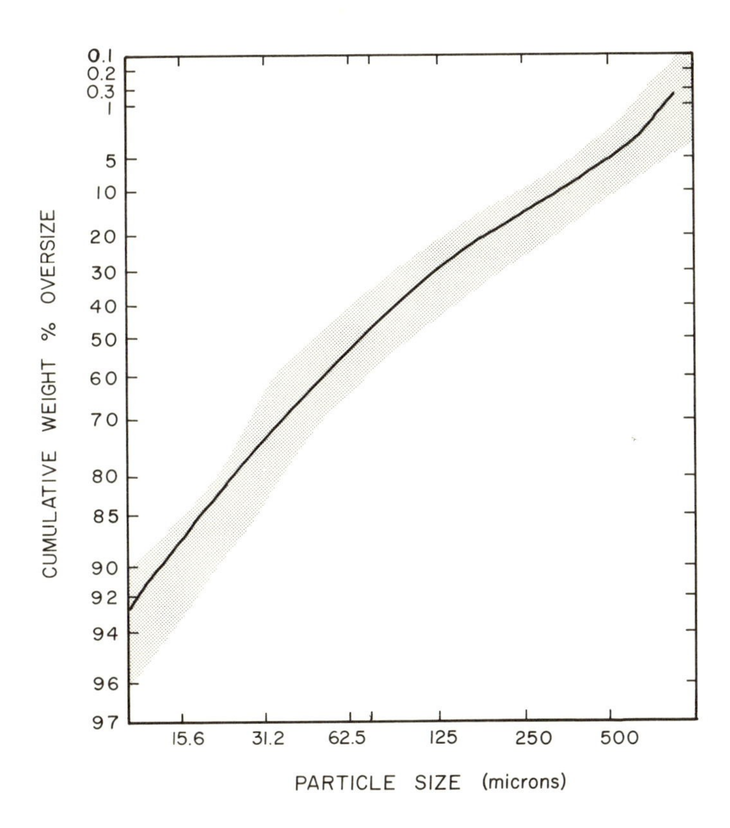

FIGURE XII-42. Size distributions of lunar fines, plotted in terms of a crushing function defined by Rosin and Rammler (1933). Shaded zone represents the range of twelve Apollo samples measured by King and others (1971). The solid line indicates the distribution for a single sample. The fact that the distributions do not form a straight line indicates that they do not observe the predicted crushing distribution.

not the case (Fig. XII-42). The deviations might be caused by agglutination and shock-lithification of small particles into larger clumps (Lindsay, 1971).

Knowledge of the composition of the lunar fines has led to important new insights concerning the reasons for telescopically observed albedo differences on the Moon. Conel and Nash (1970) have demonstrated that fusing of Apollo XI crystalline material reduces its albedo by more than one half. Utilizing this information, Adams and McCord (1971) propose that the maria owe their dark color to the presence of dark glass within the soil. Bright rays around craters are caused by the addition of crystalline fragments from depth. With time this crystalline material darkens by impact-induced vitrification. The fundamental difference between bright highlands and dark maria is probably related to different chemical compositions. Darkening by vitrification requires the presence of ilmenite and other opaques. Such minerals make up to 20 percent of the mare soils, but are apparently greatly reduced in the soils generated over highland crust.

As previously pointed out (pp. 257-259) there are competing darkening and brightening processes at work on the Moon. Although bright rays darken with time, the Surveyor determinations of a ubiquitous, thin, relatively bright surface layer less than 1 mm thick have been confirmed by Apollo investigators. This must be related in some way to bombardment by solar wind and cosmic rays or to the impact of micrometeoroids. The albedo change must occur over a relatively short time span—thousands of years; otherwise the bright layer would be disrupted by turnover (gardening) of the regolith. Hapke and others (1971) have demonstrated that many soil particles are coated with an HCl-leachable skin of semi-opaque siliceous material. Presumably this is a vapor deposit of material volatized by meteoroid impact, or perhaps it represents sputter-deposition by solar wind irradiation. Apparently the top layer of grains is continually scrubbed clean, and the volatized or sputtered material is deposited at shallow depth.

Most samples of lunar fines have been scooped up from the top few centimeters of the regolith, but some core samples have also been obtained. On the first three Apollo missions a total of seven drive core tube samples were obtained. Most of these showed internal textural and color changes, documenting a well-preserved stratigraphy. In one Apollo XII core 41 cm long, at least ten layers were identified (LSPET, 1970), and as many as sixteen depositional units may be present (Lindsay, 1971). On the Apollo XV mission a rotary percussive drill was used for the first time to obtain a core 2.4 meters in length. Preliminary examination by x-radiographs reveals 58 individual layers, differentiated on the basis of density and pebble content (LSPET, 1972).

REGOLITH

From observations of the sizes of anomalous, flat-floored craters, Shoemaker and others (1970) estimate that the regolith at the Apollo XI site ranges in thickness from 3 to 6 meters. Using relationships described on pp. 256-257 of this volume, and developed in more detail by Soderblom (1970), they determine an upper limit to the steady-state crater distribution consistent with a median thickness for the regolith of 4. 1 meters. Although their assumptions necessarily lead to the conclusion that the regolith has been turned over only once to that depth, turnover times will be reduced at lesser depths. Figure XII-43 indicates this relationship. If one assumes the age of the surface to be 3.65 billion years, the radiometrically determined age of crystalline rocks at this site, then it follows that the regolith will be churned up to a depth of 1 cm in approximately 1.7 million years. It should be recognized that this is a generalization. As demonstrated by the well preserved stratification in an Apollo XII core sample, successive build-up of the regolith by discrete ejecta-deposition events may locally be more important than gardening-disruption.

Utilizing the same crater-production function used to calculate turnover of the regolith, Shoemaker and others calculate the integrated volume of material eroded by repeated small cratering events. Added to this are the catastrophic disruption effects

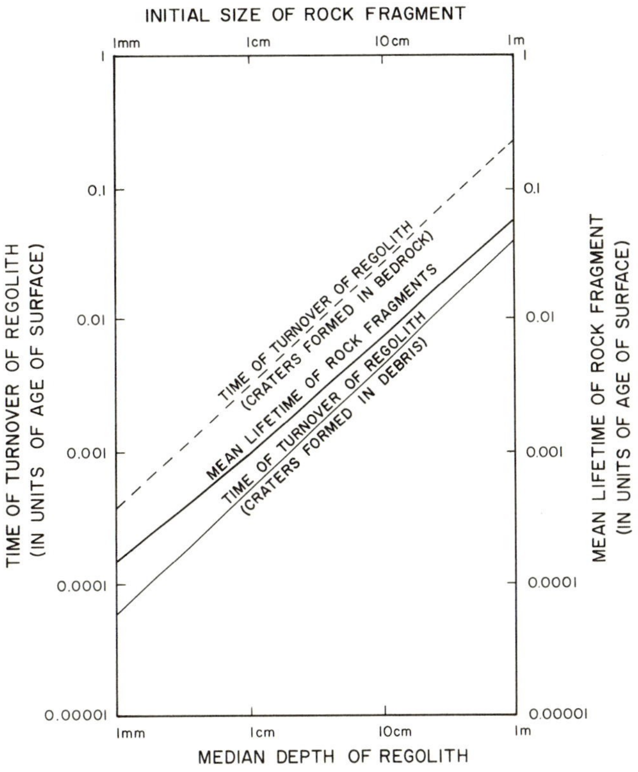

INITIAL SIZE OF ROCK FRAGMENT
1mm 1cm 10cm 1m
TIME OF TURNOVER OF REGOLITH (IN UNITS OF AGE OF SURFACE)
MEAN LIFETIME OF ROCK FRAGMENT (IN UNITS OF AGE OF SURFACE)
1 0.1 0.01 0.001 0.0001 0.00001
TIME OF TURNOVER OF REGOLITH (CRATERS FORMED IN BEDROCK)
MEAN LIFETIME OF ROCK FRAGMENTS
TIME OF TURNOVER OF REGOLITH (CRATERS FORMED IN DEBRIS)
MEDIAN DEPTH OF REGOLITH

FIGURE XII-43. Calculated times of turnover of the regolith and mean lifetimes of rock fragments, expressed as fractions of the age of the mare surface at the Apollo XI site. Craters formed in debris are proportionately deeper than those formed in bedrock. Relationships for both materials are shown. The intersection of the bedrock function with a time of 1 establishes the thickness of the regolith. For determining turnover times of the regolith at relatively shallow depths, the debris function is more accurate than the bedrock function (from Shoemaker and others, 1970).

which would accompany larger particle impact. The resulting mean lifetime of fragments is shown in Figure XII-43. For example, a fragment 10 cm in diameter will be completely comminuted in approximately 2.9 $\times$ 10^7 years.

Gault and others (1963) have calculated the dispersion of ejecta following a cratering event. This range-frequency distribution is apparently independent of impact velocity and crater size. The actual thickness of the regolith at the Apollo XI site is most consistent with a multiple ejection, or three-hop model. Half the material has been transported distances of 3.1 km or less, but 5 percent has come from distances of 100 km or more (Fig. XII-44). The Apollo XI site lies about 50 km north of a broad

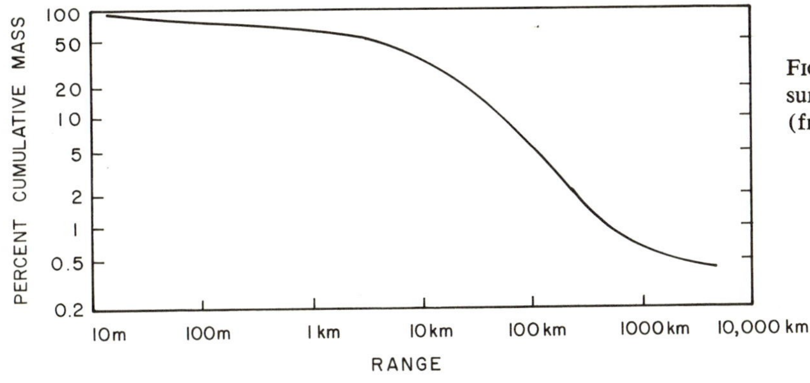

FIGURE XII-44. Dispersion of lunar surface materials by cratering ejection (from Shoemaker and others, 1970).

highland area, and, as a very general approximation, Shoemaker and others estimate that 5 percent of the regolith material is composed of highland material. Interestingly enough, this correlates well with the determination by King and others (1971) that about 5 percent of the fines are plagioclase-rich fragments.

The surface of the lunar regolith at the Apollo XI site is marked by intersecting sets of shallow grooves a fraction of a centimeter deep, about 1 cm wide, and 3 to 50 cm apart (Shoemaker and others, 1970) (Fig. XII-45). These linear grooves are remarkably pervasive, appearing at all of the first four Apollo sites. They have a slight tendency to align in northeast and northwest directions, reminding one of the "lunar grid" discussed on pp. 247-250. Shoemaker and others suggest that the grooves may be formed by drainage of soil into bedrock fractures, or by vibration of joint blocks in the bedrock. Schaber and Swann (1971) suggest that lunar rock tides associated with Earth-Moon interaction may provide sufficient energy to maintain lineament systems in the regolith. However, it is difficult to imagine fractures spaced so evenly, occurring so close together, or having

FIGURE XII-45. Linear features in Apollo XII soil viewed through the window of the LM, following landing. One prominent set of lineations trends from lower right to upper left. A less prominent set trends approximately at right angles to the first set (NASA photographs AS 12-48-7029).

such a uniform effect over sites with such disparate bedrock and regolith characteristics as Apollo XI, XII, XIV, and XV. Possibly, the grooves are formed by slight readjustments of surface particles in response to thermal contraction and expansion. The temperature range between lunar day and night is as much as 270°C. It would be remarkable if this dramatic change had no effect on the unconsolidated regolith materials. Polygonal and linear soil markings are common in terrestrial soils (Washburn, 1956). Many of these are associated with the freezing or vaporization of water, but some occur in regions with enduring subfreezing temperatures (Pewé, 1959). Even if the grooves in the regolith are formed by thermal readjustments, it remains likely that their orientation is determined by larger bedrock structures and topographic irregularities. The similar orientations of grooves and lunar-wide lineaments are too striking to be coincidental.

It should be noted that the actual contact between the regolith and underlying bedrock has not yet been sampled directly. All the crystalline rocks have been observed as boulders within the regolith. Accordingly, one cannot rule out categorically Gold's continuing assertions that the maria are underlain by a layer of dust several kilometers thick. In support of this contention, Gold and Soter (1970) point to the anomalously long reverberation of the seismic signal caused by the impact of the Apollo XII LM. They demonstrate that similar long-duration signals would be formed in a layer of dust progressively more compacted with depth. However, Latham and others (1970) maintain that the reverberation of the signal suggests only that the outer crust of the Moon is heterogeneous on a scale of several hundred meters or less to several kilometers. This heterogeneity could be caused by irregular intercalation of volcanic flows and ash deposits, by networks of lava tubes within flows, or by brecciation of rock by meteoroid impact.

Latham and others also conclude that there is no major seismic discontinuity in the vicinity of the Apollo XII site to a depth of 20 km. This is somewhat surprising. One would have thought, a priori, that the contact between mare-filling lava and premare basement would have been detected. Apparently rocks on both sides on this contact have similar seismic properties, or perhaps the contact is too irregular to be revealed in a simple two-layer model. More recently Toksöz and others (1972) have determined discontinuities in compressional velocities at 25 and 65 km in the Fra Mauro region. They tentatively interpret the layer above 25 km to be basalt, the layer between 25 and 65 km to be norite, anorthositic gabbro, or pyroxenite, and the rock below 65 km to be enriched in olivine, similar to rocks of the terrestrial mantle.

CHEMISTRY

The chemical composition of lunar samples has been studied exhaustively by numerous investigators using a variety of sophisticated techniques. More than 75 elements have been identified and, naturally enough, their abundances do not precisely duplicate those of any one terrestrial rock or meteorite. Nonetheless, a number of provocative comparisons have been made. Various investigators have concentrated their discussions on particular elements which they believe to be especially diagnostic. One group of comparisons has centered around consideration of major elements, especially Si, Al, Ti, Fe, Ca, Na, and K. Other studies have emphasized comparative distributions of rare earth elements (REE). A third area for discussion has been distribution of radioactive isotopes and their daughter products, with particular emphasis on Rb–Sr and U–Th–Pb isotopes. Finally, many studies of rare or inert gases He, Ne, Ar, Kr, and Xe, have been conducted.

Many types of meteorites have been collected on Earth. Major groups are irons, stony-irons, chondrites, and achondrites. In terms of abundance, chondrites are the most significant of all meteorites, making up 85 percent of the observed falls (Mason, 1962). Because their proportions of major metallic elements (except iron) are similar to those of the Sun, chondrites frequently have been interpreted as samples of primary, undifferentiated planetary material. In the same vein, a chondritic composition for the Moon has repeatedly been mentioned (e.g., Urey, 1952).

Clearly, the lunar rocks are completely different from the irons and stony-irons, both of which contain large amounts of metallic nickel-iron. Differences between lunar rocks and chondrites are almost as striking. For example, Al is much more abundant in

TABLE XII-2. Chemical composition, expressed as weight percents of oxides, for selected lunar, terrestrial, and meteoritic materials*

	APOLLO XI						APOLLO XII			
	Crystalline									
	Group A[1]	*Group B*[1]	*Breccia*[1]	*Fines*[1]	*"Anorthosite"*[2]	*"Granitic" Residual*[3]	*Crystalline*[4]	*Breccia*[5]	*Fines*[6]	*12013*[7]
SiO_2	40.28	40.47	41.69	42.03	46.0	76.1	40	42	42	61
TiO_2	11.88	10.32	8.87	7.50	0.3	0.5	3.7	3.4	3.1	1.2
Al_2O_3	8.95	10.41	12.35	13.84	27.3	11.7	11.2	13.2	14	12
Fe_2O_3	np	np	np	np	np	np	np	np	np	np
FeO	19.21	18.72	16.39	15.76	6.2	2.5	21.3	18.2	17	10
MnO	0.24	0.27	0.23	0.21	0.1	np	0.26	0.18	0.25	0.12
MgO	7.60	6.66	7.47	7.87	7.9	0.3	11.7	11	12	6.0
CaO	10.53	11.48	11.81	11.98	14.1	1.9	10.7	10.7	10	6.3
Na_2O	0.64	0.49	0.63	0.45	0.3	0.4	0.45	0.51	0.40	0.69
K_2O	0.31	0.09	0.15	0.14	np	6.6	0.06	0.20	0.18	2.0

* np = not present; na = not analyzed; tr = trace.
[1] Compston and others (1970). Represents average for several samples and several investigators.
[2] Wood and others (1970*b*).
[3] Roedder and Weiblen (1970).
[4] LSPET (1970). Average of 9 rocks.
[5] LSPET (1970). Average of 2 rocks.
[6] LSPET (1970). Single sample.
[7] LSPET (1970).
[8] Vinogradov (1971). Average of 4 reported analyses.

lunar samples (Table XII-2). The similarities between Apollo crystalline rocks and several types of achondrites (eucrite and howardite) are more persuasive. Ti, contained principally in the mineral ilmenite, is anomalously high for the Apollo XI rocks, but the Apollo XII rocks show values lower by a factor of three. The reasonably close match suggests that achondrites may be impact-ejected lunar fragments which have subsequently fallen to the Earth (Mason and Melson, 1970). Duke and Silver (1967) anticipated this possibility when they proposed that howardites, which characteristically have a brecciated structure, are derived from the lunar highlands and basalt-like eucrites are derived from the maria.

The Apollo crystalline rocks are commonly termed basalt in recognition of their resemblance both in composition and presumed mode of emplacement to terrestrial basalt flows. There are two broad groups of terrestrial basalts: alkali and tholeiitic. Alkali basalts are associated with many oceanic and continental volcanoes. In contrast. to Apollo basalts they have high abundances of both Na and K. Oceanic tholeiitic basalts display closer similarities to the lunar rocks. Na and K abundances are still high, but are significantly reduced relative to alkali basalts. Note that the Apollo XI basalts can also be divided into two groups on the basis of K content (Table XII-2). Group A has

	LUNA XVI	SURVEYOR		TERRESTRIAL				METEORITES		
				Basalt			Anorthositic			
	Regolith[8]	*V*[9]	*VII*[10]	*Average*[11]	*Ocean-ridge*[12]	*Ti-rich*[13]	*gabbro*[14]	*Chondrite*[15]	*Eucrite*[16]	*Tektite*[17]
O_2	41.8	46.4	46.1	49.9	49.21	41.83	50.96	38.04	48.16	73.14
iO_2	3.39	7.6	tr	1.4	1.39	7.05	0.73	0.11	0.32	0.94
l_2O_3	15.33	14.4	22.3	16.0	15.81	11.94	21.26	2.50	15.57	12.48
e_2O_3	np	np	np	5.4	2.21	7.05	1.82	np	np	—
eO	16.64	12.1	5.5	6.5	7.19	12.04	4.37	12.45	15.69	5.02
InO	0.21	na	na	0.3	0.16	0.21	na	0.25	0.31	0.13
IgO	8.78	4.4	7.0	6.3	8.53	6.47	4.49	23.84	8.41	2.00
aO	12.49	14.5	18.3	9.1	11.14	9.88	11.46	1.95	11.08	2.51
a_2O	0.34	0.6	0.7	3.2	2.71	2.43	3.17	0.98	0.45	1.45
$_2O$	0.11	na	na	1.5	0.26	0.10	0.61	0.17	0.09	2.40

[9] Turkevich and others (1969).
[10] Patterson and others (1970).
[11] Average of 161 basalts, 17 olivine diabases, 11 melaphyres, 9 dolerites (Daly, 1933).
[12] Melson and others (1968).
[13] Miyashiro and others (1970).
[14] Buddington (1939).
[15] Average of 94 analyses (Mason, 1962). Does not include Fe: 11.76 percent, FeS: 5.73 percent.
[16] Moore County eucrite (Hess and Henderson, 1949).
[17] Average of 24 analyses of Indochinites, taken from Mason (1962). All Fe reported as FeO.

K abundances very similar to those of certain ocean-ridge basalts. Ti is unusually high in Apollo XI rocks, and Fe is high in both Apollo XI and XII rocks. However, there are some examples of titanium- and iron-rich oceanic basalts collected from the mid-Atlantic ridge (Miyashiro and others, 1970).

As mentioned previously in the section on petrology, not all of the lunar crystalline rocks are basaltic. Table XII-2 demonstrates some of the chemical variations. Anorthositic fragments selected from the fines are enriched in Ca and Al and depleted in Ti and Fe. The granitic vein contained within rock 12013 is enriched in Si and K. The residual glass contained within crystalline rocks is similarly enriched in Si and K. O'Keefe (1970) has pointed out a close chemical similarity between rock 12013 and a Java tektite analyzed by Cassidy and others (1969). However, most tektites are more enriched in Si than this Java specimen. In this respect they more closely resemble the Si-rich residual glass within lunar crystalline rocks.

Numerous investigators have determined that the Apollo rocks are greatly enriched in the rare earth elements (REE) relative to potential terrestrial and meteoritic analogs. The REE are geochemically similar so that, even though their absolute abundances in two samples may be quite different, the amount of enrichment or depletion should remain constant from element to element. This relationship is confirmed for all elements except europium, which is depleted by a factor of four or more relative to other REE (Fig. XII-46). This widely-noted "europium anomaly" has been explained by preferential enrichment of Eu in plagioclase of a fractionally-crystallizing liquid, or by partial melting of a plagioclase-pyroxene-spinel mantle. In either case, the presence of a plagioclase-rich differentiate is implied. Accordingly, this is indirect evidence for the existence of an anorthositic highland crust.

The LSPET (1969) reported unexpectedly large amounts of rare gases in the breccias and fines. These observations have been confirmed by many other investigators (e.g., Hohenberg and others, 1970; Marti and others, 1970; Pepin and others, 1970). The gases can be formed in several ways: incorporation at the time of crystallization, disintegration of radioactive elements, element-spallation by cosmic-ray bombardment, and addition of solar wind components. The gases are significantly less abundant in crystalline rocks than in fines and breccias. Apparently, radiogenic sources dominate the crystalline-rock contribution, while solar wind enrichment is important for the fines and breccias. Presumably the solar-wind particles are most efficiently trapped by surface materials which are shock-melted and rapidly recrystallized.

Bombardment by high-energy, galactic cosmic rays and by lower-energy, solar wind protons and alpha particles leads to the production of disintegration products, many of them unusual radionuclides. There are several ways of studying this phenomenon. Abundances of bombardment-produced radionuclides can be directly measured, or tracks made by energetic particles penetrating lunar minerals can be artificially enhanced and then studied visually. As one might expect, a variety of complex histories are revealed for the several samples. Since cosmic rays penetrate to greater depths than solar-flare particles, a depth to time relationship can be worked out for certain samples. It is even possible to determine that certain rocks have been overturned on the surface. One rock shows high radiation effects on two opposite-facing sides, indicating an exposure time of about 2 million years for each side (Fleischer and others, 1970).

In general, measurement of radiation effects indicates that soil fragments have remained in the top 10 cm for approximately 15 million years (Fleischer and others, 1970), and have remained within a meter or so from the surface for about a billion years (Hohenberg and others, 1970). The rate of erosion due to radiation effects is slight but significant, less than 1 cm per 10 million years (Fleischer and others, 1970).

Investigators have found it useful to divide the elements into broad groups following a classification proposed by V. M. Goldschmidt in 1923. Elements are classified as

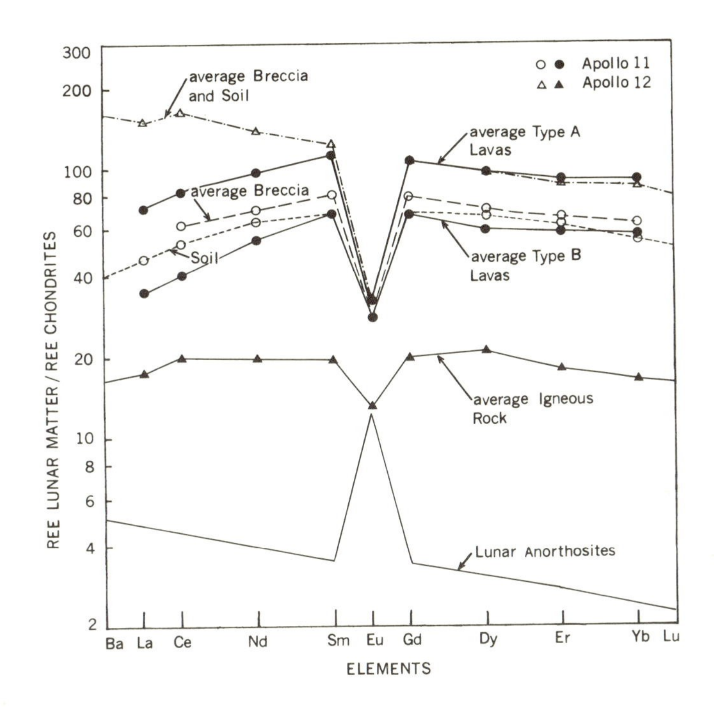

FIGURE XII-46. Chondrite normalized REE abundances for lunar materials. Average Type A (high in Rb) Apollo XI lavas, type B (low in Rb) Apollo XI lavas, and Apollo XI breccia abundances are taken from Gast and others (1970), and represent the averages of many determinations by several investigators. Abundances for Apollo XI soil, and for lunar anorthosites, are taken from Wakita and Schmitt (1970). Anorthosite determinations were made on particles selected from Apollo XI coarse fines. Note that the negative Eu anomaly in lunar basalts and associated materials is matched by a positive anomaly in anorthosite. This suggests that the Eu not present in mare basalts may be differentiated in anorthositic bodies underlying the highlands. Apollo XII lavas (not shown) closely resemble Apollo XI lavas, but Apollo XII breccias and fines are as much as ten times richer in REE. Three Apollo XII analyses are taken from Hubbard and Gast (1971).

siderophile, chalcophile, lithophile, and atmophile, according to their affinity for metallic iron, sulfides, silicates, and the atmosphere, respectively. Support for the chemical significance of the classification comes from meteorites where such a partitioning does, indeed, occur (Table XII-3). A useful way of summarizing the amount of fractiona-

TABLE XII-3. Geochemical classification of elements. Parentheses indicate a minor affinity*

Siderophile	*Chalcophile*	*Lithophile*	*Atmophile*
Fe Co Ni	S Se Te	Li Na K Rb Cs	He Ne Ar Kr Xe
Ru Rh Pd	Fe Ag Cd	Be Mg Ca Sr Ba	H N C
Os Ir Pt	Hg Tl Pb	B Al Sc Y La-Lu	
Cu Au Mo	Bi In (Mo)	Si Ti Zr Hf Th	
W Re Ge		P V Nb Ta Mn Fe	
As Sb Sn		O Cr U Zn Ga	
(Ga) (Bi)		F Cl Br I	

* Taken from Mason and Melson (1970).

tion in lunar rocks is to compare the elemental abundances with those of a Type 1 carbonaceous chondrite, a meteorite with elemental abundances that probably reflect the average composition of solar-system material, volatile elements excepted. When abundances in Apollo crystalline rocks are plotted against those of the carbonaceous chondrite, certain patterns emerge: highly enriched elements are all lithophile; highly depleted elements are mostly chalcophile, siderophile, or atmophile (Fig. XII-47). The volatile elements were probably driven off during a high temperature period late in the condensation history of the solar system and prior to aggregation of the Moon. The siderophile elements may have been fractionated in an early Moon-wide period of melting and differentiation. Conceivably chalcophile elements, in the form of sulfides, were similarly segregated in the Moon's interior. Mason and Melson (1970) have pointed out that all the enriched lithophile elements except titanium do not readily substitute for the major elements of the common lunar minerals: pyroxene, plagioclase, and ilmenite. This suggests that they might have been relatively less abundant in the original magma, but were progressively enriched in the residual liquid as fractional crystallization of the common minerals proceeded. According to this model, Apollo crystalline rocks would have formed from the late-stage residuum.

Although any detailed model for chemical differentiation of the Moon is necessarily speculative, it is certain that Apollo rocks are *not* representative of the entire Moon. Experimental work indicates that at fairly low pressures—12 kb at 1100°C—Apollo XI basalt transforms to eclogite (Ringwood and Essene, 1970). Eclogite has a density of 3.74 grams per cubic centimeter, but the mean density of the Moon is only 3.36 grams per cubic centimeter (Kaula, 1970). Accordingly, the Moon must have somehow formed a crust with composition different from that of the interior, a crust relatively enriched in lithophile elements.

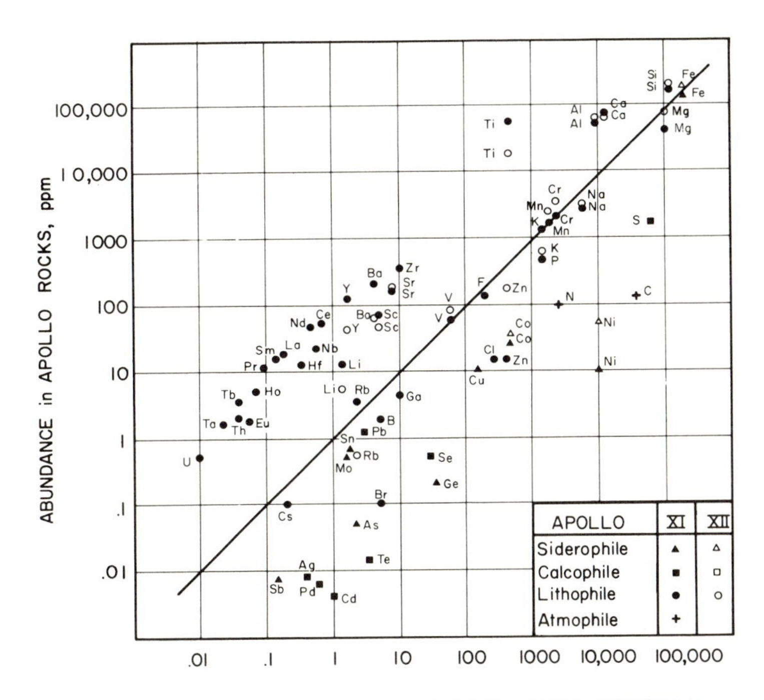

FIGURE XII-47. Plot of elemental abundances in Apollo crystalline rocks versus abundances in Type I carbonaceous chondrites. The distance of a point from the diagonal line is a measure of the relative enrichment or depletion. Note the enrichment of certain lithophile elements, and the depletion of certain siderophile and chalcophile elements in lunar rocks. Abundances of most major elements are similar in chondrites and lunar rocks. Exceptions are C, S, and Ni, which are enriched in chondrites, and Ti, which is enriched in lunar rocks. Many of the trace elements show significant preferential enrichment in chondrites or lunar rocks (Apollo XI information taken from Mason and Melson, 1970; Apollo XII information taken from LSPET, 1970).

RADIOMETRIC DATING

Among the first reports of the Apollo investigators, none received wider attention and engendered greater speculation than the radiometric age determinations. An excellent review and evaluation of these studies is provided by Wetherill (1971). Three methods have been used: uranium-thorium-lead (U–Th–Pb), rubidium-strontium (Rb–Sr), and potassium-argon (K–Ar). All methods are based on the common principle that each radioactive isotope decays according to a characteristic, invariable rate. The "half-life" is a measure of the time necessary to reduce the number of parent atoms by one half. Half-lives for U^{238}, U^{235}, Th^{232}, Rb^{87}, and K^{40} are 4.51×10^9, 0.713×10^9, 13.9×10^9, 50×10^9, and 1.30×10^9 years respectively.

In the U–Th–Pb system, different ratios are plotted against each other as follows: Pb^{206}/Pb^{204} vs. U^{238}/Pb^{204}, Pb^{207}/Pb^{204} vs. U^{235}/Pb^{204}, and Pb^{208}/Pb^{204} vs. Th^{232}/Pb^{204}. Several rocks from a presumably consanguineous suite or several mineral separates from a single rock are individually analyzed. Ideally, the plotted points fall on straight lines, or isochrons. The slope of the isochron indicates the age of the samples; the intercepts with the Pb^{206}/Pb^{204}, Pb^{207}/Pb^{204}, and Pb^{208}/Pb^{204} axes give initial values for these ratios. If an initial ratio is assumed, then a single analysis provides sufficient data to construct an isochron. This is sometimes termed a model age by Apollo investigators, alluding to the fact that a model is accepted for initial isotopic composition. (Strictly speaking, any isochron is based on a model of some sort.)

U–Pb information can also be plotted on a U–Pb evolution diagram. If Pb^{206}/U^{238} is plotted against Pb^{207}/U^{235} a theoretical "concordia curve" is established, assuming the system has remained closed since the dated event. Very often, however, an intermediate event causes lead leakage. In this case the analytically-determined points are discordant; that is, they do not plot on the concordia curve. Generally, several discordant points form a straight line. The intercepts of this line with the concordia curve indicate the times of geologic events which altered the isotope distributions.

Rb–Sr isochrons are constructed just as for U–Th–Pb. Sr^{87}/Sr^{86} is plotted against Rb^{87}/Sr^{86}. Again, the slope of the line indicates the age of the samples and the intercept with the Sr^{87}/Sr^{86} axis indicates the original ratio of these isotopes.

K–Ar ages may be roughly determined by comparing amounts of parent K^{40} to daughter Ar^{40}. This simple calculation assumes that the rock was free of Ar^{40} when formed, has retained all radiogenically produced Ar^{40}, and has not acquired additional Ar^{40} by other processes. The latter two assumptions turn out to be far from true for lunar rocks. Considerable amounts of Ar^{40} derived from the solar wind or formed by spallation are trapped in fines and microbreccias. More importantly, high-energy impact events have resulted in gas loss—even for crystalline rocks. These several effects can be calibrated by irradiating samples, so that K^{39} is converted to Ar^{39}, and then measuring the Ar^{40}–Ar^{39} ratio as the sample is progressively heated. If easily evacuated crystal sites are present, then the Ar^{39} contribution will dominate at low temperatures. At higher temperatures radiogenic Ar^{40} is released, and the Ar^{40}–Ar^{39} ratio reaches a plateau indicative of the samples' true age.

Of all the dates obtained, those for the crystalline rocks are the easiest to interpret. Six crystalline rocks from the Apollo XI site yield Rb–Sr isochrons within the range $3.65 \pm 0.06 \times 10^9$ years (Papanastassiou and others, 1970). Two subgroups are defined on the basis of different initial Sr^{87}/Sr^{86}. This suggests the existence of two magmatic reservoirs. K–Ar ages range between 4.0 and 2.2×10^9 years (Schaeffer and others, 1970). However, these ages have not been adjusted for variable Ar loss associated with shock effects or different cooling rates. Correction for the effects of these losses by the Ar^{40}/Ar^{39} technique narrows the envelope of ages to 3.52 to 3.92×10^9 years (Turner, 1970).

Eight Apollo XII crystalline rocks yield Rb–Sr ages between 3.15 and 3.35×10^9 years (Albee and others, 1971). These ages are distinctly younger than the Apollo XI ages, and indicate that mare flooding events must have occurred over a time interval of at least 500 million years. A wide range in initial Sr^{87}/Sr^{86} indicates that several different magma sources were tapped, as opposed to only two sources postulated for the high- and low-potassium rocks of Apollo XI. K–Ar ages range from 1.4 to 3.0 $\times$ 10^9 years (Schaeffer and others, 1970). As for Apollo XI rocks, the spread is largely caused by some postcrystallization loss of Ar.

The one Apollo XII granitic rock described in a previous section reveals a particularly interesting radiometric history (Albee and others, 1970). Apparently, the black basaltic part of the sample formed approximately 4.5 $\times$ 10^9 years ago from a Rb–poor system. The white, granitic "end member" was most probably derived from a Rb–rich magma which was formed about 4.5 $\times$ 10^9 years ago, and then was remelted and injected into its presently observed site about 4.0 $\times$ 10^9 years ago. This suggests the existence of granitic magmatic reservoirs which were produced during the first few tens of millions years following the Moon's formation. This early differentiation of the lunar crust was followed by remelting of the granitic rocks about four billion years ago.

Rb–Sr isochrons for four Apollo XIV basaltic rock fragments yield crystallization ages of 3.88 $\pm$ 0.04 $\times$ 10^9 years and 3.95 $\pm$ 0.04 $\times$ 10^9 years (Papanastassiou and Wasserburg, 1971*b*). Assuming that all basalts are breccia fragments from the Fra Mauro Formation, and therefore crystallized before the Imbrium impact event, the age of that event can be no greater than about 3.9 billion years.

The soil at both Apollo XI and XII sites yields a Rb–Sr model age of 4.6 $\times$ 10^9 years (Papanastassiou and others, 1970, Papanastassiou and Wasserburg, 1970). At first glance this is flagrantly inconsistent—older regolith overlying younger lavas. The rule of stratigraphic superposition would suggest just the opposite. However, there are many terrestrial examples of recently deposited sediments which yield an ancient radiometrically-determined age. A good example would be the recent sediments of the Gulf Coast which yield Paleozoic ages because the mineral constituents have been transported from Paleozoic and Precambrian terrains. During weathering and erosion the particles have been modified physically, but have resisted chemical alteration, sufficiently so to yield old ages (Hurley and others, 1961).

According to the widely accepted model for development of the lunar regolith, at least 80 or 90 percent of the soil must have been derived locally from underlying rocks. Apparently the remaining few percent came from distant source rocks and are sufficiently enriched in Rb, K, U, and Th to dominate the contribution from local sources. It is with good reason that Papanastassiou and others (1970) term this poorly understood addition a "magic component." In general one can argue that part of the soil is derived either from widespread rocks with an age of 4.6 $\times$ 10^9 years, or that it consists of a combination of rock types which, in the aggregate, represent a closed total system 4.6 $\times$ 10^9 years old. Some analytical support for the existence of a magic component is provided by a small fragment of transparent glass in a fine-grained grey groundmass, the so-called Luny Rock 1. This fragment yields an age of 4.4 $\times$ 10^9 years and is, in addition, greatly enriched in Rb (Papanastassiou and others, 1970).

The Apollo XII samples of soil and breccia are notably enriched in REE relative to the igneous rocks (Fig. XII-46). KREEP particles are even more enriched in REE so that the soil abundances can be reproduced by a two-component mixture of igneous rock and KREEP. The mixing model for particular soil samples ranges from 70 percent mare basalts, 30 percent KREEP to 30 percent basalt, 70 percent KREEP. Apollo XII soils appear to have more particles derived from distant sources. A chemically compatible mixing model consists of 70 percent low–Rb Group B basalt, 20 percent anorthosite, and 10 percent KREEP (Hubbard and Gast, 1971). According to these models KREEP is the "magic component," and was presumably derived from widespread rocks rich in K, P, and REE which were formed during the early differentiation of the Moon.

The U–Th–Pb results are more complicated and difficult to interpret than the Rb–Sr or K–Ar results. The initial isotopic composition of U, Th, Pb, and the subsequent geochemical behavior of these isotopes in lunar rocks are poorly known. In addition, analytical problems are severe, and contamination by terrestrial Pb is difficult to avoid. Nearly identical ages of 4.70×10^9, 4.67×10^9, and 4.60×10^9 are obtained for Apollo XI breccias and fines, using Pb^{206}—U^{238}, Pb^{207}—U^{235}, and Pb^{208}—Th^{232}, respectively (Tatsumoto, 1970). Isochrons calculated from four crystalline rocks for the same parent-daughter pairs give discordant ages of 3.78×10^9, 4.01×10^9, and 3.83×10^9 years. Two fine-grained crystalline rocks are slightly older. When this information is plotted on a U–Pb evolution diagram, the points lie on two chords. One, involving the fine-grained crystalline rocks, microbreccia, and fines, intersects the concordia curve at 3.8×10^9 and 4.6×10^9 years. A second chord, involving the coarser grained crystalline rocks, microbreccia, and fines, intersects at 3.4×10^9 and 4.63×10^9 years. Tatsumoto believes that the 4.6×10^9 age results from a mixture of materials compositionally approximating the lunar crust, and records the age of the crust's original formation. The younger intercepts are interpreted as recording magmatic events occurring over the time period 3.4×10^9 to 3.8×10^9 years.

The concentrations of U, Th, and Pb in Apollo XII crystalline rocks are similar to those in Apollo XI crystalline rocks, but the concentrations in Apollo XII breccias and fines are more than ten times those in crystalline rocks. All samples have extremely radiogenic lead isotopic compositions but the breccias and fines are more radiogenic than the crystalline rocks (Tatsumoto and others, 1971). This suggests that, early in the Moon's history, surficial rocks were depleted in the volatile element, lead, relative to uranium and thorium. Highly radiogenic particles within the soil contribute to apparent ages, which range from 4.3 to 4.65×10^9 years (Tatsumoto and others, 1971).

Final note should be made of the anorthositic "genesis rock" collected on the Apollo XV mission. It yields an Ar^{40}–Ar^{39} age of approximately 4.1×10^9 years (Husain and others, 1972). Inasmuch as the rock shows evidence of brecciation and metamorphic recrystallization, the measured age can best be interpreted as a younger limit for the time of the anorthosite's primary crystallization. The radiometric information strengthens the premise that differentiation of anorthositic bodies occurred very early in evolution of the lunar crust.

ORIGIN AND EMPLACEMENT OF CRYSTALLINE ROCKS

From what depth within the Moon and from what source materials were the Apollo mare basalts derived? By what processes of melting and recrystallization did they attain their present character? Three quite different answers have been proposed to these questions: first, that the lavas were produced by impact-melting of surface materials; secondly, that they are the product of differentiation within a magmatic reservoir; and thirdly, that they have been formed by partial melting of a deep zone unaccompanied by fractional crystallization.

The impact-melting hypothesis is the easiest to assess. Any surficial impact-related melting would be total. Therefore the melting would probably not be accompanied by chemical fractionation. But the Apollo lavas are highly differentiated. Of course, one can assume that the rocks were already differentiated prior to impact-melting, but then the central problem is simply sidestepped. This is not to say that the general phenomenon of impact-melting has been disproven. As described in previous chapters, lava-like features are observed in and around many craters. Furthermore, addition of impact energy to an already hot crust may trigger melting deep within the crust. However, surficial impact-related melting is inadequate to explain the overall character of the Apollo basalts.

The second proposal is that Apollo basalts are formed by fractional crystallization within a magma. It is well known that, if a silicate magma is slowly cooled, different minerals will begin to crystallize at different temperatures. In such a situation equilibrium assemblages of minerals and residual liquids are necessarily formed at specified temperatures and pressures. Appreciation of these theoretically and experimentally predictable relationships forms the cornerstone of modern igneous petrology.

O'Hara and others (1970) have demonstrated that the Apollo XI lavas are, in fact, very close to the cotectic composition—this being the composition where the residual liquid is in equilibrium simultaneously with olivine, clinopyroxene, and plagioclase feldspar. In support of this implication that the lavas are formed from residual liquid in a fractionally crystallizing magma body, they cite the experimentally-determined small temperature interval (60 to 140°C) between first melting and complete melting. In one important way, however, the Apollo lavas do *not* behave as a cotectic assemblage. When the basalts are experimentally melted and then recrystallized, not all mineral phases crystallize at the same time. Plagioclase does not appear as a primary phase until 30 to 50 percent of the liquid has crystallized as olivine, pyroxene, and ilmenite (Ringwood and Essene, 1970).

How one chooses to evaluate the approximate, but not exact, cotectic composition of Apollo lavas is largely a matter of personal taste. Ringwood (1970*a*) cites the late crystallization of plagioclase as a strong argument *against* fractional crystallization. Other authors (e.g., Hargraves and Buddington, 1970) accept the cotectic argument, maintaining that slight differences between natural and experimental conditions could account for the plagioclase anomaly.

Hargraves and Buddington (1970) point out that fractional crystallization in an undisturbed lava reservoir or lava lake should lead to a stratiform complex in which crystals settle to the bottom of the reservoir to form layers of varied composition. Stratification of this type is well known in certain terrestrial igneous bodies, notably the Skaergaard Complex of Greenland and the Stillwater Complex of Montana. Hand specimens from these complexes show a "cumulate" texture resulting from gravity settling of crystals. Hargraves and Buddington note that none of the Apollo investigators has found evidence for cumulate textures in the lunar rocks. Instead, compositional zoning in pyroxenes implies that crystals formed from a liquid equivalent in composition to that of the bulk rock in which they are found. For these several reasons Hargraves and Buddington reject the idea of local fractional crystallization and gravity settling. They consider it more likely that the mare lavas are consanguineous with anorthositic rocks presumed to underlie the highlands. The common parent magma for both rock types would be gabbroic anorthosite, with the basalts forming from subsidiary, later-stage liquids.

Ringwood (1970*a*) doubts the adequacy of any fractional crystallization scheme. He advocates the third possibility mentioned at the start of this section, namely, that the basalts were formed by partial melting of a pyroxene-rich source rock at depths of 200 to 600 km. His supporting arguments are intricately interwoven and, accordingly, are difficult to judge one strand at a time. Petrogenesis of the lunar basalts is set in a large context of the Moon's origin and early development. In brief, Ringwood maintains that major-element and trace-element contents of the Apollo rocks can be explained by partial melting which buffers the major-element compositions and causes enrichment of incompatible elements according to the degree of partial melting. With regard to the depth of generation, he maintains that, following initial widespread melting of the Moon in the vicinity of 4.6 billion years ago, the outer crust would have had ample time to cool by conduction. Only at depths in excess of 200 km would radioactive heating be adequate to raise temperatures above the melting point. Mare basalts are derived from partial melting of primary unfractionated pyroxenite which makes up the bulk of the Moon, but is overlain by a 150-km thickness of differentiated materials associated with widespread melting immediately following accretion. One can differ with Ringwood about the weight he gives or fails to give to particular pieces of evidence, but his analysis is commendable as one of the very few attempts to synthesize the work of many investigators into an integrated model of lunar development.

Discussion to this point has dealt principally with the generation of mare lavas. Some information is also available about their mode of emplacement. Murase and McBirney (1970*a*) have measured the viscosity of synthetic silicate liquid with the composition of lunar basalts. They found that the viscosity is lower than that of any previously studied volcanic rock on Earth, due principally to enrichment in Fe and Ti and depletion in Si. The distance that any lava flows is a function primarily of velocity and cooling rate. Individual flows of Columbia River basalt have extended more than 200 km on slopes of a few degrees. The velocities of low-viscosity lavas on the Moon would be about twice as great as terrestrial flows. However, with these high velocities, flow would almost certainly be turbulent, and the net horizontal velocity would be controlled by the configuration of the surface over which the lava flowed. The cooling of lunar lavas might proceed quite slowly, further favoring extensive flow. During early cooling a thin crust would be formed on top of the lava. Murase and McBirney (1970*b*) demonstrated that the thermal conductivity of synthetic lunar rock in its melting range is only half that of terrestrial basalt. This means that a thin crust would act as a very efficient insulator, and that relatively small amounts of heat would be lost by radiative processes. A final conclusion is that the slow cooling, and especially the low viscosity, would favor formation of coarse crystals, as observed in Apollo rocks. In their laboratory experiments Murase and McBirney (1970*a*) grew crystals more than 1 cm long in samples that were completely crystallized in about 30 minutes. These several conclusions provide comforting support for the pre-Apollo geologic map interpretations in which lavas are postulated to cover great areas and to form overlapping stratigraphic sequences.

One might have thought that first-hand examination of mare materials would have clarified the physical character of mascons, previously discussed on pp. 215-217. This has not proved to be the case. Several new interpretations have appeared, but none appears definitive. O'Hara and others (1970) suggested that the mass concentrations are the result of heavy minerals rich in Fe and Ti settling to the bottom of deep mare lakes. From high pressure-temperature studies of lunar basalt, Ringwood and Essene (1970) showed that the present gabbroic assemblages would be transformed to a higher density eclogite assemblage at depths in the vicinity of 100 km. This would provide the necessary excess of mass, but perhaps at depths too great to result in the observed gravity anomalies. Wood and others (1970*a*) pursued the earlier suggestion by Baldwin (1968) that the mascons result from high-density lava filling of mare basins. They point out that isostatic equilibrium would be attained when the basins were only partly filled. Further extrusion of lava would result in an excess of mass and the necessary gravity anomaly. A conceptually similar explanation has been published independently by Wise and Yates (1970). They demonstrated that the observed gravity anomaly over Mare Imbrium can be explained satisfactorily by a combination of isostatic rebound of a basement plug following basin formation, and volcanic filling. Their reconstruction is recommended by its simplicity and its detailed compatibility with the gravity anomaly as measured by Muller and Sjogren (1968).

Interpretation of mare lavas centers about a sequence of igneous events that occurred in the vicinity of 3.25 to 3.65 billion years ago. However, reference has already been made to a large-scale melting and differentiation of the Moon which occurred shortly after its accretion. Before the Apollo missions it was widely assumed that planets form at low temperatures, well below their melting points (e.g., Urey, 1969). But radiometric information from Apollo samples points to a large-scale melting and differentiation of the Moon at 4.6 billion years, shortly after its formation. Most likely this melting was precipitated by a build-up of accretional energy as small planetesimals collided to form the larger Moon. The amount of melting which could be caused by trapped accretional energy is problematical. Hanks and Anderson (1969) have examined the general relationship between accretion and temperature. Solution of the problem involves assumptions of masses, temperatures, and impact velocities for individual planetesimals, heat capacity and thermal conductivity of the accreting body's surface layer, existence of endothermic or exothermic chemical reactions, and effectiveness of energy loss by radiation.

Smith and others (1970*b*) have proposed that the entire Moon was molten shortly after its formation. Crystallization then proceeded from the interior outward, with the successive formation of a small Fe-Ni core, an olivine-rich layer of settling crystals, and a layer comprising a mixture of olivine and pyroxene. According to this model, anorthite-rich plagioclase crystallizes late in the sequence, and, because of its low density relative to the residual liquid, floats to the surface. The last liquid trapped between the solid interior and the anorthosite crust is a high-density basalt, rich in Fe and Ti. The anorthosite crust eventually grows to a thickness of 20 km, perhaps to even greater thicknesses on the farside where light material would tend to drift together due to differential tidal action from the Earth. Conversely, the heavy ferrobasaltic liquid is concentrated below relatively thin crust on the Earthside due to tidal attraction. Subsequently, when basin-forming impacts occur, the surficial crust is penetrated and the liquid escapes to form those basalts which now underlie the mare surfaces.

Ringwood (1970*a*) has drawn attention to two difficulties in this model. First, the problem of finding enough energy to melt the entire Moon is summarily dismissed by Smith and others as one of "no difficulty." Secondly, the probable time span for magma differentiation is inconsistent with Apollo results. If the Moon was extensively molten 4.6 billion years ago it would have solidified from the core outward in a few million years. But the Apollo basalts yield crystallization ages of 3.8 billion years or less. Conceivably some of the mare lavas crystallized in the first few million years of the Moon's development, and then melted and recrystallized a second time approximately 3.8 billion years ago. Papanastassiou and others (1970) have shown that Rb/Sr for some, but not all, crystalline rocks is consistent with such a sequence of events.

Whatever its failings, the model of Smith and others is notable for its attempt to explain the preferential occurrence of maria on the Earthside of the Moon. This non-random distribution was briefly acknowleged in previous chapters, but is more clearly shown in Figure XII-48. Of particular interest is the fact that, although mare materials are largely restricted to the Earthside, large basins are equally common on both sides. This suggests that the impact histories for the two sides have been approximately the same. The concentration of mare materials on the side facing the Earth can scarcely be a coincidence. It must in some way be related to the different tidal forces exerted

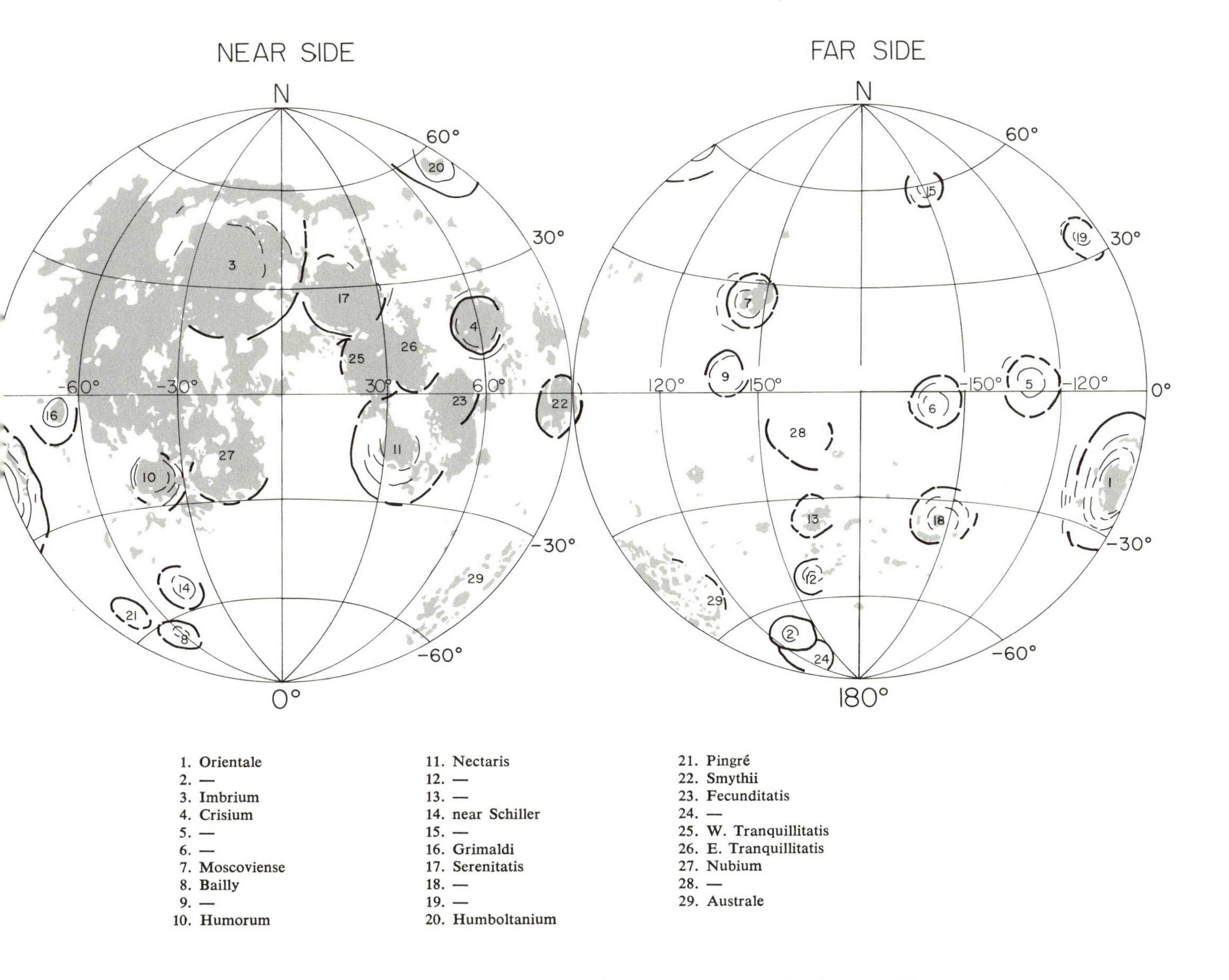

FIGURE XII-48. Distribution of maria and large circular basins 300 km or more in diameter. Mare areas are shaded. The most prominent mountain ring of each basin is shown by a heavy line; secondary rings are shown by light lines. Basin numbers correspond to names in accompanying table where basins are listed in approximate order of increasing age, Orientale youngest and Australe oldest (taken from Stuart-Alexander and Howard, 1970).

by the Earth. Conceivably this distribution might predate and be responsible for the Moon's present orientation (Nash, 1963), but it seems more likely that a thick, light crust accumulated on the farside, and a thin crust, on the nearside during an early stage of extensive melting within the sphere of the Earth's gravitational influence. Large impacts on the Earthside have probably penetrated this crust just as Smith and others suggested, either to tap subcrustal magma pools or to somehow induce melting in subcrustal rocks already close to their melting temperatures.

A model for lunar melting and differentiation somewhat similar to that of Smith and others has been independently proposed by Wood and others (1970*a*). Crystallization following widespread melting is accompanied by density segregation of major minerals. Olivine and pyroxene sink, but anorthite floats in the dense gabbroic melt and accumulates as an exterior anorthositic layer. A postulated heat source is the radioactive decay of Al^{26}, which has a relatively short half-life of 720,000 years.

In contrast to the models just mentioned, Ringwood (1970*a*) postulates that the early melting of the Moon involved only a layer approximately 150 km thick. The lower 125 km of this zone is made up of residual orthopyroxene and olivine from which all the basaltic components are removed to form an upper 25-km layer of eucritic magma. This crystallizes, as in the previous models, to form a 11-km layer of anorthosite above a 14-km layer dominantly of clinopyroxene. The final residual lqiuid is postulated to be quartzo-feldspathic and perhaps to have formed numerous intrusions within the highlands.

ORIGIN OF THE MOON

The Moon's origin is still a matter for debate, but the Apollo results do not equally favor all the competing models. The hypothesis that the Moon formed elsewhere in the solar system and was subsequently captured by the Earth remains unlikely from a dynamical point of view; it involves a two-body interaction of low probability. One of the original corollaries to this hypothesis was that the Moon is a "primary object," containing solar abundances of elements (e.g., Urey, 1952). This contention is refuted by the low abundance of volatile elements in Apollo samples. The anomaly can be explained away by assuming that volatile materials accreted differentially on the Earth and Moon during a late stage of solar nebula cooling (Singer and Banderman, 1970), but there is little empirical evidence to support the argument.

According to the binary planet hypothesis, the Earth and Moon formed in the same region from a common mixture of silicate and metal particles. If so, basalts formed from partial melting of the Earth and Moon should have closely similar compositions, but such is not the case. The fission hypothesis suffers from a similar liability. If the Moon had separated from the Earth's mantle the composition of the two bodies should show detailed similarities. Apollo results are more compatible with a modified fission hypothesis (Wise, 1969) which postulates that large tidal frictions concentrated in the outer shell of a rapidly spinning Earth led to the formation of a very high-temperature, volatilized, silicate atmosphere, and that this atmosphere subsequently condensed to form our Moon.

A "precipitation" hypothesis outlined by Ringwood (1970*b*) is suspiciously complicated, but still most persuasive in terms of supporting information from Apollo studies. The Earth is pictured as accreting in the primeval solar nebula from planetesimals resembling

Type I carbonaceous chondrites—materials with oxidized iron, carbonaceous compounds, large amounts of volatiles, and approximate primordial abundances of most elements. During early accretion the Earth's temperature is kept low by endothermic evaporation of volatiles such as H_2O. As the nucleus increases in size, impact-heating leads to the reduction of oxidized iron, the formation of a metal phase, and the formation of a primitive CO, H_2 atmosphere. With further growth and concurrent increase of temperature and reduction intensity, metals are volatilized, to be followed by silicates. During this progressive heating the primitive atmosphere grows in size until its mass is one-quarter that of the Earth and comprises 10 percent volatilized silicates. At this point in the Earth's evolution it is assumed that the amosphere is dissipated by some unknown combination of causes, possibly including sweeping out by intense solar wind, transfer of angular momentum from Earth to atmosphere, or rotational instability related to the formation of the Earth's core. Following dissipation, the silicate component of the atmosphere cools to form a ring of Earth-orbiting planetesimals which then accrete to form the Moon. Many detailed differences and similarities between terrestrial and lunar rocks can be satisfactorily explained by this fractionation model.

It is interesting to note that the entire sequence of events postulated by Ringwood probably took place over a geologically brief period of time. Noting the limited range of Sr^{87}/Sr^{86} ratio in Apollo rocks and meteorites, Papanastassiou and others (1970) estimated that the Moon separated from the solar nebula in an interval of less than 10 million years. Previously, Von Weizsäcker (1943) estimated that no more than this amount of time was necessary for dissipation of a turbulent primordial solar nebula, and for gravitational accumulation of the planets. More recently, Opik (1961) has determined that formation of the Moon by "sweeping up" of a sediment ring around the Earth could have occured in a period of less than 100 years.

Ringwood's model shares similarities with the hypotheses of other authors. Both O'Keefe (1969) and Wise (1969) suggested that the Moon is a residue of a hot terrestrial atmosphere. Opik (1955) first presented the hypothesis that the Moon formed by the coalescence of an Earth-orbiting ring of planetesimals.

A LUNAR CHRONOLOGY

It is still too early to do any more than block out a few major episodes in lunar history. The most important data points are the ages of 3.25, 3.65, and 3.9 billion years for emplacement of Apollo XII, Apollo XI, and pre-Imbrian basalts, respectively. As predicted in the last sentence of Chapter VI, determination of mare-basalt ages provides a critical "marker zone" in the stratigraphic sequence. Of the several competitive pre-Apollo models for ages of mare materials, the radiometric results correspond most closely to the arrangement shown in Figure VI-15. However, it should be noted that mare material was not emplaced over a "short period of time," even by geologic standards. This is a fact unjustifiably minimized by most Apollo investigators. For example,

Papanastassiou and Wasserburg (1970) conclude that "the flooding of the maria may have occurred only during a short interval of about 300×10^6 years duration. . . ." But here on Earth 300 million years before the present takes us back to the Pennsylvanian period, a time incidentally, before the continents had started to drift apart.

Knowledge of crystallization dates at two mare sites permits a more precise determination of cratering rates throughout lunar history. Clearly, the heavily cratered highlands cannot be explained by an extrapolation of postmare rates backward in time (Fig. XII-49). The early intense bombardment hypothesized many years ago by Urey (1952)

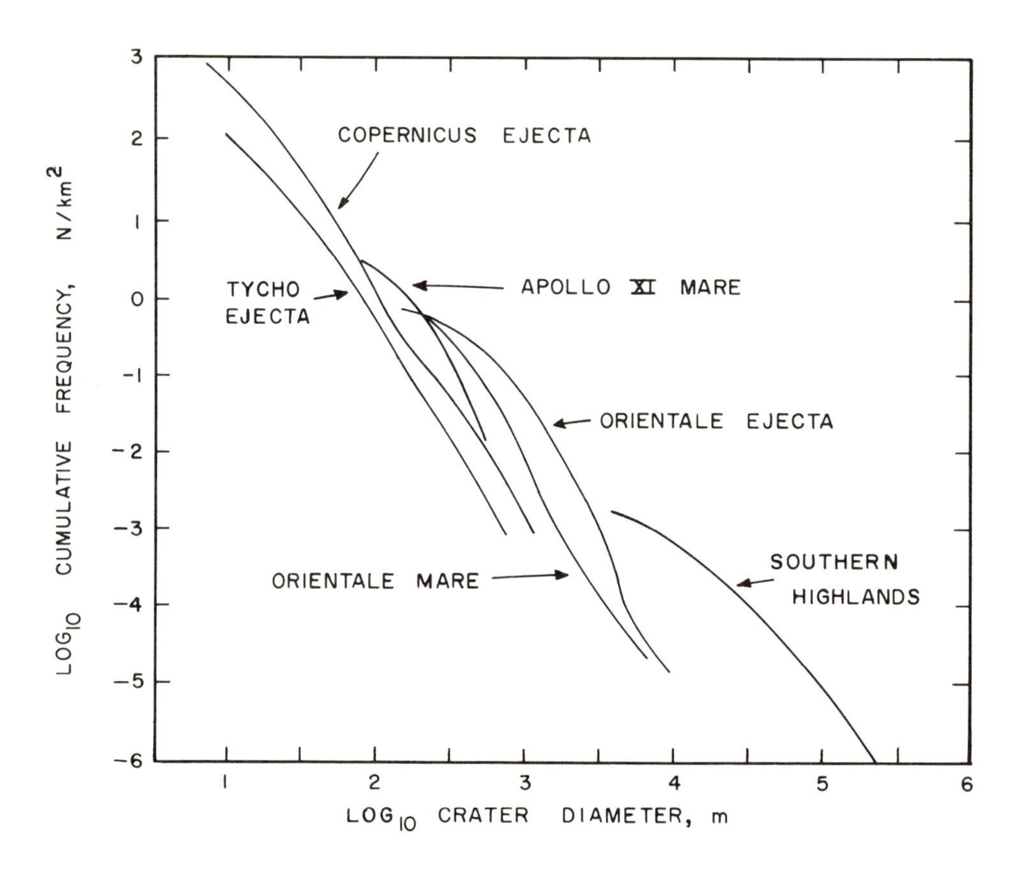

FIGURE XII-49. Crater distributions, taken from a comprehensive review paper by Gault (1970). If one assumes that the meteoroid flux has remained constant throughout lunar history, then the highlands would have an age in excess of 10^{12} years, based on the radiometrically determined age for Apollo XI mare. This is inconsistent with all other data relating to the age of the Moon and indicates that the assumption of constant flux is invalid. As an additional complexity Gault (1970) points out that the highland crater distribution probably has attained an equilibrium state. Accordingly, even if the variations in flux were known, an age measured by crater distributions would be only a minimum value.

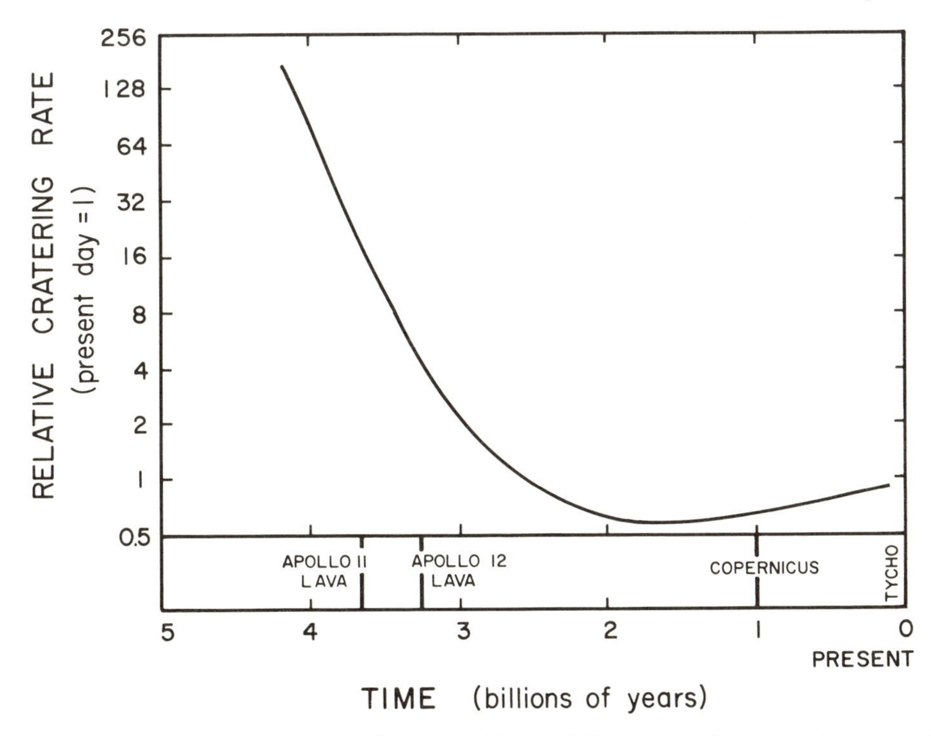

FIGURE XII-50. Empirical determination of history of lunar cratering rate, using crater-density determinations for uplands, Apollo XI site, Apollo XII site, and the region of Tycho (taken from Hartmann, 1970).

and Kuiper (1954), and more recently by Hartmann (1966), is confirmed. Hartmann (1970) notes that the decline of the cratering rate through the first half billion years of lunar history is much less than would result from impacts associated solely with a rapidly swept-up swarm of circumterrestrial particles. He suggests that the observed cratering rate may actually be the sum of three components: (1) early intense bombardment of circumterrestrial particles remaining after formation of the Moon; (2) intense bombardment by the last of the planet-forming planetesimals in low-eccentricity solar orbits. These would have collision half-lives on the order of 10^8 years, significantly greater than circumterrestrial particles; and (3) cratering throughout lunar history by sporadic asteroids and comets (Fig. XII-50).

The existence of these three populations helps to explain a phenomenon otherwise puzzling. We have already postulated that the surface of the Moon was completely molten early in its history and that the energy for that melting was provided by the infall of circumterrestrial "moonlets." This means that the impact features associated with the Moon's accretional formation would be largely obliterated by the melting which necessarily *followed* the infall of most of the circumterrestrial particles. Accordingly, many, if not most, of the craters which we now see in the differentiated highland crust may be associated with Hartmann's second set of late-stage planetesimals in solar orbit.

Even after the highland crust solidifies, there is the additional problem of its cooling until it has sufficient rigidity to maintain the nonisostatic conditions accompanying evacuation of deep craters. Concurrent with the gradual cooling of the crust there must have been a progressive change in crater modification, with the earliest craters preferentially deformed by sinking rims and rising floors (Baldwin, 1969, 1970*a,b*).

As previously discussed, radiometric data points are not restricted to times of mare emplacement. Pre-mare basalts have been dated at about 3.9 billion years. The fines give an aggregate model age of 4.6 billion years, suggesting a large-scale differentiation of the crust at that time. Individual particles within the fines give intermediate ages. One fragment (Luny Rock 1) yields a 4.4 billion-year model age. The granitic vein in rock 12013 indicates first differentiation of granite about 4.6 billion years ago and later melting 4.0 billion years ago. The Apollo XV genesis rock formed no less than 4.1 billion years ago. As more samples are returned and analyzed, it seems highly probable that a large number of additional events between 4.6 and 3.65 billion years will be identified.

The four main phases of lunar history currently identifiable are shown in Table XII-4. Figure XII-51 shows the visual appearance of the Moon just before and after Phase III, compared with the Moon's present appearance. The general distribution of geologic materials across the entire Earthside is shown in Figure XII-52.

Much remains to be known about the period of lunar history from 3.25 billion years to the present. Little can be added to what has been said in previous chapters. Evidence continues to accumulate that the Moon is, even now, an active planet. The Apollo XII seismic experiment has detected numerous Moonquakes, most of which occur at monthly intervals near times of perigee and apogee (Latham and others, 1971). This suggests they are induced by tidal stress.

Although the Moon has been modified by sporadic impact, volcanic, and seismic events throughout its history, many of the major events probably fall between 4.6 and 3.25 billion years, the same time interval for which there is no rock record on Earth. In this way the rock records of the two planets may prove to complement each other ideally, with the lunar rocks providing evidence concerning the origin and early evolution of the solar system not obtainable through study of terrestrial rocks.

TABLE XII-4. Geologic evolution of the Moon*

Stage (time in billion years before present)	*Stratigraphic system*	*Events*
I (~ 4.6)	Pre-Imbrian	a) Formation of Moon by accretional processes, probably from a circumterrestrial swarm of particles.
		b) Rapid heating, either from energy of accretion or from decay of short-lived isotopes, to produce widespread or total melting of Moon.
		c) Large-scale differentiation of Moon, probably by fractional crystallization. Formation of highland crust (dominantly anorthosite? some granite?).
II (4.6 to 3.7)	Pre-Imbrian and Imbrian	a) Infall of last circumterrestrial particles and intense bombardment by last planet-forming planetesimals in low-eccentricity solar orbit to form all major basins and many highland craters. Imbrium impact event occurred about 3.9×10^9 years B.P.
		b) Episodic igneous activity with emplacement of pre-mare basalts and intrusion of some granitic bodies.
III (3.7 to 3.2)	Imbrian	a) Partial melting of the crust, accompanied by fractional crystallization sufficient to produce mare basalts. Period of main eruptions extended over at least 400 million years.
IV (3.2 to present)	Eratosthenian and Copernican	a) Localized mare and highland volcanism (e.g., Marius Hills, Sulpicius Gallus Formation).
		b) Cratering by sporadic asteroidal fragments and cometary bodies. Associated local volcanism (e.g., Tycho).
		c) Minor seismic activity.
		d) Continuous degradation of surface materials by meteoritic impact and by radiation effects. Formation of regolith materials.

* Adapted from Lowman (1970) and Papanastassiou and Wasserburg (1971).

(a)

FIGURE XII-51. Artistic reconstructions showing the appearance of the Earthside of the Moon after formation of all the major basins and before emplacement of mare lavas (a), just after emplacement of mare lavas and before Copernican and Eratosthenian cratering (b), and as at present (c) courtesy of Don E. Wilhelms; artistic reconstructions by Donald E. Davis).

(b)

(c)

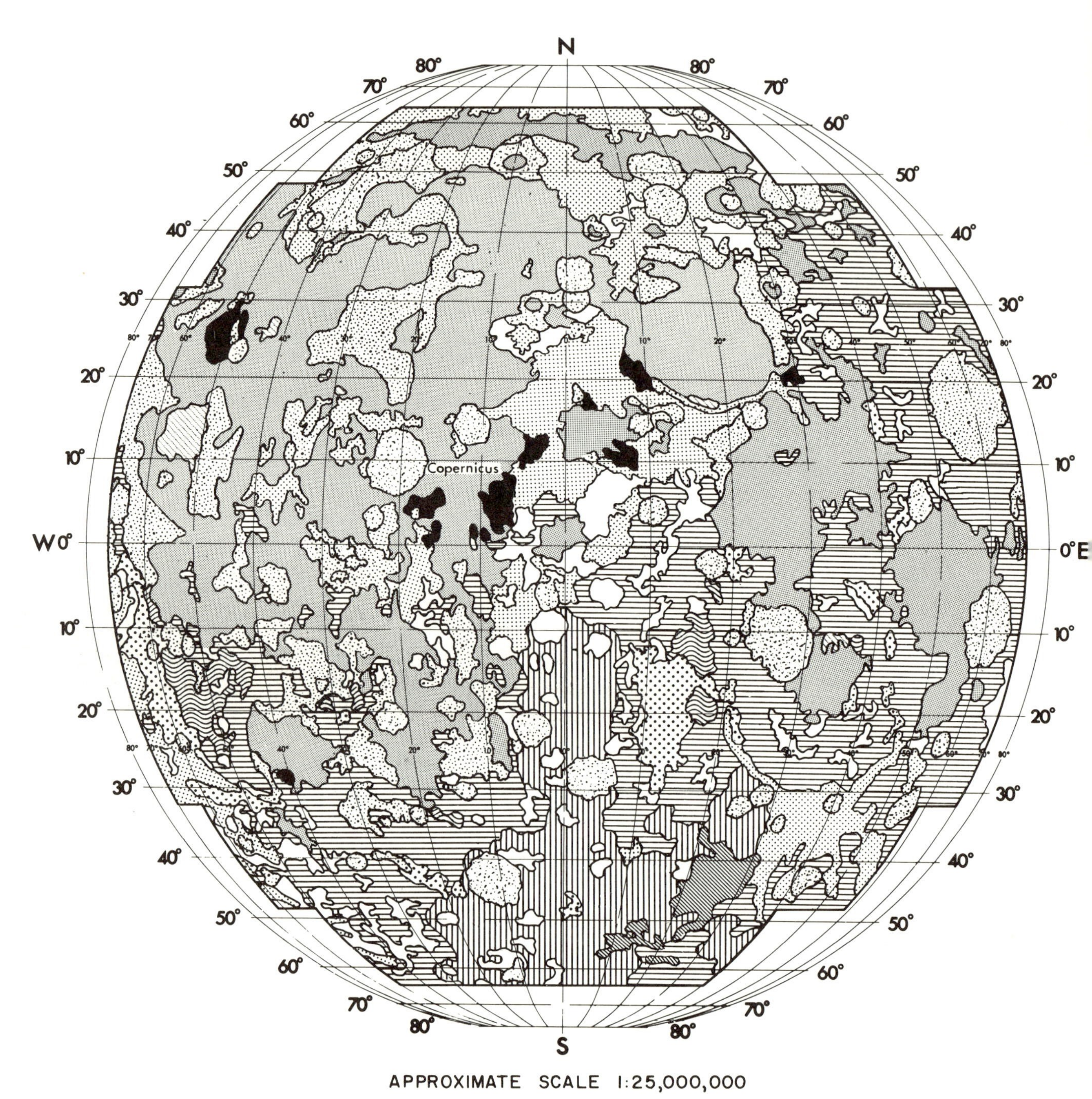

FIGURE XII-52. Generalized geologic map of the Moon's Earthside. Physiographic base is shown on the accompanying plate (courtesy of D. E. Wilhelms and J. F. McCauley).

UNIT	INTERPRETATION AND HISTORY
IMPACT CRATERS > 50 KM	Impact craters later than the Imbrium event and its associated blanket and structures. Age range from earliest Imbrian to Copernican.
MARE PLATEAUS	Mostly flows of dark to intermediate albedo in near-side multi-ring basins and smaller basin related depressions or crater floors. Mare plateaus: later aggregates of volcanic landforms, cones and domes predominate; pyroclastics and flows of same or possibly slightly different composition than maria. Younger mare: relatively thin flows, generally darker, less cratered, distribution patchy. Dark mantles: thin veneers of very dark pyroclastics; mostly cover lighter terra units, near mare-terra contacts. Light plains; more cratered than maria, fills basins and terra depressions; impact lightened old mare or compositionally distinct older volcanic fill unit. Imbrian to Copernican in age.
YOUNGER MARE	
OLDER MARE	
DARK MANTLES	
LIGHT TERRA PLAINS	
HILLY AND FURROWED TERRA	Terra modifiers. Hilly and furrowed: Mostly subdued linear structures resembling fissure cones. Hilly and pitted: rolling plains and plateau unit, many closely spaced rimless pits on surface. Embays pre-Imbrian craters but cuts Imbrium sculpture. Mostly Imbrian in age.
HILLY AND PITTED TERRA	
BASIN DEPOSITS AND STRUCTURES	Recognizable ejecta blankets and parts of encompassing structural rings and radially lineated terrain of Orientale, Imbrium and Nectaris basins. Pre-Imbrian to middle Imbrian.
CRATERED LIGHT TERRA PLAINS	Heavily cratered plains filling local depressions in terra. Most superposed craters lower Imbrian. Pre-Imbrian equivalent of light terra plains.
TERRA, UNDIVIDED	Terra undivided: blocky hills, unevenly filled depressions, segments of mantled craters. Of mixed origin, includes much interlayed ejecta and associated structural units of older basins. Terra, densely cratered: closely spaced 50-150 km craters mostly of pre-Imbrian age, least modifed part of near-side. Recognizable volcanic units subordinate. Age range from early to late pre-Imbrian.
TERRA, DENSELY CRATERED	

Appendix A

GEOLOGIC MAPS OF THE MOON AT A SCALE OF 1:1,000,000

THE following geologic maps have been published by the U.S. Geological Survey and are available as items within their Miscellaneous Investigations Series. Locations of quadrangles are shown in Figure I-11.

Quadrangle	*Author*	*Date*	*Map Number*
Aristarchus	Moore, H. J.	1965	I-465
Cassini	Page, N. J	1971	I-666
Copernicus	Schmitt, H. H., Trask, N. J., & Shoemaker, E. M.	1967	I-515
Hevelius	McCauley, J. F.	1967	I-491
J. Herschel	Ulrich, G. E.	1969	I-604
Julius Caesar	Morris, E. C. & Wilhelms, D. E.	1967	I-510
Kepler	Hackman, R. J.	1962	I-355
Letronne	Marshall, C. H.	1963	I-385
Mare Humorum	Titley, S. R.	1967	I-495
Mare Serenitatis	Carr, M. H.	1966	I-489
Mare Vaporum	Wilhelms, D. E.	1968	I-548
Montes Apenninus	Hackman, R. J.	1966	I-463
Montes Riphaeus	Eggleton, R. E.	1963	I-458
Pitatus	Trask, N. J. & Titley, S. R.	1966	I-485
Ptolemaeus	Howard, K. A. & Masursky, Harold	1968	I-566
Seleucus	Moore, H. J.	1967	I-527
Sinus Iridum	Schaber, G. G.	1969	I-602
Theophilus	Milton, D. J.	1968	I-546
Timocharis	Carr, M. H.	1965	I-462

The following geologic maps exist as U.S. Geological Survey open-file reports. They are available on interlibrary loan from the U.S. Geological Survey Library, Washington, D.C.

Quadrangle	*Author*	*Date*
Byrgius	Trask, N. J.	1965
Geminus	Grolier, M. J.	1970
Grimaldi	McCauley, J. F.	1964
Langrenus	Ryan, J. D. & Wilhelms, D. E.	1965
Macrobius	Pohn, H. A.	1965
Mare Undarum	Masursky, Harold	1965
Petavius	Wilhelms, D. E.	1965
Purbach	Holt, H. E.	1965
Rümker	Eggleton, R. E. & Smith, E. I.	1967
Schickard	Karlstrom, T.N.V.	1971
Wilhelm	Saunders, R. S.	1970

The following geologic maps are in press, scheduled for publication in 1971 or early 1972

Quadrangle	*Author*	*Map Number*
Aristoteles	Lucchitta, B. K.	I-725
Clavius	Cummings, David	I-706
Cleomedes	Casella, C. J., & Binder, A. B.	I-707
Colombo	Elston, D. P.	I-714
Eudoxus	Scott, D. H.	I-705
Fracastorius	Stuart-Alexander, D. E. & Tabor, R. W.	I-720
Hommel	Mutch, T. A. & Saunders, R. S.	I-702
Maurolycus	Scott, D. H.	I-695
Plato	M'Gonigle, J. W. & Schleicher, D. L.	I-701
Rheita	Stuart-Alexander, D. E.	I-694
Rupes Altai	Rowan, L. C.	I-690
Schiller	Offield, T. W.	I-691
Taruntius	Wilhelms, D. E.	I-722
Tycho	Pohn, H. A.	I-713

AVAILABILITY OF NASA PHOTOGRAPHS

The National Space Science Data Center, Greenbelt, Maryland 20771, was established by NASA to further the widest possible use of data obtained from space science investigations. Complete indexes are available for pictures taken on all the Orbiter and Apollo flights. Photographs can be ordered either individually or in groups, as negatives or positives.

Appendix B

INDEX MAP OF THE MOON

(*See overleaf*)

SINUS RORIS
SINUS IRIDUM
MARE IMBRIUM
PLATO
PICO
GRUITHUISEN
ARCHIMEDES
TIMOCHARIS
SHROTER'S VALLEY
ARISTARCHUS
SELEUCUS
OCEANUS
MARIUS
GALILEO
KEPLER
PROCELLARUM
COPERNICUS
ERATOSTHENES
SURVEYOR 2
REINHOLD
SURVEYOR 3
APOLLO 12
MÖSTING
RICCIOLI
GRIMALDI
SURVEYOR 1
FLEMSTEED RING
FRA MAURO
EUCLIDES
MARE COGNITUM
RANGER 7
LETRONNE
PTOLEMAEUS
ALPHONSUS
CRÜGER
GASSENDI
MARE NUBIUM
STRAIGHT WALL
DARWIN
MARE ORIENTALE
BYRGIUS
CAVENDISH
MARE HUMORUM
REGIOMONTANUS
SCHICKARD
SURVEYOR 7
TYCHO
MAGINUS
SCHILLER
CLAVIUS
NEWTON

MARE HUMBOLDTIANUM
ARISTOTELES
ALPINE VALLEY
AGASSIZ PROMONTORY
MARE SERENITATIS
RIMAE SULPICIUS GALLUS
LITTROW
DAWES
PLINIUS
PROCLUS
MARE CRISIUM
MARE MARGINUS
MARE VAPORUM
MARE TRANQUILLITATIS
JULIUS CAESAR
RANGER 6
HYGINUS RILLE
RANGER 8
SINUS MEDII
CAYLEY
SURVEYOR 5
APOLLO 11
CENSORINUS
MARE SMYTHII
LUNA 16
MESSIER
MARE FECUNDITATIS
HORROCKS
HIPPARCHUS
TAYLOR
LANGRENUS
CAPELLA
APOLLO 16
THEOPHILUS
DESCARTES
MARE NECTARIS
TACITUS
CATHERINA
PETAVIUS
WILKINS
PICCOLOMINI
ROTHMANN
RABBI LEVI
NASIREDDIN
MAUROLYCUS
NICOLAI
JANSSEN
PITISCUS
HOMMEL
MUTUS

References

Adams, J. B. (1968), Lunar and Martian surfaces: petrologic significance of absorption bands in the near-infrared: *Science*, v. 159, no. 3822, pp. 1453–1455.

Akimoto, S., and S. Yasuhiko (1969), Coesite-stishovite transition: *Jour. Geophys. Research*, v. 74, no. 6, pp. 1653–1659.

Alfvén, H. (1963), The early history of the Moon and the Earth: *Icarus*, v. 1, no. 4, pp. 357–363.

———, and G. Arrhenius (1969), Two alternatives for the history of the Moon: *Science*, v. 165, no. 3888, pp. 11–17.

American Commission on Stratigraphic Nomenclature (1961), Code of stratigraphic nomenclature: *Bull. Am. Assoc. Petroleum Geol.*, v. 45, pp. 645–665.

Anderson, E. M. (1951), *The Dynamics of Faulting*, 2nd ed.: Edinburgh, Oliver and Boyd, 206 pp.

Arking, Albert (1967), Space science, *in* R. W. Fairbridge, ed., *Encyclopedia of Atmospheric Sciences and Astrogeology*: New York, Reinhold, pp. 913–933.

Arthur, D. W. G. (1954), The distribution of lunar craters: *Jour. British Astron. Assoc.*, v. 64, no. 3, pp. 127–132.

——— (1963), Selenography, *in* B. M. Middlehurst and G. P. Kuiper, eds., *The Solar System*, Vol. 4, *The Moon, Meteorites and Comets*: Chicago, Univ. of Chicago Press, pp. 57–89.

Baldwin, R. B. (1963), *The Measure of the Moon*: Chicago, Univ. of Chicago Press, 488 pp.

——— (1964), Lunar crater counts: *Astron. Jour.*, v. 69, no. 5, pp. 377–392.

——— (1968), Lunar mascons: another interpretation: *Science*, v. 162, pp. 1407–1408.

Barrell, Joseph (1927), On continental fragmentation, and the geologic bearing of the Moon's surficial features: *Am. Jour. Sci.*, 5th series, v. 13, pp. 283–314.

Barth, T. F. W. (1950), *Volcanic Geology, Hot Springs and Geysers of Iceland*: Carnegie Inst. Washington Publ. 587, 174 pp.

Beals, C. S. (1968), On the possibility of a catastrophic origin for the great arc of eastern Hudson Bay, *in* C. S. Beals and D. A. Shenstone, eds., *Science, History and Hudson Bay*: Pub. of Dept. of Energy, Mines, and Resources, Ottawa, pp. 985–999.

Beard, D. P. (1925), Coral origin of the lunar craters: *Popular Astron.*, v. 32, no. 1, pp. 74–75.

Bowen, N. L. (1928), *The Evolution of the Igneous Rocks*: Princeton, N.J., Princeton Univ. Press, 332 pp.

Boyd, F. R., Jr. (1957), Geology of the Yellowstone Rhyolite Plateau: Ph.D. thesis, Harvard University.

———, and J. L. England (1959), Quartz-coesite transition, *in* P. H. Abelson, ed., Ann. Rpt. Director Geophysical Laboratory: *Carnegie Inst. Washington Yearbook* 58, pp. 87–88.

Brown, Harrison (1960), The density and mass distribution of meteoritic bodies in the neighborhood of the Earth's orbit: *Jour. Geophys. Research*, v. 65, no. 6, pp. 1679–1683.

Bruton, R. H., K. J. Craig, and B. S. Yaplee (1959), The radius of the Earth and the parallax of the Moon from radar range measurements on the Moon: *Astron. Jour.*, v. 64, no. 8, p. 325.

Bucher, W. H. (1963), Cryptoexplosion structures caused from without or within the earth? ("astroblemes" or "geoblemes"?): *Am. Jour. Sci.*, v. 261, pp. 597–649.

Bullard, F. M. (1962), *Volcanoes: in History, in Theory, in Eruption*: Austin, Univ. of Texas Press, 441 pp.

Cameron, W. S. (1964), An interpretation of Schröter's Valley and other lunar sinuous rills: *Jour. Geophys. Research*, v. 69, pp. 2423–2430.

Campbell, M. J., B. T. O'Leary, and Carl Sagan (1969), Moon: two new mascon basins: *Science*, v. 164, no. 3885, pp. 1273–1275.

Carder, R. W. (1962), Lunar mapping on a scale of 1:1,000,000, *in* Zdenek Kopal, and Z. K. Mikhailov, eds., *The Moon*: London, Academic, pp. 117–129.

Carlson, R. H., and W. A. Roberts (1963), *Mass Distribution and Throwout Studies*: U.S. Atomic Energy Comm. Rept. PNE-217F, 144 pp.

Carr, M. H. (1965), Geologic map of the Timocharis region of the Moon: U.S. Geol. Survey Misc. Geol. Inv. Map I-462.

——— (1966), Geologic map of the Mare Serenitatis region of the Moon: U.S. Geol. Survey Misc. Geol. Inv. Map I-489.

Chao, E. C. T. (1967), Impact metamorphism, *in* Philip Abelson, ed., *Researches in Geochemistry*, Vol. II: New York, Wiley, pp. 204–233.

———, J. J. Fahey, Janet Littler, and D. J. Milton (1962), Stishovite, SiO_2, a very high pressure new mineral from Meteor Crater, Arizona: *Jour. Geophys. Research*, v. 67, no. 1, pp. 419–421.

———, E. M. Shoemaker, and B. M. Madsen (1960), First natural occurrence of coesite: *Science*, v. 132, no. 3421, pp. 220–222.

Chapman, C. R. (1968), Interpretation of the diameter-frequency relation for lunar craters photographed by Rangers VII, VIII and IX: *Icarus*, v. 8, no. 1, pp. 1–22.

———, and R. R. Haefner (1967), A critique of methods for analysis of the diameter-frequency relation for craters with special application to the Moon: *Jour. Geophys. Research*, v. 72, no. 2, pp. 549–557.

Cloud, P. E., Jr. (1968), Atmospheric and hydrospheric evolution on the primitive earth: *Science*, v. 160, no. 3829, pp. 729–735.

Code of stratigraphic nomenclature, *see* American Commission on Stratigraphic Nomenclature.

Cohenour, Robert, and B. J. Sharp (1968), The impact theory: asteroids and the Earth-Moon system: *Geoscience News*, v. 1, no. 2, pp. 9–11, 32–34.

Cox, A., G. B. Dalrymple, and R. R. Doell (1967), Reversals of the earth's magnetic field: *Scientific American*, v. 216, pp. 44–54.

Cummings, David (1968), Preliminary geologic map of the Clavius quadrangle of the Moon: U.S. Geol. Survey open-file report.

Dana, J. D. (1846), On the volcanoes of the Moon: *Am. Jour. Sci.*, v. 2, pp. 335–353.

Darwin, G. H. (1879), On the precession of a viscous spheroid and on the remote history of the earth: *Phil. Trans. Roy. Soc.*, Part II, v. 170, pp. 447–530.

Diehl, C. H. H., and G. H. S. Jones (1967*a*), The Snowball Crater, general background information: Suffield Tech. Note No. 187, Defense Research Board of Canada, 23 pp.

———, and ——— (1967*b*), The Snowball Crater, profile and ejecta pattern: Suffield Tech. Note No. 188, Defense Research Board of Canada, 30 pp.

Dietz, R. S. (1949), The meteoritic impact origin of the Moon's surface features: *Jour. Geology*, v. 54, no. 6, pp. 359–375.

Dobar, W. I. (1966), Simulated basalt and granite magma upwelled in vacuum: *Icarus*, v. 5, pp. 399–405.

Dodd, R. T., Jr., J. W. Salisbury, and V. G. Smalley (1963), Crater frequency and the interpretation of lunar history: *Icarus*, v. 2, pp. 466–480.

Eaton, G. P. (1964), Windborne volcanic ash: a possible index to polar wandering: *Jour. Geology*, v. 72, pp. 1–35.

Eggleton, R. E. (1963), Thickness of the Apenninian Series in the Lansberg region of the Moon, *in* Astrogeol. Studies Ann. Prog. Rept., August 1961–August 1962, U.S. Geol. Survey open-file report, pp. 19–31.

——— (1964), Preliminary geology of the Riphaeus quadrangle of the Moon and definition of the Fra Mauro Formation, *in* Astrogeol. Studies Ann. Prog. Rept., August 1962–July 1963, Pt. A: U.S. Geol. Survey open-file report, pp. 46–63.

El-Baz, Farouk (1968), Classification of lunar rilles: Geol. Soc. America Ann. Meeting Program, p. 88.

Elston, D. P. (1965), Preliminary geologic map of the Colombo quadrangle of the Moon: U.S. Geol. Survey open-file report.

Elston, W. E. (1965), Rhyolite ash-flow plateaus, ring-dike complexes, calderas, lopoliths, and Moon craters, *in* H. E. Whipple, ed., *Geological Problems in Lunar Research*: Annals of the New York Acad. of Sciences, v. 123, art. 2, pp. 817–842.

Evans, J. V., and T. Hagfors (1964), On the interpretation of radar reflections from the Moon: *Icarus*, v. 3, no. 2, pp. 151–160.

———, and ———, eds. (1968), *Radar Astronomy*: New York, McGraw-Hill, 620 pp.

Faul, Henry (1966), *Ages of Rocks, Planets, and Stars*: New York, McGraw-Hill, 109 pp.

Fauth, P. (1908), *Der Mond und Hörbigers Welteislehre*: Leipzig and Berlin.

——— (1964), *Mondatlas*: Bremen, Olbers-Gesellschaft.

Fielder, Gilbert (1961), *Structure of the Moon's Surface*: New York, Pergamon, 266 pp.

——— (1963), Lunar tectonics: *Geol. Soc. London Quart. Jour.*, v. 119, no. 473, pt. 1, pp. 65–69.

——— (1965), *Lunar Geology*: London, Butterworth, 184 pp.

——— (1965*a*), Distribution of craters on the lunar surface: *Monthly Notices Royal Astron. Soc.*, v. 129, no. 5-6, pp. 351–361.

———, and A. H. Marcus (1967), Further tests for randomness of lunar craters: *Monthly Notices Royal Astron. Soc.*, v. 136, pp. 1–10.

Firsoff, V. A. (1961), *Surface of the Moon*: London, Hutchinson, 128 pp.

Fisher, R. V. (1966), Mechanism of deposition from pyroclastic flows: *Am. Jour. Sci.*, v. 264, pp. 350–363.

French, B. M., and N. M. Short, eds. (1968), *Shock Metamorphism of Natural Materials*: Baltimore, Md., Mono Book, 644 pp.

Fuller, R. E. (1927), The closing phase of a fissure eruption: *Am. Jour. Sci.*, 5th series, v. 14, pp. 228–230.

Galilei, Galileo (1610), *Nuncius Sidereus*: Padua, Bartoluzzi.

Gault, D. E., and Ronald Greeley (1968), Estimated ages for three large lunar craters (Abstract): *American Geophys. Transactions*, v. 49, no. 1, p. 273.

———, W. L. Quaide, and V. R. Oberbeck (1968), Impact cratering mechanics and structures, *in* B. M. French and N. M. Short, eds., *Shock Metamorphism of Natural Materials*: Baltimore, Md., Mono Book, pp. 87–99.

Gehrels, Thomas, T. Coffeen, and D. Owings (1964), Wavelength dependence of polarization, III—The lunar surface: *Astron. Jour.*, v. 69, no. 10, pp. 826–852.

Gerstenkorn, H. (1955), Über Gezeitenreibung beim Zweikörperproblem: *Z. Astrophys.* v. 36, pp. 245–274.

Gilbert, G. K. (1893), The Moon's face, a study of the origin of its features: *Philos. Soc. Washington Bull.*, v. 12, pp. 241–292.

——— (1896), The origin of hypotheses, illustrated by the discussion of a topographic problem: *Science*, new series, v. 3, pp. 1–12.

Gilvarry, J. J. (1960), Origin and nature of lunar surface features: *Nature*, v. 188, no. 4754, pp. 886–891.

——— (1969), What are the Mascons? *Saturday Review*, v. 52, no. 23, 54–57.

Glasstone, Samuel, ed. (1950), *The Effects of Atomic Weapons*: Los Alamos, N.M., Los Alamos Sci. Lab., U.S. Atomic Energy Comm., 456 pp.

Goetz, A. F. H. (1968), Differential infrared lunar emission spectroscopy: *Jour. Geophys. Research*, v. 73, no. 4, pp. 1455–1466.

Gold, T. (1966), The Moon's surface, *in* W. N. Hess, D. H. Menzel, and J. A. O'Keefe, eds., *The Nature of the Lunar Surface, Proceedings of the 1965 IAU-NASA Symposium*: Baltimore, Md., Johns Hopkins, pp. 107–124.

——— (1969), Apollo 11 observations on a remarkable glazing phenomenon on the lunar surface: *Science*, v. 165, no. 3900, pp. 1345–1349.

Goodacre, Walter (1910), A new map of the Moon, 60″ in diameter, divided into 25 sections, each 13″ square, enclosed in portfolio with index: London.

Green, Jack (1962), The geosciences applied to lunar exploration, *in* Zdenek Kopal and Z. K. Mikhailov, eds., *The Moon*: London, Academic, pp. 169–257.

Gressly, Amanz (1838), Observations géologiques sur le Jura Soleurois (Part I, pp. 1–112): Schweizer. Gesell. gesamten Naturwiss. Neue Denkschr. (Soc. helvétique sci. nat. Nouv. Mem.), Band 2 (part 6).

Gruithuisen, F. v.P. (1924), *Entdeckung vieler deutlichen Spuren der Mondbewohner*: Nürnberg, Germany.

Hackman, R. J. (1961), Photointerpretation of the lunar surface: *Photogram. Eng.*, v. 27, no. 3, pp. 377–386.

——— (1962), Geologic map of the Kepler region of the Moon: U.S. Geol. Survey Misc. Geol. Inv. Map I-355.

——— (1964), Stratigraphy and structure of the Montes Apenninus quadrangle of the

Moon, *in* Astrogeol. Studies Ann. Prog. Rept., August 1962–July 1963, Pt. A: U.S. Geol. Survey open-file report, pp. 1–8.

——— (1966), Geologic map of the Montes Apenninus region of the Moon: U.S. Geol. Survey Misc. Geol. Inv. Map I-463.

Hackman, R. J., and A. C. Mason (1961), Engineer special study of the surface of the Moon: U.S. Geol. Survey Misc. Inv. Map I-351 (3 maps and expl.).

Hagfors, T. (1966), Review of radar observations of the Moon, *in* W. N. Hess, D. H. Menzel, and J. A. O'Keefe, eds., *The Nature of the Lunar Surface: Proceedings of the 1965 IAU-NASA Symposium*: Baltimore, Md., Johns Hopkins, pp. 229–240.

———, and J. V. Evans (1968), Radar studies of the Moon, *in* J. V. Evans and T. Hagfors, eds., *Radar Astronomy*: New York, McGraw-Hill, pp. 219–246.

Handy, R. L., and D. T. Davidson (1953), On the curious resemblance between fly ash and meteoritic dust: Proc. Iowa Acad. Sci., v. 60, pp. 373–379.

Hapke, Bruce (1968), Lunar surface: Composition inferred from optical properties: *Science*, v. 159, no. 3810, pp. 76–79.

Hartmann, W. K. (1963), Radial structures surrounding lunar basins, I—The Imbrium system: Arizona Univ. Lunar and Planetary Lab. Commun., v. 2, no. 24, pp. 1–15.

——— (1964), Radial structures surrounding lunar basins, II—Orientale and other systems, conclusions: Arizona Univ. Lunar and Planetary Lab. Commun., v. 2, no. 36, pp. 175–191.

——— (1965), Terrestrial and lunar flux of large meteorites in the last two billion years: *Icarus*, v. 4, no. 2, pp. 157–165.

——— (1966), Early lunar cratering: *Icarus*, v. 5, no. 4, pp. 406–418.

——— (1968), Lunar crater counts, VI—The young craters, Tycho, Aristarchus, and Copernicus: Arizona Univ. Lunar and Planetary Lab. Commun., v. 7, no. 3, pp. 145–156.

———, and G. P. Kuiper (1962), Concentric structures surrounding lunar basins: Arizona Univ. Lunar and Planetary Lab. Commun., v. 1, no. 13, pp. 51–66.

———, and F. G. Yale (1968), Lunar crater counts, IV—Mare Orientale and its basin system: Arizona Univ. Lunar and Planetary Lab. Commun., v. 7, no. 117, pp. 131–138.

Hess, W. N., D. H. Menzel, and J. A. O'Keefe (1966), *The Nature of the Lunar Surface, Proceedings of the 1965 IAU-NASA Symposium*: Baltimore, Md., Johns Hopkins, 320 pp.

Howard, K. A., and T. W. Offield (1968), Shatter cones at Sierra Madera, Texas: *Science*, v. 162, no. 3850, pp. 261–265.

Hunt, C. B. (1956), Cenozoic geology of the Colorado Plateau: U.S. Geol. Survey Prof. Paper 279, 99 pp.

Hunt, G. R., J. W. Salisbury and R. K. Vincent (1968), Lunar eclipse: infrared images and an anomaly of possible internal origin: *Science*, v. 162, no. 3850, pp. 252–254.

Hutton, James (1788), Theory of the Earth: *Roy. Soc. Edin. Trans.*, v. 1, pp. 209–304.

———(1795), *Theory of the Earth, With Proofs and Illustrations*, two volumes: Edinburgh, W. Creech.

Irving, E. (1964), *Paleomagnetism and its Application to Geological and Geophysical Problems*: New York, Wiley, 399 pp.

Katterfield, G. N. (1967), Types, ages and origins of lunar ring structures: statistical and comparative geologic approach: *Icarus*, v. 6, pp. 360–380.

Kay, Marshall, and E. H. Colbert (1965), *Stratigraphy and Life History*: New York, Wiley, 736 pp.

Khabakov, A. V. (1962), Characteristic features of the relief of the Moon: Basic problems of the genesis and sequence of development of lunar formations, *in* A. V. Markov, ed., *The Moon, A Russian View* (translated from the Russian edition of 1960 by Royer & Royer, Inc.): Chicago, Univ. of Chicago Press, pp. 247–303.

King, P. B. (1959), *The Evolution of North America*: Princeton, N.J., Princeton Univ. Press, 190 pp.

Koestler, Arthur (1959), *The Sleepwalkers—A History of Man's Changing Vision of the Universe*: New York, Macmillan, 624 pp.

Kopal, Zdenek (1966), *An Introduction to the Study of the Moon*: Dordrecht, Holland, D. Reidel, 464 pp.

———, and Z. K. Mikhailov, eds. (1962), *The Moon: Proceedings of the International Astronomical Union Symposium 14, Leningrad, 1960*: London, Academic, 571 pp.

———, Josef Klepesta and T. W. Rackham (1965), *Photographic Atlas of the Moon*: New York, Academic, 277 pp.

Kreiter, T. J. (1960), Dating lunar surface features by using crater frequencies: *Publ. Astron. Soc. Pacific*, v. 72, pp. 393–398.

Krumbein, W. C., and L. L. Sloss (1963), *Stratigraphy and Sedimentation*, 2nd. ed.: San Francisco, Freeman, 660 pp.

Kuiper, G. P. (1951), On the origin of the solar system, Chapt. 8 *in* J. A. Hynek, ed., *Astrophysics*, New York, McGraw-Hill, pp. 357–424.

———, ed. (1960), *Photographic Lunar Atlas*: Chicago, Univ. of Chicago Press.

——— (1960*a*), Orthographic atlas of the Moon—Supplement no. 1 to the *Photographic Lunar Atlas*, comp. by D. W. G. Arthur and E. A. Whitaker: Tucson, Univ. of Arizona Press, 30 pp.

——— (1966), Interpretation of Ranger VII records: Arizona Univ. Lunar and Planetary Lab. Commun., v. 4, no. 58, pp. 1–70.

———, R. G. Strom, and R. S. LePoole (1966), Interpretation of Ranger records, Pt. I, Sinuous rilles, sec. 3 of *Ranger VIII and IX*, Pt. 2—*Experimenters' Analyses and Interpretations*: Jet Propulsion Lab. Tech. Rept. 32-800, California Inst. Technology, pp. 199–210.

———, W. A. Whitaker, R. G. Strom, J. W. Fountain, and S. M. Larson (1967), Consolidated lunar atlas—Supplements no. 3 and 4 to the U.S.A.F. photographic lunar atlas: Arizona Univ. Lunar and Planetary Lab. Centr. no. 4.

Kulp, J. L. (1961), Geologic time scale: *Science*, v. 133, no. 3459, p. 1105–1114.

Kuno, Hisashi, T. Ishikaua, Y. Katsui, K. Yagi, M. Yamasaki, and S. Taneda (1964), Sorting of pumice and lithic fragments as a key to eruptive and emplacement mechanism: *Japanese Jour. Geology and Geography*, v. 35, pp. 223–238.

Lamar, D. L., and P. M. Merifield (1968), Cambrian fossils and origin of the Earth-Moon system: *Bull. Geol. Soc. Am.* v. 78, no. 11, pp. 1359–1368.

Larsen, E. S., Jr., and W. Cross (1956), Geology and petrology of the San Juan region, southwestern Colorado: U.S. Geol. Survey Prof. Paper 258, 303 pp.

Lenham, A. P. (1964), The distribution of lunar craters: *Jour. British Astron. Assoc.*, v. 74, no. 5, pp. 182–185.

Levin, Ellis, D. D. Viele, and L. B. Eldrenkamp (1968), The Lunar Orbiter missions to the Moon: *Scientific American*, v. 218, no. 5, pp. 58–78.

Lingenfelter, R. E., S. J. Peale, and G. Schubert (1968), Lunar rivers: *Science*, v. 161, pp. 266–269.

Longwell, C. R., ed. (1949), Sedimentary facies in geologic history: *Geol. Soc. America Mem.* 39, 171 pp.

Lunar Sample Preliminary Examination Team (1969), Preliminary examination of lunar samples from Apollo 11: *Science*, v. 165, no. 3899, pp. 1211–1227.

Lyubarski, K. A. (1960): *Missiles and Rockets*, v. 6, no. 9, p. 30.

McCauley, J. F. (1964), The stratigraphy of the Mare Orientale region of the Moon, *in* Astrogeol. Studies Ann. Prog. Rept., August 1962–July 1963, Pt. A: U.S. Geol. Survey open-file report, pp. 86–98.

——— (1967*a*), The nature of the lunar surface as determined by systematic geologic mapping, *in* S. K. Runcorn, ed., *Mantles of the Earth and Terrestrial Planets*: London, Interscience, pp. 431–460.

——— (1967*b*), Geologic map of the Hevelius region of the Moon: U.S. Geol. Survey Misc. Inv. Map I-491.

——— (1968*a*), Geologic results from the lunar precursor probes: *American Institute Aeronautics and Astronautics Jour.*, v. 6, no. 10, pp. 1991–1996.

——— (1968*b*), Preliminary photogeologic map of the Orientale basin region, *in* G. E. Ulrich, Advanced systems traverse research project report: U.S. Geol. Survey Interagency Rept., Astrogeology 7, pp. 32–33.

——— (1968*c*), Preliminary small-scale geologic map of the Marius Hills region, *in* T. N. V. Karlstrom, J. F. McCauley, and G. A. Swann, Preliminary exploration plan of the Marius Hills region of the Moon: U.S. Geol. Survey Interagency Report: Astrogeology 5, 42 pp.

——— (1969), Geologic map of the Alphonsus GA region of the Moon: U.S. Geol. Survey Misc. Inv. Map I-586 (RLC 15).

———, and Harold Masursky (1968), The bedded white sands at Meteor Crater, Arizona: Abstract of paper presented at 31st ann. meeting, Meteoritical Society.

McConnell, R. K., Jr., L. A. McClaine, D. W. Lee, J. R. Aronson, and D. U. Allen (1967), A model for planetary igneous differentiation: *Reviews of Geophysics*, v. 5, no. 2, pp. 121–171.

MacDonald, G. A., and J. P. Eaton (1964), Hawaiian volcanoes during 1955: U.S. *Geological Survey Bull.* 1171, 170 pp.

MacDonald, G. J. F. (1964), Origin of the Moon: past and future: *Science*, v. 145, no. 3635, pp. 881–890.

McGetchin, T. R. (1968), The Moses Rock Dike: geology, petrology, and mode of emplacement of a kimberlite-bearing breccia dike, San Juan County, Utah: Ph.D. thesis, California Institute of Technology.

Mackin, J. H. (1961), A stratigraphic section in the Yakima Basalt and the Ellensburg Formation in south-central Washington: Washington Div. Mines and Geology Rept. Inv. 19, 45 pp.

Mackin, J. H. (1969), Origin of lunar maria: *Bull. Geol. Soc. Am.* v. 80, no. 5, pp. 735–748.

Marcus, A. H. (1964), A stochastic model of the formation and survival of lunar craters, I—Distribution of diameter of clean craters: *Icarus*, v. 4, no. 5–6, pp. 460–472.

——— (1966*a*), A stochastic model of the formation and survival of lunar craters, II—Approximate distribution of diameter of all observable craters: *Icarus*, v. 5, no. 2, pp. 165–177.

——— (1966*b*), A stochastic model of the formation and survival of lunar craters, III—Filling and disappearance of craters: *Icarus*, v. 5, no. 2, pp. 178–189.

——— (1966*c*), A stochastic model of the formation and survival of lunar craters, IV—On the randomness of crater centers: *Icarus*, v. 5, no. 2, pp. 190–200.

——— (1966*d*), A stochastic model of the formation and survival of lunar craters, V—Approximate diameter distribution of primary and secondary craters: *Icarus*, v. 5, no. 6, pp. 590–605.

——— (1967), A stochastic model of the formation and survival of lunar craters, VI—Initial depth, distribution of depths, and lunar history: *Icarus*, v. 6, no. 1, pp. 56–74.

Markov, A. V., ed. (1962), *The Moon, a Russian View* (translated from the Russian edition of 1960 by Royer & Royer, Inc.): Chicago, Univ. of Chicago Press, 391 pp.

Marshall, C. H. (1963), Geologic map of the Letronne region of the Moon: U.S. Geol. Survey Misc. Inv. Map I-385.

Mason, C. C. (1969), Particle size distribution of lunar surface material: *Bull. Geol. Soc. Am.*, v. 80, pp. 587–594.

Masursky, Harold (1964), A preliminary report on the role of isostatic rebound in the geologic development of the lunar crater, Ptolemaeus, *in* Astrogeol. Studies Ann. Prog. Rept., July 1963–July 1964, Pt. A: U.S. Geol. Survey open-file report, pp. 102–134.

Matumoto, Tadaiti (1943), The four gigantic caldera volcanoes of Kyushu: *Japanese Jour. Geology and Geography*, v. 19, spec. no., 57 pp.

Maxwell, J. C. (1959), Turbidite, tectonic and gravity transport, northern Apennine Mountains, Italy: *Bull. Amer. Assoc. Petroleum Geologists*, v. 43, no. 11, pp. 2701–2719.

Middlehurst, B. M. (1967), An analysis of lunar events: *Reviews of Geophys.*, v. 5, pp. 173–189.

Miller, T. G. (1965), Time in stratigraphy: *Paleontology*, v. 8, no. 1, pp. 113–131.

Mills, G. A. (1968), Absolute coordinates of lunar features, II: *Icarus*, v. 8, no. 1, pp. 90–116.

Milton, D. J. (1968), Geologic map of the Theophilus quadrangle of the Moon: U.S. Geol. Survey Misc. Geol. Inv. Map I-546.

Minnaert, M. (1961), Photometry of the Moon, *in* G. P. Kuiper and B. M. Middlehurst, eds., *The Solar System*, Vol. 3, *Planets and Satellites*: Chicago, Univ. of Chicago Press, pp. 213–245.

Moore, H. J. (1965), Geologic map of the Aristarchus region of the Moon: U.S. Geol. Survey Misc. Geol. Inv. Map I-465.

——— (1966), Craters produced by missile impact, *in* Astrogeol. Studies Ann. Prog. Rept., July 1965–July 1966, Pt. B., Crater Investigations: U.S. Geol. Survey open-file report, pp. 79–106.

——— (1967), Geologic map of the Seleucus quadrangle of the Moon: U.S. Geol. Survey Misc. Geol. Inv. Map I-527.

Moore, J. G. (1967), Base surge in recent volcanic eruptions: *Bull. Volcanologique*, v. 30, pp. 337–363.

Morris, E. C., and D. E. Wilhelms (1967), Geologic map of the Julius Caesar quadrangle of the Moon: U.S. Geol. Survey Misc. Geol. Inv. Map I-510.

Müller, G., and G. Veyl (1957), The birth of Nilahue, a new maar type volcano at Rininalue, Chile: Twentieth Internat. Geol. Cong. Rept., Sec. 1, pp. 375–396.

Muller, P. M., and W. L. Sjogren (1968), Mascons: lunar mass concentrations: *Science*, v. 161, no. 3842, pp. 680–684.

Murray, J. (1876), On the distribution of volcanic debris over the floor of the ocean: *Proc. Royal. Soc. Edinburgh*, v. 9, pp. 247–261.

Mutch, T. A. and R. S. Saunders (1968), Preliminary geologic map of the Hommel quadrangle of the Moon: U.S. Geol. Survey open-file report.

Newell, N. D. (1966), Problems in geochronology: *Acad. Nat. Sci. Philadelphia Proc.*, v. 118, no. 3, pp. 63–89.

Nininger, H. H. (1956), *Arizona's Meteor Crater*: Denver, Colo., American Meteorite Laboratory, 232 pp.

Nordyke, M. D., and M. M. Williamson (1965), *The Sedan Event*: U.S. Atomic Energy Comm. Rept. PNE-242F, 103 pp.

Oberbeck, V. R., and W. L. Quaide (1968), Genetic implications of lunar regolith thickness variations: *Icarus*, v. 9, pp. 446–465.

Ocampo, S. (1949), El origen de los circos lunares: *Revista Astronómica* (Buenos Aires), pp. 76–86.

Offield, T. W. (1966), Structure of the Triesnecker-Hipparchus region, *in* Astrogeol. Studies Ann. Prog. Rept., July 1965–July 1966, Pt. A: U.S. Geol. Survey open-file report, pp. 133–154.

——— (1968), Preliminary geologic map of the Schiller quadrangle of the Moon: U.S. Geol. Survey open-file report.

———, and H. A. Pohn (1969), Lunar crater morphology and relative age determination of lunar geologic units, Part 2, applications: U.S. Geol. Survey Interagency Rept., Astrogeology 13, 35 pp.

O'Keefe, J. A. (1966), Lunar ash flows, *in* W. N. Hess, D. H. Menzel, and J. A. O'Keefe, eds., *The Nature of the Lunar Surface: Proceedings of the 1965 IAU-NASA Symposium*: Baltimore, Md., Johns Hopkins, pp. 259–266.

——— (1968), Isostasy on the Moon: *Science*, v. 162, pp. 1405–1406.

Opik, E. J. (1960), The lunar surface as an impact counter: *Monthly Notices Royal Astron. Soc.*, v. 120, no. 5, pp. 404–411.

Ovechkin, N. K., L. S. Librovich, Bobkova, N. N. (1961), Stratigraphic classification and terminology: Rept. of the XXI Session, Int. Geol. Congress, Part XXV, pp. 31–33.

Patterson, C., R. G. Tilton, and M. G. Inghram (1955), Age of the Earth: *Science*, v. 121, no. 3134, pp. 69–75.

Peterson, N. V., and E. A. Groh, eds. (1965), *State of Oregon Lunar Geological Field Conference Guide Book*: State of Oregon Dept. of Geology and Mineral Industries, 51 pp.

Pettengill, G. H., and T. W. Thompson (1968), A radar study of the lunar crater Tycho at 3.8 cm and 70 cm wavelengths: *Icarus*, v. 8, no. 3, pp. 457–471.

Pickering, W. H. (1903), *The Moon*: New York, Doubleday, Page, 103 pp.

Playfair, John (1802), *Illustrations of the Huttonian Theory of the Earth*: Edinburgh, W. Creech.

Pohn, H. A. (1963), New measurements on steep lunar slopes: *Publ. Astron. Soc. Pacific*, v. 75, no. 443, pp. 186–187.

———, and T. W. Offield (1969), Lunar crater morphology and relative age determination of lunar geologic units: U.S. Geological Survey Interagency Rept., Astrogeology 13, 35 pp.

Quaide, W. L. (1965), Rilles, ridges and domes—clues to maria history: *Icarus*, v. 4, pp. 374–389.

———, and V. R. Oberbeck (1968), Thickness determinations of the lunar surface layer from lunar impact craters: *Jour. Geophys. Research*, v. 73, no. 6, pp. 5247–5270.

Rackham, T. W. (1967), A comparison of lunar photography from space probes and ground-based observatories: *Icarus*, v. 6, pp. 440–443.

Ranger VII Experimenters' Analyses and Interpretations (1965): Tech. Rept. No. 32–700, Pt. II, Jet Propulsion Lab., Pasadena, Calif.

Ranger VIII and IX Experimenters' Analyses and Interpretations (1966): Tech. Rept. No. 32-800, Pt. II, Jet Propulsion Lab., Pasadena, Calif.

Ray, R. G. (1960), Aerial photographs in geologic interpretation and mapping: U.S. Geol. Survey Prof. Paper 373, 230 pp.

Richards, W. D. (1964), *Geologic Study of the Sedan Nuclear Crater*: U.S. Atomic Energy Comm. Rept. PNE-240F, 43 pp.

Rittmann, A. (1939), Threngslaborgir-line isländische Eruptions-Spalte am Myvatn: *Natur u. Volk*, v. 69, pp. 275–289.

——— (1962), *Volcanoes and Their Activity* (translated from the second German edition by E. A. Vincent): New York, Wiley, 305 pp.

Roberts, G. L. (1968), Photoelectric scanning of the Moon: *Icarus*, v. 9, no. 2, pp. 253–280.

Roberts, W. A. (1964), Notes on the importance of shock crater lips to lunar exploration: *Icarus*, v. 3, no. 4, pp. 342–347.

——— (1967), Stratigraphic chronology—a problem in extraterrestrial manned geological exploration: *Icarus*, v. 6, pp. 427–433.

Roddy, D. J. (1968), The Flynn Creek Crater, Tennessee, *in* B. M. French and N. M. Short, eds., *Shock Metamorphism of Natural Materials*: Baltimore, Md., Mono Book, pp. 291–322.

Ronca, L. B. (1966), An introduction to the geology of the Moon: *Proc. Geol. Assoc. London*, v. 77, pp. 101–126.

Rowan, L. C. (1965), Preliminary geologic map of the Rupes Altai quadrangle of the Moon: U.S. Geol. Survey open-file report.

Runcorn, S. K. (1962), Convection in the Moon: *Nature*, v. 195, pp. 1150–1151.

Ruskol, E. L. (1962), The origin of the Moon, *in* Zdenek Kopal and Z. K. Mikhailov, eds., *The Moon*: London, Academic, pp. 149–156.

Saari, J. M., and R. W. Shorthill (1966), Infrared and visual images of the eclipsed Moon of December 19, 1964: *Icarus*, v. 5, no. 6, pp. 635–659.

Sagan, Carl (1961), Organic matter and the Moon: Natl. Acad. Sci.—Natl. Res. Council, Publ. 757, 49 pp.

Saunders, R. S. (1968), Problems for geologic investigations of the Orientale region of the

Moon, *in* G. E. Ulrich, Advanced systems traverse research project report, U.S. Geol. Survey Interagency Rept., Astrogeology 7, pp. 30–55.

———, and T. A. Mutch (1967), Preliminary geologic map of ellipse Orb. III-9-5 and vicinity: U.S. Geol. Survey open-file report.

Schmidt, J. F. J. (1878), *Die Charte der Gebirge des Montes*: Berlin, Dietrich Reimer.

Schmidt, R. A., and K. Keil (1966), Electron microprobe study of spherules from Atlantic Ocean sediments: *Geochim. et Cosmochim. Acta*, v. 30, pp. 471–478.

Schmincke, H. U. (1967), Stratigraphy and petrography of four upper Yakima Basalt flows in south-central Washington: *Bull. Geol. Soc. Am.*, v. 78, pp. 1385–1422.

Schmitt, H. H., N. J. Trask, and E. M. Shoemaker (1967), Geologic map of the Copernicus quadrangle of the Moon: U.S. Geol. Survey Misc. Geol. Inv. Map I-515.

Schumm, S. A. (1969), Experimental studies on the formation of lunar surface features by gas emission—a preliminary report: U.S. Geological Survey Interagency Rept., Astrogeology 16, 22 pp.

———, and D. B. Simons (1969), Lunar rivers or coalesced crater chains? *Science*, v. 165, no. 3889, p. 201.

Shoemaker, E. M. (1960), Penetration mechanics of high velocity meteorites, illustrated by Meteor Crater, Arizona: Intern. Geol. Cong. XXI Session, v. 18, pp. 418–434.

——— (1962), Interpretation of lunar craters, *in* Zdenek Kopal, ed., *Physics and Astronomy of the Moon*: London, Academic, pp. 283–359.

——— (1964), The Geology of the Moon: *Scientific American*, v. 211, no. 6, pp. 38–47.

——— (1966), Progress in the analysis of the fine structure and geology of the lunar surface from Ranger VIII and IX photographs, *in Ranger VIII and IX Experiments' Analyses and Interpretations* JPL Tech. Rept. No. 32-800, Part 2, pp. 275–284.

———, R. M. Batson, H. E. Holt, E. C. Morris, J. J. Rennilson, and E. A. Whitaker (1968), Television observations from Surveyor III: *Jour. Geophys. Research*, v. 73, no. 12, pp. 3989–4044.

———, ———, ———, ———, ———, and ——— (in press), Observations of the lunar regolith and the Earth from the television camera on Surveyor VII: *Jour. Geophys. Research.*

———, and R. J. Hackman (1962), Stratigraphic basis for a lunar time scale, *in* Zdenek Kopal and Z. K. Mikhailov, eds., *The Moon*: London, Academic, pp. 289–300.

———, ———, and R. E. Eggleton (1962), Interplanetary correlation of geologic time, *in Advances in the Astronautical Sciences*, Vol. 8: New York, Plenum, pp. 70–89.

———, ———, ———, and C. H. Marshall (1962), Lunar stratigraphic nomenclature, *in* Astrogeol. Studies Semiann. Prog. Rept., February 1961–August 1961: U.S. Geol. Survey open-file report, p. 114–116.

Short, N. M. (1965), A comparison of features characteristic of nuclear explosion craters and astroblemes, *in* H. E. Whipple, ed., *Geological Problems in Lunar Research*, Annals of New York Acad. Sci., v. 123, art. 2, pp. 573–616.

——— (1966*a*), Shock-lithification of unconsolidated rock materials: *Science*, v. 154, pp. 382–384.

——— (1966*b*), Shock processes in geology: *Jour. Geol. Education*, v. 14, pp. 149–166.

Shorthill, R. W., and J. M. Saari (1966), Recent discoveries of hot spots on the lunar surface, *in* W. N. Hess, D. H. Menzel, and J. A. O'Keefe, eds., *The Nature of the Lunar Surface: Proceedings of the 1965 IAU-NASA Symposium*: Baltimore, Md., Johns Hopkins, pp. 215–228.

Simon, Ivan (1966), *Infrared Radiation*: Princeton, N.J., Van Nostrand, 119 pp.

Simpson, G. G. (1963), Historical science, *in* C. C. Albritton, Jr., ed., *The Fabric of Geology*: Reading, Mass., Addison-Wesley, pp. 24–48.

Smith, R. L. (1960*a*), Ash flows: *Bull. Geol. Soc. Am.*, v. 71, pp. 795–842.

——— (1960*b*), Zones and zonal variations in welded ash flows: U.S. Geol. Survey Prof. Paper 354-F, 10 pp.

——— (1966), Terrestrial calderas, associated pyroclastic deposits, and possible lunar applications, *in* W. N. Hess, D. H. Menzel, and J. A. O'Keefe, eds., *The Nature of the Lunar Surface: Proceedings of the 1965 IAU-NASA Symposium*: Baltimore, Md., Johns Hopkins, pp. 241–258.

Smith, William (1820), A new geological map of England and Wales with the inland navigations, exhibiting the districts of coal: London, J. Carey.

Spurr, J. E. (1944), *Geology Applied to Selenology*, Vol. I, *The Imbrian Plain Region of the Moon*: Lancaster, Pa., Science Press, 112 pp.

——— (1945), *Geology Applied to Selenology*, Vol. II, *The Features of the Moon*: Lancaster, Pa., Science Press, 318 pp.

——— (1948), *Geology Applied to Selenology*, Vol. III, *Lunar Catastrophic History*: Concord, N.H., Rumford, 253 pp.

——— (1949), *Geology Applied to Selenology*, Vol. IV, *The Shrunken Moon*: Concord, N.H., Rumford.

Stearns, H. T., and G. MacDonald (1946), Geology and ground-water resources of the island of Hawaii: Bull. 9, Hawaii Division of Hydrography, 363 pp.

Strom, R. G. (1964), Analysis of lunar lineaments, I—Tectonic map of the moon: Arizona Univ. Lunar and Planetary Lab. Commun., v. 2, no. 39, pp. 205–221.

Surveyor I Mission Report (1966), Part II, Scientific data and results: Tech. Rept. 32-1023, Jet Propulsion Lab., Pasadena, Calif.

Surveyor III Mission Report (1967*a*), Part II, Scientific results: Tech. Rept. 32-1177, Jet Propulsion Lab., Pasadena, Calif.

Surveyor V Mission Report (1967*b*), Part II, Science results: Tech. Rept. 32-1246, Jet Propulsion Lab., Pasadena, Calif.

Surveyor VI Mission Report (1968*a*), Part II, Science results: Tech. Rept. 32-1262, Jet Propulson Lab., Pasadena, Calif.

Surveyor VII Mission Report (1968*b*), Part II, Science results: Tech. Rept. 32-1264, Jet Propulsion Lab., Pasadena, Calif.

Surveyor Project Final Report (1968*c*), Part II, Science results: Tech. Rept. 32-1265, Jet Propulsion Lab., Pasadena, Calif.

Surveyor Program, OSSA, *compilers* (1966), Surveyor I, a preliminary report: NASA Spec. Pub. 126, Washington, D.C., 39 pp.

Surveyor Program, OSSA, *compilers* (1967*a*), Surveyor III, a preliminary report: NASA Spec. Pub. 146, Washington, D.C., 159 pp.

Surveyor Program, OSSA, *compilers* (1967*b*), Surveyor V, a preliminary report: NASA Spec. Pub. 163, Washington, D.C., 161 pp.

Surveyor Program, OSSA, *compilers* (1968*a*), Surveyor VI, a preliminary report: NASA Spec. Pub. 166, Washington, D.C., 165 pp.

Surveyor Program, OSSA, *compilers* (1968*b*), Surveyor VII, a preliminary report: NASA Spec. Pub. 173, Washington, D.C., 303 pp.

Thom, T. W., Jr. (1923), The relation of deep-seated faults to the surface structural features of central Montana: *Bull. Am. Assoc. Petroleum Geol.*, v. 7, pp. 1–13.

Thompson, T. W., and R. B. Dyce (1966), Mapping of radar reflectivity at 70 centimeters: *Jour. Geophys. Research*, v. 71, no. 20, pp. 4843–4853.

Titley, S. R. (1967), Geologic map of the Mare Humorum region of the Moon: U.S. Geol. Survey Misc. Geol. Inv. Map I-495.

———, and R. E. Eggleton (1964), Description of an extensive hummocky deposit around the Humorum basin, *in* Astrogeol. Studies Ann. Prog. Rept., July 1963–July 1964, Pt. A: U.S. Geol. Survey open-file report, pp. 85–89.

Trask, N. J. (1965), Preliminary report on the geology of the Byrgius quadrangle of the Moon, *in* Astrogeol. Studies Ann. Prog. Rept., July 1964–July 1965, Pt. A: U.S. Geol. Survey open-file report, pp. 1–8.

——— (1966), Size and spatial distribution of craters estimated from the Ranger photographs, in *Ranger VIII and IX, Pt. II, Experimenters' Analyses and Interpretations*: JPL Tech. Rept. No. 32-800, pp. 252–262.

———, and L. C. Rowan (1967), Lunar Orbiter photographs: some fundamental observations: *Science*, v. 158, no. 3808, pp. 1529–1535.

Tsuya, H., I. Murai, and Y. Hosoya (1958), Size characteristics of the pumice deposits distributed in the vicinity of Komoro on the southwest foot of Asana volcano: Tokyo Univ. Earthquake Research Inst. Bull., v. 36, pt. 3, pp. 413–429.

Turkevich, A. L., E. J. Franzgrote, and J. H. Patterson (1968*a*), Chemical analysis of the Moon at the Surveyor VI landing site: preliminary results: *Science*, v. 160, no. 3832, pp. 1108–1110.

——— (1968*b*), Chemical analysis of the Moon at the Surveyor VII landing site: preliminary results: *Science*, v. 162, no. 3849, pp. 117–118.

——— (1969), Chemical composition of the lunar surface in Mare Tranquillitatis: *Science*, v. 165, no. 3890, pp. 277–279.

Tyler, G. L., 1968, Oblique-scattering radar reflectivity of the lunar surface: preliminary results from Explorer 35: *Jour. Geophys. Research*, v. 73, no. 24, pp. 7609–7620.

Tyrrell, G. W. (1937), Flood basalts and fissure eruption: *Bull. Volcanologique*, ser. 2, no. 1, pp. 89–111.

U.S. Geological Survey (1963), Astrogeologic Studies Annual Progress Report, August 25, 1961–August 24, 1962, Pt. A, Lunar and planetary investigations; Pt. B, Crater investigations; Pt. C, Cosmochemistry and petrography; Pt. D, Studies for space flight program: U.S. Geol. Survey open-file report, 4 vols.

——— (1964*a*), Astrogeologic Studies Annual Progress Report, August 25, 1962–July 1, 1963, Pt. A, Lunar and planetary investigation; Pt. B, Crater and solid-state investigations;

Pt. C, Cosmochemistry and petrography; Pt. D, Studies for space flight program: U.S. Geol. Survey open-file report, 4 vols.

——— (1964*b*), Astrogeologic Studies Annual Progress Report, July 1, 1963–July 1, 1964, Pt. A, Lunar and planetary investigations; Pt. B, Crater and solid-state investigations; Pt. C, Cosmic chemistry and petrology: U.S. Geol. Survey open-file report, 3 vols.

——— (1965), Astrogeologic Studies Annual Progress Report, July 1, 1964–July 1, 1965, Pt. A, Lunar and planetary investigations; Pt. B, Crater investigations; Pt. C, Cosmic chemistry and petrology: U.S. Geol. Survey open-file report, 3 vols.

——— (1966), Astrogeologic Studies Annual Progress Report, July 1, 1965–July 1, 1966, Pt. A, Lunar and planetary investigations; Pt. B, Crater investigations; Pt. C, Cosmic chemistry and petrology; Pt. D, Space flight investigations: U.S. Geol. Survey open-file report, 4 vols.

Urey, H. C. (1962), The origin of the Moon and its relationship to the origin of the solar system, *in* Zdenek Kopal and Z. K. Mikhailov, eds., *The Moon*, New York, Academic, pp. 133–148.

——— (1968), Mascons and the history of the Moon: *Science*, v. 162, pp. 1408–1410.

——— (1969), The contending moons: *Astronautics and Aeronautics*, v. 7, no. 1, pp. 37–41.

Van Houten, F. B. (1952), Sedimentary record of Cenozoic orogenic and erosional events, Big Horn Basin, Wyo.: Wyoming Geol. Assoc. 7th Ann. Field Conf. Guidebook, pp. 74–79.

Velikovsky, Immanuel (1950), *Worlds in Collision*: New York, Doubleday, 401 pp.

von Bandat, H. F. (1962), *Aerogeology*: Houston, Texas, Gulf, 350 pp.

Wadia, D. N. (1953), *Geology of India*, 3rd. ed.: London, MacMillan, 531 pp.

Waters, A. C. (1961), Stratigraphic and lithologic variations in the Columbia River basalt: *Am. Jour. Sci.*, v. 259, pp. 583–611.

——— (1967), Moon craters and Oregon volcanoes: 1967 Condon Lecture, Oregon State System of Higher Education, 70 pp.

Watson, Kenneth (1968), Photoclinometry from spacecraft images: U.S. Geol. Survey Prof. Paper 599B, 10 pp.

Wells, A. K. and J. F. Kirkaldy (1966), *Outline of Historical Geology*: London, Thomas Murby, 503 pp.

Wells, J. W. (1963), Coral growth and geochronometry: *Nature*, v. 197, pp. 948–950.

Wentworth, C. K., and G. A. MacDonald (1953), Structures and forms of basaltic rocks in Hawaii: U.S. Geol. Survey Bull. 994, 98 pp.

Wheeler, H. E., and H. A. Coombs (1967), Late Cenozoic Mesa Basalt sheet in northwestern United States: *Bull. Volcanologique*, v. 31, pp. 21–44.

Whipple, F. L. (1963), *Earth, Moon and Planets*, rev. ed.: Cambridge, Mass., Harvard Univ. Press, 278 pp.

Whipple, H. E., ed. (1965), *Geological Problems in Lunar Research*, Annals of New York Academy of Sciences, v. 123, art. 2, 1257 pp.

Whitaker, E. A. (1966), The surface of the Moon, *in* W. N. Hess, D. H. Menzel, and J. A. O'Keefe, eds., *The Nature of the Lunar Surface: Proceedings of the 1965 IAU-NASA Symposium*: Baltimore, Md., Johns Hopkins, pp. 79–98.

———, G. P. Kuiper, W. K. Hartmann, and L. H. Spradley, eds. (1964), *Rectified Lunar*

Atlas, Supplement no. 2 to the *Photographic Lunar Atlas*: Tucson, Univ. of Arizona Press, 30 pp.

Wilhelms, D. E. (in press), Summary of lunar stratigraphy—telescopic observations: U.S. Geol. Survey Prof. Paper 599.

——— (1965), Fra Mauro and Cayley Formations in the Mare Vaporum and Julius Caesar quadrangles, *in* Astrogeol. Studies Ann. Prog. Rept., July 1964–July 1965, Pt. A: U.S. Geol. Survey open-file report, pp. 13–28.

———, and N. J. Trask (1965), Polarization properties of some lunar geologic units, *in* Astrogeologic Studies Ann. Prog. Rept., July 1964–July 1965, Pt. A: U.S. Geol. Survey open-file report, pp. 63–80.

Wilkins, H. P. (1933), Great 200″ map of the Moon: Linwood, 22 Bradford St., Llanelly, Im Selbstverlag.

———, and Patrick Moore (1955), *The Moon*: New York, MacMillan, 388 pp.

Williams, Howell (1941), Calderas and their origin: Univ. Calif. Pub. Bull. Dept. Geol. Sci., v. 25, no. 6, pp. 239–346.

Wilshire, H. G., and K. A. Howard (1968), Structural patterns in central uplifts of crypto-explosive structures as typified by Sierra Madera: *Science*, v. 162, no. 3850, pp. 258-261.

Wise, D. U. (1963), An origin of the Moon by rotational fission during formation of the Earth's core: *Jour. Geophys. Research*, v. 68, pp. 1547–1554.

Wise, W. S. (1969), Geology and petrology of the Mt. Hood area: a study of High Cascade volcanism: *Bull. Geol. Soc. Am.*, v. 80, no. 6, pp. 969–1006.

Young, G. A. (1965), *The Physics of the Base Surge*: Silver Spring, Md., U.S. Naval Ordinance Lab., 284 pp.

Young, James (1933), Preliminary report of a statistical investigation of the diameters of lunar craters: *Jour. British Astron. Assoc.*, v. 3, no. 5, pp. 201–209.

Zeller, E. J., and L. B. Ronca (1967), Space weathering of lunar and asteroidal surfaces: *Icarus*, v. 7, no. 3, pp. 372–379.

SUPPLEMENTARY REFERENCES

Adams, J. B., and T. B. McCord (1971), Alteration of lunar optical properties: age and composition effects: *Science*, v. 171, no. 3971, pp. 567–571.

Adler, I., J. Trombka, J. Gerard, P. Lowman, L. Yin, and H. Blodgett (1972), Preliminary results from the S-161 X-ray fluorescence experiment: Proceedings, 1972 Lunar Science Conf., Jan. 10-13, Houston, Texas.

Albee, A. L., D. S. Burnett, A. A. Chodos, E. L. Haines, J. C. Huneke, D. A. Papanastassiou, F. A. Podosek, G. P. Russ III, and G. J. Wasserburg (1970), Mineralogic and isotopic investigations of lunar rock 12013: *Earth Planetary Sci. Letters*, v. 9, pp. 137–163.

Anderson, A. T., Jr., A. V. Crewe, J. R. Goldsmith, P. B. Moore, J. C. Newton, E. J. Olsen, J. V. Smith, and P. J. Wyllie (1970), Petrologic history of Moon suggested by petrography, mineralogy, and crystallography: *Science*, v. 167, no. 3918, pp. 587–590.

Baldwin, R. B. (1968), Lunar mascons: another interpretation: *Science*, v. 162, no. 3860, pp. 1407–1408.

——— (1969), Asbolute ages of the lunar maria and large craters: *Icarus*, v. 11, pp. 320–331.

——— (1970*a*), Absolute ages of the lunar maria and large craters II. The viscosity of the Moon's outer layers: *Icarus*, v. 13, pp. 215–225.

——— (1970*b*), Summary of arguments for a hot Moon: *Science*, v. 170, no. 3964, pp. 1297–1300.

Buddington, A. F. (1939), Adirondack igneous rocks and their metamorphism: *Geol. Soc. Amer. Mem. 7.*

Carr, M. H. (1972), Sketch map of the region around candidate Littrow landing sites: Apollo XV Prelim. Science Rept., NASA Spec. Pub., in press.

Cassidy, W. A., B. P. Glass, and B. C. Heezen (1969), Physical and chemical properties of Australasian microtektites: *Jour. Geophys. Research*, v. 74, no. 4, pp. 1008–1025.

Chao, E.C.T., O. B. James, J. A. Minkin, J. A. Boreman, E. D. Jackson, and C. B. Raleigh (1970), Impact metamorphic effects in lunar samples from Tranquillity Base: *Geochim. Cosmochim. Acta*, Suppl. 1, v. 34, pp. 287–314.

Compston, W., B. W. Chappell, P. A. Arriens, and M. J. Vernon (1970), The chemistry and age of Apollo 11 lunar material: *Geochim. Cosmochim. Acta*, Suppl. 1, v. 34, pp. 1007–1027.

Conel, J. E., and D. B. Nash (1970), Spectral reflectance and albedo of Apollo 11 lunar samples: effects of irradiation and vitrification and comparison with telescopic observations: *Geochim. Cosmochim. Acta*, Suppl. 1, v. 34, pp. 2013–2024.

Daly, R. A. (1933), *Igneous Rocks and the Depths of the Earth*: New York, McGraw-Hill, 598 pp.

Duke, M. B., and L. T. Silver (1967), Petrology of eucrites, howardites, and mesosiderites: *Geochim. Cosmochim. Acta*, v. 31, pp. 1637–1666.

Eggleton, R. E. (1965), Geologic map of the Riphaeus Mountains region of the Moon: U.S. Geol. Survey Misc. Geol. Inv. Map I-458.

Fleischer, R. L., E. L. Haines, R. E. Hanneman, H. R. Hart, Jr., J. S. Kasper, E. Lifshin, R. T. Woods, and P. B. Price (1970), Particle track X-ray, thermal, and mass spectrometric studies of lunar material: *Science*, v. 167, no. 3918, pp. 568–571.

Frondel, C., C. Klein, Jr., J. Ito, and J. C. Drake (1970), Mineralogical and chemical studies of Apollo 11 lunar fines and selected rocks: *Geochim. Cosmochim. Acta*, Suppl. 1, v. 34, pp. 445–474.

Ganapathy, R., R. R. Keays, J. C. Laul, and E. Anders (1970*a*), Trace elements in Apollo 11 lunar rocks: implications for meteorite influx and origin of moon: *Geochim. Cosmochim. Acta*, Suppl. 1, v. 34, pp. 1117–1142.

———, R. R. Keays, and E. Anders (1970*b*), Apollo 12 lunar samples: trace elements analysis of a core and the uniformity of the regolith: *Science*, v. 170, no. 3957, pp. 533–535.

Gast, P. W., N. J. Hubbard, and H. Wiesmann (1970), Chemical composition and petrogenesis of basalts from Tranquillity Base: *Geochim. Cosmochim. Acta*, Suppl. 1, v. 34, pp. 1143–1164.

Gault, D. E. (1970), Saturation and equilibrium conditions for impact cratering on the lunar surface: criteria and implications: *Radio Science*, v. 5, no. 2, pp. 273-291.

Gault, D. E., E. M. Shoemaker, and H. J. Moore (1963), Spray ejected from the lunar surface by meteroid impact: NASA Tech. Note D-1767.

Gold, Thomas, and Steven Soter (1970), Apollo 12 seismic signal: indication of a deep layer of powder: *Science*, v. 169, no. 3950, pp. 1071–1075.

Greeley, Ronald (1971), Lunar Hadley Rille: considerations of its origin: *Science*, v. 172, no. 3984, pp. 722–725.

Hanks, T. C., and D. L. Anderson (1969), The early thermal history of the Earth: *Phys. Earth Planet. Interiors*, v. 2, pp. 19–29.

Hapke, B. W., W. A. Cassidy, E. N. Wells, and A. J. Cohen (1971), Analyses of optical coatings on Apollo fines: Proceedings, 1971 Lunar Science Conf., Jan. 11-14, Houston, Texas.

Hargraves, R. B., and A. F. Buddington (1970), Analogy between anorthosite series of the Earth and Moon: *Icarus*, v. 13, pp. 371–382.

Hartmann, W. K. (1966), Early lunar cratering: *Icarus*, v. 5, pp. 406–418.

——— (1970), Lunar cratering chronology: *Icarus*, v. 13, pp. 299–301.

Helsley, C. E. (1971), Evidence for an ancient lunar magnetic field: *Geochim. Cosmochim. Acta*, Suppl. 2, pp. 2485–2490.

Hess, H. H., and E. P. Henderson (1949), The Moore County meteorite: a further study with comment on its primordial environment: *Am. Mineral.*, v. 34, pp. 494–507.

Hohenberg, C. M., P. K. Davis, W. A. Kaiser, R. S. Lewis, and J. H. Reynolds (1970), Trapped and cosmogenic rare gases from stepwise heating of Apollo 11 samples: *Geochim. Cosmochim. Acta*, Suppl. 1, v. 34, pp. 1283–1310.

Howard, K. A. (1971), Geologic map of part of the Apennine-Hadley region of the Moon, Apollo 15 pre-mission map: U.S. Geol. Survey Misc. Geol. Inv. Map. I-723.

Hubbard, N. J., and P. W. Gast (1971), Chemical composition and origin of nonmare lunar basalts: *Geochim. Cosmochim. Acta*, Suppl. 2, pp. 999–1020.

Hurley, P. M., D. G. Brookins, W. H. Pinson, S. R. Hart, and H. W. Fairbairn (1961), K-Ar age studies of Mississippi and other river sediments: *Geol. Soc. America Bull.*, v. 72, pp. 1807–1816.

Husain, Liaquat, O. A. Schaeffer, and J. F. Sutter (1972), Age of a lunar anorthosite: *Science*, v. 175, no. 4020, pp. 428–430.

Kaula, W. M. (1970), The gravitational field of the Moon: *Science*, v. 166, no. 3913, pp. 1581–1588.

King, E. A., Jr., J. C. Butler, and M. F. Carman, Jr. (1971), The lunar regolith as sampled by Apollo 11 and 12: grain size analyses, model analyses, origins of particles: *Geochim. Cosmochim. Acta*, Suppl. 2, pp. 737–746.

Kosofsky, L. J., and Farouk El-Baz (1970), *The Moon as Viewed by Lunar Orbiter*: NASA Spec. Pub. 200.

Kovach, R. L., J. S. Watkins, and Tom Landers (1971), Active seismic experiment: Apollo 14 Prelim. Science Rept., NASA SP-272, pp. 163–174.

Kuiper, G. P. (1954), On the origin of the lunar surface features: *Proc. Natl. Acad. Sci. U.S.*, v. 40, p. 1096.

Latham, G., M. Ewing, J. Dorman, D. Lammlein, F. Press, N. Toksöz, G. Sutton, F. Duennebier, Y. Nakamura (1971), Moonquakes: *Science*, v. 174, no. 4010, pp. 687–692.

Latham, G., M. Ewing, J. Dorman, F. Press, N. Toksöz, G. Sutton, R. Meissner, F. Duennebier, Y. Nakamura, R. Kovach, and M. Yates (1970), Seismic data from man-made impacts on the Moon: *Science*, v. 170, no. 3958, pp. 620–626.

Lindsay, J. F. (1971), Sedimentology of the Apollo 11 and 12 lunar soils: *Jour. Sedimentary Petrology*, v. 41, pp. 780–797.

Lowman, P. D, Jr. (1969), *Lunar Panorama*: Zurich, Weltflugbild, 132 pp.

——— (1970), The geologic evolution of the Moon: rpt. X-644-70-381, Goddard Space Flight Center, Md.

Lunar Sample Preliminary Examination Team (1969), Preliminary examination of samples from Apollo 11: *Science*, v. 165, no. 3899, pp. 1211–1227.

——— (1970), Preliminary examination of lunar samples from Apollo 12: *Science*, v. 167, no. 3923, pp. 1325–1339.

——— (1971), Preliminary examination of lunar samples from Apollo 14: *Science*, v. 173, no. 3998, pp. 681–693.

——— (1972), The Apollo 15 lunar samples: a preliminary description: *Science*, v. 175, no. 4020, pp. 363–375.

Marti, K., G. W. Lugmair, and H. C. Urey (1970), Solar wind gases, cosmic-ray spallation products and the irradiation history of Apollo 11 samples: *Geochim. Cosmochim. Acta*, Suppl. 1, v. 34, pp. 1357–1368.

Mason, Brian (1962), *Meteorites*: New York, Wiley, 274 pp.

———, and W. G. Melson (1970), *The Lunar Rocks*: New York, Wiley, 179 pp.

McKay, D. S., W. R. Greenwood, and D. A. Morrison (1970), Origin of small lunar particles and breccia from Apollo 11 site: *Geochim. Cosmochim. Acta*, Suppl. 1, v. 34, pp. 673–694.

———, D. A. Morrison, U. S. Clanton, G. H. Ladle, and J. F. Lindsay (1971), Apollo 12 soil and breccia: *Geochim. Cosmochim. Acta*, Suppl. 2, pp. 755–773.

Melson, W. G., G. T. Thompson, and T. H. Van Andel (1968), Volcanism and metamorphism in the mid-Atlantic ridge, 22°N. latitude: *Jour. Geophys. Research*, v. 73, no. 18, pp. 5925–5941.

Meyer, Charles, Jr., Robin Brett, N. J. Hubbard, D. A. Morrison, D. S. McKay, F. K. Aitken, H. Takeda, and E. Schonfeld (1971), Mineralogy, chemistry, and origin of the KREEP component in soil samples from the Ocean of Storms: *Geochim. Cosmochim. Acta*, Suppl. 2, pp. 393–411.

Middlehurst, B. M. (1967), An analysis of lunar events: *Reviews of Geophys.*, v. 5, pp. 173–189.

Miyashiro, A., F. Shido, and M. Ewing (1970), Crystallization and differentiation in abyssal tholeiites and gabbros from mid-oceanic ridges: *Earth Planetary Sci. Letters*, v. 7, pp. 361–365.

Muller, P. M., and W. L. Sjogren (1968), Mascons: lunar mass concentrations: *Science*, v. 161, no. 3842, pp. 680–684.

Murase, Tsutomu, and A. R. McBirney (1970*a*), Viscosity of lunar lavas: *Science*, v. 167, no. 3924, pp. 1491–1493.

———, and A. R. McBirney (1970*b*), Thermal conductivity of lunar and terrestrial igneous rocks in their melting range: *Science*, v. 170, no. 3954, pp. 165–167.

Nash, D. B. (1963), On the distribution of lunar maria and the synchronous rotation of the Moon: *Icarus*, v. 1, pp. 372-373.

Offield, T. W., and H. A. Pohn (1970), Lunar crater morphology and relative age determination—Part 2. Applications, *in* Geological Survey research 1970: U.S. Geol. Survey Prof. Paper 700-C, pp. C163–169.

O'Hara, M. J., G. M. Bigger, S. W. Richardson, C. E. Ford, and B. G. Jamieson (1970), The nature of seas, mascons, and the lunar interior in the light of experimental studies: *Geochim. Cosmochim. Acta*, Suppl. 1, v. 34, pp. 695–710.

O'Keefe, J. A. (1969), Origin of the Moon: *Jour. Geophys. Research*, v. 74, no. 10, pp. 2758–2767.

——— (1970), Tektite glass in Apollo 12 samples: *Science*, v. 168, no. 3936, pp. 1209–1210.

Opik, E. J. (1955), The origin of the Moon: *Irish Astron. Jour.*, v. 3, pp. 245–248.

——— (1961), Tidal deformations and the origin of the Moon: *Astron. Jour.*, v. 66, pp. 60–67.

Papanastassiou, D. A., and G. J. Wasserburg (1970), Rb–Sr ages from the Ocean of Storms: *Earth and Planet. Science Letters*, v. 8, pp. 269–278.

———, and G. J. Wasserburg (1971*a*), Lunar chronology and evolution from Rb–Sr studies of Apollo 11 and 12 samples: *Earth Planetary Sci. Letters*, v. 11, pp. 37–62.

———, and G. J. Wasserburg (1971*b*), Rb–Sr ages of igneous rocks from the Apollo 14 mission and the age of the Fra Mauro Formation: *Earth Planetary Sci. Letters*, v. 12, pp. 36–48.

———, G. J. Wasserburg, and D. S. Burnett (1970), Rb–Sr ages of lunar rocks from the Sea of Tranquillity: *Earth and Planet. Science Letters*, v. 8, pp. 1–19.

Patterson, J. H., A. L. Turkevich, E. J. Franzgrote, T. E. Economou, and K. P. Sowinski (1970), Chemical composition of the lunar surface in a terra region near the crater Tycho: *Science*, v. 168, no. 3933, pp. 825–828.

Pepin, R. O., L. E. Nyquist, D. Phinney, and D. C. Black (1970), Rare gases in Apollo 11 lunar material: *Geochim. Cosmochim. Acta*, Suppl. 1, v. 34, pp. 1435–1454.

Péwé, T. L. (1959), Sand-wedge polygons (tesselations) in the McMurdo Sound region, Antarctica—a progress report: *American Jour. Sci.*, v. 257, pp. 545–552.

Pohn, H. A., and T. W. Offield (1970), Lunar crater morphology and relative age determination—Part 1. Classification, *in* Geological Survey research 1970: U.S. Geol. Survey Prof. Paper 700-C, pp. C153–C162.

Ringwood, A. E. (1970*a*), Petrogenesis of Apollo 11 basalts and implications for lunar origin: *Jour. Geophys. Research*, v. 75, no. 32, pp. 6453–6479.

——— (1970*b*), Origin of the Moon—The precipitation hypothesis: *Earth and Planetary Sci. Letters*, v. 8, no. 2, pp. 131–140.

———, and E. Essene (1970), Petrogenesis of Apollo 11 basalts, internal constitution and origin of the Moon: *Geochim. Cosmochim. Acta*, Suppl. 1, v. 34, pp. 769–800.

Roedder, E., and P. W. Weiblen (1970), Lunar petrology of silicate melt inclusions, Apollo 11 rocks: *Geochim. Cosmochim. Acta*, Suppl. 1, v. 34, pp. 801–838.

Rosin, P., and E. Rammler (1933), The laws governing the fineness of powdered coal: *Jour. Institute Fuel*, v. 7, pp. 29–36.

Schaber, G. G., and G. A. Swann (1971), Surface lineaments at the Apollo 11 and Apollo 12 landing sites: *Geochim. Cosmochim. Acta*, Suppl. 2, pp. 27–38.

Schaeffer, O. A., J. G. Funkhouser, D. D. Bogard, and J. Zähringer (1970), Potassium-argon ages of lunar rocks from Mare Tranquillitatis and Oceanus Procellarum: *Science*, v. 170, no. 3954, pp. 161–162.

Schmitt, H. H., Gary Lofgren, G. A. Swann, and Gene Simmons (1970), The Apollo 11 samples: introduction: *Geochim. Cosmochim. Acta*, Suppl. 1, v. 34, pp. 1–54.

———, and R. L. Sutton (1971), Stratigraphic sequence for samples returned by Apollo Missions 11 and 12; Proceedings, 1971 Lunar Science Conf., Jan. 11-14, Houston, Texas.

Sclar, C. B. (1970), Shock metamorphism of lunar rocks and fines from Tranquillity Base: *Geochim. Cosmochim. Acta*, Suppl. 1, v. 34, pp. 849–864.

Shoemaker, E. M., M. H. Hait, G. A. Swann, D. L. Schleicher, G. G. Schaber, R. L. Sutton, D. H. Dahlem, E. N. Goddard, and A. C. Waters (1970), Origin of the lunar regolith at Tranquillity Base: *Geochim. Cosmochim. Acta*, Suppl. 1, v. 34, pp. 2399–2412.

Short, N. M. (1970), Evidence and implications of shock metamorphism in lunar samples: *Geochim. Cosmochim. Acta*, Suppl. 1, v. 34, pp. 865–872.

Singer, S. F., and L. W. Bandermann (1970), Where was the Moon formed?: Science, v. 170, no. 3956, pp. 438–440.

Smith, J. V., A. T. Anderson, R. C. Newton, E. J. Olsen, A. V. Crewe, M. S. Isaacson, D. Johnson, and P. J. Wyllie (1970*a*), Petrologic history of the Moon inferred from petrography, mineralogy, and petrogenesis of Apollo 11 rocks: *Geochim. Cosmochim. Acta*, Suppl. 1, v. 34, pp. 897–926.

———, A. T. Anderson, R. C. Newton, E. J. Olsen, and P. J. Wyllie (1970*b*), A petrologic model for the Moon based on petrogenesis, experimental petrology, and physical properties: *American Jour. Sci.*, v. 78, no. 4, pp. 381–405.

Soderblom, L. A. (1970), A model for small-impact erosion applied to the lunar surface: *Jour. Geophys. Research*, v. 75, no. 14, pp. 2655–2661.

Stuart-Alexander, D. E., and K. A. Howard (1970), Lunar maria and circular basins—a review: *Icarus*, v. 12, pp. 440–456.

Sutton, R. L., R. M. Batson, K. B. Larson, J. P. Schafer, R. E. Eggleton, and G. A. Swann (1971), Documentation of the Apollo 14 samples: U.S. Geol. Survey Interagency Rept. Astrogeology-28, 37 pp.

Swann, G. A., N. J. Trask, M. H. Hait, and R. L. Sutton (1971), Geologic setting of the Apollo 14 samples: *Science*, v. 172, no. 3998, pp. 716–719.

Tatsumoto, M. (1970), Age of the moon: an isotopic study of U–Th–Pb systematics of Apollo 11 lunar samples—II: *Geochim. Cosmochim. Acta*, Suppl. 1, v. 34, pp. 1596–1612.

———, R. J. Knight, and B. R. Doe (1971), U–Th–Pb systematics of Apollo 12 lunar samples: *Geochim. Cosmochim. Acta*, Suppl. 2, pp. 1521–1546.

Toksöz, M. N., F. Press, K. Anderson, A. Dainty, G. Latham, M. Ewing, J. Dorman, D. Lammlein, G. Sutton, F. Duennebier, and Y. Nakamura (1972), Velocity structure and properties of the lunar crust: Proceedings, 1972 Lunar Science Conf., Jan. 10-13, Houston, Texas.

Trask, N. J. (1970), Geologic maps of early Apollo landing sites: Supp. to U.S. Geol. Survey Misc. Geol. Inv. Maps I–616 and I–627.

Turkevich, A. L., E. J. Franzgrote, and J. H. Patterson (1969), Chemical composition of the lunar surface in Mare Tranquillitatis: *Science*, v. 165. no. 3890, pp. 277–279.

Turner, G. (1970), Argon-40/argon-39 dating of lunar rock samples: *Geochim. Cosmochim. Acta*, Suppl. 1, v. 34, pp. 1665–1684.

Urey, H. C. (1952), *The Planets, Their Origin and Development*: New Haven, Yale Univ. Press, 245 pp.

——— (1969), Early temperature history of the Moon: *Science*, v. 165, no. 3899, p. 1275.

Vinogradov, A. P. (1971), Preliminary data on lunar ground brought to Earth by automatic probe "Luna-16": *Geochim. Cosmochim. Acta*, Suppl. 2, pp. 1–16.

Von Weizsäcker, C. F. (1943), Über die Entstehung des Planetensystems: *Zeitschrift Astrophysik*, v. 22, no. 5, pp. 319–355.

Wakita, Hiroshi, and R. A. Schmitt (1970), Lunar anorthosites: rare-earth and other elemental abundances: *Science*, v, 170, no. 3961, pp. 969–974.

Warner, J. L. (1971), Lunar crystalline rocks: petrology and geology: *Geochim. Cosmochim. Acta*, Suppl. 2, pp. 469–480.

Washburn, A. L. (1956), Classifications of patterned ground and review of suggested origins: *Geol. Soc. America Bull.*, v. 67, pp. 823–866.

Wetherill, G. W. (1971), Of time and the Moon: *Science*, v. 173, no. 3995, pp. 383–392.

Wilshire, H. G., and E. D. Jackson, Petrology and stratigraphy of the Fra Mauro Formation at the Apollo 14 site: unpublished manuscript.

Wise, D. U. (1969), Origin of the moon from the earth: some new mechanisms and comparisons: *Jour. Geophys. Research*, v. 74, no. 25, pp. 6034–6045.

———, and M. T. Yates (1970), Mascons as structural relief on a lunar "Moho": *Jour. Geophys. Research*, v. 75, no. 2, pp. 261–268.

Wood, J. A., J. S. Dickey, Jr., U. B. Marvin, and B. N. Powell (1970*a*), Lunar anorthosites and a geophysical model of the moon: *Geochim. Cosmochim. Acta*, Suppl. 1, v. 34, pp. 965–988.

———, U. B. Marvin, B. N. Powell, and J. S. Dickey, Jr. (1970*b*), Mineralogy and petrology of the Apollo 11 lunar samples: *Smithsonian Astrophys. Lab. Special Rept.* 307.

Wu, S. S. C., F. J. Schafer, R. Jordan, G. M. Nakata, and J. L. Derick (1972), Photogrammetry of Apollo 15 photography: Apollo 15 Prelim. Science Rept., NASA Spec. Pub., in press.

Index

(References to pages on which there are figures appear in italic type.)

Index